THE HEPATOCYTE REVIEW

THE HEPATOCYTE REVIEW

Edited by

MICHAEL N. BERRY

and

ANTHONY M. EDWARDS

SPRINGER-SCIENCE+BUSINESS MEDIA, B.V.

Library of Congress Cataloging-in-Publication Data

The hepatocyte review / edited by Michael N. Berry and Anthony M. Edwards.
p. cm.
Includes bibliographical references and index.
ISBN 978-90-481-5402-9 ISBN 978-94-017-3345-8 (eBook)
DOI 10.1007/978-94-017-3345-8
1. Liver cells. 2. Cell culture. I. Berry, Michael N. II. Edwards, Anthony M.

QP185 .H485 2000
611'.0181--dc21

99-086813

ISBN 978-90-481-5402-9

Printed on acid-free paper

Contents

Acknowledgements

The Editors are grateful to the many scientists who advised us on what topics this book should cover. We are especially obliged to those who then volunteered to contribute one or more chapters. Particular thanks go to Miss Emma Faehrmann who helped prepare the manuscript for electronic processing.

Preface

It is thirty years since the technique of high-yield preparation of isolated hepatocytes, by collagenase perfusion of the liver, was published. The original method described by Berry and Friend has undergone many minor modifications by other workers, and the two-step procedure introduced by Seglen in 1976 has become the most frequent way to prepare hepatocyte suspensions. An important development introduced by Bissell in 1973 was the use of the cells as the first step in monolayer culture. The availability of the isolated hepatocyte preparation as cells in suspension or culture has undoubtedly facilitated research on the liver. This was emphasised in our book, published (with Dr. Greg Barritt) in 1990, which described in detail methods of preparation and the properties of the isolated hepatocytes. It also discussed the usefulness of the preparation for the study of intermediary and xenobiotic metabolism, calcium ion transport, and the growth and differentiation of hepatocytes in culture. The book also touched briefly on a range of specialised techniques, including perifusion, subcellular fractionation, transplantation, cryopreservation and measurement of intracellular pH. Although standard procedures for the manipulation of hepatocytes have not changed a great deal in ten years, they have undoubtedly been refined. This applies particularly to hepatocyte culture techniques, cryopreservation, and even to preparation of hepatocyte suspensions, where it is now feasible to use purified enzymes. There is also much more emphasis on the use and study of human hepatocytes, particularly in the field of pharmacology and therapeutics.

The present volume places a much larger stress on hepatocyte biology. Despite the relatively short time since our previous publication, the problem has been not what to include but rather what to leave out. In addressing this we have tried to provide a volume that would be of interest to a wide variety of researchers. To this end, we have devoted Part 1 primarily to methodology with discussion of relevant hepatocyte biology. Part 2 consists of a series of up-to-date reviews covering topical areas of hepatic biology where the use of isolated hepatocytes is making important contributions to knowledge. These are written by experts in the field and extensively referenced. Parts 3 and 4 address applications that were in their infancy ten years ago. The chapters in these Parts are either in the form of comprehensive reviews or are descriptions of the subject by individuals with hands on experience. They cover the use of hepatocytes for pharmacological and toxicological studies, as well as their employment as therapeutic tools. The therapeutic applications described are hepatocyte transplantation for acute hepatic failure, gene therapy and the use of hepatocytes in the bioartificial liver.

It is our hope that this book will prove a valuable source of information not only to those commencing studies with isolated hepatocytes but also to more advanced researchers. It is gratifying that the isolated hepatocyte preparation has made a substantial contribution to the understanding of the biochemistry, physiology and cell biology of the normal and diseased liver over the past thirty years. This book is intended to enhance that contribution.

Abbreviations

3MC	3-methylcholanthrene
3-MPA	3 mercaptopicolinate
4-HNE	4-hydroxyalkenals
7-EC	7-ethoxycoumarin
AAV	adeno-associated virus
AcNPV	*autographa californica*, nuclear polyhedrosis virus
ActD	actinomycin-D
Ah	aryl hydrocarbon
ALF	acute liver failure
AMC-BAL	Academic Medical Centre Bioartificial Liver
APAP	acetaminophen
APR	acute phase response
AST	aspartate aminotransferase
BAL	Bioartificial Liver
BB	bromobenzene
BSA	bovine serum albumin
CAD	caspase activated deoxyribonuclease
CCCM	cell conditioning and cryopreservation medium
CCM	cell culture medium
CDKs	cyclin-dependent serine/threonine kinases
CDM	chemically defined media
CDNB	1-chloro-2,4-dinitrobenzene
CL_H	hepatic clearance
CL_{int}	intrinsic clearance
CM	cytomix
CMV	Cytomegalus virus
cNOS	expressed nitric oxide synthase
CPP	cerebral perfusion pressure
CYP	cytochrome P450
DAG	diacylglycerol
DCF	dichlofluorescein
DG	double gel
DHAP	dihydroxyacetonephosphate
DISC	death inducing signalling complex
DME/F12	Dulbecco's modified Eagle's and Ham's F-12
DMEM	Dulbecco`s modified Eagles medium
DMN	dimethylnitrosamine
DMSO	dimethylsulphoxide
DNIC	dinitrosyl iron complex
DPG	diphosphoglycerate
DTT	dithiothreitol
E2	estradiol-17β
ECM	extracellular matrix
ECOD	7-ethyoxycoumarin-o-deethylase
EDTA	ethylenediaminetetraacetic acid
EGF	epidermal growth factor
EPR	electron paramagnetic resonance
ER	estrogen receptor
ER	extraction ratio

EROD	ethoxyresorufin-O-deethylase
ETF	Electron-transferring protein
F16P	fructose 1,6-biphosphate
FADD	Fas Associated Death Domain
FAH	fumarylacetoacetate hydrolase
FasL	Fas ligand
FCS	fetal calf serum
FGF	fibroblast growth factor
FHF	fulminant hepatic failure
FMB	flat membrane bioreactor
GalN	D-galactosamine
GAP	glyceraldehyde 3-phosphate
GBC	glass bead chromatography
GDH	glutamate dehydrogenase
GFP	green fluorescent protein
GP	glycerophosphate
GPDH	glycerophosphate dehydrogenase
GPI	glycosyl-phosphatidylinositol
GSH	glutathione
GSK-3	glycogen synthase kinase-3
GST	glutathione S-transferase
hAAT	human alpha-1-antitrypsin
HBDH	3-hydroxybutyrate dehydrogenase
HB-EGF	heparin-binding EGF
HbsAg	hepatitis B surface antigen
HC	hydrocortisone
HGF	hepatocyte growth factor
HGH	Hepatocyte growth medium
HLH	helix-loop-helix
HSP70	heat shock protein-70
HSV-1	herpes simplex virus-1
ICP	intracranial pressure
IL-1	interleukin-1
IMV	inferior mesenteric vein
iNOS	inducible nitric oxide synthase
$InsP_3$	inositol 1,4,5-trisphosphate
IPGs	inositol phosphoglycans
IRS	insulin receptor substrates
KHB	Krebs-Henseleit buffer
LDH	lactate dehydrogenase
LDL	low density lipoprotein
LECs	liver epithelial cells
LIS	Complete liver ischemia
LMW	low-molecular-weight
LPS	lipopolysaccharide
LSS	liver support system
M-A	malate-aspartate
MCV	Medical College of Virginia
MDA	malondialdehyde
MDH	malate dehydrogenase
MILCA	matrix-induced liver cell aggregates

MMS	methyl methanesulphonate
MNNG	*N*-methyl-*N'*-nitro-*N*-nitrosoguanine
MPT	mitochondrial permeability transition
mRNA	messenger RNA
MT	metallothionein
MTT	3-(4,5-dimethylthiazol-2-yl)-2,5-diphenyl tetrazolium bromide
NAPQI	N-acetyl-p-benzoquinone imine
NAR	Nagase analbuminemic rat
NDMA	*N*-nitrosodimethylamine
NMR	nuclear magnetic resonance spectroscopy
NO	nitric oxide
NOS	nitric oxide synthase
NPC	non-parenchymal cell
OLT	orthotopic liver transplant
OTC	ornithine transcarbamoylase
PAHs	polycyclic aromatic hydrocarbons
PARP	poly(ADP-ribose) polymerase
PBBs	polybrominated biphenyls
PBS	phosphate buffered saline
PC	phosphatidylcholine
PCBs	polychlorinated biphenyls
PCN	pregnenolone 16α-carbonitrile
PEG	polyethylene glycol
PEPCK	phosphoenolpyruvate carboxykinase
pHA	para-hydroxyamphetamine
PI	phosphatidylinositol
PI-PLC	phosphoinositide-specific phospholipase C
PKA	protein kinase
PKC	protein kinase C
PLC	phospholipase C
PLD	phospholipase D
PNLP	phenylephrine
Ptdlns	Phosphotidylmositol
RIF	rifampicin
RLEC	rat liver epithelial cells
ROS	reactive oxygen species
RT-PCR	reverse transcriptase-polymerase chain reaction
RV_d	regulatory volume decrease
RV_i	regulatory volume increase
SF	scatter factor
SG	Second Generation
SG	single gel
SH	sulphhydryl
SH2	src-homology-2
SNAP	S-nitroso-N-acetyl-penicillamine
SOD	superoxide dismutase
SSDB	single strand DNA breakage
TAA	thioacetamide
TBA	thiobarbituric acid
TBARS	TBA reactive substance
TBT	tri-*n*-butyltin

TCDD	2,3,7,8-tetrachloro-dibenzo-p-dioxin
TEM	transmission electron microscopy
TGFα	transforming growth factor alpha
TGFβ	transforming growth factor beta
TNFα	tumor necrosis factor alpha
TNFR1	type 1 TNF receptor
TO	Tryptophan dioxygenase
TRADD	TNF receptor associated death domain
UDS	unscheduled DNA synthesis
UM	umbelliferone
uPA	urokinase-type plasminogen activator
UW	University of Wisconsin
VTG	Vitellogenin
WHHL	Watanabe heritable hyperlipidemic

Chapter 1

Isolated hepatocytes: forty years on

Michael N. Berry
Department of Medical Biochemistry, Flinders University School of Medicine, GPO Box 2100, Adelaide, 5001, Australia

Key words: calcium ions, collagenase, damaged hepatocytes, elastase, incubation temperature, isolated hepatocyte preparation, hepatocyte structure, liver perfusion, purified enzymes

1.1 Introduction

When forty years ago, I made my first attempt at preparing isolated hepatocytes it would have been quite impossible to imagine the extraordinary diversity of studies for which they would be employed. Indeed, many of the topics covered in this book were not even part of the body of scientific knowledge at that time, and it is pleasing to know that much new knowledge concerning the liver has come to light as a consequence of the availability of the isolated hepatocyte preparation. In this chapter I will give a brief history of the development of methods for isolated hepatocyte preparation. In Chapter 21 the results of some of the metabolic studies of our group are described in the context of macromolecular organisation.

1.2 The development of the method for high-yield preparation of isolated hepatocytes

1.2.1 Preparation of damaged cell suspensions

In 1958, two years after graduating from Medical School at Otago University in New Zealand, and having completed my hospital residency, I took up a Junior Lectureship in the Department of Biochemistry. The Head of the Department, Professor Norman Edson suggested that I carry out some studies on hepatic metabolism in the NZO obese mouse, using isolated hepatocytes prepared by a technique [1] based on that of Anderson [2]. In

M.N. Berry and A.M. Edwards (eds.), The Hepatocyte Review, 1–9.

this method for cell separation, the liver is distended under pressure with a Ca^{2+}-free medium such as 0.3 M sucrose solution, buffered with Tris to pH 7.4. The swollen liver is cut into small pieces and gently homogenized by hand. This treatment does not produce a homogenate, as one might expect, but rather a suspension of isolated hepatocytes, which can be readily separated from cell debris by a 2 min centrifugation at 50 × *g*.

Preparing the cell suspension was straightforward, but the cells, when suspended in an extracellular medium had, as others had found, no apparent metabolic activity other than the ability to vigorously oxidize succinate. For the next 18 months I attempted without success to demonstrate additional metabolic properties. The paradox was that many workers were at this time revealing the metabolic properties of isolated mitochondria. Why then were isolated hepatocytes so intractable? It finally occurred to me that the suspending medium might be the critical factor. Isolated mitochondria were routinely suspended in an "intracellular" type medium, whereas it seemed only reasonable to suspend whole cells in an extracellular medium containing Ca^{2+} and phosphate. Nevertheless, I decided to use the same type of high K^+ medium used by those working with mitochondria, and had instant success [1]. The explanation proved straightforward; cells prepared by the high-pressure perfusion technique had gross damage to their plasma membranes and were essentially nothing more than a particulate preparation. Their mitochondria took on the form of isolated mitochondria, and the membranes of the endoplasmic reticulum were converted into microsomes [3]. Hence the need for an intracellular medium. It was only too apparent that there was no possibility of preparing intact cells by this technique.

1.2.2 Structure within damaged cells

Despite my disappointment, this initial introduction to research had a major impact on my future directions. Looking at these damaged flattened hepatocytes under the phase-contrast microscope, I noted that none of their mitochondria showed the Brownian motion possessed by isolated mitochondria that formed part of the debris in the surrounding medium. Clearly, there must be an invisible scaffold holding them in place, not observable even under the electron microscope [3]. I also noted that the mitochondria adopted definite patterns within the cell [1]. Thus, an appreciation of the existence of intracellular organisation was established early in my career, as was the determination to find a method for preparing *intact* isolated hepatocytes.

1.3 Preparation of intact suspensions of hepatocytes

Nine years passed, the first five of which I was fortunate enough to spend in the Department of Professor Hans Krebs in Oxford, first as a D.Phil. student, and then as a worker in his team. In this role I visited Hans Schimassek in Marburg and learned the technique of rat liver perfusion. Krebs' interests had turned to gluconeogenesis, and at that time the only *in vitro* hepatic preparation that could synthesize glucose was the perfused liver. Soon after setting up the perfusion system, I took up a one year appointment in Pittsburgh and then spent two years in London at the Postgraduate Medical School, before being attracted to an Associate Professorship position at the University of California in San Francisco.

1.3.1 Successful attempts at intact hepatocyte preparation by Howard et al.

Prior to taking up my appointment in San Francisco, I was asked to referee a paper submitted to the Journal of Cell Biology by a post-doctoral scientist, Roger Howard, who had devised a technique for isolating hepatocytes. This was achieved by injecting crude collagenase and hyaluronidase into the main vessels of the rat liver, then slicing the liver and incubating the slices in a Ca^{2+}-free medium. Collagenase had been previously used by Rodbell to prepare suspensions of adipocytes [4]. The yield of hepatocytes was only a few per cent of the original starting material, but the electron micrographs provided convincing evidence that the isolated cells were intact [5]. I strongly recommended that the paper be published, and, having arrived in San Francisco, took the first opportunity to visit Roger, who was based at Stanford. He generously took me through the technique, and within a few days I was preparing intact hepatocytes, though unfortunately in yields too low for them to be useful for metabolic studies.

1.3.2 Perfusion of the liver

It was not till several months later that it occurred to me to set up a liver perfusion system and try out the effects of putting collagenase in the perfusion medium. The outcome was more than I could possibly have hoped for, and I can still clearly remember the feeling of exhilaration as I looked down the microscope – a rare feeling indeed for a basic scientist! I then teamed up with Dan Friend, an electron microscopist of the highest order, so that our paper, when it appeared in 1969, demonstrated not only that high yields of isolated hepatocytes could be prepared from rat liver but also the mechanisms involved in cell separation [6]. Nevertheless, the paper remained relatively unnoticed till 1972 when I visited Oxford for 2 months to work with Hans Krebs. He, not surprisingly, was sceptical. If liver slices could not make glucose, why should isolated hepatocytes be capable of doing so? He asked me to demonstrate their ability to synthesize urea and I was extremely despondent when the experiment, using cells incubated in a phosphate buffer, was a complete failure. But Professor Krebs himself came to the rescue. He questioned me concerning what medium I had employed and sent me away to repeat the experiment, this time using a bicarbonate buffer and 5% CO_2 in the gas phase. Given a physiological medium the hepatocytes duly obliged, synthesising urea at a rate equivalent to that seen *in vivo*. From that day Professor Krebs promoted the isolated hepatocyte preparation enthusiastically, and its popularity rapidly spread. For my part I learned another lesson, namely to use physiological media in my experiments whenever feasible [7].

1.4 Modifications to the original technique for hepatocyte isolation

1.4.1 The "two-step" procedure

Although numerous variations to the original isolation technique have been introduced during the 30 years that my colleagues and I have successfully continued to use the original high-yield method, the only significant modification has been that introduced by Seglen [8]. On the assumption that it was necessary to wash Ca^{2+}out of the liver, before the hepatocytes could be detached from each other, he recommended a two-step procedure in which the liver is first perfused with Ca^{2+}-free medium for about 10 min with or without EDTA. The medium is allowed to run to waste and a collagenase-containing medium with up to 5 mM Ca^{2+} is substituted, the rationale being that the enzyme is Ca^{2+}-dependent and therefore relatively high concentrations of the ion should be included in the second step of the perfusion.

This modification, in one manifestation or another, has long been favoured by most other workers, as the articles in this book illustrate. However, we have found repeatedly that the one-step and two-step methods yield virtually identical results in terms of yield and time required for isolated cell preparation. It would seem that the use of two perfusates of different composition (or three separate solutions if EDTA must be flushed from the liver) introduces an unnecessary degree of complexity to the procedure. Furthermore, the flushing of a 10 g liver with several hundred ml of medium introduces the possibility of leaching small molecular weight metabolites, such as amino-acids from the organ. If EDTA is employed, there is also a risk of depleting the cells of trace metals such as Zn^{2+}, which bind to the chelating agent much more strongly than do Ca^{2+} [9]. Nevertheless, provided the hepatocytes are placed in a balanced saline, bicarbonate-buffered medium at the time of their final wash, we have found their carbohydrate metabolism to be identical as determined by measurement of rates of glycolysis and gluconeogenesis (unpublished observations).

1.4.2 The role of calcium ions

The modifications introduced by Seglen were based on the understandable misconception that it was essential to flush out Ca^{2+} with a large volume of medium to break down cellular adhesiveness. But as was originally observed, the initial wash-out of blood with about 50 ml of medium is sufficient to lower Ca^{2+} concentrations to a level that allows the desmosomes to cleave [6]. This irreversible cleavage takes less than 10 min and, since collagenase does not interfere with this process, there is no reason why the enzyme should not be present from the start. Addition of Ca^{2+} is not required for collagenase activation because the enzyme contains sufficient bound $Ca^{2+}$for full activity. As the efficacy of the one-step method demonstrates, these bound Ca^{2+} do not interfere with desmosomal cleavage. However, in recent years, we have included 1 mM Ca^{2+} in the cell washing medium immediately after liver digestion because of evidence that the sooner the hepatocytes are re-exposed to physiological concentrations of Ca^{2+}, the better their metabolic performance [10].

1.5 Preparation of cell suspensions with purified enzymes

A very recent modification of the original technique, introduced by Ronal MacGregor of Kansas State University, promises to be of much greater significance [11]. MacGregor has sought an answer to the question of why crude collagenase is effective, whereas attempts to prepare isolated hepatocytes with purified enzyme yield very disappointing results. It was shown early on that collagenase was an essential component of the digestive cocktail and that hyaluronidase was unnecessary, but what the extra ingredients were that facilitated liver digestion by a crude collagenase preparation has remained uncertain for 30 years. MacGregor in a series of systematic experiments discovered that the combination of purified collagenase and elastase was much more effective than collagenase alone in the isolation of parathyroid cells. He went on to demonstrate that this also applied in the case of hepatocyte isolation (personal communication). Since there is little elastin in the liver, it would appear that elastase acts at a different site to that attacked by collagenase.

1.5.1 Preparation of isolated hepatocytes with purified enzymes

Ronal MacGregor and I, together with my colleague John Phillips, have collaborated to devise an optimal procedure for hepatocyte isolation using purified collagenase and elastase, and are preparing a protocol for publication, which differs little from the methods used with crude enzyme preparations. Cell isolation can be carried out on the basis of the original Berry and Friend method [6]. We recommend gassing 180 ml of Ca^{2+}-free Krebs bicarbonate medium with carbogen and flushing a 10 g rat liver with 80 ml of this medium over a 2 min period to wash out blood. A few ml of medium containing collagenase and elastase and 25 μM Ca^{2+} is then added to the reservoir and a recirculating perfusion through the liver established, which results in its digestion within 8-15 min depending on the concentration of enzymes employed. A yield equivalent to that achieved with 50 mg collagenase per 100 ml medium can be obtained with 5000 units of collagenase and 10 units of elastase. Maximal yields, representing at least 25% more cells than can be obtained with crude collagenase, are attained by including 10000 units of collagenase with 10 units of elastase in the medium.

Cell integrity as measured by Trypan blue exclusion is regularly over 92% and can exceed 95%. Addition of Ca^{2+} is unnecessary for the digestive process, but we include trace amounts on the assumption that it is desirable for some external Ca^{2+} to be present, though the concentration must be limited to allow desmosomal cleavage. Metabolic rates of gluconeogenesis and ureogenesis are enhanced 5-10 %, although this may merely reflect the much smaller proportion of damaged cells in the suspension. It seems reasonable to anticipate that surface damage to the hepatocytes will be less, and in keeping with this, attachment to plastic by cells plated for culture is facilitated (MacGregor, personal communication).

The importance, and possible site, of elastase action is illustrated by the observation that if the amount of this enzyme is reduced below a critical level, regardless of the quantity of collagenase employed, a good yield of cells can be obtained but 30% or more fail to exclude Trypan blue. This implies that the desmosomes do not cleave completely and raises the possibility that elastase action may be involved in desmosome cleavage. This possibility is under investigation.

For those workers committed to the two-step method, an initial perfusion of 8 min can be carried after the initial flushing of blood and prior to addition of enzymes and Ca^{2+}. However, we recommend that this perfusion be performed by recirculation of the medium, rather than by allowing the perfusate to flow to waste. In our hands pre-perfusion for 8-10 min prior to enzyme addition lengthens the overall process by a few minutes but the outcome in terms of cell yield and integrity appears to be the same as in the one-step procedure.

1.6 Use of hepatocytes in the bioartificial liver

1.6.1 Incubation temperatures for isolated hepatocytes in suspension

Many contributors to this book have made use of cultured hepatocytes, a technique that was introduced by Bissell [12] as early as 1973. For short term metabolic studies, however, our preference has been to use cells in suspension, a procedure which offers the opportunity to carry out 60 or more different incubations in a single experiment. This in turn permits the concomitant examination, at frequent intervals over prolonged time periods, of the effects of concentration gradients of substrates, inhibitors or hormones. In fact, the robust nature of isolated hepatocytes in suspension enables experiments lasting at least 3-4 h when incubations are carried out at 37°C. Furthermore, cells incubated at 25°C for 24 h, exhibit close to normal rates of glucose and urea synthesis when subsequently incubated at 37°C (Table 1).

Table 1. Urea synthesis and cellular ATP concentration in hepatocytes maintained in suspension for 24 h at 25°C

*J*urea µmol/min/g		[ATP] µmol/g	
0 h	24 h	0 h	24 h
2.79 ± 0.13	2.17 ± 0.13	2.44 ± 0.09	1.84 ± 0.08

Hepatocytes from 24 h fasted rats were maintained in suspension at 25°C for 24 h in a bicarbonate-buffered medium containing 2.25% BSA and 30 mM glucose. These cells were subsequently incubated at 37°C as previously described [11] in the presence of 5mM lactate, 10 mM ammonium ions and 2 mM ornithine. Results are presented as the mean ± SEM of 12 separate experiments.

Because an incubation temperature of 25°C reduces metabolic activity by no more than 60% (Table 2), there is a realistic possibility that hepatocytes in suspension might be incorporated into a Bioartificial Liver (BAL). At present, all operational or experimental BAL's rely on cells adhered to a matrix (see also Chapters 30-34); but the use of cells in suspension would provide much greater cell mass and would greatly simplify BAL design, whilst enhancing the possibility of creating Distribution Centres in various countries, strategically placed for allocating hepatocytes to end-users.

Table 2. Effect of temperature on the rates of glucose and urea synthesis by hepatocytes

Temp	*J*glucose	*J*urea
	µmol/min/g	
25°C	0.50 ± 0.03	1.04 ± 0.05
37°C	1.42 ± 0.04	2.26 ± 0.16

Hepatocytes from fasted rats were incubated as previously described [11] at 25°C or 37°C with 8 mM lactate, 1 mM pyruvate, 2 mM asparagine and 1.75 mM palmitate for the measurement of gluconeogenesis. Ureogenesis was determined with 8 mM Lactate, 1 mM pyruvate, 2 mM asparagine, 10 mM ammonium ions and 2 mM ornithine. The maximal rates of formation during a 40-min incubation are presented as the mean ± SEM of 8 experiments.

1.7 Conclusions

While isolated hepatocytes have for many years been a routine tool for experimental studies, examination of their potential importance for therapeutic uses is much more recent. Several chapters in the book are devoted to the possibilities of gene transplantation for inborn errors of metabolism, and hepatocyte transplantation, or use in the bioartificial liver for treatment of fulminant hepatic failure. These treatments are in their infancy but, as the authors describe, promising results are being obtained. Other chapters describe useful techniques for hepatocyte preparation storage or manipulation and still others concentrate on basic biology. The Editors hope that this wide variety of information will prove of value and interest to readers, whether they are just commencing in this field or have been preparing hepatocyte suspensions for many years.

Acknowledgments

The work carried out in Adelaide has been supported by the NH&MRC, the Drug and Alcohol Services Council and the Flinders Medical Centre Foundation.

References

1. Berry MN. Metabolic properties of cells isolated from adult mouse liver. J Cell Biol 1962; 15: 1-8.
2. Anderson NG. Mass isolation of whole cells from rat liver. Science 1953; 117: 627-628.
3. Berry MN and Simpson FO. Fine structure of cells isolated from adult mouse liver. J Cell Biol 1962; 15: 9-17.
4. Rodbell M. Metabolism of isolated fat cells. J Biol Chem 1964; 239: 375-380.
5. Howard RB, Christensen AK, Gibbs FA and Pesch LA. The enzymatic preparation of isolated intact parenchymal cells from rat liver. J Cell Biol 1967; 35: 675-684.
6. Berry MN and Friend DS. High-yield preparation of isolated rat liver parenchymal cells: a biochemical and fine structural study. J Cell Biol 1969; 43: 506-520.
7. Berry MN, Werner HV and Kun E. Effects of bicarbonate on intercompartmental reducing-equivalent translocation in isolated parenchymal cells from rat liver. Biochem J 1974; 140: 355-361.
8. Seglen PO. Preparation of rat liver cells. I. Effect of Ca^{2+} on enzymatic dispersion of isolated, perfused liver. Exp Cell Res 1972; 74: 450-454.
9. Dawson RMC, Elliott DC, Elliott WH and Jones KM. Data for Biochemical Research. 3rd ed. Oxford: Oxford University Press, 1986.
10. Berry MN, Edwards AM and Barritt GJ. Isolated Hepatocytes Preparation, Properties and Application. Amsterdam: Elsevier, 1991.
11. Ferzandi TR and MacGregor RR. Thapsigargin shifts the Ca set point of parathyroid cells to lower extracellular [Ca]. Endocrine 1997; 7: 361-366.
12. Bissell DM, Hammaker LE and Meyer UA. Parenchymal cells from adult rat liver in nonproliferating monolayer culture. 1. Functional studies. J Cell Biol 1973; 59: 722-734.

Chapter 2

Isolation and culture of human hepatocytes

María José Gómez-Lechón and José V. Castell
Unidad de Hepatología Experimental, Centro de Investigación, Hospital Universitario La Fe. Avda. ampanar 21, 46009-Valencia. Spain

Key words: cell density, culture media, hepatocyte proliferation, human hepatocytes, liver biopsy, liver perfusion

2.1 Isolation of human hepatocytes from liver biopsies

Small tissue samples obtained in the course of liver surgery (*i.e.* for trauma, tumour resections, cholecystectomy etc.) or from tissue resected for pathological examination are available on a regular basis, and constitute a reliable source of viable human hepatocytes. The enzymatic perfusion of liver through the blood vessels is the procedure of choice to obtain human hepatocytes for culturing. Another key factor is the use of appropriate enzymes to weaken cell-cell junctions and liberate cells with minimal membrane damage [1,2]. On the basis of these considerations, a two-step microperfusion technique for isolating hepatocytes has been successfully applied to very small samples of human liver [3]. The key steps towards obtaining high yields of viable, minimally membrane-damaged, metabolically competent cells are as follows:

1. The biopsy must be wedge-shaped and have only one cut surface accessible to blood vessels;
2. The tissue must be immediately cooled to avoid warm ischemia, and processing of the tissue sample should take place within the first three hours;
3. Adequate microcannulation of the small blood vessels (the biopsy should have at least one) to ensure efficient perfusion of the whole tissue sample;
4. Efficient oxygenation of the perfusion buffer during cell isolation;
5. Pretesting of collagenase to select the most appropriate type and batch [3,4]. Pre-testing is strongly recommended since commercial collagenase is available with different grades of purity, containing minor amounts of other proteolytic enzymes that also contribute to cell-cell dissociation and rapid, complete digestion of the liver biopsy, thus reducing perfusion

M.N. Berry and A.M. Edwards (eds.), The Hepatocyte Review, 11–15.

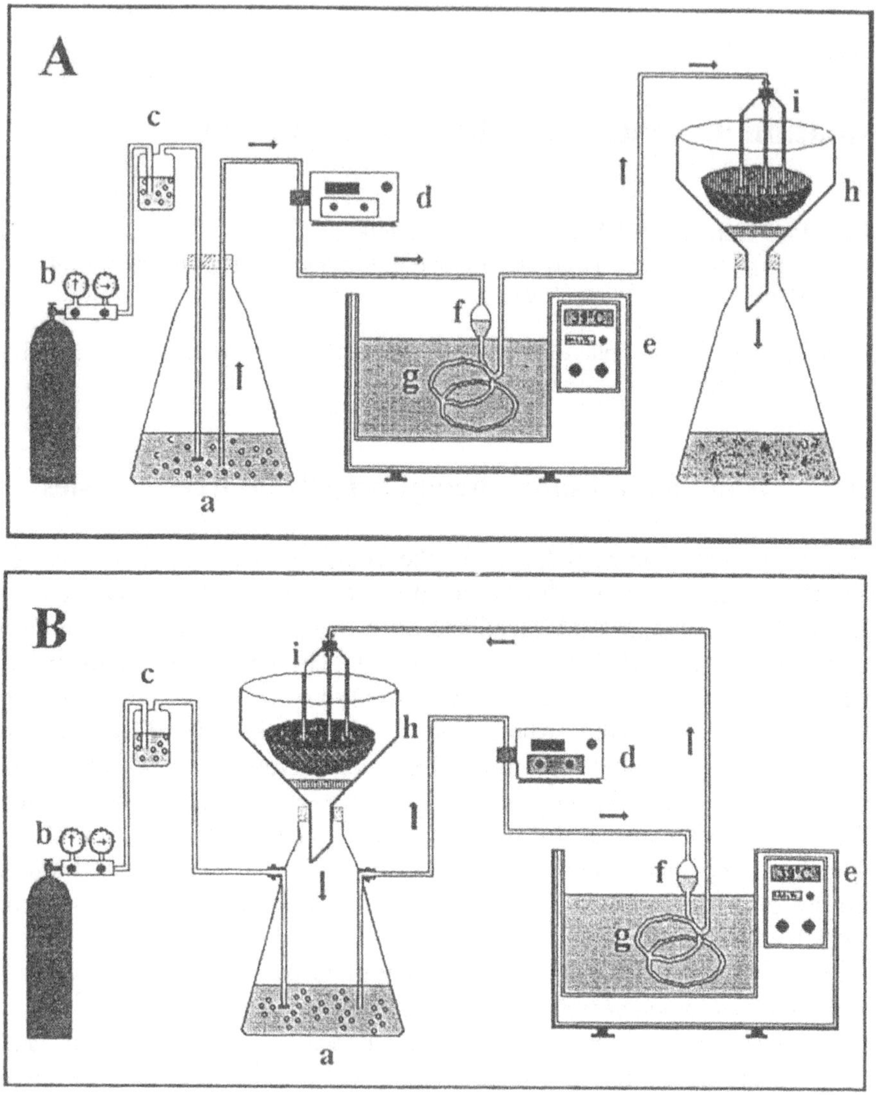

Figure 1. Procedure and devices for liver biopsy perfusion. The microperfusion equipment consists of: a) a flask containing perfusion solution; b) carbogen; c) a washing carbogen flask; d) a peristaltic pump; e) a water bath; f) a bubble-trapping container; g) a warming coil; h) a funnel to hold the liver specimen and i) perfusing silicon catheters (0.75 mm diameter). (A) During the first step, the tissue is extensively washed by non-recirculating perfusion with a calcium-free buffer. Warmed (37°C) and continuously gassed solutions (95% O_2: 5% CO_2) are gently pumped through microcatheters into the liver biopsy with the aid of a peristaltic pump. The effluent solution is drained off. (B) The second step is a recirculating perfusion with the same buffer, without EGTA but with calcium and collagenase. Tube connections need to be changed in the course of perfusion (after addition of collagenase) to allow the solution coming from the liver sample to be gassed and recirculated back to the biopsy.

time and increasing the percentage of viable cells [3,4] (see also description of use of purified collagenase and elastase in Chapter 1.)

Hepatocytes are isolated from surgical liver biopsies (2-5 g) by a two-step perfusion technique to digest the liver sample (Figure 1). The liver sample is placed on the holding device and one to four silicon catheters (0.75 mM diameter) are inserted into the lumen of the venous orifices on the cut surface. The liver tissue sample is first extensively washed with 250 ml of perfusion buffer (137 mM NaCl, 2.68 mM KCl, 0.7 mM Na_2HPO_4.12 H_2O, 10 mM HEPES, 10 mM glucose, 0.5 mM EGTA, pH 7.4) at a flow rate of 10 ml/min/catheter, which is then allowed to flow to waste. In a second stage, an EGTA-free perfusion buffer containing 5 mM $CaCl_2$, previously reported to be effective for human liver [5], and collagenase (0.5 mg/ml; approximately 0.20 U/ml) is recirculated at the same flow rate for 45-60 min until sufficient softening of the tissue is observed. The perfusion buffers are continuously oxygenated with carbogen and delivered to the liver at 37°C. Once the enzymatic digestion is completed, the liver tissue is transferred to a petri dish and the cells are gently dispersed with a blunt spatula. The cell suspension is filtered through gauze and there is no need for additional mechanical disruption of the tissue. The cellular suspension is centrifuged for 2 min at $100 \times g$ and the pellet is resuspended with ice-cold culture medium and again centrifuged. The pellet is finally resuspended in culture medium, and cell viability is assessed by means of a dye exclusion test (*i.e.* 0.4% Trypan blue [3]). Viability should be around 85-95%. Typical yields are $23 \pm 3.5 \times 10^6$ intact hepatocytes/g liver, although this value depends to a large extent on the size of the biopsy, accessibility of vessels and efficiency of perfusion.

2.2 Culture of human hepatocytes in chemically defined media

Active selective attachment of viable hepatocytes to plastic culture plates and long-term survival of these cells require the surface of dishes to be coated with adhesive extracellular matrix components such as fibronectin (3 μg/cm^2) [3] or collagen [6,7]. The culture medium used in our studies is Ham F-12/Leibovitz L-15 (1:1 v/v) supplemented with 2% adult calf serum, 10 mM glucose, 50 U/ml penicillin, 50 μg/ml streptomycin, 0.2% bovine serum albumin and 10^{-8} M insulin. Other culture media are also currently used by researchers, *i.e.* Ham F-12/Williams' E, (1:1 v/v) [3], and Williams' E [7,8].

Cell density strongly influences the proliferative response of cultured hepatocytes. Adult human hepatocytes do not proliferate even in the presence of growth stimulating factors when they are cultured at high density or near confluency [6-8]. Only when hepatocytes are cultured at sub-confluency and in the presence of growth factors do they exhibit a proliferative response as assessed by DNA synthesis and cyclin A accumulation [6,7,9-12]. To assess proliferative responses, hepatocytes were seeded on fibronectin-coated plates at a low density of $2\text{-}4 \times 10^4$ cells/cm^2 suspended in an appropriate volume of culture medium. This cell density is also used by other authors [6-8] and it produces maximal DNA synthesis by human hepatocytes stimulated with growth factors.

One hour after cell seeding the medium is aspirated to remove unattached cells and new medium is added. Twenty-four hours after cell plating, cells are shifted to serum-free medium supplemented with 10^{-8} M insulin and 10^{-8} M dexamethasone. Under these culture conditions cells survive up to 10-12 days but do not divide in unstimulated conditions [3,8].

Acknowledgements

The authors wish to thank the financial support of European Union (projects AIR2-CT93-0860 and BIO4-CT96-0052) and the ALIVE Foundation.

References

1. Le Rumeur E, Guguen-Guillouzo C, Beaumont C, Saunier A and Guillouzo A. Albumin secretion and protein synthesis by cultured diploid and tetraploid rat hepatocytes separated by elutriation. Exp Cell Res 1983; 157: 247-254.
2. Strom SC, Jirtle RL, Jones RS et al. Isolation, culture, and transplantation of human hepatocytes. J Natl Cancer Inst 1982; 68: 771-778.
3. Gómez-Lechón MJ, Ponsoda X, Fabra R, Trullenque R and Castell JV. Isolation, culture and use of human hepatocytes in drug research. In: Castell JV and Gómez-Lechón MJ, Eds. *In vitro* methods in pharmaceutical research. London: Academic Press, 1997: 129-154.
4. Blaauboer BJ, Boobis AR, Castell JV, Coecke S et al. The practical applicability of hepatocyte cultures in routine testing. The report and recommendations of ECVAM Workshop 1. ATLA 1994; 22: 231-241.
5. Guguen-Guillouzo C, Campion JP, Brissot P, Glaise B, Launois B, Bourel M and Guillouzo A. High yield preparation of isolated human adult hepatocytes by enzymatic perfusion of the liver. Cell Biol Int Rep 1982; 6: 625-628.
6. Blanc P, Etienne H, Daujat M, Fabre I, Zindy F, Domergue J, Astre C, Saint Aubert B, Michel H and Maurel P. Mitotic responsiveness of cultured adult human hepatocytes to epidermal growth factor, Transforming growth factor α, and human serum. Gastroenterology 1992; 102: 1340-1350.
7. Ismail T, Howl J, Wheatley M, McMaster P, Neuberger JM and Strain AJ. Growth of normal human hepatocytes in primary cultures. Effect of hormones and growth factors on DNA synthesis. Hepatology 1991; 14: 1076-1082.

8. Strain AJ, Ismail T, Tsubouchi H, Arakaki N, Hishida T, Kitamura H, Daikuhara Y and McMaster P. Native and recombinant human hepatocyte growth factors are highly potent promoters of DNA synthesis in both human and rat hepatocytes. Clin Invest 1991; 87: 1853-1857.
9. Gómez-Lechón MJ, Castell JV, Guillen I, O'Connor E, Nakamura T, Fabra R and Trullenque R. Effects of hepatocyte growth factor in the growth and metabolism of human hepatocytes in primary culture. Hepatology 1995; 21: 1248-1254.
10. Michalopoulos G, Cianciulli HD, Novotny AR, Kligerman AD, Strom SC and Jirtle RL. Liver regeneration studies with rat hepatocytes in primary culture. Cancer Res 1982; 42: 4673-4682.
11. Nakamura T, Tomita Y and Ichiara A. Density-dependent growth control of adult rat hepatocytes in primary culture. J Biochem 1983; 94: 1020-1035.
12. Greuet J, Pichard L, Ourlin JC, Bonfils C, Domargge J, Le Truet P and Maurel P. Effect of cell density and epidermal growth factor on the inducible expression of CYP3A4 genes in human hepatocytes in primary culture. Hepatology 1997; 25: 1166-1175.

Chapter 3

Isolation and culture of porcine hepatocytes

Hans G. Koebe, Sascha A. Pahernik and Friedrich W. Schildberg
Department of Surgery, Klinikum Grosshadern, University of Munich, Marchioninistr. 15, D-81377 Munich, Germany

Key words: collagen gel, collagen preparation, culture method, double gel configuration, hepatocyte isolation, liver lobe cannulation, slaughterhouse liver, percoll, porcine hepatocytes

3.1 Introduction

Isolated hepatocytes have increasingly become an important research tool in the medical, pharmacological, biological, and biochemical sciences so that the demand for viable cell material has increased. Cell-lines could produce sufficient numbers of hepatocytes; however, uncertainty exists with respect to their stability and how well their properties compare with those of primary cells. In bioartificial liver applications the use of cell lines has been advocated [1] as well as challenged [2]. In most cases, hepatocytes for various kinds of experiments have been isolated from small vertebrates, such as rats. Other animal species have been used as donors in order to achieve large-scale isolation of liver cells. The choice of animal reflects not only well-recognized biological differences between various animal species [3], but is also a consequence of practical considerations in individual laboratory settings.

3.2 Should donor tissue be from the slaughterhouse or the operating theatre?

In this chapter, after a brief description of cell isolation from whole pigs, we discuss the procedure for preparing isolated porcine hepatocytes under laboratory conditions [4] and then focus on the technical aspects and requirements that allow successful pig hepatocyte isolation from slaughterhouse organs [5]. We see two major advantages for this latter approach. First, it does not require surgical facilities for large animal operations, and thus saves time, money and animal lives. Second, it provides

M.N. Berry and A.M. Edwards (eds.), The Hepatocyte Review, 17–26.

large quantities of intact cells for research groups, predictably and early in the morning, which enables further processing the same day.

3.3 Preparation of pig hepatocytes from organs of living donors

3.3.1 Requirements

1. Operation facilities for large animals, surgical and anaesthetic equipment
2. Experience in large animal anaesthesia
3. Approval for animal experiments by local authorities
4. Pigs
5. Perfusion device with appropriate tubing, connectors
6. Medium 199 or other cell culture medium (CCM)
7. Trypan blue Solution (0.05%, v/v)
8. Perfusion buffer A: Phosphate buffered saline (PBS) without calcium/magnesium + 34 μM EDTA
9. Perfusion buffer B: PBS without calcium/magnesium + 5.5 mM glucose + 26 mM sodium bicarbonate
11. Perfusion buffer C: PBS with calcium/magnesium and penicillin/streptomycin
12. Perfusion buffer D: Medium 199 + 26 mM sodium bicarbonate + penicillin/streptomycin/amphotericin + 0.08% (w/v) collagenase
13. Perfusion buffer E: Medium 199 + 20 mM HEPES
14. Centrifuge
15. Microscope

3.3.2 Perfusion technique

Hepatocyte isolation protocols based on *ex vivo* perfusion techniques have been described by various authors with only minor modifications [6-11]. In general, all techniques resemble a scale-up of the collagenase perfusion techniques usually described for small vertebrates [12]. (With every new batch of collagenase, the optimal concentration in the perfusate as well as the requisite perfusion time should be determined experimentally). Piglets with an average body weight of 30 kg are usually recommended as donors. Pigs are anaesthetized intramuscularly with 0.01 mg/kg body wt atropine and 50 mg/kg metomidate, and intravenously with 15 mg/kg thiopental. A mixture of isoflurane/N_2O is then administered by inhalation in combination with 0.01 mg/kg Fentanyl™. Laparotomy is then performed, the liver prepared and the portal vein cannulated. Following the incision and/or cannulation of the inferior vena cava, *in situ* perfusion of the liver is commenced. The liver is perfused with buffer A at 4°C and then removed and placed in a pre-cooled dish. Perfusion is continued with buffer A, warmed to 37°C and equilibrated, as are all other solutions, with 5% CO_2, 95% O_2, for a further 15 min at a flow of 2 litres/min at a mean pressure of 25 mM H_2O. The isolated organ is then placed in the perfusion device and perfusion continued with buffer B for 15 min, followed by Buffer C for 2 min, and collagenase-containing Buffer D for 40 min. In order to reduce the required amount of collagenase, Buffer D should be recirculated. Finally, the liver is perfused with Buffer E for 2 min at 4°C.

Next, the liver lobe is carefully dissected by hand, using a scalpel blade, (this takes about 4 min) and the resulting cell suspension is filtered through two nylon meshes (grid sizes 500 and 100 μm, respectively). The resulting cell suspension is washed twice with Medium E at $10 \times g$ for 5 min at 4°C. The average cell yield after isolation is about 68 ± 11% of the organ's volume with an average cell viability of 65 ± 14% [4].

3.4 Hepatocyte isolation from slaughterhouse organs

3.4.1 Requirements

The following protocol for the preparation of cells from "living donors" has been described by Gerlach *et al.* [4] and was modified by us for the successful harvesting of viable liver cells from slaughterhouse organs:

1. Perfusion device (pump, tubing, connectors, oxygenator)
2. Collagenase
3. Dulbecco's Modified Eagles Medium (DMEM) with 25 mM glucose or other CCM
4. Leibovitz L-15 Medium
5. Trypan blue Solution (0.05% v/v)
6. Perfusion buffer containing 154 mM NaCl, 5.6 mM KCl, 5 mM glucose, 25 mM sodium bicarbonate and 20 mM HEPES, pH 7.4
7. Washing buffer containing 120 mM NaCl, 6.2 KCl, 0.9 mM $CaCl_2$, 10 mM HEPES and 0.2% (w/v) bovine albumin, pH 7.4
8. Centrifuge
9. Microscope

3.4.2 Perfusion technique

For isolation of porcine hepatocytes from slaughterhouse organs we use Landrace piglets (both sexes, 4-8 weeks old, weighing 15-35 kg) which are slaughtered at the local slaughterhouse. We found it essential to harvest cells from young animals because pig livers of larger weight from older animals need a higher collagenase activity and longer digestion times to liberate parenchymal cells from the organ. This increases the rate of cell death and damage during isolation and reduces both cell viability and specific functions. Piglets that have undergone the usual slaughtering process (electric shock and death through exsanguination), are immersed in a boiling water bath for 3 min, cleaned, and thereafter the abdominal cavity is opened by butchers and the liver is excised. After transport in plastic bags on ice to the laboratory (20 ± 10 min), the left medial liver lobe is carefully excised (Figure 1). The liver can be perfused *in toto* via the portal vein or – if lesser cell numbers are sufficient – a perfusion of a single lobe may be performed. Since the cannulation of a central vessel is of crucial importance in order to reach the largest possible portion of a liver section, the lobe has to be cut out of the organ far to the dorsal surface and close to the gallbladder on the opposite side.

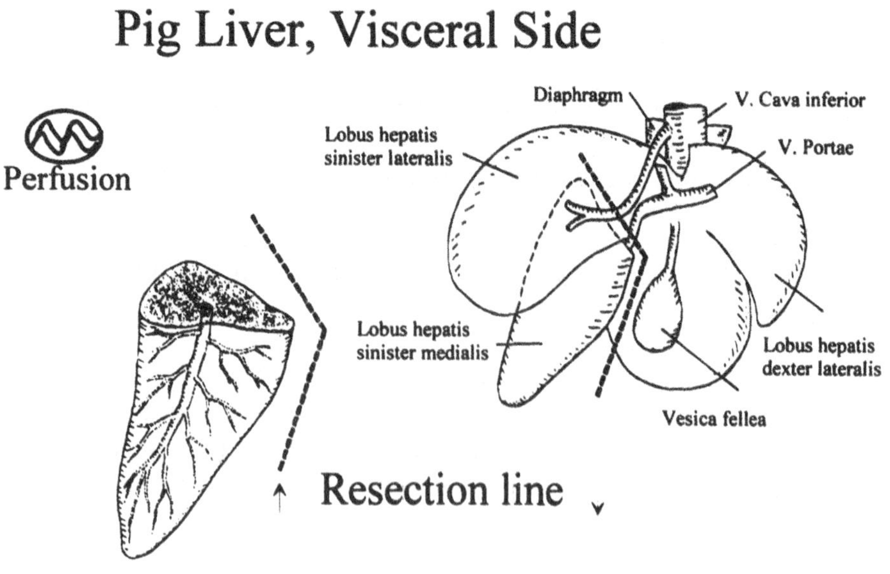

Figure 1. Illustration of the resection of the left medial liver lobe and its cannulation via the left hepatic vein.

3.4.3 General considerations

The anatomical site of perfusion as well as the following details concerning perfusion buffers derive from practical considerations and our own experience during the process of optimising the protocol. The tissue is perfused via the left hepatic vein (ramus ventralis). This vessel collects blood from all parts of the left medial lobe, whereas the portal system is more difficult to track down to the corresponding part of the liver. The hepatic vein runs within the liver lobe, whereas branches from the portal vein are situated directly under the liver capsule. Thus, cannulas can be retained far more easily within the hepatic vein branch than within the portal vein. The cannulation of one central vessel was of considerable benefit as compared to a multi-vessel procedure with respect to the cell yield and was far easier to standardize. Following cannulation of the main-stem branch of the left hepatic vein and retrograde perfusion, complete digestion of the left medial liver lobe, assessed by macroscopical appearance, is achieved. This is in contrast to multi-vessel cannulation where larger tissue areas regularly remain out of reach of the perfusate. Retrograde perfusion, although apparently unphysiological, results in cell yields comparable to reported data from the literature, with a high percentage of viable cells and consistent results. Perfusion of the whole liver is very expensive due to high volumes of perfusion buffers and the amount of collagenase needed. Additionally, we found complete tissue perfusion was achieved more efficiently by performing a single lobe perfusion rather than a whole organ perfusion.

3.4.4 Details of technique

Perfusion via the left hepatic vein for 10 min is performed with oxygenated buffer containing 1 mM EDTA, and 40 µg/litre dexamethasone (37°C, 50 ml/min), and is followed by a perfusion under identical conditions but without dexamethasone (Figure 2). Next, the liver lobe is perfused without EDTA for 2 min, followed by a recirculating perfusion with buffer containing 0.05% collagenase (w/v) and 5 mM calcium chloride at identical temperature and flow rates (37°C, 50 ml/min) for 16 ± 2 min, depending on the enzymatic activity of the collagenase batch used.

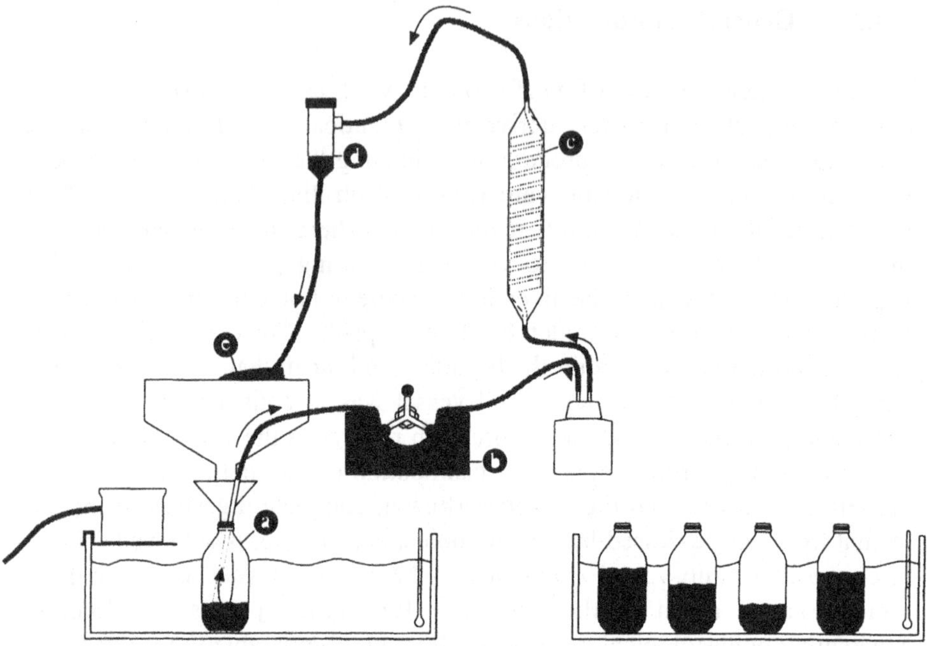

Figure 2. Schematic illustration of the perfusion system. The perfusion solutions are recirculated in a system containing a perfusate heated in water bath (a), peristaltic pump (b), temperature unit (c), bubble trap (d), and a tray for liver tissue (e).

The liver lobe is carefully dissected by hand, using a scalpel blade, and the resulting crude homogenate shaken for 5 min at 37°C in the presence of the final perfusate (0.05% collagenase (w/v), 5 mM calcium chloride). The resulting cell suspension is filtered through two nylon meshes (grid size 210 and 70 µm, respectively). Washing buffer (4°C) is added to the filtered suspension to a final volume of 500 ml, followed by three washing procedures in the same buffer at $17 \times g$ for 5 min. In order to obtain a purified hepatocyte suspension, a Percoll™ gradient centrifugation with 10.8 ml of Percoll™, 1.2 ml of 10× concentrated DMEM and 17 ml of cell suspension is performed at $76 \times g$ for 5 min (Pharmacia LKB Biotechnology, Uppsala, Sweden). Next, the resulting cell pellets are washed twice with washing buffer at $17 \times g$ for 5 min. Hepatocytes are resuspended in Leibovitz L-15-Medium (Serva Feinbiochemica GmbH) containing 0.2% (w/v) albumin. Cell viability is tested by Trypan blue exclusion. Nonparenchymal cells, as judged by size (<10 µm in diameter) and morphology (non-polygonal) are <2%. Aliquots of the resulting cell suspension are checked for DNA and total protein content to confirm results of cell counting.

3.4.5 Cell yield

The average yield (n=110) of viable hepatocytes after Percoll™ gradient centrifugation is $1.92 \pm 0.5 \times 10^9$ (viability as assessed by Trypan blue exclusion: $93 \pm 2\%$) per liver lobe ($1.0 \pm 0.3 \times 10^7$ per g tissue). The average number of (rat) hepatocytes per g tissue *in vivo* is around 1.5×10^8. Thus, at the present time we harvest about 10% of all hepatocytes of the perfused liver lobe as viable cells. We did not encounter any problems with respect to bacterial or fungal contamination as monitored by macroscopical findings and microbiological assays in control cultures.

3.5 *Ex vivo* - versus slaughterhouse organ cell isolation

The aforementioned isolation protocol seems applicable for most laboratories even if they are separated by longer distances from the local slaughterhouse. We did not observe any influence of different time lags (range 35 to 60 min) on cell isolation or culture results. Compared to published results from other groups who perform isolations mainly on living animals, no adverse effects were seen that might have been caused by isolating cells from cadaver livers with respect to cell morphology or function. Apart from the ethical point of view, *i.e.* the reduction of animal experiments, hepatocytes from slaughtered piglets are easier to obtain without the requirement for large animal operation facilities. Furthermore, experiments on large animals are widely restricted and force scientists to look for alternative ways to obtain biological material. Further advantages include the reduction of costs and labor, as well as the reliable availability of liver tissue early in the morning, allowing experiments on cells the same day.

In order to improve the methodology for the isolation of viable hepatocytes from slaughterhouse organs, the procedure has to be easy to apply, since perfusion of liver tissues within slaughterhouses is only rarely applicable. Whole organs can be perfused that way, but large collagenase volumes are needed, which are expensive, and in most cases not necessary. We suggest harvesting hepatocytes from single liver lobes, and perfusing the tissue via a central vessel. This technique supplies sufficient cell material in a reliable fashion for *in vitro* experiments or for testing bioartificial liver devices in small vertebrate models.

In conclusion, cell harvesting from slaughterhouse pig livers yields viable hepatocytes, thus representing an unlimited resource of cell material for related fields of research, including technologies for hybrid artificial liver devices. It should be preferred to *ex vivo* preparations, wherever applicable.

3.6 Porcine hepatocytes in primary cell culture

Hepatocytes, in general, do not proliferate *in vitro* and thus need extracellular matrix components to attach and to survive in culture. The classical culture configuration was introduced by Michalopoulos and requires collagen as an extracellular matrix component [13]. Collagen can be obtained commercially or can be prepared from biological material, such as fibrous tissues. In our laboratory, collagen (Type I) is prepared from rat tail tendons by a procedure modified from that of Elsdale [14]. Tendons are dissected from rat tails (about 1.75 g per tail) and stirred overnight at 4°C in 1 litre of 3% (v/v) acetic acid. The solution is filtered through cheese cloth and centrifuged at 10,960 × *g* for 2 h at 4°C. The supernatant is precipitated with 0.2 vol of 30% NaCl (w/v). The solution is centrifuged at 1750 × *g* for 30 min at 4°C. After washing (0.6% acetic acid/5% NaCl w/v) the resulting pellets are dissolved in 200 ml of 0.6% acetic acid and stirred overnight. The collagen is then dialysed against 2 litre of 1 mM HCl. After addition of 0.3% chloroform (v/v) the collagen is stirred again for 48 hours at 4°C. Aliquots are lyophilized and weighed to determine collagen content (about 100 mg/rat tail). This preparation yields Type I collagen, mainly in its native, triple-helical, rather than cross-linked, form.

For pig hepatocytes we use the following CCM composition which has shown good results in preliminary tests: phenol red-free DMEM which is supplemented with insulin (125 mU/ml), hydrocortisone (60 ng/ml), glucagon (10 ng/ml), gentamycin (100 μg/ml), and penicillin (100 mU/ml). In order to support attachment of cells to the surface of the matrix, we add 5% Fetal Calf Serum (FCS) to the medium for the first 24 hours.

3.6.1 Standard culture technique

In standard culture technique, extracellular matrix (ECM) proteins like collagen, fibronectin, laminin, or ECM preparations such as Matrigel™ or Vitrogen™, are usually added to the culture dishes. We use collagen as the ECM component. Culture dishes are precoated with a cooled collagen solution. This solution is prepared by mixing 1 part 10 × DMEM, pH 7.4 and 9 parts stock collagen solution at a concentration of 2.2 mg/ml. Collagen on culture dishes is allowed to gel at room temperature for 30 min before cells are distributed on culture dishes.

3.6.2 Double gel configuration

For long-term cultures of hepatocytes we recommend a three dimensional extracellular matrix configuration which allows the maintenance of morphological viability and functional activity of hepatocytes [15]. Hepatocytes are sandwiched in this configuration between two layers of collagen gel (see Chapter 8). The advantages of this system for *in vitro* investigations on liver cells have been reported by several authors using various donor species. The Double Gel Technique helps to form an extracellular protein environment that closely mimics the *in vivo* architecture. Polarity of cells cultured in the collagen gel sandwich configuration resembles the *in vivo* situation of the liver acinus. Using this culture technique physiological activity of a variety of hepatocyte specific functions are maintained over several weeks [15]. The total content of DNA for 3×10^6 hepatocytes decreased from 69 ± 4 µg in cell suspension to 52 ± 2 µg per culture dish at day 3 after isolation, and remained stable over the rest of the culture period, indicating a stable culture phase after an initial cell loss. Alanine aminotransferase release showed peak values (43 µU/hr/µg DNA) immediately after isolation, probably because of cellular damage due to enzymatic digestion during the isolation process, and remained constant at low values (4 µU/hr/µg DNA), indicating a low cell loss thereafter. CYP 1A1-dependent activity, as measured by the deethylation of ethoxycoumarin, slightly decreased during culture with a mean activity of 4.5 ± 0.5 nmol/h/µg DNA. Albumin synthesis increased during the culture period to stable values (4.5 ± 0.8 ng/h/µg DNA) for at least 11 days.

References

1. Sussman NL, Chong MG, Koussayer T, He DE, Schang TA et al. Reversal of fulminant hepatic failure using an extracorporeal liver assist device. Hepatology 1992; 16: 60-65.
2. Muller TE and Jauregui HO. Letters to the editor. Artif Organs 1993; 17: 44-45.
3. Oldham HG, Standring P, Norman SJ, Blake TJ, Beattie I et al. Metabolism of temelastie (SK&F 93944) in hepatocytes from rat, dog, cynomolgus monkey and man. Drug Metab Dispos 1990; 18: 146-152.
4. Gerlach J, Kloppel K, Schon MR, Brombacher J, Courtney JM, Unger J and Neuhaus P. Comparison of pig hepatocyte isolation using intraoperative perfusion without warm ischemia and isolation of cells from abattoir organs after warm ischemia. Artif Organs 1993; 17: 950-953.
5. Koebe HG, Pahernik SA, Sproede M, Thasler WE and Schildberg FW. Porcine hepatocytes from slaughterhouse organs. An unlimited resource for bioartificial liver devices. ASAIO J 1995; 41: 189-193.

6. Donini A, Corno V, Lavaroni S, Pellerito R, Bigi L et al. *Ex vivo* preparation of porcine hepatocytes for use in bioartificial hepatic support systems. Transplant Proc 1997; 29: 1948-1949.
7. Chen SC, Mullon C, Kahaku E, Watanabe F, Hewitt W et al. Treatment of severe liver failure with a bioartificial liver. Ann N Y Acad Sci 1997; 831: 350-360.
8. Joly A, Desjardins JF, Fremond B, Desille M, Campion JP et al. Survival, proliferation, and functions of porcine hepatocytes encapsulated in coated alginate beads: a step toward a reliable bioartificial liver. Transplantation 1997; 63: 795-803.
9. Lazar A, Peshwa MV, Wu FJ, Chi CM, Cerra FB and Hu WS. Formation of porcine hepatocyte spheroids for use in a bioartificial liver. Cell Transplant 1995; 3: 259-268.
10. Naik S, Trenkler D, Santangini H, Pan J and Jauregui HO. Isolation and culture of porcine hepatocytes for artificial liver support. Cell Transplant 1996; 5: 107-115.
11. Naruse K, Sakai Y, Nagashima I, Jiang GX, Suzuki M and Muto T. Comparisons of porcine hepatocyte spheroids and single hepatocytes in the non-woven fabric bioartificial liver module. Int J Artif Organs 1996; 19: 605-609.
12. Seglen PO. Preparation of isolated rat liver cells. Methods Cell Biol 1976; 13: 29-83.
13. Michalopoulos G and Pitot HC. Primary culture of parenchymal liver cells on collagen membranes. Exp Cell Res 1975; 94: 70-78.
14. Elsdale T and Bard J. Collagen substrata for studies on cell behaviour. J Cell Biol 1972; 54: 626-637.
15. Dunn JCY, Tompkins RG and Yarmush ML. Long-term *in vitro* function of adult hepatocytes in a collagen sandwich configuration. Biotechnology Progress 1991; 7: 237-245.

Chapter 4

Preparation and properties of isolated hepatocytes from sheep

Michael G. Clark and Michelle A. Vincent
Division of Biochemistry, Medical School, University of Tasmania, GPO Box 252-58 Hobart, 7001, Australia

Key words: caudate lobe perfusion, gluconeogenesis, hepatocyte isolation from ruminants, metabolic differences between ruminants and non-ruminants, nitrogen metabolism, ruminant metabolism, short-chain fatty acids

4.1 Preparation

4.1.1 Introduction

The advent of a reliable method for the preparation in high-yield of viable hepatocytes from the rat, developed by Berry and Friend [1], opened up many new studies that otherwise could only be addressed using liver slices (inferior) or perfused liver (ignoring vascular effects). The situation for metabolic studies in larger animals such as the sheep, was even more a problem as sheep liver perfusions were technically cumbersome. Thus the idea of adapting the method of Berry and Friend [1] to prepare lamb, and then later, sheep liver cells was very attractive. Such a method had the potential to allow investigators an opportunity to study the metabolic changes that occurred in ovine liver as weaning proceeded and metabolism changed from non-ruminant to ruminant.

In 1974 Ivan Jarrett and Owen Filsell, long time researchers on sheep metabolism at the CSIRO Division of Human Nutrition in Adelaide, suggested that the problem could be solved by perfusing the excised caudate lobe with a buffer containing collagenase, rather than the far more expensive and cumbersome alternative of perfusing the entire liver. The caudate lobe of the sheep's liver, which is a finger-like structure that represents only approximately 4.5% of the total organ, appeared suitable provided that retrograde perfusion through its major large vein could be successfully accomplished. A paper detailing this approach was published in 1976 [2].

M.N. Berry and A.M. Edwards (eds.), The Hepatocyte Review, 27–36.

Essentials of the method were as follows: The liver of an anesthetized lamb was exposed by a lateral abdominal incision on the right flank of the animal and the caudate lobe removed by a single transverse cut. The lobe was placed on an appropriately sized porous tray in a perfusion apparatus at 37°C with the portal vein uppermost. A tapered, stepped, cannula, attached to the perfusion line, was then fitted into the portal vein blood vessel and the perfusion commenced. After discarding the blood, the venous effluent was allowed to return through the porous tray and funnel into the perfusion system. With practice, cannulation of the caudate lobe from the sheep could be completed within 20 s.

The method for tissue disaggregation was essentially as described for rat liver [1] with a few modifications [3]. These included the use of Ca^{2+}-free buffer with 1.5% w/v gelatin as the protein support. Details for disaggregation perfusion involved recirculating perfusion with 160 ml of buffer for a total of 35 min at a flow rate of 40-60 ml/min and a pressure head of 20 cm of water. Gassing with 95%O_2:5% CO_2 (carbogen) at 3 litre per min ensured that the pH stayed at 7.4 ± 0.2 and that the entering perfusate oxygen tension was 400 ± 50 mm Hg. Collagenase 60 mg (0.375 mg/ml final concentration), was introduced in the last 20 min of perfusion.

Although digestion was usually extensive after the perfusion, it was found preferable to mince the digested lobe, filter off the undigested material through nylon mesh, (pore size approximately 400 μm) and then incubate the filtrate with fresh collagenase-containing buffer for an additional 15 min at 37°C while continuously gassing. Cells were harvested at the end of the incubation by filtering through two layers of nylon mesh (0.1 mm pore size.) followed by centrifugation at 100 × *g* for 75 s with two washings. Using this method yields of approximately 30% of the initial wet weight of the lobe were obtained as cells.

4.1.2 Current method of preparation

Subsequent studies led to minor changes. The method that is widely used today [4-15] involves the following steps:

1. Perfusion of the caudate lobe or a piece of another lobe. This is perfused within a perfusion apparatus near to where the sheep is slaughtered. Alternatively, if these facilities are unavailable then the lobe or portion of liver may be excised at a regional abattoir, cannulated and perfused by a syringe to wash out blood and replace it with the primary buffer. The primary buffer preferred by most researchers is Ca^{2+}-free, and comprises a physiological salts solution, such as Krebs Henseleit

medium [16], containing 0.2-0.6 mM EGTA, but no collagenase. The temperature is kept close to 37°C during transport. Once the liver portion or lobe is connected to the perfusion apparatus, this first perfusion is continued at 37°C for 5-20 min. It is assumed that the period of time that is chosen for this perfusion is dependent on final outcome of yield and viability, but 10 min is usually sufficient for desmosome cleavage at 37°C.

2. Perfusion with collagenase- and Ca^{2+}-containing buffer at 37°C. The concentration of collagenase ranges from 0.3 to 1 mg per ml and that of Ca^{2+} from 0.8 to 5 mM. Investigators have decided on their own recipes but in general, the higher the concentration of collagenase and Ca^{2+} (up to 1 mg per ml and 5 mM, respectively) the shorter the time required for digestion. The minimum perfusion time appears to be 10 min.
3. Removal of digested from undigested material. The requirement of a collagenase incubation after perfusion [2] has now been circumvented by using Ca^{2+}-containing buffer in (2) above, thus ensuring a complete digestion. Some groups have adopted use of a stainless steel comb to free the hepatocytes from the digested lobe (*e.g.* [13]). The hepatocytes are then filtered through a 200 μm pore size nylon filter mesh and collected by centrifugation at 100 × *g* for 90 s. Cells are finally suspended and incubated in Ca^{2+}- (1.27-2.5 mM) and albumin- (0.75-2.5%) containing, Krebs Henseleit bicarbonate medium that has been gassed with carbogen.

It is worthy of note that perfusion of the caudate lobe of the liver of cows and part of the left lobe of goats has been used successfully to prepare hepatocytes from these species [15]. There is also a method for preparing sheep hepatocytes from liver slices [17,18]

4.2 Properties

4.2.1 Integrity

Trypan blue exclusion is routinely used, and is claimed by many to indicate sustained viability of greater than 90% over an extended period of storage on ice (often 3 h). Some others are prepared to accept lower Trypan blue exclusion values (*e.g.* 65-85%, [19]). Few groups have reported the extent of contamination by non-hepatocytes, although this may be as much as 15% [20].

Other indicators of integrity that have been used are:

1. Oxygen uptake rate [2]
2. Gluconeogenesis from propionate ± NH_4Cl [2,11,21,22]
3. Ureagenesis from propionate and NH_4Cl [4]
4. Lactate dehydrogenase leakage [2]
5. Cellular content of ATP [2,8,23]
6. Linearity of phenylalanine incorporation into protein [13]

The literature cited for each of these integrity tests gives well defined ranges that researchers new to hepatocyte isolation can use for reference.

4.2.2 Changes in hepatocyte size with age of sheep

As shown by Clark *et al.* [2], maturation of lambs to sheep results in a considerable increase in hepatocyte size. This can be a problem for the calculation of metabolic rates when the number of cells is used rather than wet or dry weight. It was noted that for 1-week-old lambs there were approximately 9×10^8 cells per g dry weight of cell suspension with a mean diameter of about 15μm. By close to 60 days of age, cell size had increased to a plateau, with approximately 4.5×10^8 cells per g dry weight and a mean diameter of about 25μm (4). Since this value is similar to that for liver cells from 24 h-starved adult rat, guinea pig and mouse of 4.1×10^8, 3.4×10^8 and 3.3×10^8 cells per g dry weight, respectively, it is clear that lamb liver cells at 40 days of age or younger are appreciably smaller and some caution is necessary in interpreting results. The increase in cell size may merely reflect the maturation process or the development of ruminant from non-ruminant metabolism, as the latter also takes place over the interval from 7 to 60 days.

4.2.3 Comparison with perfused liver

A method for *in situ* perfusion of neonatal ovine liver has been described for the study of chemical clearance [24]; and perfused caudate lobe has been used to study the effects of pregnancy on fatty acid metabolism [25]. Adult sheep liver perfusion, also *in situ*, has been used to study effects of glucagon on the rate of utilisation of simultaneously infused propionate and amino acids [26], and to study the fate of amino acids in isolated livers of developing sheep [27]. However, direct comparison between metabolic properties of isolated sheep hepatocytes and perfused sheep liver appears limited to our own study [2], which addressed only gluconeogenesis. The rates of glucose formation from 10 mM propionate were compared for animals deprived of food for 24 h. Net glucose formation from unlabelled propionate was 1.76 and 1.95 μmol/min/g dry weight for perfused liver and hepatocytes, respectively. Similarly, the corresponding values for ^{14}C glucose formation from [^{14}C]propionate were 1.43 and 2.06 μmol/min/g dry weight [2]. Thus the rates were comparable between the two preparations. The absolute rate of glucose formation from 10 mM propionate has been reported by several research groups and found to range from 0.81 [11], through 2.10 [10] to 3.0 μmol/min/g dry weight [7]. Cells from the median lobe of the lamb also showed rates of glucose formation of 2.02 μmol/min/g dry weight [2]. Interestingly, gluconeogenesis from 10 mM propionate, when a perfused caudate lobe preparation was used, gave similar figures [5].

4.2.4 Gluconeogenic rates

Ovine hepatocytes have been used by a number of workers to compare gluconeogenic rates for a variety of substrates. These studies are important because ruminants are in a state of continuous gluconeogenesis as the amount of glucose available for intestinal absorption is limited due to the ruminal fermentation of dietary carbohydrate. In addition, the main gluconeogenic substrates utilized can vary depending on the physiological state of the animal. Propionate is the major gluconeogenic substrate in fed sheep, but this may change when food is scarce. Several research groups have compared the relative gluconeogenic potential of a variety of substrates with hepatocytes from lambs (which are non-ruminant), adult sheep (ruminant) and sheep which are essentially equivalent to 24 h-starved rats (*i.e.* adult sheep that have been deprived of food for 4 days). Some interesting findings have emerged that contrast with comparable studies using rat hepatocytes. For suckling lambs (1-20 days post partum; non-ruminant) the order of substrates (all 10 mM) that gave the highest rates of glucose synthesis was: galactose> lactate> pyruvate> propionate> fructose> alanine [2]. As the animals became ruminant the order changed so that hepatocytes from a fed sheep showed: fructose> propionate> galactose> lactate> pyruvate> glutamate> alanine. [2,11]. Depriving the adult sheep of food for 4 days markedly affected the rate of gluconeogenesis from propionate when tested with incubated hepatocytes. For example, Lomax *et al* [10] reported that the rate declined from 338.2 ± 44.0 to 124.6 ± 9.5 nmol/h/g dry weight when fed sheep were starved for this period. Thus while starvation decreased the ability to convert propionate to glucose, starvation enhanced the handling of lactate. Demigne *et al* [28] reported that hepatocytes from fed and 72 h-starved sheep gave rates of [^{14}C]propionate (2.5 mM) gluconeogenesis of 0.73 ± 0.06 and 0.48 ± 0.03 μmol/min/g cell, respectively. In contrast, rates of gluconeogenesis from 2.5 mM lactate for fed and starved sheep were 0.03 ± 0.01 and 0.18 ± 0.02 μmol/min/g cells, respectively. Again these rates confirm that propionate rather than lactate is the preferred gluconeogenic substrate for fed sheep [9].

Comparison of hepatocytes from starved sheep and rats show that while the sheep adapts to starvation by increasing its potential for handling lactate, the capacity does not reach that of the rat (*e.g.* 0.18 ± 0.02 and 0.70 ± 0.06 μmol/min/g cells, for sheep and rat, respectively) [28]. Another difference between rat and sheep hepatocytes is the peak rate of utilisation of propionate. For sheep hepatocytes this occurs at 5 mM, while for rat the peak rate is observed at only 1.0-1.2 mM propionate [28]. This reflects the differing capacity to handle propionate, a fermentation product from the rumen. Other than the specific case for propionate, the order of substrate preference for gluconeogenesis does not differ substantially between rat and sheep hepatocytes from starved animals: fructose> dihydroxyacetone> pyruvate> lactate [29].

4.2.5 Responses to catecholamines, glucagon and insulin

There seems little doubt that hepatocytes from sheep possess appropriate receptors [22] and respond well to each of these hormones. Studies by Morand *et al.*, [30] have shown that glycogenolysis is stimulated by catecholamines involving an increase in cytosolic concentrations of both cAMP and Ca^{2+}. However, in contrast to rat hepatocytes, glucagon was not found to induce an increase in cytosolic Ca^{2+}. Even so, glucagon was more efficient than epinephrine in promoting the production of cAMP and the relatively small increase in cAMP (*cf.* rat hepatocytes) by glucagon was able to induce a marked activation of phosphorylase *a*. In addition, the α_1- and β-adrenergic agonists (phenylephrine and isoproterenol), were strong activators of phosphorylase *a* [30]. Glycogen synthase activity was decreased by glucagon or α_1 adrenergic agents and increased by insulin. Time course studies showed that glycogen synthase inactivation by epinephrine or glucagon did not require prior activation of phosphorylase [14]. Furthermore, insulin's activating effect on glycogen synthase was independent of the phosphorylase *a* level. Insulin also appeared to antagonize the effects of glucagon (but not α_1 agonists) on glycogen metabolism in isolated sheep hepatocytes by counteracting the rise in cAMP.

In other studies Faulkner and Pollock [9] reported that neither glycogenolysis nor gluconeogenesis (from lactate and pyruvate or propionate) was stimulated by vasopressin. This contrasts with rat hepatocytes that respond well to vasopressin. In general, it would appear that hormonal stimulation of propionate gluconeogenesis from hepatocytes of adult sheep [9] is less than for lambs [2,11]. Insulin has been reported to inhibit glucagon-mediated gluconeogenesis from lactate plus pyruvate or fructose [2] and activate carnitine palmitoyltransferase I in mitochondria from insulin-treated sheep hepatocytes [6]. It is also worthy of note that stimulation of gluconeogenesis is retained in hepatocytes isolated from lambs that have undergone prior treatment with glucagon [5] and that hepatocytes in primary culture maintain their responsiveness to glucagon and insulin [19].

4.2.6 Effects of fatty acids on gluconeogenesis

Short-chain fatty acids, such as propionate, are important for the ruminant as they provide the carbon skeleton for gluconeogenesis. In addition, short-chain fatty acids such as acetate and butyrate, also produced by microbial fermentation in the rumen, may influence rates of gluconeogenesis. Pyruvate carboxylase is allosterically activated by acetyl-CoA and butyryl-CoA [31] and, these fatty acids may stimulate gluconeogenesis by increasing the concentrations of the corresponding CoA derivatives. Thus the general question of regulation of ovine hepatic gluconeogenesis by fatty acids has been explored by several groups. Butyrate appears not to be a gluconeogenic substrate for ovine liver cells (confirmed for lambs that were non-ruminant, [2]) and thus can be used to assess the effects on the rate of gluconeogenesis from other substrates. Butyrate (2 mM) resulted in a marked stimulation of glucose formation from galactose, with a much reduced effect from propionate [2]. However, this stimulatory effect of butyrate on gluconeogenesis from propionate may be restricted to hepatocytes from young non-ruminant lambs [2] as several other authors, using hepatocytes from sheep, have reported butyrate to inhibit gluconeogenesis from this substrate [6,8,10,11]. Interestingly, neither oleate [8], nor acetate [11] had any effect as stimulants of gluconeogenesis from propionate. This would seem at odds with findings of Chow and Jesse [6] who, in an elegant study, showed that tetradecylglycidic acid, an inhibitor of carnitine palmitoyltransferase I, inhibited glucose formation from propionate. Furthermore they showed that neither butyrate nor caproate were able to replace octanoate in overcoming the inhibition. This led the authors to conclude that β-oxidation was required to support gluconeogenesis from propionate and that intermediates from the metabolism of butyrate were responsible for the inhibition of gluconeogenesis. Overall, this suggests that oxidation of some of the propionate is providing energy for its own conversion to glucose.

4.2.7 Nitrogen metabolism

Ruminants, even when fed, depend on hepatic glucose formation, largely from propionate. But fed sheep also can absorb large amounts of ammonia from the gastrointestinal tract and this is converted by the liver to urea. This has the potential to deplete amino acid reserves, as some of the urea nitrogen originates from amino acids. When sheep are deprived of food, loss of amino acids is unavoidable as amino acids become important gluconeogenic precursors and ureagenesis is required to remove ammonium ions captured by transamination. A number of workers have studied interactions between ammonia and propionate that influence the partitioning of amino acids between oxidation and gluconeogenesis by ovine hepatocytes. It is clear that ammonia detoxification (conversion of absorbed ammonia to urea) by sheep hepatocytes increases amino acid deamination [4,12,32]. Others have shown that propionate markedly increased total ammonia utilisation for ureagenesis and diminished the rate of urea synthesis from endogenous sources [21]. Comparative studies indicate that the basal rate of ammonia utilisation by sheep hepatocytes is much lower than in rat hepatocytes. In addition, ammonia metabolism is markedly enhanced by butyrate and in contrast to rat liver cells, also by propionate [8]. The only study reporting a stimulatory effect of glucagon on ureagenesis involved primary cultures of sheep hepatocytes [19].

4.2.8 Lipid turnover

Isolated sheep liver cells convert palmitate to ketone bodies and this is increased when cells are derived from starved animals [23]. In addition, propionate was found to be a potent inhibitor of ketogenesis; it increased palmitate esterification, while causing a reduction in long-chain acyl carnitine concentrations and the cellular content of CoA. Taken together the findings point to a role for propionate or one of its metabolites as a regulator specific for ruminant hepatocyte metabolism. Thus, stimulation of ketogenesis during starvation in the sheep may relate to decreased availability of antiketogenic propionate. Carnitine palmitoyltransferase I activity appears to be essential for gluconeogenesis from propionate and is inhibited by prior incubation of hepatocytes with insulin [6].

4.3 Summary

Intact hepatocytes from lambs and sheep are proving to be a useful preparation for addressing metabolic issues of this species. By far the most successful method of preparation involves an initial perfusion of the caudate lobe.

References

1. Berry MN and Friend DS. High-yield preparation of isolated rat liver parenchymal cells: a biochemical and fine structural study. J Cell Biol 1969; 43: 506-520.
2. Clark MG, Filsell OH and Jarrett IG. Gluconeogenesis in isolated intact lamb liver cells. Effects of glucagon and butyrate. Biochem J 1976; 156: 671-680.
3. Zahlten RN and Stratman FW. The isolation of hormone-sensitive rat hepatocytes by a modified enzymatic technique. Arch Biochem Biophys 1974; 163: 600-608.
4. Mutsvangwa T, Buchanan-Smith JG and McBride BW. Effects of ruminally degradable nitrogen intake and *in vitro* addition of ammonia and propionate on the metabolic fate of L-[1-14C]alanine and L-[15N]alanine in isolated sheep hepatocytes. J Anim Sci 1997; 75: 1149-1159.
5. Ali AM and Jois M. Uptake and metabolism of propionate in the liver isolated from sheep treated with glucagon. Br J Nutr 1997; 77: 783-793.
6. Chow JC and Jesse BW. Interactions between gluconeogenesis and fatty acid oxidation in isolated sheep hepatocytes. J Dairy Sci 1992; 75: 2142-2148.
7. Cremin JDJ, Drackley JK, Hansen LR, Grum DE, Odle J and Fahey GCJ. Effects of glycine and bovine serum albumin on inhibition of propionate metabolism in ovine hepatocytes caused by reduced phenolic monomers. J Anim Sci 1995; 73: 3009-3021.
8. Demigne C, Yacoub C, Remesy C and Fafournoux P. Propionate and butyrate metabolism in rat or sheep hepatocytes. Biochim Biophys Acta 1986; 875: 535-542.
9. Faulkner A, and Pollock HT. Effects of glucagon and α- and β-agonists on glycogenolysis and gluconeogenesis in isolated ovine hepatocytes. Biochim Biophys Acta 1990; 1052: 229-234.
10. Lomax MA, Donaldson IA and Pogson CI. The effect of fatty acids and starvation on the metabolism of gluconeogenic precursors by isolated sheep liver cells. Biochem J 1986; 240: 277-280.
11. Looney MC, Baldwin RL and Calvert CC. Gluconeogenesis in isolated lamb hepatocytes. J Anim Sci 1987; 64: 283-294.
12. Luo QJ, Maltby SA, Lobley GE, Calder AG and Lomax MA. The effect of amino acids on the metabolic fate of 15NH4Cl in isolated sheep hepatocytes. Eur J Biochem 1995; 228: 912-917.
13. McBride BW and Early RJ. Energy expenditure associated with sodium/potassium transport and protein synthesis in skeletal muscle and isolated hepatocytes from hyperthyroid sheep. Br J Nutr 1989; 62: 673-682.
14. Morand C, Redon C, Remesy C and Demigne C. Non-hormonal and hormonal control of glycogen metabolism in isolated sheep liver cells. Int J Biochem 1990; 22: 873-881.
15. van 't Klooster GA, Blaauboer BJ, Noordhoek J and van Miert AS. Sulfadimidine metabolism *in vitro*: II. Comparative studies in cultured rat, goat, sheep and cattle hepatocytes. J Vet Pharmacol Ther 1993; 16: 454-461.
16. Dawson RMC, Elliott DC, Elliott WH and Jones KM. Data for Biochemical Research. Oxford: Clarendon Press, 1969.
17. Ash R, Elliott KR and Pogson CI. Preparation and utilization of isolated hepatocytes for the study of the sheep liver metabolism. Proc Nutr Soc 1976; 35: 29A.

18. Ash R and Pogson CI. Preparation and biochemical characterisation of isolated parenchymal cells from adult sheep liver. Biochim Biophys Acta 1977; 496: 475-483.
19. Luo QJ, MacRae JC and Scislowski PW. Characterization of sheep hepatocytes in primary culture. Int J Biochem 1993; 25: 719-723.
20. Townsend SF, Thureen PJ, Hay WWJ and Narkewicz MR. Development of primary culture of ovine fetal hepatocytes for studies of amino acid metabolism and insulin-like growth factors. In Vitro Cell Dev Biol Anim 1993; 29A: 592-596.
21. Garwacki S, Wiechetek M, Karlik W, Souffrant WB and Krawielitzki K. Effect of propionate on the utilization of nitrogen from 15NH4Cl for urea synthesis in hepatocytes isolated from sheep liver. Int J Biochem 1990; 22: 1185-1188.
22. Gill RD and Hart IC. Properties of insulin and glucagon receptors on sheep hepatocytes: a comparison of hormone binding and plasma hormones and metabolites in lactating and non-lactating ewes. J Endocrinol 1980; 84: 237-247.
23. Lomax MA, Donaldson IA and Pogson CI. The control of fatty acid metabolism in liver cells from fed and starved sheep. Biochem J 1983; 214: 553-560.
24. Ring JA, Ghabrial H, Ching MS, Potocnik S, Shulkes A, Smallwood RA and Morgan DJ. Hepatic uptake and excretion of [14C]sodium taurocholate by the isolated perfused fetal sheep liver. Biochem Pharmacol 1994; 48: 667-674.
25. Butler SM, Faulkner A, Zammit VA and Vernon RG. Fatty acid metabolism of the perfused caudate lobe from livers of fed and fasted non-pregnant and fasted late pregnant ewes. Comp Biochem Physiol 1988; 91: 25-31.
26. Gill W, Mitchell GEJ, Boling JA, Tucker RE, Schelling GT and DeGregorio RM. Glucagon influence on gluconeogenesis and oxidation of propionic acid and threonine by perfused ovine liver. J Dairy Sci 1985; 68: 2886-2894.
27. Savan PM, Jeacock MK and Shepherd DA. The fate of amino acids in isolated perfused livers of developing sheep. Br J Nutr 1979; 42: 467-476.
28. Demigne C, Yacoub C, Morand C and Remesy C. Interactions between propionate and amino acid metabolism in isolated sheep hepatocytes. Br J Nutr 1991; 65: 301-317.
29. Clark MG, Kneer NM, Bosch AL and Lardy HA. The fructose 1,6-diphosphatase-phosphofructokinase substrate cycle. A site of regulation of hepatic gluconeogenesis by glucagon. J Biol Chem 1974; 249: 5695-5703.
30. Morand C, Yacoub C, Remesy C and Demigne C. Characterization of glucagon and catecholamine effects on isolated sheep hepatocytes. Am J Physiol 1988; 255: R539-R546
31. Keech DB and Utter MF. Pyruvate carboxylase. II. Properties. J Biol Chem 1963; 2609-2617.
32. Mutsvangwa T, Buchanan-Smith JG and McBride BW. Interactions between ruminal degradable nitrogen intake and *in vitro* addition of substrates on patterns of amino acid metabolism in isolated ovine hepatocytes. J Nutr 1996; 126: 209-218.

Chapter 5

Elasmobranch hepatocytes

Nazzareno Ballatori[1], Gregory C. Connolly[1] and James L. Boyer[2]
[1]Department of Environmental Medicine, University of Rochester School of Medicine, Rochester, NY 14642 and [2]Liver Center, Department of Medicine, Yale University School of Medicine, New Haven, CT 06520.

Key words: bile secretion, fish liver, hepatocyte isolation, hepatocyte membrane polarity, metabolism, osmoregulation, skate hepatocytes

5.1 Unique properties of elasmobranch hepatocytes

The elasmobranchs are an ancient group of fishes represented by approximately 800 extant species, and occupying environments ranging from polar oceans to tropical seas and from surface waters to the abyss [1]. Some species are exclusively marine, others are found in fresh water ecosystems, and still others are capable of moving freely between these two media. This diverse group of organisms ranges from bottom feeders to warm-bodied predators, and includes sharks, rays, and skates. One of the elasmobranch species, the little skate (*Raja erinacea*), has been used extensively to study hepatic function. Skates are relatively abundant in many marine environments and are easily captured by net, but are not heavily harvested due to their relatively low commercial value. In captivity, skates are easy to handle and maintain in a healthy state.

5.1.1 Advantages of skate hepatocytes

Hepatic function in the skate has been studied in several experimental models, including whole animal, perfused liver, isolated hepatocytes, and subcellular organelles. Of these experimental models, freshly isolated skate hepatocytes in suspension appear to be the most versatile system. Skate hepatocytes have been utilized by several investigators to examine multiple biochemical and functional parameters. They offer several advantages when compared with other hepatocyte preparations (Table 1). As described in detail below, skate hepatocytes are relatively easy to isolate and incubate, and they exhibit high structural and metabolic viability [2]. For example,

M.N. Berry and A.M. Edwards (eds.), The Hepatocyte Review, 37–47.

Trypan blue exclusion is greater than 98% shortly after isolation and is maintained for extended incubation periods. Moreover, there is no significant release of lactate dehydrogenase (LDH) or cell K^+ during this same time interval [2]. By comparison, rat hepatocytes are much less resistant to damage with 10-20% of the isolated cells staining with Trypan blue and 20% of the LDH released shortly after isolation [3,4]. Skate hepatocyte stability also exceeds that of hepatocytes from carp (*Cyprinus carpio*), goldfish (*Carassius auratus*), and trout (*Salmo gairdneri*) where Trypan blue staining accounts for about 5% of all cells, and where approximately 10% of the LDH is normally released shortly after isolation [5-7].

Table 1. Characteristics of skate hepatocytes

- **Easy to isolate and incubate for long periods**

- **High structural and metabolic integrity:**
  - Trypan blue exclusion > 98%
  - No significant release of lactate dehydrogenase or cell K^+
  - O_2 consumption is stable for up to 8 h; is not stimulated by succinate
  - ATP and GSH levels are maintained and GSH/GSSG is high
  - Maintainance of a large negative plasma membrane potential
  - Transport and accumulate amino acids (L-alanine, methylaminoisobutyric acid), organic anions (xenobiotics, bile acids and alcohols, bumetanide), and rubidium

- **Poikilothermic**

- **Low oxygen demand**

- **Maintainance of membrane polarity**

- **Specialized features:**
  - Primitive evolutionary position
  - Unique osmotic strategy
  - Low incidence of neoplasia

5.1.2 Structural integrity

An additional advantage of isolated skate hepatocytes is that they can be obtained in clusters and thus retain structural integrity and polarity, a feature that is usually lost in other isolated hepatocyte preparations. Electron microscopy has demonstrated that approximately 80% of a typical skate cell suspension consists of three to five associated hepatocytes surrounding a single bile canaliculus [2]. These clusters of cells have intact tight junctions and desmosomes, and retain their apical and basolateral structural and functional polarity. Preliminary studies with long-term cultures of skate hepatocytes indicate that membrane polarity can be maintained for at least 5 days [8]. This preservation of a functional primary bile secretory unit is most likely related to the anatomy of the liver cell plates, which in skate liver are two to three cells thick, as opposed to one cell thick in mammals.

5.1.3 Oxygen consumption

Oxygen consumption by isolated skate hepatocytes is remarkably stable for 8 hours after isolation, suggesting no significant deterioration of metabolic activity over this time period, and it is similar to that observed in the isolated perfused intact skate liver [2]. A 55% stimulation of oxygen consumption by 2,4-dinitrophenol indicates that these isolated hepatocytes retain a satisfactory degree of respiratory control. Moreover the failure of succinate to stimulate oxygen consumption is an indication of the excellent functional integrity of the skate hepatocyte plasma membrane in these isolated cell preparations. By comparison, the ratio of succinate-stimulated respiration to endogenous respiration of isolated rat hepatocytes increases approximately 1.5-fold a few hours after isolation [9]. Elasmobranch hepatocytes also maintain a large negative plasma membrane potential, as well as high ATP and GSH levels, and GSH/GSSG ratios [10-13].

5.1.4 Additional attributes

Elasmobranch fish have additional attributes that can be exploited to study several areas of interest, including biochemical pathways, transport mechanisms, and cell volume regulation. Elasmobranchs provide a primitive biological model from which considerable insight can be gained into the evolution and regulation of metabolic processes. Because of their efficient osmoconforming strategy, elasmobranchs provide an ideal system for studying cell volume regulation and the interaction of xenobiotics with this process. Osmoregulation in skates relies heavily on their ability to produce and retain unusually high concentrations of urea (as high as 680 mM [14]), levels which would be disruptive to the structure and function of proteins and other cellular components in most other organisms. This disruptive effect of urea is compensated at least in part by the presence of high concentrations of protective osmolytes such as betaine, taurine, and polyols. It is also of interest that elasmobranchs have the lowest incidence of neoplasia of any vertebrate group [15,16].

5.2 Methods of skate hepatocyte isolation and incubation

5.2.1 Animals

Studies in our laboratory have been carried out with skates caught in the Gulf of Maine off Mount Desert Island, and delivered by boat to the Mount Desert Island Biological Laboratory. Skates are usually captured by net and kept in circulating seawater tanks during delivery to the laboratory.

Overall hepatocyte function is optimal if hepatocytes are isolated from freshly captured skates, although the animals may be kept for an extended period (>4 weeks) in large tanks equipped with circulating water. Water composition and temperature should mimic that of the animal's normal environment. If the skates are used within 2-3 days of capture, feeding is not usually required, but appropriate food (*e.g.* shrimp or squid) should be provided for longer periods of captivity. Male skates are preferred over females, because of the unusually high lipid content in hepatocytes from female skates, which makes it difficult to sediment these cells by centrifugation. Adult male *R. erinacea* are ~1 kg in weight with a liver weight of 15-30 g, a convenient size for perfusion during hepatocyte isolation.

5.2.2 Isolation and incubation protocols

The skate hepatocyte isolation protocol was devised in our laboratory by Smith and co-workers [2], based on a collagenase perfusion method described for mammalian hepatocytes [4]. The following is a description of the protocol currently utilized by our group at the Mount Desert Island Biological Laboratory.

Skates are anaesthetized by pentobarbital sodium (5 mg/kg) administration via the caudal vein. The animal's gills are perfused with 12-14°C seawater through latex rubber tubing secured to the opercula. The skate is inverted onto an operating board, with the abdominal cavity facing the investigator. The abdominal cavity is opened by midline and transabdominal incisions, tributaries to the portal vein are ligated, and bile is aspirated from the gallbladder. The skate gallbladder is embedded within the hepatic tissue, and cannot readily be excised without damaging the liver. The portal vein is cannulated with a plastic cannula (3 cm of PE-240 tubing) attached to a 3-4 cm segment of latex rubber tubing filled with heparinized, Ca^{2+}-Mg^{2+}-free elasmobranch Ringer solution (in mM, 270 NaCl, 4 KCl, 0.5 Na_2SO_4, 1 KH_2PO_4, 8 $NaHCO_3$, and 350 urea). The distal end of the portal vein cannula is sealed with a glass bead. The cannula is secured to the portal vein with two 3-0 silk sutures. In order to prevent blood from clotting within the sinusoids, the liver is immediately flushed with 25-40 ml of heparinized, Ca^{2+}-Mg^{2+}-free, elasmobranch Ringer solution at 10-15°C. This solution is delivered via a 60 ml syringe and blunt-end 16-gauge needle which is inserted into the portal vein cannula after temporarily removing the occluding glass bead. Care must be taken to avoid introduction of air bubbles which will occlude portal flow.

The liver is then quickly excised from the abdominal cavity by transecting the hepatic vein sinuses above both the right and left lobes and cutting the proximal portal vein remnant, hepatic artery, and bile duct. The liver is transferred to a temperature-regulated perfusion dish (15°C in our experiments), and is immediately perfused at a rate of 30 ml/min with heparinized, Ca^{2+}-Mg^{2+} -free elasmobranch Ringer solution maintained at 15°C. After 10 min of single-pass perfusion, a recirculating perfusion with collagenase (0.07-0.1%) in complete elasmobranch Ringer solution is initiated. Complete elasmobranch Ringer contains (in mM), 270 NaCl, 4 KCl, 0.5 Na_2SO_4, 1 KH_2PO_4, 8 $NaHCO_3$, 350 urea, 2.5 $CaCl_2$, 3 $MgCl_2$, 5 D-glucose, and 5 *N*-2-hydroxyethylpiperazine-*N*'-2-ethanesulphonic acid (HEPES)-tris(hydroxymethyl)aminomethane (Tris) at pH 7.5. The collagenase perfusion is continued for 5-10 min or until gentle pressure on the surface of the liver results in a permanent depression. At this point the liver is transferred to a beaker containing 80 ml of complete elasmobranch Ringer solution supplemented with deoxyribonuclease II (100 U/ml) (this does not appear to be essential) and the liver capsule and connective tissue are torn apart with forceps. The resulting cell suspension is filtered sequentially through gauze and 60 μm nylon mesh, and is then centrifuged at $100 \times g$ for two min. The hepatocytes to be used for short-term incubation are washed twice in complete elasmobranch Ringer solution (15°C) and resuspended at a concentration of approximately 20-40 mg wet weight/ml (~$2\text{-}4 \times 10^6$ cells/ml) in complete elasmobranch Ringer solution.

The cells are preincubated at 15°C in 50 ml polypropylene tubes for

20-30 min in room air, before the start of each experiment, being maintained in suspension by gentle agitation of the tubes. Skate hepatocytes and hepatocyte clusters also readily attach, within 10-30 min, to plastic culture dishes or collagen I-, or collagen IV-coated glass cover slips, allowing conventional and fluorescent microscopic evaluation of the cells [17]. We are currently characterising the properties of cells cultured for several days on plastic dishes at 12°C, and although initial results are quite promising, theses studies are not yet complete.

5.3 Applications

5.3.1 Biochemical pathways

Isolated skate hepatocyte suspensions have been employed in the study of several biochemical pathways, including gluconeogenesis, ketone body synthesis and oxidation, and methylamine and amino acid metabolism, as reviewed by Ballantyne [18].

The liver is the primary site of gluconeogenesis in elasmobranch fishes [19], and studies have characterized the relative contribution of a variety of precursors to the production of glucose in isolated skate hepatocytes [10,19]. Of the substrates tested, glycerol as precursor elicits the highest rates of glucose production, followed by the amino acids serine and alanine [19]. Lactate as precursor resulted in the lowest gluconeogenic rates [19]. Our studies indicate that alanine is not a major gluconeogenic precursor in isolated skate hepatocytes [10].

Elasmobranchs differ from most other vertebrates in the importance of "ketone bodies", such as 3-hydroxybutyrate and acetoacetate, as a source of energy during non-starvation conditions. Their increased reliance on these metabolites is most likely attributed to the absence of plasma albumin, the major nonesterified fatty acid binding protein in the blood, and the resultant inability to transport lipid carbon from the liver to extrahepatic tissue in the form of nonesterified fatty acids. The use of ketone bodies is an effective alternative because they are much more soluble than nonesterified fatty acids, and therefore do not require plasma carrier proteins. Isolated skate hepatocytes have been used to support the notion that the liver is essentially a net exporter of ketone bodies for use by other tissues. Mommsen and Moon [19] have demonstrated that 3-hydroxybutyrate is a poor oxidative substrate in isolated skate hepatocytes, and that oleate and leucine are strong precursors for the synthesis of ketone bodies in these same isolated hepatocyte incubations.

Isolated hepatocytes from *R. erinacea* have also been used to examine the contribution of several amino acids, including alanine, serine, and leucine, as oxidative substrates in the skate liver [18,19]. Similar rates of CO_2 production were observed for the three amino acids tested [19].

Skate hepatocytes maintain high intracellular concentrations of taurine, β-alanine, and sarcosine as a means of counteracting the disruptive effects of unusually high concentrations of urea (350 mM) on protein and cellular structure. Studies by Ballantyne *et al.* [20], demonstrating that hypoosmotic stress results in increased oxidation of sarcosine in isolated hepatocyte suspensions, reflect a secondary role of these amino acids in elasmobranchs.

5.3.2 Transport mechanisms

Skate hepatocyte preparations are a particularly attractive experimental model for transport studies because the isolated cells maintain their ultrastructure and membrane polarity, closely resembling the *in vivo* condition [2,17,21].

This preserved polarized bile secretory unit has been used to study bile salt uptake and efflux mechanisms [17,22-26], and to characterize ATP receptors [27]. Miller *et al.* [17] used skate hepatocyte clusters to identify and characterize an active microtubule-dependent secretion mechanism for a fluorescent bile-salt derivative.

5.3.3 Cytotoxic mechanisms

Toxicological research employs a multitude of approaches and models to identify hazardous substances and to examine mechanisms of toxicological responses in biological systems. *In vitro* models, especially those that closely resemble the more complex *in vivo* condition, offer several advantages for research on cytotoxic mechanisms. These benefits include better control over experimental conditions, the elimination of biological factors which may interfere with mechanistic studies, reduced variability, and repeated and controlled sampling over time.

Isolated skate hepatocytes provide a powerful *in vitro* model system for the study of cytotoxic mechanisms [11,12,28-32]. They have been used to demonstrate that mercury impairs cell volume regulation at multiple sites, including volume regulatory taurine channels and Na^+ and K^+ permeability pathways [28]. The mechanisms of mercury-induced intracellular Ca^{2+} signalling have also been characterized using skate hepatocyte preparations [33]. Studies of mercury-induced cell injury in hepatocytes of *R. erinacea* extend studies in other systems indicating that mercury has many effects on plasma membrane permeability, including inhibition of amino acid transport, decreased Na^+-K^+-ATPase activity as reflected by ^{86}Rb uptake, impairment of volume regulatory mechanisms, and increased membrane permeability to monovalent cations [11]. The effects of nocodazole, a microtubule disrupting agent, on the structure and function of the cytoskeleton in isolated skate hepatocytes have been investigated [21].

5.3.4 Cell volume regulation

Because of their efficient osmoregulatory mechanisms, elasmobranch tissues provide an excellent *in vitro* model system for studying volume regulation [34-37]. Under normal seawater conditions, elasmobranchs maintain osmotic balance with their environment by retaining high concentrations of NaCl and nitrogenous compounds (urea and methylamines) in extracellular body fluids, and high intracellular concentrations of organic solutes, including taurine, β-alanine, betaine, sarcosine, and urea. When exposed to dilute seawater, the hypoosmotic stress leads initially to an increase in cell volume, but this is corrected by the release and/or metabolism of intracellular solutes, and by branchial and renal elimination of specific solutes. Whereas in cells of most terrestrial vertebrates, including liver cells [38,39], regulatory volume decrease (RVD) is accomplished largely by efflux of K^+ and Cl^-, elasmobranchs rely largely on intracellular organic solutes for volume adjustments, although they also release K^+ and Cl^-. Taurine is one of the major intracellular osmolytes in marine organisms [34,37,40,41]. Intracellular concentrations of this amino acid in marine organisms are as high as 190 mM [42]. In skate hepatocytes taurine is present in relatively high concentrations, approximately 65 mM [43], whereas plasma taurine levels are only 0.13 mM [44].

The intracellular signals that mediate volume-sensitive taurine release have not yet been identified for any tissue, of any species. However studies with skate hepatocyte suspensions have identified some of the key properties of this volume regulatory mechanism. Taurine efflux is mediated by a membrane channel that requires intracellular ATP or a non-hydrolysable ATP analogue, and which is inhibited by mercury and sulfhydryl reagents [12,28-30,45]. The channel is localized to the basolateral membrane of the liver cell [30], and is anion selective with a taurine/chloride permeability ratio of 0.17 [46].

5.4 Perspectives

The inability to maintain adult hepatocytes in a fully differentiated state in cell culture for prolonged periods has prevented wider use of hepatocyte cell culture systems. In this review we describe our experiences and those of others, with the preparation and experimental use of hepatocytes isolated from the livers of elasmobranchs, primarily the little skate, *Raja erinacea*. The majority of these studies have been performed with freshly isolated cells or short-term (less than one day) cell culture systems, and the evidence to date suggests that they are more stable preparations than their more widely used mammalian counterparts. These encouraging results suggest that additional effort should be directed towards the examination of long-term hepatocyte culture systems from lower vertebrates.

References

1. Evans DH, Ed. The Physiology of Fishes. Boca Raton: CRC Press, 1997.
2. Smith DJ, Grossbard M, Gordon ER and Boyer JL. Isolation and characterization of a polarized isolated hepatocyte preparation in the skate Raja erinacea. J Exp Zool 1987; 241: 291-296.
3. Barnabei O, Leghissa G and Tomasi V. Hormonal control of the potassium level in isolated rat liver cells. Biochim Biophys Acta 1974; 362: 316-325.
4. Seglen PO. Preparation of isolated liver cells. In: Prescott DM, Ed. Methods in Cell Biology Vol 13. New York: Academic Press, 1976: 29-83.
5. French CJ, Mommsen TP and Hochachka PW. Amino acid utilization in isolated hepatocytes from rainbow trout. Eur J Biochem 1981; 113: 311-317.
6. Saez L, Goicoechea O, Amthauer R and Krauskopf M. Behavior of RNA and protein synthesis during the acclimitization of the carp. Studies with isolated hepatocytes. Comp Biochem Physiol 1982; 72B: 31-38.
7. van Waarde A and Kesbeke F. Nitrogen metabolism in goldfish, Carassius auratus (L.). Comp Biochem Physiol 1981; 70B: 499-507.
8. Ballatori N, Barnes DM, Seward D, O'Connell D, Toure J, Dobak L, Henson JH, Runnegar M, Miller DS, Xie Z and Boyer JL. A long-term primary culture model of hepatocytes isolated from the little skate, Raja erinacea. Bull MDI Biol Lab 1998; 37: 85-86.
9. Baur H, Kasperek S and Pfaff E. Criteria of viability of isolated liver cells. Hoppe-Seyler's Z Physiol Chem 1975; 356: 827-838.
10. Ballatori N and Boyer JL. Characteristics of L-alanine uptake in freshly isolated hepatocytes of elasmobranch Raja erinacea. Am J Physiol 1988; 254: R801-R808.
11. Ballatori N, Shi C and Boyer JL. Altered plasma membrane ion permeability in mercury-induced cell injury: studies in hepatocytes of elasmobranch Raja erinacea. Toxicol Appl Pharmacol 1988; 95: 279-291.

12. Ballatori N, Truong AT, Jackson PS, Strange K and Boyer JL. ATP depletion and inactivation of an ATP-sensitive taurine channel by classic ion channel blockers. Mol Pharmacol 1995; 48: 472-476.
13. Ballatori N, Wondergem R and Boyer JL. Effects of L-alanine on membrane potential, potassium (86Rb) permeability and cell volume in hepatocytes from Raja erinacea. Biochim Biophys Acta 1988; 946: 261-269.
14. Browning J. Urea levels in plasma and erythrocytes of the southern fiddler skate, Tryogogorhina fasciata guanerius. J Exp Zool 1978; 203: 325-330.
15. Harshbarger JC. Role of the registery of tumours in lower animals in the study of environmental carcinogenesis in acquatic animals. Ann NY Acad Sci 1977; 298: 280-289.
16. Harshbarger JC, Charles AM and Spero PM. Collection and analysis of neoplasms in sub-homeothermic animals from a phyletic point of view. In: Dawe CJ, Harshburger JC, Kondo S, Sagimura T and Takayama S, Eds. Phyletic Approaches to Cancer. Tokyo: Jap Soc Sci Press, 1981: 357-384.
17. Miller DS, Fricker G, Schramm U, Henson JH, Hager DN et al. Active microtubule-dependent secretion of a fluorescent bile salt derivative in skate hepatocyte clusters. Am J Physiol 1996; 270: G887-G896.
18. Ballantyne JS. Jaws: The inside story. The metabolism of elasmobranch fishes. Comp Biochem Physiol 1997; 118B: 703-742.
19. Mommsen TP and Moon TW. The metabolic potential of hepatocytes and kidney tissue in the Little skate, Raja erinacea. J Exp Zool 1987; 244: 1-8.
20. Ballantyne JS, Moyes C and Moon TW. Osmolarity affects oxidation of sarcosine by isolated hepatocytes and mitochondria from euryhaline elasmobranch. J Exp Zool 1986; 238: 267-271.
21. Henson JH, Capuano S, Nesbitt D, Hager DN, Nundy S et al. Cytoskeletal organization in clusters of isolated polarized skate hepatocytes: structural and functional evidence for microtubule-dependent transcytosis. J Exp Zool 1995; 271: 273-284.
22. Blumrich M, Petzinger E and Boyer JL. Characterization of bumetanide transport in isolated skate hepatocytes. Am J Physiol 1993; 265: G926-G933.
23. Fricker G, Dubost V, Finsterwald K and Boyer JL. Characteristics of bile salt uptake into skate hepatocytes. Biochem J 1994; 299: 665-670.
24. Fricker G, Wossner R, Drewe J, Fricker R and Boyer JL. Enterohepatic circulation of scymnol sulfate in an elasmobranch, the little skate (Raja erinacea). Am J Physiol 1997; 273: G1023-G1030.
25. Fricker G, Hugentobler G, Meier PJ, Kurz G and Boyer JL. Identification of a single sinusoidal bile salt uptake system in skate liver. Am J Physiol 1987; 253: G816-G822.
26. Smith DJ, Grossbard M, Gordon ER and Boyer JL. Taurocholate uptake by isolated skate hepatocytes: effect of albumin. Am J Physiol 1987; 252: G479-G484.
27. Nathanson MH and Mariwalla K. Characterization and function of ATP receptors on hepatocytes from the little skate Raja erinacea. Am J Physiol 1996; 270: R561-R570.
28. Ballatori N and Boyer JL. Disruption of cell volume regulation by mercuric chloride is mediated by an increase in sodium permeability and inhibition of an osmolyte channel in skate hepatocytes. Toxicol Appl Pharmacol 1996; 140: 404-410.
29. Ballatori N and Boyer JL. Taurine transport in skate hepatocytes. II. Volume activation, energy, and sulfhydryl dependence. Am J Physiol 1992; 262: G451-G460.
30. Ballatori N, Simmons TW and Boyer JL. A volume-activated taurine channel in skate hepatocytes: membrane polarity and role of intracellular ATP. Am J Physiol 1994; 267: G285-G291.

31. Blazka ME, Yoshida M and Shaikh ZA. Comparison of cadmium, mercury and calcium accumulations by isolated hepatocytes of the small skate (Raja erinacea) and rat. Comp Biochem Physiol - C: Comp Pharmacol Toxicol 1992; 101: 631-639.
32. Simmons TW, Hinchman CA and Ballatori N. Polarity of hepatic glutathione and glutathione S-conjugate efflux, and intraorgan mercapturic acid formation in the skate. Biochem Pharmacol 1991; 42: 2221-2228.
33. Nathanson MH, Mariwalla K, Ballatori N and Boyer JL. Effects of Hg^{2+} on cytosolic Ca^{2+} in isolated skate hepatocytes. Cell Calcium 1995; 18: 429-439.
34. Chamberlin ME and Strange K. Anisosmotic cell volume regulation: a comparative view. Am J Physiol 1989; 257: C159-C173.
35. Forster RP and Goldstein L. Intracellular osmoregulatory role of amino acid and urea in marine elasmobranchs. Am J Physiol 1976; 230: 925-931.
36. Hoffmann EK and Simonsen LO. Membrane mechanisms in volume and pH regulation in vertebrate cells. Physiol Rev 1989; 69: 315-382.
37. King PA and Goldstein L. Organic osmolyes and cell volume regulation in fish. Mol Physiol 1983; 4: 53-66.
38. Boyer JL, Graf J and Meier PJ. Hepatic transport systems regulating pH_i, cell volume, and bile secretion. Ann Rev Physiol 1992; 54: 415-438.
39. Corasanti JG, Gleeson D and Boyer JL. Effects of osmotic stresses on isolated rat hepatocytes I. Ionic mechanisms of cell volume regulation. Am J Physiol 1990; 258: G290-G298.
40. Finchman DA, Wolowyk MW and Young JD. Volume sensitive taurine transport in fish erythrocytes. J Membr Biol 1987; 96: 45-56.
41. Gilles R. Intracellular free amino acids and cell volume regulation during osmotic stresses. In: Jorgenson CB and Skadhauge E, Eds. Osmotic and Volume Regulation. A. Benzon Symposium XI. Munksgaard: 1978: 470-491.
42. Costa CJ, Pierce SK and Warrren MK. The intracellular mechanisms of salinity tolerance in polychaetes: Volume regulation by isolated Glycera dibranchiata red coelomocytes. Biol Bull 1980; 159: 626-655.
43. Ballatori N and Boyer JL. Taurine transport in skate hepatocytes. I. Uptake and efflux. Am J Physiol 1992; 262: G445-G450.
44. Boyd TA, Cha CJ, Forster RP and Goldstein L. Free amino acids in tissue of the skate Raja erinacea and the stingray Dasyatis sabina: effects of environmental dilution. J Exp Zool 1977; 199: 435-442.
45. Ballatori N and Boyer JL. ATP regulation of a swelling-activated osmolyte channel in skate hepatocytes. J Exp Zool 1997; 279: 471-475.
46. Jackson PS, Churchwell N, Ballatori N, Boyer JL and Strange K. Swelling-activated anion conductance in skate hepatocytes: regulation by Cl^- and ATP. Am J Physiol 1996; 270: C57-C66.

Chapter 6

Isolation and cultivation of teleost hepatocytes

Thomas Braunbeck[1] and Helmut Segner[2]
[1] *Aquatic Ecology and Ecotoxicology Group, Department of Zoology, University of Heidelberg, Im Neuenheimer Feld 230, D-69120 Heidelberg, Germany, and* [2] *Department of Chemical Ecotoxicology, UFZ Centre for Environmental Research, Permoserstr. 15, D-04318 Leipzig, Germany*

Key words: culture conditions, fish, isolation techniques, hepatocyte attachment, primary culture, substrata, teleost hepatocytes

6.1 Introduction

After the original description by Berry and Friend in 1969 of the high yield isolation of rat liver parenchymal cells by collagenase perfusion [1], the use of isolated hepatocytes was soon established as an ideal system for the study of many aspects of hepatic metabolism [2]. In 1976, Birnbaum and colleagues [3] adapted the technique of hepatocyte isolation for studies of hormone-stimulated glycogenolysis in goldfish cells. Since then, isolated fish hepatocytes have been used intensively in fresh suspensions and in primary culture as a powerful, yet simple tool for the investigation of piscine hepatic functions. Cultured hepatocytes have proved to be well suited for applications not only in basic fish physiology, but also in the fields of pharmacology and environmental toxicology [4-7].

Initial studies utilized isolated fish hepatocytes over only limited time periods [7], but in the mid-1980's methods for long-term culture of fish hepatocytes were reported. Klaunig and co-workers [8,9] were the first to establish a system for the cultivation of rainbow trout and channel catfish hepatocytes for a period of up to two weeks. Since then, there have been about 250 reports on isolation and culture of teleost hepatocytes, and primary cultures of teleost hepatocytes have been used to investigate a variety of aspects of hepatic structure and function including cell morphology, temperature acclimation, cell senescence, establishment of cell-to-cell interactions, lipoprotein, protein and carbohydrate metabolism, nutrient metabolism and endocrine regulation, ammoniogenesis, energy metabolism, vitellogenin synthesis, hormone receptor expression, processes of endocytosis, CYP-related metabolism, induction of metallothioneins and

M.N. Berry and A.M. Edwards (eds.), The Hepatocyte Review, 49–71.

heat-shock proteins, mutagenesis, pharmacological processes and general toxicological effects. Since standard laboratory procedures for isolation and cultivation of teleost hepatocytes have recently been reviewed in detail [4,5], the following review will focus on recent developments in fish hepatocyte culture systems and on more general aspects of the isolation of piscine liver cells.

6.2 Selection of fish species as donors of hepatocytes

To date, rainbow trout (*Oncorhynchus mykiss*) has received by far the most attention among the 30 fish species studied with respect to the isolation and cultivation of hepatocytes. More than 60% out of approximately 250 reported investigations have focused on rainbow trout hepatocytes, and methodological studies especially have been devoted to rainbow trout (Table 1). Selection of species for hepatocyte preparations has been based on a number of criteria. Most likely, the major reason for selecting rainbow trout is that it is, apart from zebrafish, the best-studied fish species [10]; moreover, it can easily be obtained from local hatcheries and can be held in most laboratories without trouble. Another consideration in the selection of species may be an interest in a specific metabolic function of a particular species of fish. For instance, antifreeze glycoproteins and acclimation to very low water temperatures have been studied in Antarctic *Pagothenia borchgrevinki* by O'Grady as well as Haschemeyer and colleagues [11,12].

Usually, fish of an average size of 100-300 g are preferred; the use of fish of smaller size (4-10 cm) usually requires special perfusion and surgical techniques as well as pooling of hepatocytes from several individuals [3,13]. The use of small aquarium fish for hepatocyte isolation has been limited due to the size; however, recently techniques for hepatocyte isolation have been developed for such species as well [14].

6.3 Cell isolation techniques

6.3.1 Selection of perfusion media

The vast majority of techniques used for the isolation of fish hepatocytes can be traced back to the two-step collagenase-perfusion technique followed by mechanical disruption, as was originally developed for the isolation of mammalian hepatocytes [1,15,16]. After an initial pre-perfusion step to clear the liver of blood, the extracellular matrix is enzymatically digested by perfusion with collagenase-containing media. After enzymatic dissociation,

Table 1. Selected methodological studies with isolated fish hepatocytes

Species / Subject of the study	References
Rainbow trout (*Oncorhynchus mykiss*)	
As a model in toxicological research	[4,46,118]
Hepatocyte cultures for *in vitro* carcinogenicity studies	[8,9]
As a model in physiological research	[7,118]
Separation of different populations of hepatocytes	[5,73,107]
Isolation of biliary epithelial cells	[67]
Comparison of different cell culture substrates	[49,99]
Long-term culture of hepatocytes	[113]
Senescence of hepatocytes in culture	[29,119]
Comparison of monolayer and aggregate cultures	[64,120]
Use of trout serum as a coating of culture dishes	[121]
Atlantic salmon (*Salmo salar*)	
Preparation of Atlantic salmon hepatocytes	[84]
Sockeye salmon (*Oncorhynchus nerka*)	
Preparation of sockeye salmon hepatocytes	[84]
Japanese eel (*Anguilla japonica*)	
Preparation of isolated cells from Japanese eel	[88,101,122]
Carp (*Cyprinus carpio*)	
Isolation of hepatocytes from common carp	[54,55]
Channel catfish (*Ictalurus punctatus*)	
Preparation of hepatocytes from channel catfish	[8,123]
Black bullhead catfish (*Ictalurus melas*)	
Separation of different hepatocyte populations	[74]
Striped bass (*Morone saxatilis*)	
Striped bass hepatocytes as a model in toxicology	[56]

some authors include a post-perfusion step in order to rapidly remove any collagenase prior to mechanical separation, washing of hepatocytes and eventual transfer to the incubation medium.

6.3.2 Composition of the pre-perfusion medium

For the initial pre-perfusion, the liver is cannulated - usually *via* the portal vein - and cleared of blood. The composition of the pre-perfusion medium varies with the fish species [7]. Most frequently, Hank's buffered salt solutions are used for this purpose; however, Ca^{2+}- Mg^{2+}-free phosphate-buffered saline [17,18], Leibowitz L-15 [7,19], Ca^{2+}-free Krebs-Ringer [20-22], Krebs-Henseleit buffer [23,24], as well as specialized buffers such as Ca^{2+}-free "*Salmo* buffer" [25] or Ca^{2+}-free "teleost Ringer" [26,27] have successfully been used. The pH of the buffers is dependent on the fish species; a range of 7.3 to 8.0 has been reported. Moon and co-workers [7] suggested the use of lower hydrogen ion concentrations (pH 7.6) than employed in mammalian studies (pH 7.4).

Addition of a chelating agent such as EDTA or EGTA to remove Ca^{2+} (and thus irreversibly break the Ca^{2+}-dependent cell junctions [28]) may considerably improve subsequent cell yield. Although glucose has been recommended as an energy source [7], it may be omitted in the pre-perfusion step [10,29]. Finally, heparin may be added to reduce blood clotting [30-42]. However, application of heparin has also been reported to cause undesired side effects [7].

In our own laboratory, the liver of rainbow trout is cleared by *in situ* pre-perfusion with well-oxygenated Ca^{2+}- and Mg^{2+}-free Dulbecco's medium [29] adjusted to pH 7.55, equivalent to the pH in fish blood [7], and supplemented with 2 mM EDTA; the pre-perfusion medium is discarded after single use. The temperature of the perfusion buffers depends on the species used, as well as on the acclimation temperature. In general, most fish studies have been performed at room temperature, although some experiments have been done at 0-4°C [11,12].

6.3.3 Enzymatic digestion of the liver

In the second step, one or several enzymes are added to the perfusion medium for the purpose of degrading extracellular connective tissue, thus facilitating dispersal of the cells. Although in theory, removal of Ca^{2+} should be sufficient to loosen the desmosomal connections between hepatocytes, degradative enzymatic activity appears to be necessary for the digestion of the extracellular matrix [4]. Usually, collagenase (EC 3.4.24.3) obtained from crude extracts of *Clostridium histolyticum* is selected, which digests intercellular collagen [4-6]. Since collagenase preparations differ in composition between batches and possess variable proteolytic side activities [43], the choice of the collagenase lot has proved to be critical. Moreover, different collagenase batches may have variable efficacy for different species. An example of the damaging effects of crude collagenase is given by Segner and co-workers [44], who showed that the number of membrane-bound insulin receptors of isolated carp hepatocytes critically depends on the collagenase type used for cell isolation. For the isolation of teleost hepatocytes, collagenase types I, II and IV have generally been applied; in addition, in our own laboratories collagenase types D and H (Clostridiopeptidase A) have successfully been used [6,29]. Normally, collagenase concentrations range between 0.30 and 0.50 mg/ml [5,45,46]. The composition of the collagenase buffer is usually similar to that of the pre-perfusion except for the absence of chelators [29]. The perfusion medium is also well-oxygenated and buffered close to the pH optimum of collagenase (pH 7.5).

Ca^{2+} are usually added to the isolation medium, since collagenase requires Ca^{2+} for enzymatic activity [28]. Although Ca^{2+} ions are essential for collagenase activation [16], the need to add Ca^{2+} is uncertain since crude collagenase preparations may contain sufficient enyme-bound Ca^{2+}. Moreover, Ca^{2+} ions have repeatedly been reported to decrease cell quality and to lead to cell clumping and low cell yields [5,7]. Since Mg^{2+} compete with Ca^{2+}, Mg^{2+} has frequently been excluded from the enzyme medium.

In rare exceptions, hyaluronidase [47,48], protease [49], elastase and nagarase [5] or trypsin [47] have been used to dissociate fish hepatocytes. (See discussion of use of purified collagenase and elastase in Chapter 1) As an alternative to enzymatic digestion, non-enzymatic isolation of liver cells (by EDTA perfusion) has been successfully employed for at least two teleosts, sea raven [50] and brown bullhead catfish [5]. Although this approach may be particularly advantageous when the use of collagenase and its protease side-activities has to be avoided, *e.g.* in membrane and receptor studies, this method appears less robust and reproducible than collagenase perfusion [6].

6.3.4 Dissection techniques

Although many anesthetics for fish are available, the anesthetics of choice are ethyl-3-aminobenzoate [29,51] and tricain methanesulphonate (MS-222, [4]). Other anesthetics used include urethane, chlorobutanol, or 2-phenoxyethanol [5]. It should be noted, however, that choice of anesthetic may influence experimental results obtained subsequently with isolated cells [52]. Physical methods such as decapitation and stunning by a blow to the head have also been used to incapacitate fish, especially when smaller species are used, from which the liver is immediately excised.

Following anesthesia, fish are disinfected by swabbing with 70% ethanol and placed ventral or lateral side up on a surgical board. The abdominal and pericardial cavities are opened by a midline incision from the anus to the pectoral fins. A second incision along the lateral line is made from the urogenital pore to the pectoral girdle thus exposing the heart and abdominal cavity. In species such as salmonids, which have a well-developed portal vein, a blunt cannula is inserted into the hepatic portal or intestinal vein, and an orthograde perfusion is performed [9,29]. In order to avoid excessive pressure, blood and perfusate are released by ventrally severing the Ductus Cuvieri, the Sinus venosus or the atrium of the heart. For species such as many cyprinid or flatfish species, which lack a distinct portal vein, but have multiple venous inputs into the liver, a perfusion can be performed through the bile duct [53] or the Arteria coeliaca [54,55]. Alternatively, the liver is perfused in the retrograde direction via the major hepatic vein by means of a cannula placed in the heart [56]. So far, no obvious differences have emerged between hepatocytes prepared by orthograde or by retrograde perfusion techniques [5].

Depending on the temperature as well as the species and size of the fish, the times for pre-perfusion and isolation perfusion have been reported to vary between 1 and 5 min, and 15 and 90 min, respectively. The perfusion may be performed completely *in situ*, or the liver may be removed between the first and second perfusion step and placed either on a platform [57], a perfusion apparatus based on a modified laboratory support [29], or a watch glass [48] in order to facilitate the recirculation of the collagenase solution. The individual choice depends on factors such as ease of cannulation or liver removal, accessibility of vessels and liver, rapidity of blood clotting, location of the gall bladder, length of the bile duct etc. During perfusion, the liver may be cleared of the gall bladder and adhering connective tissue.

Once the liver tissue has become soft and malleable, enzymatic digestion is followed by mechanical disruption with a stainless steel comb, razor blades, scissors, forceps or by carefully scraping the cells from the connective tissue with a sterile transfer pipette.

6.3.5 Removal of debris from freshly isolated hepatocyte suspensions

To remove cell aggregates and connective tissue remnants, the crude cell suspension is gently massaged through several filter screens of varying mesh sizes (usually 50 to 250 μm). This clean-up may be effected by passing the freshly isolated cell suspensions through sterile nylon screening [58], gauze [29,54], plankton netting [7] or a combination of the above [4]. Hepatocytes are collected in buffer, washed repeatedly by low-speed centrifugation (60-120 × *g*) for 2 to 10 min, depending on species, cell size, and lipid content) and resuspended in the final culture medium. Hepatocytes isolated from species with excessive lipid storage require longer centrifugation times. The washing buffers used are normally similar to those used for the pre-perfusion, but more complete solutions such as Minimal Essential Medium [41], Leibowitz L15 [9] or RPMI 1640 [56] may be used.

6.3.6 Alternative isolation procedures

Several investigators have used alternatives to the two-step perfusion procedure outlined above. Bhattacharya and co-workers [59] as well as Plisetskaya and colleagues [60,61] perfused the livers of Atlantic salmon with HEPES-buffered saline, then excised the livers. The livers were cut into pieces and incubated in a 0.5% collagenase solution in a shaking water bath until digested. Especially for studies with small fish such as zebrafish [62] or mummichog [13], but also for smaller specimens of larger species such as carp [54], yellow perch [63], coho salmon [61] or goldfish [3], the liver is excised without any perfusion, minced and placed in a medium with a digestive enzyme. Cells are dispersed by shaking or magnetic stirring. In a modified procedure, juvenile zebrafish are cleared of blood with Ca^{2+}- and Mg^{2+}-free saline by total body perfusion via the heart, and the liver is then cut into pieces and immersed in collagenase-containing buffer solution [14].

Table 2. Composition of fish liver tissue (after [123], modified)

Component	Relative Volume (%)				Isolation	Suspected pathological role
	Rainbow trout (*Oncorhynchus mykiss*)	**Golden ide** (*Leuciscus idus melanotus*)	**Female zebrafish** (*Danio rerio*)	**Male zebrafish** (*Danio rerio*)		
Stroma	5.3	8.2	7.6	8		
Hepatocytes	80.1	81.6	81.5	82	Low-speed pellet: 60 - 120 × g	
Non-hepatocytes	14.6	10.2	10.9	10	High-speed pellet: > 600 × g	
Extracellular spaces	94	6.8	7.5	7		
Biliary cells	1.2	2.1	1.8	1.9		Hepato-carcinogenesis [67]
Endothelial cells	2.2	1	1.5	1.3		Dioxin-sensitive [66]
Fat-storing cells	1.3	0.2	0.2	0.2		Spongiosis hepatis [34]
Macrophages	0.6	0.1	*	*		
Reference	[123] after [53]	[123]	Own unpublished data and recalculated from [20]			

* Relative volume of macrophages is highly variable, depending on physiological status, rate of parasitic infection and degree of chemical contamination.

6.3.7 Cellular components of primary liver cell suspensions

Cell suspensions obtained after perfusion of teleost fish liver contain a variety of cell types (Table 2). Hepatocytes clearly represent the predominant cell type, occupying more than 80% of the liver volume [64,65]. Non-hepatocytic elements in fish liver include biliary epithelial cells, sinusoidal endothelial cells, persinusoidal fat-storing cells, as well as macrophages. As in the intact liver of fish [51], the relative proportion of macrophages, which may functionally replace the Kupffer cells present in mammalian liver, is particularly variable depending on the physiological status, health (rate of infection, etc.) and the degree of environmental stress to the donor fish. Although the relative number of non-hepatocytes is small, they are believed to participate in a number of important toxic or neoplastic processes: Endothelial cells are particularly responsive to the action of dioxin-like xenobiotics [66], biliary cells have been considered to be involved in neoplastic alterations of the liver [67], and fat-storing cells are thought to play an important role in the etiology of the so-called "spongiosis hepatis" [68]. Moreover, recent research indicates that the role of non-parenchymal cells for the function and differentiation of parenchymal cells has been considerably underestimated [69]. In fish, the study of the non-hepatocytes has been almost completely neglected, and it is hoped that the work published recently by Blair and co-workers [67] stimulates further research in this important and interesting field of teleost liver physiology and pathology [6].

The liver parenchyma in mammals displays metabolic zonation [70], *i.e.* hepatocytes from distinctive zones of the liver parenchyma express different metabolic functions and structures. However, consideration of such microheterogeneity appears to be less important in teleost systems, because there is no unequivocal evidence available so far that a typical metabolic zonation comparable to that of mammalian liver exists in fish liver [71-76].

In species possessing an intrahepatic exocrine pancreas (*e.g.* carp), exocrine pancreatic cells are isolated together with liver cells, if perfusion is carried out via arterial vessels. Thus, approximately 10% of the isolated cells from carp liver are pancreatic cells; they may be removed by density gradient centrifugation [54], or they may be cultured together with the hepatocytes [77]. The presence of pancreatic cells remains without negative effects on cultured hepatocytes, probably because the pancreatic cells release only negligible amounts of proteolytic enzymes into the medium [6].

6.3.8 Viability control and cell yield

An aliquot of the final cell suspension is counted on a hemocytometer, and the viability of the cells is usually assessed by Trypan blue exclusion [9,78,79]. Other methods of viability control include various dye exclusion or dye absorption assays, physiological parameters such as liberation of cytosolic enzymes, maintenance of Na^+/K^+ gradients, ATP levels, or succinate oxidation [80]. Only recently have ultrastructural criteria been proposed as a sensitive indicator of hepatocyte viability [10,25,29,81-83].

Cell yields from liver perfusion vary with species [7,84], strain, sex [65,72,85,86], and age of the donor fish; yields range from 30% to 90% of initial liver weight [5]. Following collagenase perfusion, the average yield of hepatocytes from most fish species is about 10^8 viable cells per g liver with maximum cell recoveries of $1\text{-}6 \times 10^8$ per g [29]. The nutritional status of the donor fish can also significantly influence the cell yield [6]. Cell recovery after isolation by mincing and immersion in collagenase is usually lower and amounts to a maximum $1\text{-}5 \times 10^7$ cells per g liver.

6.4 Culture conditions for isolated hepatocytes

For fresh isolates of hepatocytes, a large variety of solutions ranging from simple salines to complete media has been employed, and in most cases, no attempts to systematically improve conditions for the maintenance of hepatocytes have been made. Since, however, the composition of the incubation medium may be crucial for the performance of hepatocytes in primary culture, the following discussion will focus on the selection of incubation media. Culture conditions for isolated hepatocytes largely depend on experimental designs and the species of the donor fish. Use of densities between 1×10^5 to 2×10^6 cells/ml have been reported. The temperature of primary cultures has been varied from 0°C for cultures of hepatocytes of Antarctic species [11,12] to 30°C for hepatocytes from carp [87], Japanese eel (*Anguilla japonica*, [88]) and black bullhead catfish (*Ictalurus melas*, [74]). Rainbow trout hepatocytes are usually cultivated between 10 and 20°C.

6.4.1 Incubation media

A large variety of incubation media including Krebs-Henseleit medium, Tyrode's buffer, Hank's and Eagle's Minimum Essential Media, to more complex media such as Leibowitz L15, Williams' Medium E, Ham's F-12 Medium, Medium 199 or RPMI 1640 and special "*Salmo* buffer" [89] have been utilized (for composition of the culture medium used in our own laboratory, see [29]). Although systematic investigations on the influence of incubation media on hepatocellular longevity and the preservation of liver-specific functions in primary hepatocyte cultures are lacking, the incubation medium is likely to strongly influence the performance ofhepatocytes in primary culture. An excellent example that illustrates the importance not only of the incubation medium, but also of serum supplements is the synthesis of vitellogenin in primary cultures of rainbow trout hepatocytes [90-93]. In our own laboratory, the secretion rate of vitellogenin was found to be significantly higher in F-12 Ham's medium supplemented with 25 % Ultraculture (Bio Whittaker, Walkersville, Maryland, USA), when compared to non-supplemented Medium 199.

Hepatocyte suspensions have frequently been kept in a shaking water bath and continuously gassed with CO_2 and O_2, if media are buffered with bicarbonate. However, when bicarbonate-buffered media are used, the low bicarbonate concentrations *in vivo* and concomitant low pCO_2 values should be kept in mind [94]. Thus, bicarbonate concentrations should be kept low (4-8 mM), and CO_2-poor gas phases (0.5-1%, remainder O_2) should be chosen [7]. In fact, major differences in metabolite flux have been reported at low and high bicarbonate concentrations [95]. In complex media, bicarbonate can be replaced by zwitterionic buffers such as HEPES [46]. Osmolality conditions may differ significantly between cultures of hepatocytes from freshwater and marine fish. Depending on serum osmolality, that of media for marine fish hepatocytes may need to be adjusted by increasing the final content of NaCl [96].

6.4.2 Cell attachment

Attachment of fish (especially rainbow trout) hepatocytes to commercially available culture dishes was found to be difficult [49]. Generally it is held that monolayer cultures of fish hepatocyte last significantly longer than hepatocytes held in suspension [4]. In the early studies with rainbow trout and channel catfish [8,9] only 50% of the cells loosely attached to the uncoated culture dishes; yet, after 14 days of incubation, cell viability was 63.4 and 40.2% at 4 and 10°C, respectively. In contrast, the longest incubation time reported for catfish hepatocytes in suspension culture is 24 h [97]. Other species that have been used successfully as sources of hepatocytes for monolayer culture are antarctic fish [11,12], carp [98], and oyster toadfish [47].

Several approaches to improve cell seeding and settling have been explored. In serum- and hormone-free medium, for instance, Koban *et al.*

[41] were able to study temperature acclimatisation of catfish hepatocytes in monolayer culture on plates coated with catfish liver biomatrix for 20 days, and Lipsky *et al.* [99] described 93% attachment of isolated rainbow trout hepatocytes to culture plates coated with a commercially available extracellular matrix. Collagen-coating of dishes in the presence of rainbow trout serum resulted in fairly firm attachment of rainbow trout hepatocytes, with optimum conditions at 10°C with 10% serum [100]. Hayashi and Ooshiro [101] were successful in the attachment of Japanese eel hepatocytes to plastic culture dishes coated with horse fibronectin, and Blair and colleagues found good adhesion on plates coated with a fish skin extract [64]. In a comparative study, Maitre *et al.* [102] were successful in maintaining rainbow trout hepatocytes in monolayer culture on dishes without any pretreatment for 10 to 12 days; however, cell attachment was slow and remained fairly weak. Coating of the dishes with fetal bovine serum, collagen or fibronectin failed to improve attachment [102]. In contrast to Mommsen and Lazier [103], who described stable cultivation of salmon hepatocytes on Falcon Primaria™ plates (Becton & Dickinson Labware) for 9 days, Blair and co-workers [64] reported only transient attachment of isolated rainbow trout hepatocytes to Falcon plates for 24 h with subsequent detachment, but found good adhesion for up to six days on culture plates coated with an extract of rainbow trout skin.

In our own experience, the major means of ensuring a high rate of permanent adhesion of rainbow trout hepatocytes is to reduce the stress exerted on hepatocytes during the isolation procedure, to follow a strict temperature regime (maximum temperature 14°C) and to allow cells to settle firmly on the plastic dishes for at least 24 h [104].

Only recently, there has been good evidence that aggregate (spheroid) cultures of rainbow trout hepatocytes may provide stable expression of liver-specific functions including the expression of the estradiol receptor and synthesis of vitellogenin as well as both phase I and II biotransformation enzymes [105,106].

For studies in which bathing media have to be changed rapidly or time-dependent changes in cellular behavior or output are to be recorded, perifusion of hepatocytes immobilized in minicolumns packed with a support such as Biogel-P4 [5], is the method of choice. For perfusion, settling of hepatocytes and firm attachment are, of course, indispensable prerequisites. A detailed review of perifusion techniques has recently been given by Ottolenghi and colleagues [107].

6.5 Preparation of hepatocytes for ultra-structural investigations

For morphological purposes such as electron microscopic inspection of hepatocyte morphology and ultrastructure, cells can be cultured on uncoated Thermanox™ tissue culture slides [108] or uncoated plastic petri dishes [10,29]. In the former case, all steps of the fixation procedure can be carried out *in situ*, *i.e.* on the coverslips. In the case of petri dishes, cells need to be resuspended, and all fixation steps are performed in small centrifuge tubes. Cells have to be carefully sedimented at 50 × *g* after each step; supernatants are removed by aspiration. Fixation is preferentially accomplished by immersion for 30 min in ice-cold 1.5% glutaraldehyde and 1.5% formaldehyde (freshly prepared from paraformaldehyde) in 0.1 M sodium phosphate buffer (pH 7.6) containing 2.5% polyvinylpyrrolidone (PVP). Following brief rinsing in 0.1 M cacodylate buffer (pH 7.6), fixation is continued in 2.5% glutaraldehyde in 0.1 M sodium cacodylate buffer (pH 7.6) containing 4% PVP and 0.05% $CaCl_2$ for 30 min at 4°C. After rinsing in cacodylate buffer, hepatocytes are postfixed for 30 min at 4°C with 1% osmium ferrocyanide (1:1 mixture of 2% aqueous osmium tetroxide and 3% $K_4[Fe(CN)_6]$). After osmication, suspended cells are suspended in liquid agar (2 g/40 ml), and after solidification small slices of the agar-cell mixture can be further processed. After repeated rinsing in 0.1 M cacodylate and 0.05 M maleate buffers (pH 5.2), specimens are stained *en bloc* with 1% uranyl acetate in maleate buffer for at least 1 h at 4°C, dehydrated in a graded series of ethanol, and embedded in Spurr's medium in small beem capsules turned upside down. Agar-embedded cell suspensions may be cut directly; in the case of hepatocytes cultured on Thermanox™ tissue culture slides, the coverslips may be easily broken off by brief immersion in liquid nitrogen.

6.6 Senescence of isolated hepatocytes in primary culture

During prolonged monolayer culture of rainbow trout, *e.g.* the number of viable cells tends to gradually decline almost linearly after five to eight days of culture. In an in-depth study on the senescence of rainbow trout hepatocytes in culture, Braunbeck and Storch [29] investigated the ultrastructural appearance of cells in monolayer culture over a period of 8 days. Freshly isolated hepatocytes appeared as solitary spherical cells with numerous microvilli at the outer surface (Figure 1).

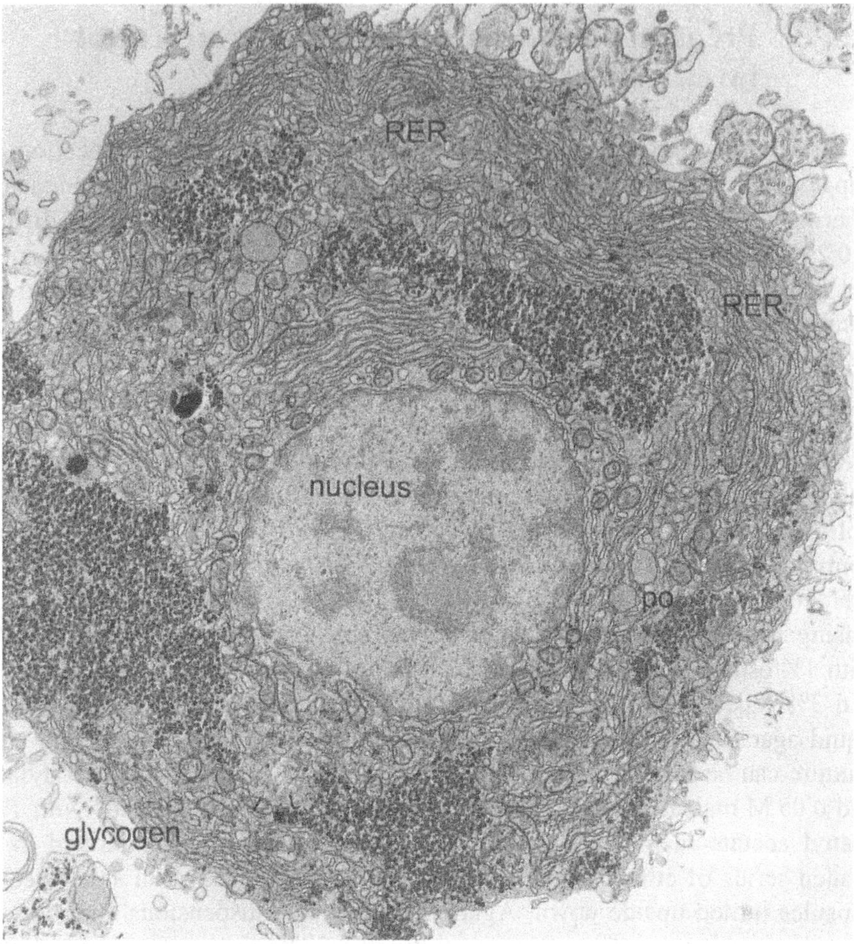

Figure 1. Electron micrograph of isolated rainbow trout hepatocytes. Except for abundant slender microvilli protruding from the plasmalemma, the cytological appearance of isolated rainbow trout (*Oncorhynchus mykiss*) hepatocytes immediately after isolation resembles that of hepatocytes *in vivo*. Freshly isolated rainbow trout hepatocytes are rich in glycogen deposits and display a clear-cut cytoplasmic compartmentation into large glycogen areas and organelle-containing cytoplasm (*cf.* [86, 115-117]). Lamellae of the rough endoplasmic reticulum (RER) show intensive stacking, mitochondria (mi) are in intimate relationship to cisternae of RER and accumulate around the nucleus and along the edges of RER piles. Most peroxisomes (po) aggregate between RER stacks and glycogen fields, and part of the Golgi cisternae appear slightly fenestrated, but are clearly active in secreting very low density lipoprotein (VLDL) particles. Reprinted with permission from [29]. 11,800×

Except for a 30% reduction in cell size due to stress-induced glycogen reduction, the ultrastructure of freshly isolated hepatocytes closely resembled that of rainbow trout hepatocytes *in vivo*. Within one day of culture, about 60-80% of the isolated hepatocytes sedimented to form a monolayer attached to the uncoated culture dishes (Nunclon^{TM} dishes, Nunc), whereas up to 20% remained in suspension forming hepatocyte aggregates. From day 1 to day 5, hepatocytes in primary culture displayed only minor cytological alterations [29,64]. Excellent cytoplasmic compartmentation, restoration of hepatocytic glycogen stores, high secretion rates of very low density lipoproteins by dictyosomes, establishment of cell-to-cell contacts, restitution of cellular polarity and the epithelial character of the cells, as well as new formation of bile canaliculi, documented recovery of the hepatocytes in primary culture. From day 5, an increasing number of cells and cell aggregates detached from the substrate [29,64], and cell senescence was indicated by a marked increase in ultrastructural changes.

Concomitant biochemical studies demonstrated almost constant activities of metabolic key enzymes [109], and hepatocytes preserved inducibility of CYPIA as determined by incubation with β-naphthoflavone [110,111]. Induction of CYP by β-naphthoflavone could be drastically stimulated by use of hepatocytes isolated from pre-induced rainbow trout. Comparative studies on the senescence of isolated hepatocytes from European eel revealed continuous cytological modifications over the entire culture period [109], thus indicating the need for further improvement of culture conditions.

6.7 Perspectives

Given that hepatocytes have been isolated from only 30 out of some 25,000 to 30,000 fish species [112], these cells provide a vast potential not only to study a wealth of different cellular adaptations characteristic of specific fish, but also to explore a broad spectrum of fundamental processes typical of cellular systems in general. The tremendous morphological, physiological and ecological variability of fishes should, therefore, be given much more attention than has occurred in the last two decades.

For the use of cultured hepatocytes as an experimental model, the differentiated state of the cells *in vitro* is of major importance. Except for viability data, surprisingly little information on the morphological and physiological differentiation status of piscine hepatocytes in primary culture is available [6]. For short-term culture (up to 5 or 8 days), monolayer systems seem appropriate to conserve liver-specific gene expression. However, to prevent phenotypic changes of piscine liver cells during prolonged culture, techniques seem to be required that mimic more closely the *in vivo* environment of hepatocytes. For this purpose, the use of aggregate cultures provides an interesting model that should be studied more intensively.

Continued investigation into isolation and maintenance of non-hepatocytic cell types also seems most promising, since these cell types are present in all primary cell suspensions [64,67,113]. The role and importance of non-parenchymal cells to the function of parenchymal cells has generally been neglected or underestimated [5], although there is good evidence for the establishment of cell-to-cell-contacts and interaction between fish hepatocytes and non-hepatocytic cells [67,113]. Therefore, the study of cocultures between parenchymal and non-parenchymal cells deserves particular attention [114].

As a consequence of the growing interest in the use of isolated fish hepatocytes for both physiological and toxicological research, efforts will be made to develop cultures from new species. Given the vast number of fish species, in combination with the abundance of physiological adaptations and specialisations in fish, careful selection of the donor fish species will provide access to a multitude of specialized functions. Especially, small aquarium fish species will have to be given more consideration as donors of hepatocytes, since they are easier to breed under full genetic control and to manipulate than larger species such as rainbow trout. This will require the development and evaluation of more sophisticated surgical techniques, and determination of the optimal composition of initial as well as enzyme perfusion media; incubation techniques and incubation media will have to be re-examined. Although systematic evaluation of these parameters requires a significant investment of time and effort, knowledge of the species (such as temperature, pH, and osmolality requirements) will greatly facilitate the development of optimal culture systems [4].

References

1. Berry MN and Friend DS. High yield preparation of isolated rat liver parenchymal cells. J Cell Biol 1969; 43: 506-520.
2. Berry MN, Edwards AM and Barritt GJ. Isolated hepatocytes - preparation, properties and application. In: Burdon RH and van Knipfenberg PH, Eds. Laboratory techniques in biochemistry and molecular biology. Amsterdam: Elsevier 1991: 1-460.
3. Birnbaum MJ, Schultz J and Fain JN. Hormone-stimulated glycogenolysis in isolated goldfish hepatocytes. Am J Physiol 1976; 231: 191-197.
4. Baksi SM and Frazier JM. Review. Isolated fish hepatocytes - model systems for toxicology research. Aquat Toxicol 1990; 16: 229-256.
5. Mommsen TP, Moon TW and Walsh PJ. Hepatocytes: isolation, maintenance and utilization. In: Hochachka PW and Mommsen TP, Eds. Biochemistry and molecular biology of fishes. Vol. 3. Analytical techniques. Amsterdam: Elsevier, 1994: 355-372.

6. Segner H. Isolation and primary culture of teleost hepatocytes. Comp Biochem Physiol C 1999; in press.
7. Moon TW, Walsh PJ and Mommsen TP. Fish hepatocytes: a model metabolic system. Can J Fish Aquat Sci 1985; 42: 1772-1782.
8. Klaunig JE. Establishment of fish hepatocyte cultures for use in *in vitro* carcinogenicity studies. Natl Cancer Inst Monogr 1984; 65: 163-173.
9. Klaunig JE, Ruch RJ and Goldblatt JP. Trout hepatocyte culture: Isolation and primary culture. *In Vitro* Cell Dev Biol 1985; 21: 221-228.
10. Braunbeck T. Cytological alterations in isolated hepatocytes from rainbow trout (*Oncorhynchus mykiss*) exposed to 4-chloroaniline. Aquat Toxicol 1993; 25: 83-110.
11. O'Grady SM, Clarke A and De Vries AL. Characterization of glycoprotein antifreeze biosynthesis in isolated hepatocytes from *Pagothenia borchegrevinki*. J Exp Zool 1982; 220: 179-189.
12. Haschemeyer AEV and Mathews RW. Temperature dependency of protein synthesis in isolated hepatocytes of antarctic fish. Physiol Zool 1983; 56: 78-87.
13. Moerland TS and Sidell BD. Characterization of metabolic carbon flow in hepatocytes isolated from thermally acclimated killifish *Fundulus heteroclitus*. Physiol Zool 1981; 54: 379-389.
14. Schnurstein A, Leist E, Froschauer A et al. Activation of indirectly acting genotoxic substances in primary cell cultures of liver and gill cells from zebrafish (*Brachydanio rerio*). Verh Dtsch Zool Ges 1997; 90.1: 384.
15. Seglen PO. Preparation of rat liver cells. I. Effects of Ca^{2+} on enzymatic dispersion of isolated, perfused liver. Exp Cell Res 1972; 74: 450-454.
16. Seglen PO. Preparation of rat liver cells. III. Enzymatic requirements for tissue dispersion. Expl Cell Res 1973; 82: 391-398.
17. Gagné F and Blaise C. Evaluation of the genotoxicity of environmental contaminants in sediments to rainbow trout hepatocytes. Environ Toxicol Water Qual 1995; 10: 217-229.
18. Gagné F and Blaise C. Flow cytometry measurement of mixed function oxidase activity and cell viability in rainbow trout (*Oncorhynchus mykiss*) hepatocytes: method development. Environ Toxicol Water Qual 1996; 11: 53-63.
19. Donohue M, Baldwin LA, Leonard DA, Kostecki PT and Calabrese EJ. Effect of hypolipidemic drugs gemfibrozil, ciprofibrate, and clofibric acid on peroxisomal β-oxidation in primary cultures of rainbow trout hepatocytes. Ecotox Environ Safety 1993; 26: 127-132.
20. Hazel JR and Sellner PA. Fatty acid and sterol synthesis by hepatocytes of thermally acclimated trout (*Salmo gairdneri*). J Exp Zool 1979; 209: 105-113.
21. Steward AR, Elmarakby SA, Stuart KG, Kumar S and Sikka HC. Metabolism of 2-acetylaminofluorene by hepatocytes isolated from rainbow trout. Toxicol Appl Pharmacol 1995; 130: 188-196.
22. Duan C, Hanzawa N, Takeuchi Y et al. Use of primary cultures of salmon hepatocytes for the study of hormonal regulation of insulin-like growth factor I expression *in vitro*. Zool Sci 1993; 10: 473-480.
23. Parker RS, Morrissey MT, Moldeus P et al. The use of isolated hepatocytes from rainbow trout (*Salmo gairdneri*) in the metabolism of acetaminophen. Comp Biochem Physiol 1981; 70B: 631-633.
24. Rabergh CMI, Ziegler K, Isomaa B, Lipsky MM and Eriksson JE. Uptake of taurocholic acid and cholic acid in isolated fish hepatocytes from rainbow trout. Am J Physiol 1994; 267: G380-G386.

25. Andersson T and Förlin L. Spectral properties of substrate-cytochrome P-450 interaction and catalytic activity of xenobiotic metabolizing enzymes in isolated rainbow trout liver cells. Biochem Pharmacol 1985; 34: 1407-1413.
26. Renaud JM and Moon TW. Characterization of gluconeogenesis in hepatocytes isolated from the American eel, *Anguilla rostrata* LeSueur. J Comp Physiol 1980; 135B: 115-125.
27. Renaud JM and Moon TW. Starvation and metabolism of hepatocytes isolated from the American eel, *Anguilla rostrata* LeSueur. J Comp Physiol 1980; 135B: 127-137.
28. Seglen PO. Preparation of isolated liver cells. Meth Cell Biol 1976; 13: 29-59.
29. Braunbeck T and Storch V. Senescence of hepatocytes isolated from rainbow trout (*Oncorhynchus mykiss*) in primary culture: an ultrastructural study. Protoplasma 1992; 170: 138-159.
30. Walsh PJ. Lactate uptake by toadfish hepatocytes: passive diffusion is sufficient. J Exp Biol 1987; 130: 295-304.
31. Porthé-Nibelle J and Lahlou B. Mechanisms of glucocorticoid uptake by isolated hepatocytes of the trout. Comp Biochem Physiol 1981; 69B: 425-433.
32. Williams EE and Hazel JR. Restructuring of plasma membrane phospholipids in isolated hepatocytes of rainbow trout during brief *in vitro* cold exposure. J Comp Physiol 1995; 164B: 600-608.
33. Nishimoto M, Yanagida GK, Stein JE, Baird WM and Varanasi U. The metabolism of benzo[a]pyrene by English sole (*Parophrys vetulus*): comparison between isolated hepatocytes *in vitro* and liver *in vivo*. Xenobiotica 1992; 22: 949-961.
34. Bianchini L, Fossat B, Porthe-Nibelle J, Ellory JC and Lahlou B. Effects of hyposmotic shock on ion fluxes in isolated trout hepatocytes. J Exp Biol 1988; 137: 303-318.
35. Hazel JR. The incoporation of unsaturated fatty acids of the n-9, n-6 and n-3 families into individual phospholipids by isolated hepatocytes of thermally acclimated trout, *Salmo gairdneri*. J Exp Zool 1983; 227: 167-176.
36. Goeger DE, Shelton DW, Hendricks JD and Bailey GS. Mechanisms of anti-carcinogenesis by indole-3-carbinol: effect on the distribution and metabolism of aflatoxin B1 in rainbow trout. Carcinogenesis 1986; 7: 2025-2031.
37. Coulombe RA, Bailey GS and Nixon JE. Comparative activation of aflatoxin B1 to mutagens by isolated hepatocytes from rainbow trout (*Salmo gairdneri*) and coho salmon (*Oncorhynchus kisutch*). Carcinogenesis 1984; 5: 29-33.
38. Bailey GS, Taylor MJ and Selivonchick DP. Aflatoxin B1 metabolism and DNA binding in isolated hepatocytes from rainbow trout (*Salmo gairdneri*). Carcinogenesis 1982; 3: 511-518.
39. Jobling S and Sumpter JP. Detergent components in sewage effluent are weakly estrogenic to fish: an *in vitro* study using rainbow trout (*Oncorhynchus mykiss*) hepatocytes. Aquat Toxicol 1993; 27: 361-372.
40. French CJ, Mommsen TP and Hochachka PW. Amino acid utilization in isolated hepatocytes from rainbow trout. Europ J Biochem 1981; 113: 311-317.
41. Koban M. Can cultured teleost hepatocytes show temperature acclimation? Am J Physiol 1986; 250: R211-R220.
42. Mommsen TP and Walsh PJ. Evolution of urea synthesis in vertebrates: the piscine connection. Science 1989; 243: 72-75.
43. Garrison JC and Haynes RC. Hormonal control of glucogenolysis and gluconeogenesis in isolated rat liver cells. J Biol Chem 1973; 248: 5333-5343.

44. Segner H, Böhm R and Kloas W. Binding and bioactivity of insulin in primary cultures of carp (*Cyprinus carpio*) hepatocytes. Fish Physiol Biochem 1993; 11: 411-420.
45. Baksi SM. A fish hepatocyte model for the investigation of the effects of trihalomethanes. Water chlorination: chemistry, environmental impact and health effects 1990;
46. Pesonen M and Andersson T. Fish primary hepatocyte cultures; an important model of xenobiotic metabolism and toxicity studies. Aquat Toxicol 1997; 39: 253-267.
47. Kelly JJ and Maddock MB. *In vitro* induction of unscheduled DNA synthesis by genotoxic carcinogens in the hepatocytes of the toadfish (*Opsanus tau*). Arch Environ Contam Toxicol 1985; 14: 555-563.
48. Walsh PJ, Moon TW and Mommsen TP. Interactive effects of acute changes in temperature and pH on metabolism in hepatocytes from the sea raven *Hemitripterus americanus*. Physiol Zool 1985; 58: 727-735.
49. Rabergh CM, Kane AS, Reimschuessel R et al. Viability and induction of tyrosine aminotransferase in rainbow trout hepatocytes cultured on laminin and polylysine in serum-free medium. Meth Cell Sci 1995; 17: 207-215.
50. Foster GD and Moon TW. Hormonal sensitivity and response in sea raven hepatocytes: changes with fasting and collagenase exposure. Can J Zool 1993; 71: 1755-1762.
51. Braunbeck T. Cytological alterations in fish hepatocytes - *in vivo* and *in vitro* biomarkers of environmental contamination. In: Braunbeck T, Hinton DE and Streit B, Eds. Fish ecotoxicology. Experientia, Suppl. Ser. Birkhäuser, Basel, 1998: 61-140.
52. Puceat M, Garin D and Freminet A. Inhibitory effects of anaesthesia with 2-phenoxyethanol as compared to MS 222 on glucose release in isolated hepatocytes from rainbow trout (*Salmo gairdneri*). Comp Biochem Physiol 1989; 94A: 221-224.
53. Morrison H, Young P and George S. Conjugation of organic compounds in isolated hepatocytes from a marine fish, the plaice, *Pleuronectes platessa*. Biochem Pharmacol 1985; 34: 3933-3938.
54. Bouche G, Gas N and Paris H. Isolation of carp hepatocytes by centrifugation on a discontinuous ficoll gradient. A biochemical and ultrastructural study. Biol Cellul 1979; 36: 17-24.
55. Segner H, Scholz S and Böhm R. Carp (*Cyprinus carpio*) hepatocytes in primary culture: morphology and metabolism. Actes du Colloque (Ed l'IFREMER) 1995; 18: 77-82.
56. Baksi SM and Frazier JM. A fish hepatocyte model for the investigation of the effects of environmental contaminants. Mar Environ Res 1988; 243: 141-145.
57. Seibert H. Viability control and oxygen consumption of isolated hepatocytes from thermally acclimated rainbow trout (*Salmo gairdneri*). Comp Biochem Physiol 1985; 80B: 677-683.
58. Dannevig BA and Berg T. Endocytosis of galactose-terminated glycoproteins by isolated liver cells of the rainbow trout (*Salmo gairdneri*). Comp Biochem Physiol 1985; 82B: 683-688.
59. Bhattacharya S, Plisetskaya E, Dickhoff WW and Gorbman A. The effects of estradiol and triiodothyronine on protein synthesis by hepatocytes of juvenile coho salmon (*Oncorhynchus kisutch*). Gen Comp Endocrin 1985; 57: 103-109.
60. Plisetskaya EM, Fabbri E, Moon TW et al. Insulin binding to isolated hepatocytes of Atlantic salmon and rainbow trout. Fish Physiol Biochem 1993; 11: 401-409.

61. Plisetskaya E, Bhattacharya S, Dickhoff WW et al. The effect of insulin on amino acid metabolism and glycogen content in isolated liver cells of juvenile coho salmon, *Oncorhynchus kisutch*. Comp Biochem Physiol 1984; 78A: 773-778.
62. Deventer K. Detecting of genotoxic effects on cells of liver and gills of *B. rerio* by means of single cell gel electrophoresis. Bull Environ Contam Toxicol 1996; 56: 911-918.
63. Foster GD, and Moon TW. Hypometabolism with fasting in the yellow perch (*Perca flavescens*): A study of enzymes, hepatocyte metabolism, and tissue size. Physiol Zool 1991; 64: 259-275.
64. Blair JB, Miller MR, Pack D, Barnes R, Teh SJ and Hinton DE. Isolated trout liver cells: establishing short-term primary cultures exhibiting cell-to-cell interactions. *In Vitro* Cell Dev Biol 1990; 26: 237-249.
65. Hampton JA, Lantz RC and Hinton DE. Functional units in rainbow trout (*Salmo gairdneri*, Richardson) liver: III. Morphometric analysis of parenchyma, stroma, and component cell types. Am J Anat 1989; 185: 58-73.
66. Husoy AM, Myers MS, Willis ML, Collier TK, Celander M and Goksoyr A. Immunohistochemical localization of CYP1A and CYP3A-like isoenzymes in hepatic and extrahepatic tissues of Atlantic cod *(Gadus morhua* L.), a marine fish. Toxicol Appl Pharmacol 1994; 129: 294-308.
67. Blair JB, Ostrander GK, Miller MR and Hinton DE. Isolation and characterization of biliary epithelial cells from rainbow trout liver. *In Vitro* Cell Dev Biol 1995; 31: 780-789.
68. Couch JA. Spongiosis hepatis: chemical induction, pathogenesis, and possible neoplastic fate in a teleost fish model. Toxicol Pathol 1991; 19: 237-250.
69. Skett P. Problems in using isolated and cultured hepatocytes for xenobiotic metabolism/metabolism-based toxicity testing - solutions? Toxicol *in vitro* 1994; 8: 491-504.
70. Buhler R, Lindros KO, Nordling A, Johansson I and Ingelman-Sundberg M. Zonation of cytochrome P-450 isozyme expression and induction in rat liver. Eur J Biochem 1992; 204: 407-412.
71. Schaer M, Maly LP and Sasse D. Histochemical studies on metabolic zonation of the liver in the trout (*Salmo gairdneri*). Histochem 1985; 83: 147-151.
72. Hampton JA, McCuskey PA, McCuskey RS and Hinton DE. Functional units in rainbow trout (*Salmo gairdneri*) liver: I. Arrangement and histochemical properties of hepatocytes. Anat Rec 1985; 213: 166-175.
73. Mommsen TP, Danulat E, Gavioli ME et al. Separation of enzymatically distinct populations of trout hepatocytes. Can J Zool 1991; 69: 420-426.
74. Ottolenghi C, Ricci D, Gavioli ME et al. Separation of two populations of fish hepatocytes by digitonin infusion: some metabolic patterns and hormonal responsiveness. Can J Zool 1991; 69: 427-435.
75. Segner H and Braunbeck T. Hepatocellular adaptation to extreme nutritional conditions in ide, *Leuciscus idus melanotus* L. (Cyprinidae). A morphofunctional analysis. Fish Physiol Biochem 1988; 5: 79-97.
76. Mommsen TP and Walsh PJ. Metabolic and enzymatic heterogeneity in liver of the ureogenic teleost *Opsanus beta*. J Exp Biol 1991; 156: 407-418.
77. Vogt G and Segner H. Spontaneous formation of intercellular bile canaliculi and hybrid biliary-pancreatic canaliculi in co-culture of hepatocytes and exocrine pancreas cells from carp. Cell Tissue Res 1997; 289: 191-194.

78. Patterson MK. Measurement of growth and viability of cells in culture. In: Jakoby WB and Pastan IH, Eds. Meth Enzymol 1979; 58: 141-152.
79. Tyson CA and Green CE. Cytotoxicity measures: choices and methods. In: Rauckman EJ and Padilla GM, Eds. The isolated hepatocyte: use in toxicology and xenobiotic biotransformations. Orlando: Academic Press, 1987: 119-158.
80. Klaassen CD and Stacey NH. Use of isolated hepatocytes in toxicity assessment. In: Plaa G and Hewitt WR, Eds. Toxicology of the liver. New York: Raven Press, 1982: 147-179.
81. Zahn T and Braunbeck T. Cytologic evidence for malachite green toxicity in isolated rainbow trout (*Oncorhynchus mykiss*) hepatocytes. Toxicol *In Vitro* 1995; 9: 729-741.
82. Zahn T, Arnold H and Braunbeck T. Cytological and biochemical response of R1 cells and isolated hepatocytes from rainbow trout (*Oncorhynchus mykiss*) to subacute *in vitro* exposure to disulfoton. Exp Toxicol Pathol 1996; 48: 47-64.
83. Zahn T, Hauck C, Holzschuh J and Braunbeck T. Acute and sublethal toxicity of seepage water from garbage dumps to permanent cell cultures and primary cultures of hepatocytes from rainbow trout (*Oncorhynchus mykiss*): a novel approach to environmental risk assessment. Zbl Hyg 1995; 196: 455-479.
84. Mommsen TP. Comparative gluconeogenesis in hepatocytes from salmonid fishes. Can J Zool 1986; 64: 1110-1115.
85. Langer M. Histologische Untersuchungen an der Teleosteerleber. I. Der Aufbau des Leberparenchyms. Z mikr-anat Forsch 1979; 93: 829-848.
86. Braunbeck T, Storch V and Nagel R. Sex-specific reaction of liver ultrastructure in zebrafish (*Brachydanio rerio*) after prolonged sublethal exposure to 4-nitrophenol. Aquat Toxicol 1989; 14: 185-202.
87. Macartney AI, Tiku PE and Cossins AR. An isothermal induction of δ^9-desaturase in cultured carp hepatocytes. Biochim Biophys Acta 1996; 1302: 207-216.
88. Hayashi S and Ooshiro Z. Primary culture of eel hepatocytes in serum-free medium. Bull Jap Soc Sci Fish 1986; 52: 1641-1651.
89. Andersson T and Koivusaari U. Influence of environmental temperature on the induction of xenobiotic metabolism by β-naphthoflavone in rainbow trout, *Salmo gairdneri*. Toxicol Appl Pharmacol 1985; 80: 43-50.
90. Islinger M, Pawlowski S, Hollert H et al. Measurement of vitellogenin-mRNA expression in primary cultures of rainbow trout hepatocytes in a non-radioactive dot blot/RNAse protection-assay. Sci Total Environ 1999; 233: 109-122.
91. Vaillant C, Le Guellec C, Pakdel F and Valotaire Y. Vitellogenin gene expression in primary culture of male rainbow trout hepatocytes. Gen Comp Endocrinol 1988; 70: 284-290.
92. Peyon P, Baloche S, Burzawa and Gerard E. Synthesis of vitellogenin by eel (*Anguilla anguilla* L.) hepatocytes in primary culture: requirement of 17 β-estradiol-priming. Gen Comp Endocrinol 1993; 91: 318-329.
93. Pelissero C, Flouriot G, Foucher JL, Bennetau B, Dunogues J, Le Gac F and Sumpter JP. Vitellogenin synthesis in cultured hepatocytes; an *in vitro* test for the estrogenic potency of chemicals. J Steroid Biochem Mol Biol 1993; 44: 263-272.
94. Albers C. Acid-base balance. In: Hoar WS and Randall DJ, Eds. Fish physiology. New York: Academic Press, 1970: 173-213.
95. Mommsen TP and Suarez RK. Control of gluconeogenesis in rainbow trout hepatocytes: role of pyruvate branch point and phosphoenolpyruvate-pyruvate cycle. Mol Physiol 1984; 6: 9-18.

96. Sigel MM and Beasley AR. Trypsin C. Marine teleost tissues. In: Kruse PF and Patterson MD, Eds. Tissue culture methods and applications. New York: Academic Press, 1973: 12-14.
97. Baird WM, Smolarek TA, Plakunov I et al. Formation of benzo[a]pyrene-7,8-dihydrodiol glucuronide is a major pathway of metabolism of benzo[a]pyrene in cell cultures from bluegill and brown bullhead. Symposium on Toxic Chemicals and Aquatic Life. Seattle: 1986.
98. Saez L, Goicoechea O, Amthauer R and Krauskopf M. Behaviour of RNA and protein synthesis during the acclimatization of the carp. Studies with isolated hepatocytes. Comp Biochem Physiol 1982; 72B: 31-38.
99. Lipsky MM, Sheridan TR, Bennett RO et al. Comparison of trout hepatocyte culture on different substrates. *In Vitro* Cell Dev Biol 1986; 22: 360-362.
100. Kocal T, Crane TL, Quinn BA, Ferguson HW and Hayes MA. Degradation of extracellular thymidine by cultured hepatocytes from rainbow trout (*Salmo gairdneri*). Comp Biochem Physiol 1988; 91B: 557-561.
101. Hayashi S and Ooshiro Z. Primary culture of the freshly isolated liver cells of the eel. Bull Jpn Soc Sci Fish 1985; 51: 765-771.
102. Maitre JL, Valotaire Y and Guguen-Guillouzo C. Estradiol-17β stimulation of vitellogenin synthesis in primary culture of male rainbow trout hepatocytes. *In Vitro* Cell Dev Biol 1986; 22: 337-343.
103. Mommsen TP and Lazier CB. Stimulation of estrogen receptor accumulation by estradiol in primary cultures of salmon hepatocytes. FEBS Lett 1986; 195: 269-271.
104. Braunbeck T. Protracted senescence - a fundamental process of toxicity in isolated fish hepatocytes? Verh Dtsch Zool Ges 1997; 90.1: 345.
105. Flouriot G, Pakdel F, Ducouret B and Valotaire Y. Influence of xenobiotics on rainbow trout liver estrogen receptor and vitellogenin gene expression. J Mol Endocrinol 1995; 15: 143-151.
106. Cravedi J-P, Paris A, Monod G et al. Maintenance of cytochrome P450 content and phase I and phase II enzyme activities in trout hepatocytes cultured as spheroidal aggregates. Comp Biochem Physiol 1996; 113C: 241-246.
107. Ottolenghi C, Puviani AC, Cooper GA et al. Perifusion of hepatocytes. In: Hochachka PW and Mommsen TP, Eds. Biochemistry and molecular biology of fishes. Vol. 3. Analytical techniques. Amsterdam: Elsevier, 1994: 387-397.
108. Vogt G, Böhm R and Segner H. Mimosine-induced cell death and related chromatin changes. J Submicroc Cytol Pathol 1994; 26: 319-330.
109. Braunbeck T. Strukturelle und funktionelle Veränderungen in Hepatocyten von Fischen als Biomarker für die Belastung durch Umweltchemikalien. Habilitation Thesis, Dept. of Zoology, University of Heidelberg, 1994.
110. Hauck C, Zahn T and Braunbeck T. Culture of liver-derived rainbow trout cells and their biotransformation capacity. Verh Dtsch Zool Ges 1993; 86: 190.
111. Braunbeck T, Hauck C, Scholz S et al. Mixed function oxygenases in cultured fish cells: contributions of *in vitro* studies to the understanding of MFO induction. Z Angew Zool 1996; 81: 55-72.
112. Nelson JS. Fishes of the world. New York: Wiley & Sons, 1994.
113. Ostrander GK, Blair JB, Stark BA et al. Long-term primary culture of epithelial cells from rainbow trout (*Oncorhynchus mykiss*) liver. *In Vitro* Cell Dev Biol Anim 1995; 31: 367-378.

114. Scholz S, Oulmi Y, Braunbeck T et al. Viability and differential function of rainbow trout hepatocytes in primary culture: the influence of culture conditions. *In Vitro* Cell Dev Biol 1998; 34: 762-771.
115. Braunbeck T, Gorgas K, Storch V et al. Ultrastructure of hepatocytes in golden ide (*Leuciscus idus melanotus* L.; Cyprinidae: Teleostei) during thermal adaptation. Anat Embryol 1987; 175: 303-313.
116. Braunbeck T, Storch V and Bresch H. Species-specific reaction of liver ultrastructure of zebrafish (*Brachydanio rerio*) and trout (*Salmo gairdneri*) after prolonged exposure to 4-chloroaniline. Arch Environ Contam Toxicol 1990; 19: 405-418.
117. Segner H and Braunbeck T. Environmental adaptation of the cyprinid teleost *Leuciscus idus melanotus*: Changes in liver composition and structure during the winter season. J Exp Zool 1990; 255: 171-185.
118. Singh Y, Cooke JB, Hinton DE et al. Trout liver slices for metabolism and toxicity studies. Drug Metab Dispos 1996; 24: 7-14.
119. Braunbeck T. Protracted senescence - a fundamental process of toxicity in isolated fish hepatocytes? Cell Biol Toxicol 1996; 31.
120. Flouriot G, Vaillant G, Salbert G et al. Monolayer and aggregate cultures of rainbow trout hepatocytes: long-term and stable liver-specific expression in aggregates. J Cell Sci 1993; 105: 407-416.
121. Kocal T, Quinn BA, Smith IR et al. Use of trout serum to prepare primary attached monolayer cultures of hepatocytes from rainbow trout (*Salmo gairdneri*). *In Vitro* Cell Dev Biol 1988; 24: 304-308.
122. Hayashi S and Ooshiro Z. Preparation of isolated cells from eel liver. Bull Jpn Soc Sci Fish 1978; 44: 499-503.
123. Campbell JW, Aster PL, Casey CA et al. Preparation and use of fish hepatocytes. In: Harris RA and Cornell NW, Eds. Isolation, characterization and use of hepatocytes. New York: Elsevier, 1983: 31-40.

Chapter 7

Conditions for growth of hepatocytes in culture

Anthony M. Edwards[1] and George K. Michalopoulos[2]
[1]*Department of Medical Biochemistry, Flinders University, GPO Box 2100, Adelaide, South Australia, 5001.* [2]*Department of Pathology ,University of Pittsburgh Medical School, Pittsburgh, Pennsylvania, USA, 1526*

Key words: cell cycle, cell density, culture media, differentiation, DMSO, EGF, extracellular matrix, HGF, Matrigel, metal ions, nicotinamide, non-parenchymal cell, nutrients, rat hepatocytes, small hepatocytes, stellate cells, substrata, transferrin, transcription factors

7.1 The context for studies of hepatocyte growth in culture

Investigators have a variety of reasons for studying the factors that determine the ability of hepatocytes to grow, and for pursuing some of these studies utilising the relatively defined and easily-manipulated conditions of cell culture. There are intriguing intellectual and technical challenges in dissecting out the factors that determine patterns of hepatocyte growth, death and differentiation - the cell lineage relationships, the relevant humoral and nutritional factors and the interactions with extracellular matrix and with non-hepatocyte cell types present in the liver. Some notable issues in biology and medicine underlie efforts to address these challenges. These include a need to understand the ability of generally healthy liver to rapidly replace lost tissue via liver regeneration; to understand normal patterns of hepatocyte growth and their control and the perturbations of these processes relevant to liver carcinogenesis; and to understand factors relevant to regeneration and repair when liver is damaged by toxins (including alcohol) or by virus infection. In addition, a better understanding of hepatocyte growth (and its relationship to cell differentiation or death) could assist efforts to replace defective liver function either with transplanted hepatocytes or with extracorporeal bioartificial livers. For example, ability to manipulate growth *in vitro* could increase options for generating normal or genetically-modified hepatocytes for transplantation or for developing bioartificial liver devices with sufficient cell mass, appropriate tissue architecture and extended function necessary for successful application in replacing the function of failing human livers.

M.N. Berry and A.M. Edwards (eds.), The Hepatocyte Review, 73–96.

This chapter focuses on the conditions that support short- and longer-term growth of hepatocytes and on the nature of cells capable of proliferation. The emphasis is on primary cultures for which the starting population is predominantly mature rat hepatocytes and on relatively simple culture systems which should prove useful for a variety of studies on hepatocyte growth. Advances in the last decade, in particular, are considered.

7.2 Some key issues in studies of hepatocyte growth: an overview

When procedures for the isolation and culture of hepatocytes for studies on growth were reviewed a decade ago (Chapters 10 and 11 in [1]), there were a number of common elements in the approaches of different laboratories to studies on growth *in vitro*. Most laboratories focussed on the *in vitro* behaviour of relatively pure populations of mature (rat) hepatocytes. These were obtained by similar collagenase perfusion procedures, following which single hepatocytes in the resulting cell suspensions were separated from most other liver-derived cells by washing steps, with low-speed (about $50 \times g$) centrifugation. In general, the aim was to minimize contamination with other cell types and, by choice of culture conditions, to minimize any subsequent contributions (or growth) of the 2-5% non-hepatocytes typically remaining in the population plated into culture dishes [1,2].

A decade ago, numerous variations of culture conditions had been tested in efforts to achieve multiple rounds of DNA synthesis and mitosis in hepatocyte cultures. In general, it was believed that culture conditions which tended to preserve normal hepatocyte differentiated function and those which promoted growth were quite distinct. For growth studies, hepatocytes were typically plated at relatively low density ($1\text{-}2 \times 10^4$ cells/cm^2) on simple substrata, such as culture dishes coated with a film of type I collagen, and maintained with complex media containing insulin, EGF and sometimes glucocorticoid and other hormones. In these conditions, mature hepatocytes could be induced to undergo one, and sometimes two, rounds of DNA synthesis. While mitoses could be seen in a substantial number of cultured cells, measurements of total cell or nucleus numbers after 5 days, generally showed only modest increases (50 - 70%) over numbers on day 1 in culture (reviewed in [1], pp 265-278).

Thus, some key issues or questions faced researchers a decade ago in their efforts to overcome the apparent limitations on hepatocyte growth *in vitro*; and in efforts to relate findings *in vitro* to the characteristics of hepatocyte growth in intact liver *in vivo* during development or in regeneration or repair. These issues or questions fall into three broad domains which might be briefly summarized as follows.

7.2.1 Providing essential requirements for growth in culture

It is now clear from a variety of studies, such as experiments with serial transplantation of mature hepatocytes to repopulate the livers of transgenic mice [3] that adult hepatocytes, or a substantial fraction of them, are capable of many generations of clonal growth. Earlier failure to achieve more than limited proliferation in culture suggested that essential factors to support growth *in vitro* remained to be identified. These might include essential growth factors, macro- or micronutrients or aspects of cell contact with extracellular matrix. From studies of the sequence of events during liver regeneration [4,5], and recognition that there are multiple points of regulation in the cell cycle of mammalian cells, it seems likely that multiple, distinct signals are important in "priming" a hepatocyte for growth and then allowing progress through G_1, S, G_2 and M phases of the cell cycle. There is still much to learn about the influence of growth factors, nutrients and cell contacts at different stages of the cycle.

7.2.2 The nature of the cell population in culture

Early efforts to achieve hepatocyte growth in culture concentrated on the mature hepatocytes present in adult liver. These adult cells are heterogeneous in that some are mononucleate and diploid but many are binucleate with levels of ploidy up to at least octapoid [6-8]. It appears that proliferating hepatocytes are predominantly diploid [6-8] (and discussion below). Also, greater stimulation of DNA synthesis can be achieved in cultured hepatocytes from young, rather than older rats, apparently reflecting differences in signal transduction down-stream from growth factor receptors and/or in expression of genes regulating the cell cycle [9,10]. Such observations suggest that differentiated hepatocytes in adult liver may exist in subpopulations with differing ability to proliferate. On the other hand, most hepatocytes appear to proliferate during liver regeneration [4,11], so it seems important to establish the extent to which any existing subpopulations in the liver can undergo transitions to forms with limited or continuing capacity to proliferate.

In a wider context, differentiated hepatocytes represent one end-point of a cell lineage [11], possibly extending from multipotential stem cells in bone marrow [12]. It is known that during development, multipotential hepatoblasts differentiate to yield both hepatocytes and ductular cells and that related stem-like ("oval") cells can appear in regenerating liver when proliferation of mature hepatocytes is suppressed (reviewed in [11]). It is of great importance to clarify relationships, *in vivo* or in culture, between fully differentiated (non-growing) hepatocytes, populations of hepatocytes capable of proliferation in culture and "oval" or other stem-like/progenitor cells. It is increasingly clear that "oval" and other stem-like cells, like hepatoblasts, may differentiate to either hepatocytes or ductular cells [11] but the extent to which mature hepatocytes may revert to earlier stages in the hepatocyte lineage is only recently becoming apparent. Culture systems, established with relatively defined total, or fractionated, populations of hepatocytes and allowing close control of cell environment, have great potential for further clarifying such relationships.

Since hepatocytes, either in resting or regenerating mode in the intact liver, co-exist with other cell types (endothelial, Kupffer, stellate and others) and with the extracellular matrix to which these cell types contribute, a further issue for clarification is the extent to which hepatocyte function is influenced or determined by these other cell types via direct cell contacts, or soluble factors such as cytokines, or production of matrix components.

7.2.3 Differentiation versus growth in culture

Ten years ago, most *in vitro* studies - with a few possible exceptions [13,14] - seemed to support the view that culture conditions (substrata, co-culture interactions and some media components) which favoured growth and those which favoured differentiated hepatocyte function were almost mutually exclusive (reviewed in [1], pp 265-278). In the liver *in vivo* however, transitions between resting, fully differentiated hepatocytes and proliferating hepatocytes in regenerating liver do not involve dramatic changes in the relationship of hepatocytes to other hepatocytes or non-parenchymal cells or major changes in the extracellular matrix, although of course subtle changes in matrix contacts and/or cell-cell communication are likely to be important [15-18]. Similarly during regeneration, there are significant changes in gene expression but much of the differentiated program of mature liver cells is preserved [4,5,19]. Aided by better understanding of the transcription factors important in controlling hepatocyte growth and differentiation [4,20,21], culture systems have the potential for exploring possible growth/differentiation relationships - and for identifying key controlling factors - as mature hepatocytes acquire ability to grow or proliferating cells of the hepatocyte lineage stop growing and differentiate.

These broad areas of investigation, and conclusions about experimental approaches for achieving hepatocyte growth in short-, and longer- term studies in the light of a decade of progress, are considered in more detail in the sections below.

7.3 Providing essential requirements for growth in culture

Hepatocytes in the intact liver have a complex relationship with extracellular matrix and other cell types, and are presented with a complex (and changing) mixture of components via the blood or by diffusion from surrounding cells. There are thus many possible variables in identifying essential requirements for growth. Earlier studies focussed predominantly on mature hepatocytes over a few days in primary culture and concentrated on identifying factors necessary to maximize the ability of hepatocytes to undergo at least an initial round of DNA replication since this was most amenable to study (reviewed in [1], Chapter 12).

Work in the last decade has included many studies on growth extending beyond 4 or 5 days (to weeks or months) and has involved greater focus on the transitions between adult hepatocytes and proliferating, multipotential cells which may (re)exhibit hepatocyte differentiation under appropriate conditions as discussed below. It seems likely that hepatocyte growth can be achieved by a number of different paths involving different combinations of starting cell population, substrata and media. There are still uncertainties about relationships between individual medium components and processes leading to appearance of the proliferating cells in the culture systems of different laboratories. Nevertheless, there are now complex media which on simple substrata, clearly support extended proliferation of hepatocytes or of closely-related cells in a hepatocyte lineage [22-25] (and see Chapter 14) and there is a better understanding of the individual medium components important to growth in several experimental systems.

7.3.1 Media

Although the basic culture media used with supplements by different laboratories have varied [*e.g.* 22-26], Dulbecco's modified Eagle's Medium (DMEM) has provided a base medium in some extended series of studies successful in achieving long-term hepatocyte growth [22,23,25] (and see Chapter 14).

It seems that shifts in nutrient levels *in vivo* can have a significant role in priming hepatocytes for growth [4] so that nutrients may have regulatory significance as well as a role in supporting processes such as protein or nucleic acid synthesis. However, unequivocal *in vitro* requirements for individual nutrients have been difficult to define. Apart from the tedium of systematically varying individual components, it seems clear that the effects of one amino acid, for instance, depend on the combination of other amino acids present [27]. Previous ideas on macronutrient (carbohydrate, amino acid), micronutrient (vitamins, metal ions) and serum protein requirements (reviewed in ref [1], pp 292-318) have been refined in the last decade. Various effective growth media contain galactose and/or glucose as an energy source: Block *et al* [23] found that inclusion of glucose (2 g/L), with or without galactose, was important to support growth over 14 days. Media such as DMEM have been supplemented with proline [23] (and see Chapter 14) and glutamine [23]. Block *et al* [23] suggested that arginine (in addition to ornithine) may be required when proliferating hepatocytes lose differentiation functions such as urea cycle enzymes that would allow conversion of ornithine to arginine. Serum albumin is a desirable addition to growth media [23,24].

Cable and Isom [24] examined trace metal requirements in a DMSO-containing culture system and showed that addition of Zn^{2+}, Cu^{2+} and Fe^{2+} to medium containing EGF and 2% DMSO (and apo-transferrin) induced previously quiescent hepatocytes to proliferate, suggesting that these metal ions have a key role in supporting growth. Block *et al* [23] include Zn^{2+} and Cu^{2+} in their Hepatocyte Growth Medium (HGM) but reported that iron-saturated diferric transferrin rather than Fe^{2+} and apotransferrin was required to sustain growth.

In 1991, Mitaka *et al* [26] demonstrated a striking effect of nicotinamide (2.5 to 10 mM) on the ability of EGF-treated primary cultures to sustain DNA synthesis beyond an initial round of replication. Nicotinamide has been included in media for a series of subsequent studies focused on clonal growth of "small hepatocytes" [22,25,26,28-31] (and see Chapter 14), and was also found to be an important component in the HGM of Block *et al* [23] where it contributed to sustained replication and division of "proliferating hepatocytes". At least slow growth of hepatocytes has been observed in the absence of nicotinamide [24,32] but it seems to be important in facilitating a transition from mature to "small" [28] or "proliferating" [23] hepatocytes capable of sustained growth.

Studies in which "small hepatocytes" were the key proliferating cell have indicated that the long-acting vitamin C derivative, ascorbic acid-2-phosphate (0.2 - 1.0 mM) should be included in media to support growth [22,25,29-31] but this was not essential in other culture systems [23,24].

Addition of DMSO, previously considered to be a tool only forpreserving differentiated function (reviewed in [1], pp 315-316) now appears to provide one way of sustaining hepatocyte longer-term survival and growth [22,24,28-30] (and see Chapter 14) despite inhibitory effects on DNA synthesis in newly-established primary cultures. DMSO may contribute to the preservation of cultures in various ways. It seems clear that DMSO is growth inhibitory rather than stimulatory [24,28,29,31,32]. Its inhibitory effect appears to be more complete on freshly isolated hepatocytes than on the proliferating "small hepatocyte" population which emerges in the culture systems of Mitaka [28] (and see Chapter 14) or Tateno and Yoshizato [29]. Once "small hepatocytes" are growing clonally, DMSO prevents colonies from detaching over time [31] and could help preserve the growing small cells, the mature hepatocytes also present and/or prevent over-growth or other adverse effects of non-parenchymal cells present in the cultures.

The Michalopoulos group showed that 2% DMSO reversibly inhibited DNA synthesis in primary cultures stimulated with either EGF or HGF [33] and that cycles of DMSO addition and removal stimulated multiple cycles of DNA synthesis but with no net increase in the number of cells [34].

The Isom laboratory add 2% DMSO with EGF to primary cultures immediately after attachment. In the absence of key metal ions, the DMSO/EGF combination contributes to long-term maintenance of relatively well-differentiated, non-proliferating cells and, as above, removal of DMSO allows a burst of DNA synthesis but not an increase in cell numbers [35]. Inclusion of Fe^{2+}, Cu^{2+} and Zn^{2+} in the culture medium from day 1 (or addition after a period in culture) allows slow proliferation over 2 to 3 months [24]. Thus DMSO confers advantages for the survival and differentiation of the overall cell population without totally blocking proliferation of large or small hepatocytes present in these cultures, given an appropriate combination of growth factors and nutrients.

Hepatocyte growth *in vitro* is totally dependent on addition of growth factors. Essentially all growth media include insulin (0.5 - 5 μg/ml) and dexamethasone (0.1 μM), the latter presumably in an important permissive role (*e.g.* [23]). Most growth studies have used EGF as a mitogen. HGF and TGF-α alone are each at least as effective as EGF in stimulating DNA replication and an increase in total DNA over 15 days in culture: more sustained increases in DNA synthesis and total DNA are achieved with EGF and HGF in combination [23]. While numerous other factors have been shown to influence hepatocyte DNA synthesis in culture (see Chapters 15 and 17), none appears to be essential for cell growth *in vitro*.

More recent studies with cultured hepatocytes have begun to dissect out nutritional and growth factor requirements, as well as other factors of regulatory significance, at various points in the hepatocyte cell cycle (see Chapters 15 and 17). It is clear that EGF or TGF-α or HGF is required for cells to pass a restriction point in G_1 phase of the cell cycle. The nature of growth factors or further signals required at other points of control in the cell cycle is less clear. HGF presumably has additional regulatory role(s), given its effects on cell morphology and differentiation, which are distinct from those of EGF [23,36], and its generally potent effects on hepatocyte growth [37].

Freshly isolated hepatocytes have now been shown in many laboratories to be "primed" to respond to growth factors by the events of cell isolation *i.e.* they have moved from G_0 to G_1 as shown by expression of various immediate early genes [4] (and see Chapter 17). Thus once cells have attached to substrata such as collagen, no added growth factors other than EGF (or TGF-α or HGF) with insulin are necessary to evoke the initial round of DNA synthesis. It may be that these same factors have a role in preventing cultured cells re-entering G_0 and/or in stimulating the transition of resting cells from G_0 to G_1, at least when signals from cell-cell contact or cell-matrix contact are also consistent with growth. However, studies on liver regeneration *in vivo* strongly suggest that signals other than HGF/TGF-α/EGF are necessary to "prime" hepatocytes to respond to these growth factors [4,5]: it is likely that changes in hepatocyte cytoskeleton organisation have some role (*e.g.* [18,38,39]) but there is also a strong case for involvement of TNF-α and IL6 in "priming", consistent with a rapid, transient rise in TNF-α immediately after partial hepatectomy [40] (and references therein). While TNF-α has been shown to stimulate DNA synthesis in cultured rat [40] or mouse [41,42] hepatocytes, its role in combination with other growth factors *in vitro* is yet to be characterized in detail. Recent work suggests that TNF-α may be mitogenic only in combination with serum [43] or growth factors [44].

External signals (if any) necessary for hepatocytes to move through G_2 and M phase of cell cycle after completing replication are unclear, as are requirements for completing cytokinesis. Given the effect of nicotinamide in sustaining hepatocyte growth, it is tempting to suggest that it may have permissive or regulatory effects late in the cell cycle. Controls on mitosis or cytokinesis are beginning to be characterized at the molecular level (*e.g.* [39]) and may prove to be important. Various studies have suggested that sustained hepatocyte growth may occur more readily in mononucleate, diploid cells rather than polyploid hepatocytes [6-8,20,23], although both diploid and polyploid hepatocytes, when serially transplanted in transgenic mice, apparently have high growth potential [3].

7.3.2 Cell density and substrata

Ten years ago, it was considered that even limited growth of hepatocytes in culture required plating at relatively low cell density on a substratum that allowed cell spreading, rather than substrata (such as Matrigel) that favoured retention of cuboidal morphology and differentiated function. As improvements in media have made possible the longer-term growth of hepatocytes (or their proliferating, multipotential forms), plating cells at relatively low (subconfluent) density on simple substrata [22,23,26,28-30] remains the simplest, but not the only, approach to achieving growth. While convenient for experimental studies on the hepatocyte cell cycle and its control, these culture conditions are very different from those in which hepatocytes grow *in vivo* during liver regeneration. Much of the normal liver architecture, where hepatocytes are in close contact with each other and with a complex extracellular matrix, is retained throughout the regeneration process [4,5,15]. Thus it would seem that it should be possible to elicit hepatocyte growth in culture at high cell density and in the presence of a complex substatrum.

Where cultures containing freshly isolated hepatocytes with few non-parenchymal cells have been studied, appearance of proliferating cells has generally been favoured by low [23] to moderate [25,26,28-31] plating densities. Mitaka *et al* [28] found that appearance of proliferating "small hepatocytes" was optimal with moderate, rather than high or low, plating densities; so in this context specific cell-cell contacts may be important. In the culture system described by Cable and Isom [24] involving maintenance of relatively well-differentiated hepatocytes in EGF and 2% DMSO, inclusion of key metal ions Zn^{2+}, Cu^{2+} and Fe^{2+}, either from the outset or after some weeks in culture, stimulated slow proliferation of hepatocytes in areas of continuous monolayer within overall subconfluent cultures. In all these relatively simple culture systems, growth slows and cells tend to exhibit greater differentiation as cells reach confluence and/or pile up [22-26,28-31]. In culture systems involving more complex combinations of hepatocytes, non-parenchymal cells and extracellular matrix, hepatocytes can proliferate to form multiple layers where close contact between cells does not block growth [46,47]. Thus recent findings suggest that cell density is only one factor influencing proliferative potential and that hepatocytes in close contact can be induced to proliferate under appropriate conditions *in vitro*.

With recently-plated hepatocyte primary cultures, simple substrata such as type I collagen films [23,26,29] or gels [23] are adequate to sustain growth and collagens type I or IV or fibronectin or laminin were found to be equally effective in supporting growth [23]. With appropriate growth media containing EGF or HGF, an initial proliferative phase has also been observed for hepatocytes cultured in collagen-sandwich configuration [23,36]. However, the rate of proliferation was much less than in cultures on collagen-coated plastic [23] and after a period of proliferation, culture in a collagen-sandwich with HGF favoured differentiation towards a ductular phenotype [23] (see further discussion below).

The effect of a complex extracellular matrix may depend on its composition. Some time ago, Enat *et al* [13] reported that a liver bio-matrix extract supported growth, but Matrigel (an extract of the EHS mouse sarcoma which has high laminin content along with a mix of collagens, including type IV, with various glycosaminoglycans and proteoglycans), has been shown in a variety of studies to favour differentiated expression and block growth (*e.g.* [21]). Matrigel overlay of actively proliferating hepatocytes in monolayer cultures stopped growth and favoured a return of the cultures to a mature hepatocyte phenotype [23]. However, in a more complex three dimensional culture system, in which hepatocytes and non-parenchymal cells between them synthesize extensive extracellular matrix, it *was* possible to observe high levels of proliferation even when Matrigel was added to the cultures [46].

Variations on conventional (unstirred) monolayer culture may enhance growth. Takeshita *et al* [48] embedded hepatocytes in collagen gel in polyester sponge discs and maintained them in HGM in spinner baskets. Continuous perfusion with circulating medium resulted in higher DNA synthesis than in stationary cultures. Similarly maintenance of hepatocytes on collagen-coated beads in HGM in roller bottles allowed extensive growth of relatively well-differentiated hepatocytes [46].

Thus, simple substrata are adequate to support hepatocyte growth *in vitro* but the presence of either a single component such as collagen in more complex architecture (sandwich) or the presence of more complex matrix components do not necessarily preclude *in vitro* growth, given an appropriate balance of other regulatory factors.

7.3.3 Non-parenchymal cells

As discussed below, primary hepatocyte cultures always contain some contaminating non-parenchymal cells and, if desired, cell preparation methods may be varied to increase the numbers of these cells [22,25]. Early studies in which hepatocytes were cocultured with liver-derived non-parenchymal cells or other cell types suggested that in general, coculture favoured differentiation rather than growth (see [1], pp 321-327). In culture systems where substantial growth of mature hepatocytes has been achieved in the initial few days in primary culture [23,26,27], it seems unlikely that the small numbers of stellate or other non-parenchymal cells present are either necessary for, or contribute to, the initial cycles of hepatocyte growth.

The contributions of non-parenchymal cells to longer term growth are more difficult to quantify. In monolayer cultures maintained with growth factors, with [22,25] or without serum [23], various non-parenchymal cells persist and grow. Block *et al* [23] reported focal growth of "fibroblasts" which favoured differentiation of adjacent hepatocytes. Studies by Bissell's group have demonstrated that most of the "fibroblastic" cells appearing in aged cultures derive from hepatic stellate cells, present in the initial population added to culture dishes [49]. Media containing serum [22,25] support proliferation of stellate [22,25] and liver epithelial cells [22], while DMSO suppresses their growth. In these systems, overgrowth of non-parenchymal cells or production by them of inhibitory factors such as TGF-β may exert confounding or negative effects on hepatocyte growth (Chapter 14). Tateno and Yoshizato [25] suggested that stellate cells and hepatocytes co-operate in production of extracellular matrix which supports growth and subsequent differentiation of small hepatocytes.

In complex culture systems, where the numbers of non-parenchymal cells have been enriched during cell isolation [47] or by growth in culture [46] there is presumably a major contribution of stellate cell-derived fibroblasts and other cells to extensive deposition of extracellular matrix, and possibly production of cytokines, which may set up localized regions favouring growth or hepatocytic- or ductular-differentiation.

7.4 The nature of the initial cell population and emerging populations in primary culture

In efforts to understand liver cell growth under a variety of circumstances *in vivo* and *in vitro*, enormous interest has centered on the relationship between mature, fully differentiated hepatocytes, which do not normally grow, and various populations with liver stem cell-like properties, which grow but are less differentiated. The great progress in understanding liver cell lineage relationships has been comprehensively reviewed [11]. While it is fairly clear that growing stem-cell-like populations may differentiate under appropriate conditions to yield mature hepatocytes or ductular cells [11], a key goal of studying hepatocyte growth in culture has been to establish the extent to which the reverse process can occur. This Section focuses on the cells placed into, or emerging in, primary cultures; on the extent to which hepatocyte populations pre-existing in the liver at the time of cell isolation differ in growth potential; on the extent to which the original population(s) can undergo transition to new population(s) of proliferating hepatocytes; and on the nature of proliferating cells which emerge in culture under different conditions.

Evidence from studies of liver regeneration *in vivo* suggests that after partial hepatectomy, most hepatocytes replicate their DNA and divide at least once. This is true regardless of location in the liver lobule, although periportal hepatocytes enter S-phase earlier. Most hepatocytes are probably capable of repeated cell division if repeated hepatectomies are performed. Likewise bile duct epithelial cells proliferate to extend intrahepatic ducts. Thus for both hepatocytes and ductular cells. new cells appear to arise mainly by division of existing mature cells, not by growth and differentiation of stem cells (reviewed in [11]). These observations suggest that it should be possible to observe at least a few cycles of proliferation in most of the mature hepatocyte population in culture, if appropriate conditions can be devised. Indeed, as discussed below observations from a number of laboratories suggest that at least in the initial few days in primary culture, a high proportion of the original population of rat hepatocytes go through up to 3 cycles of replication.

There is also evidence from conducting repeated hepatectomies *in vivo* [50] or transplantation studies (*e.g.* [3]) that at least some hepatocytes in liver are capable of completing many cell cycles. As discussed below, cells capable of sustained growth *can* emerge in primary cultures, although findings differ on the fraction of mature hepatocytes from which these arise. The discussion below focuses on rat hepatocytes. Sustained growth of hepatocytes from mice [51] and humans [23,52,53] (see also Chapter 16) has also been reported.

7.4.1 Cell types present in populations isolated from adult liver by collagenase perfusion

Cells obtained from collagenase dispersion of rat livers and then sedimented by centrifugation at approximately $50 \times g$ are predominantly mature, large hepatocytes about 35-45 μm in diameter in a monolayer. About 3% of the $50 \times g$ cell pellet is made up of "non-parenchymal" cells, namely stellate cells, endothelial cells and small numbers of Kupffer cells, bile duct epithelial cells and "liver epithelial cells" (see for instance[2,23] and discussion in ref [1] pp 242-245). As discussed by Mitaka and Sato in Chapter 14, the total liver cell suspension also contains some "small hepatocytes" (diameter around 20 μm in a monolayer). These cells decrease as a proportion of the original hepatocyte population with age of donor rats (ranging from 6% in 4 week old rats down to less than 0.5% in livers of aged rats).

Higher proportions of non-parenchymal cells and of small hepatocytes (relative to large hepatocytes) can be recovered by changes to the washing and centrifugation steps during cell isolation. Mitaka and co-workers (see Chapter 14) and others [25,30] have used the supernatant from an initial $50 \times g$ centrifugation to enrich for small hepatocytes; and Michalopoulos *et al* [46] used a related approach to isolate a "non-parenchymal cell" population containing mainly stellate, endothelial and bile duct cells with about 5% small hepatocytes.

7.4.2 Some key studies on proliferating hepatocytes in culture

In their 1991 report describing the influence of nicotinamide on hepatocyte growth (discussed above), Mitaka and co-workers [26] in Pitot's laboratory used an initial population prepared by collagenase perfusion and $50 \times g$ centrifugation and plated onto collagen-coated dishes at a moderate density (about 6×10^4 cells/cm^2). The initial population in culture contained mononucleate and binucleate hepatocytes (40% diploid nuclei, and about 55% tetraploid in [26]; these percentages varied in other studies [8]). In the presence of EGF and 10 mM nicotinamide, about half of the cells were in S-phase between 48 and 72 hours corresponding to an initial peak of DNA synthesis; and slightly more than half the cells were in S-phase between 72 and 96 hours corresponding to a second peak of DNA synthesis in the total culture. Double-labelling studies showed that more than 17% of cells replicated their DNA twice during the first 4 days in culture. In a subsequent study, Mitaka *et al* [28] distinguished two phases of growth in the presence of EGF and nicotinamide. During the first phase, on days 2-4, more than half of the original population, including hepatocytes with both diploid and tetraploid nuclei, synthesize DNA and divide. Total DNA increases by about 130% and cell numbers by about 80% in 4-5 days [28]. In the second phase, after day 4, in the presence of EGF and nicotinamide, some cells continue to proliferate forming colonies of diploid, mononucleate "small hepatocytes" which retain many features of hepatocyte morphology and differentiated function. The growing colonies then coexist with mature hepatocytes in culture. Appearance of mononucleate diploid cells in these cultures contrasts with a progressive shift towards multinucleate cells which stop proliferating, in cultures with EGF but lacking nicotinamide [28]. Thus the presence of nicotinamide may allow cells to complete cytokinesis to form diploid cells with further proliferative potential.

The proliferating "small hepatocytes" emerge after freshly isolated hepatocytes complete a small number of cell cycles (Mitaka's Phase 1). Since "small hepatocyte" colonies appeared only after plating at moderate, but not high or low density, transition to small hepatocytes might require specific cell contacts [28]. During Phase 1, proliferation of large hepatocytes and the subsequent appearance of colonies is prevented if DMSO is added [28,29]. Addition of 1-2% DMSO *after* day 4 (*i.e.* Mitaka's Phase 2), initially delays small hepatocyte proliferation but then allows more sustained colony growth [28,29]. In medium with nicotinamide, EGF and DMSO, the small cell colonies continued dividing for a few days, then the cells enlarged and became difficult to distinguish from mature (large) hepatocytes [28]. The small cells exhibited further growth if subcultured and their growth was enhanced by addition of 10% fetal bovine serum and ascorbic acid 2-phosphate [31]. Tateno and Yoshizato reported similar findings [29] and further showed that when cultures were initiated with a non-parenchymal cell/small hepatocyte-enriched population and maintained with the nicotinamide/EGF/DMSO/serum/ascorbic acid 2-phosphate medium, small hepatocyte colonies grew rapidly for several days and then more slowly, eventually exhibiting mature hepatocyte or ductular differentiation [30]. Synthesis of extracellular matrix by stellate cells present in the cultures was considered to contribute to survival and differentiation [25,30].

The clonigenic small hepatocytes in these culture systems are considered to derive from a small fraction of the original hepatocyte population (0.5% to 6% depending on rat age) [28] (see also Chapter 14) designated Type I cells - small hepatocytes with high proliferative capacity. These are considered to be distinct from Type II hepatocytes, (fully differentiated with limited proliferative capacity) and Type III hepatocytes (fully differentiated cells which have lost the ability to divide). These studies, discussed in more detail in Chapter 14, suggest that many hepatocytes after isolation can complete one or two cell cycles but that a specific, minor subfraction of adult hepatocytes has the greatest potential for longer-term clonal growth.

Studies in the Michalopoulos laboratory have led to somewhat different findings. In 1996, Block *et al* [23] reported studies with rat hepatocytes, prepared by standard collagenase perfusion and $50 \times g$ centrifugation and plated at relatively low density (10^4 cells/cm^2) on type I collagen gels or films. When maintained in HGM, DNA synthesis in these cultures increased markedly after 2-3 days, so that during each of the intervals, days 4-6 and 8-10, about 80% of cells replicated their DNA. During the initial 6-10 days, the majority of mature hepatocytes in culture apparently underwent initial cycles of replication and division. By 9 days, total DNA in cultures increased about fourfold and by day 15, about twelvefold. This represents a more extensive expansion (from a lower initial density) than seen in the Mitaka and Tateno systems. Rates of DNA synthesis slowed as cells reached confluence on days 12-15.

From the normal compact hepatocyte morphology seen in cultures on day 1, most cells adopted irregular shapes with many projections by day 4, presumably reflecting the effect of HGF/scatter-factor. From day 6 the cells lost projections, became smaller and lost cytoplasmic granularity. In contrast to the Mitaka and Tateno systems, the morphological changes occur in the overall culture population, not a minor subset. Cells exhibiting clonal growth between days 1 and 10 do not appear to come from a distinct subgroup of hepatocytes at day 1 (small or large, mono- or bi-nucleate for instance). Gene expression studies indicate that the small proliferating hepatocytes in the system of Block *et al* [23] lose many aspects of differentiated hepatocyte function and express genes contributing to growth and some markers typical of bile duct epithelium. These observations suggest that many mature hepatocytes have undergone a transition to bipotential hepatoblast-like cells. Their ability to redifferentiate is discussed in the next section. These hepatoblast-like cells in monolayer cultures appear to retain less hepatocyte differentiation than in the "small hepatocytes" of the Mitaka and Tateno laboratories. However it is notable that the pattern of differentiation of proliferating hepatocytes in HGM is influenced by the culture conditions. For example, compared with unstirred monolayers, there was greater preservation of differentiated morphology and function with hepatocytes (and some non-parenchymal cells) growing into multiple layers on collagen-coated beads in rollerbottle cultures [46].

Thus in HGM, it appears that a substantial fraction of the mature hepatocyte population complete initial cell cycles and that many of their progeny undergo a progressive transition to hepatoblast-like, small, morphologically-simple cells capable of clonal growth or redifferentiation, depending on the culture environment. Factors which may be important to achieving sustained growth include EGF, nicotinamide, dexamethasone and trace metals (similar to the media of Mitaka and Tateno groups) as well as inclusion of glucose and arginine as nutrients, diferric iron-saturated transferrin (rather than apotransferrin and Fe^{2+}) and the inclusion of HGF in combination with EGF and permissive nutrients. HGF alone resembles EGF in stimulating the appearance of small cell colonies [54]. It is possible that partly dedifferentiated "small" hepatocytes are unable to convert ornithine to arginine and so grow less readily in media lacking arginine. In some way the combination of factors in HGM may allow a larger fraction of mature hepatocytes to undergo transition to small, proliferative hepatocytes, which perhaps represent a less-differentiated stage in a continuum from stem cells to mature hepatocytes (or bile duct cells) than that represented by "small hepatocytes".

A third culture system in which slow, but sustained hepatocyte growth has been achieved was described by Cable and Isom [24]. Rat hepatocytes prepared by the standard procedure were plated on collagen-coated dishes at a medium density, about 4×10^4 cells/cm^2. Cells maintained with defined medium (RPCD) supplemented with 2% DMSO and EGF retain most of the morphological and functional features of differentiated hepatocytes for months but do not grow. Addition to the medium of Fe^{2+}, Cu^{2+} and Zn^{2+} ions from day 1 initially had little effect on hepatocyte numbers but from about 14 to 40 days in culture, stimulated an increase in cell numbers to about 3 times the day-14 numbers or numbers in cultures maintained without the metal ions. When tested at intervals during the period of cell growth up to 40 days, 6-7% of cells were found to be in S-phase during each 16-hour test interval.

The cultures in this system after some days, contain both large (35-45 μm in the monolayer) and small (10-15 μm) hepatocytes. Except for size, the large and small cells are similar in ultrastructure. In the presence of the metal ions, both large and small cells synthesize DNA; and metal ion-stimulated proliferation does not change the proportions of large and small cells. Growth is not in colonies: DNA synthesis occurs in cells randomly distributed through the monolayer. The nature of the proliferating cells in this system appears to be different from the small "proliferating hepatocytes" of Block *et al* [23] which were less differentiated and grew in colonies.

The proliferating cells of Cable and Isom [24] resemble "small hepatocytes" to the extent that there is substantial/partial preservation of differentiated structure and function. However slower growth of both large and small cells at random rather than in colonies in the Cable and Isom cultures suggest a different route to proliferation.

Taken together the observations in different culture systems [22-26] suggest that most mature hepatocytes can be induced to undergo a few cell divisions, but the fraction of these cells able to continue growing, and properties of the proliferating cells, is dependent on culture conditions. The Cable and Isom cultures [24] may represent a "minimal change" system where essential requirements for growth are present but a substantial shift to less-differentiated hepatocytes with enhanced proliferative capacity is blocked by constant presence of DMSO. In the presence of nicotinamide and growth factors (but without DMSO) mature hepatocytes may complete one or more cycles necessary for transition to a state (or states?) with capacity for continuing growth. In the conditions of Mitaka *et al* [22,26] (without HGF and EGF in combination?) the transition (or change in the pattern of gene expression) may be modest and restricted to "predisposed" hepatocytes, whereas conditions used in the Michalopoulos laboratory may promote a more substantial dedifferentiation, allowing many cells to establish longer-term proliferative potential.

It might be noted that proliferating hepatocytes in the experimental systems discussed above appear to be distinct from "liver epithelial cells". These small cells with clear cytoplasm, possibly derived from stem cells in the intact liver, have been recognized in primary cultures for many years. They grow in the presence of serum and may yield propagable cell-lines which have been studied in detail (see [11] for review). Mature hepatocytes have never been observed to give rise to these cells [11]. However, liver epithelial cells clearly have aspects in common with proliferating hepatocytes since they [11], like "oval" cells [11,55], can be induced to exhibit features of hepatocyte or ductular phenotype under appropriate conditions *in vivo* or in culture.

7.5 Growth and differentiation relationships in liver cell cultures

The focus of this chapter is primarily on conditions for hepatocyte growth in culture. A consideration of various aspects of hepatocyte differentiation is closely linked with an understanding of what is required for hepatocytes to grow and of the factors that determine the balance between growth and differentiation. Relevant aspects include: features of the less-fully differentiated phenotypes of different populations of proliferating hepatocytes in culture; the detailed conditions which promote redifferentiation, and the changes in gene expression underlying transitions between differentiation and growth of mature hepatocytes in culture. Detailed consideration of these topics is beyond the scope of the chapter. Nevertheless some general comments seem relevant.

A common view of a decade ago [1] - that hepatocyte growth in culture could only be achieved under conditions that led to dramatic loss of hepatocyte differentiated function - has been considerably refined by more recent work. In all the systems where relatively extensive growth of rat hepatocyte-derived cells has been achieved [22-24,46,47], the proliferating cells are small (relative to hepatocytes in the liver), predominantly diploid and, to varying degrees, exhibit a simpler morphology than mature cells and reduced expression of characteristic markers of hepatocyte differentiation. At the same time, various aspects of thehepatocyte phenotype are preserved, including expression of key transcription factors (eg [23]). After a period of clonal growth at least some of the proliferating cells express markers consistent with the phenotype of bipotential hepatoblasts or stem-like cells, indicating that some [22,25] or many [23] mature hepatocytes may reverse the process of terminal differentiation to earlier stage(s) in the hepatocyte cell lineage.

The extent of "dedifferentiation" associated with acquiring proliferative capacity seems to depend strongly on culture conditions and is generally less in the presence of DMSO [22,24,25,28-30] or in three-dimensional cultures (as opposed to simple monolayers) [48], especially where non-parenchymal cells contribute to a complex extracellular matrix and perhaps provide soluble factors [46,47]. That "dedifferentiation" of mature hepatocytes to yield growing cells is not just an aberration of cell culture, is strongly suggested by the ability of the proliferating populations to (re)acquire either ductular or mature hepatocyte phenotypes.

In the monolayer culture systems established with hepatocytes prepared in the standard way [23,26,28], proliferating cells tend to assume features of mature hepatocytes as cultures reach confluence and pile up. In the system of Block *et al* [23] this was observed mainly in regions associated with focal growth of stellate cells/fibroblasts. In cultures initiated with populations enriched in non-parenchymal cells [22,25,30,47], the presence of significant numbers of stellate cells (together with DMSO) apparently enhances appearance of different groups of cells with mature hepatocyte or ductular differentiation, after a period of hepatocyte growth.

The findings of Block *et al* [23] add significantly to an understanding of factors influencing differentiation of proliferating hepatocytes. They showed that addition of complex, laminin-rich extracellular matrix extract (Matrigel) to proliferating hepatocytes in HGM led to mature hepatocyte differentiation. Alternatively culture of hepatocytes in a type I collagen gel sandwich with HGF (but omitting EGF) led, after a period of proliferation, to ductular differentiation.

Clearly, extracellular matrix and the non-parenchymal cells associated with its production have a defining influence on hepatocyte growth and differentiation. Recent studies have centered on the role of transcription factors in this interplay. Hepatocytes maintained in HGM with EGF and HGF lose most markers of mature hepatocyte function (such as albumin expression) [23]. Liver-associated transcription factors such as HNF1, HNF3 and HNF4 are preserved but levels of C/EBPα decrease while activity of AP1 and NFκB increase. Transcription factors STAT3 and STAT5 are also activated [23,56], a new low molecular weight isoform of HNF4 appears and there is a marked increase in nuclear levels of the transcription factors

COUP TFI and COUP TFII. All these changes are reversed by addition of matrix [57]. These findings suggest that a complex signalling pathway, originating from cell-matrix contacts, is involved in modifying expression of specific transcription factors.

Interestingly, the conditions that favour hepatocytic or ductular redifferentiation of proliferating hepatocytes in primary cultures have elements in common with culture conditions which induce differentiation of "oval" cells or liver epithelial cell lines [11,55].

As an extension of the findings in simpler monolayer culture systems, the laboratories of Michalopoulos [46] and Mitaka [47] have recently described complex culture systems in which relatively well-differentiated hepatocytes and associated non-parenchymal cells grow into three-dimensional structures in which various features of intact liver structure and function develop and are then preserved for extended periods. These findings have exciting implications for the development of bioartificial livers: as our understanding of the factors which favour growth or specific paths of differentiation is further refined, it should become possible to generate mixed-cell, liver-like complex cultures of substantial size, with stable function and with characteristics selected "to order".

7.6 Cultured hepatocytes in studies on growth regulation: past and future

Over the last two decades, findings with hepatocyte cultures have had a direct impact on understanding various aspects of liver regeneration and growth. The demonstration that EGF stimulated DNA synthesis in hepatocytes *in vitro* led to studies establishing EGF receptor ligands (including EGF and TGF-α) as having a role in liver regeneration. The same system led to isolation and characterisation of HGF (reviewed in [4,5,58]).

Hepatocyte cultures have allowed systematic studies on the actions and interactions in growth control of various comitogens and correlation of *in vitro* findings with effects *in vivo*: comitogens studied in this way include norepinephrine [59], estrogens [60], xenobiotics and liver tumour promoters [61] and a variety of other agents (see also Chapter 17). Related studies are currently unravelling possible actions of TNF-α [40-44]. Actions of inhibitory factors significant in growth control, such as TGF-β, have also been characterized in culture and related to likely roles *in vivo* (eg,[62,63].)

Hepatocyte cultures have provided an important tool for investigating the signal transduction mechanisms mediating actions of inhibitors, mitogens and comitogens and for investigating points in the cell cycle at which they exert regulatory effects, as discussed in Chapters 15-17. Further studies should be facilitated by the conditions, discussed in the preceding Sections, that allow hepatocytes to complete multiple cell cycles.

Findings of the last decade, identifying conditions which support sustained growth of hepatocytes or differentiation along particular paths, have already added substantially to understanding of some of the issues of liver biology and medicine listed in Section 7.1, but many mysteries remain. Hepatocyte cultures will undoubtedly continue to play a key role as a model for unravelling regulation of the complex events underlying growth, differentiation and death in differentiated cells.

References

1. Isolated hepatocytes: Preparation, properties and applications. Berry MN, Edwards AM and Barritt GJ, Eds. Amsterdam: Elsevier, 1991.
2. Seglen PO. Preparation of isolated liver cells. Methods Cell Biol 1976; 13: 29-83.
3. Overturf K, Al-Dhalimy M, Ou C-N, Finegold M and Grompe M. Serial transplantation reveals the stem-cell-like regenerative potential of adult mouse hepatocytes. Amer J Pathol 1997; 151: 1273-1280.
4. Bucher NLR and Farmer SR. Liver regeneration following partial hepatectomy: genes and metabolism. In: Liver growth and repair. Strain AJ and Diehl AM, Eds. London: Chapman Hall, 1998, pp 3-27.
5. Fausto N and Webber EM. Liver regeneration. In: The Liver: Biology and Pathobiology (3rd ed.). Arias IM, Boyer JL, Fausto N, Jakoby WB, Schachter DA and Shafritz DA, Eds. New York: Raven Press, 1994, pp 1059-1084.
6. Mossin L, Blankson H, Huitfeldt H and Seglen PO. Ploidy-dependent growth and binucleation in cultured rat hepatocytes. Exp Cell Res 1994; 214: 551-560.
7. Seglen PO. DNA ploidy and autophagic protein degradation as determinants of hepatocellular growth and survival. Cell Biol Toxicol 1997; 13: 301-315.
8. Rajvanshi P, Liu D, Ott M, Gagandeep S, Schilsky ML and Gupta S. Ploidy-based fractionation of rat hepatocyte subpopulations with varying metabolic potential, proliferative capacity and retroviral gene transfer efficiency. Exp Cell Res 1998; 244: 405-419.
9. Roth GS. Age changes in signal transduction and gene expression. Mech Ageing Dev 1997; 98: 231-238.
10. Sawada N, Kojima T, Obata H, Saitoh M, Isomura K, Kokai Y, Satoh M and Mori M. P21 waf-1/cip-1/sdi-1 is expressed at G1 phase in primary culture of hepatocytes from old rats, presumably preventing the cells from entering the S-phase of the cell cycle. Biochem Biophys Res Commun 1996; 228: 819-824.
11. Coleman WB and Grisham JW. Epithelial stem-like cells of the rodent liver. In: Liver growth and repair. Strain AJ and Diehl AM, Eds. London: Chapman Hall, 1998, pp 50-99.
12. Petersen BE, Bowen WC, Patrene KD, Mars WM, Sullivan AK et al. Bone marrow as a potential source of hepatic oval cells. Science 1999; 284: 1168-1170.

13. Enat R, Jefferson DM, Nelson RM, Gatmaitan Z, Leinward LA and Reid LM. Hepatocyte proliferation *in vitro*: its dependence on the use of serum-free, hormonally defined medium and substrata of extracellular matrix. Proc Natl Acad Sci USA 1984; 81: 1411-1415.
14. Shimaoka S, Nakamura T and Ichihara A. Stimulation of growth of primary cultured adult rat hepatocytes without growth factors by coculture with non-parenchymal liver cells. Exp Cell Res 1987; 172: 228-242.
15. Maher JJ. The extracellular matrix in liver regeneration. In: Liver Growth and Repair. Strain AJ and Diehl AM, Eds. London: Chapman Hall, 1998, 451-464.
16. Kim T-H, Mars WM, Stolz DB, Petersen BE and Michalopoulos GK. Extracellular matrix remodelling at the early stages of liver regeneration in the rat. Hepatology 1997; 26: 896-904.
17. Fladmark KE, Gjertsen BT, Molven A, Mellgren G, Vintermyr OK and Doskeland SO. Gap junctions and growth control in liver regeneration and in isolated rat hepatocytes. Hepatology 1997; 25: 847-855.
18. Liu ML, Mars WM, Zarnegar R and Michalopoulos GK. Collagenase pretreatment and the mitogenic effects of hepatocyte growth factor and transforming growth factor-α in adult rat liver. Hepatology 1994; 19: 1521-1527.
19. Tygstrup N, Jensen SA, Krog B, Dietranjelo A and Shafritz DA. Expression of messenger RNA for liver functions following 70% and 90% hepatectomy. J Hepatol 1996; 25: 72-78.
20. Taub R. Liver regeneration 4: transcriptional control of liver regeneration. FASEB J 1996; 10: 413-427.
21. Rana BD, Mischoulon Y, Xie Y, Bucher NL and Farmer SR. Cell-extracellular matrix interactions can regulate the switch between growth and differentiation in rat hepatocytes: reciprocal expression of C/EBPα and immediate-early growth response transcription factors. Mol Cell Biol 1994; 14: 5858-5869.
22. Mitaka T, Mizuguchi T, Sato F, Mochizuchi C and Mochizuki Y. Growth and maturation of small hepatocytes. J Gastroent Hepatol 1998; 13 (suppl): S70-S77.
23. Block GD, Locker J, Bowen WC, Petersen BE, Katyal S et al. Population expansion, clonal growth and specific differentiation patterns in primary cultures of hepatocytes induced by HGF/SF, EGF and TGF-α in a chemically defined (HGM) medium. J Cell Biol 1996; 132: 1133-1149.
24. Cable EE and Isom HC. Exposure of primary rat hepatocytes in long-term DMSO culture to selected transition metals induces hepatocyte proliferation and formation of duct-like structures. Hepatology 1997; 26: 1444-1457.
25. Tateno C and Yoshizato K. Growth and differentiation of adult rat hepatocytes regulated by the interaction between parenchymal and non-parenchymal liver cells. J Gastroent Hepatol 1998; 13 (suppl): S83-S92.
26. Mitaka T, Sattler CA, Sattler GL, Sargent LM and Pitot HC. Multiple cell cycles occur in rat hepatocytes cultured in the presence of nicotinamide and epidermal growth factor. Hepatology 1991; 13: 21-30.
27. Li Y, Sattler GL and Pitot HC. The effect of amino acid composition of serum-free medium on DNA synthesis in primary hepatocyte cultures in the presence of epidermal growth factor. *In Vitro* Cell Dev Biol Anim 1995; 31: 867-870.
28. Mitaka T, Mikami M, Sattler GL, Pitot HC and Mochizuki Y. Small cell colonies appear in the primary culture of adult rat hepatocytes in the presence of nicotinamide and epidermal growth factor. Hepatology 1992; 16: 440-447.

29. Tateno C and Yoshizato K. Long-term cultivation of adult rat hepatocytes that undergo multiple cell divisions and express normal parenchymal phenotypes. Amer J Pathol 1996; 148: 383-392.
30. Tateno C and Yoshizato K. Growth and differentiatioon in culture of clonigenic hepatocytes that express both phenotypes of hepatocytes and biliary epithelial cells. Amer J Pathol 1996; 149: 1593-1605.
31. Mitaka T, Kojima T, Mizuguchi T and Mochizuki Y. Subculture of proliferating adult rat hepatocytes in medium supplemented with nicotinamide and EGF. *In Vitro* Cell Dev Biol Anim 1996; 32: 469-477.
32. Mitaka T, Norioka K and Mochizuki Y. Redifferentiation of proliferated rat hepatocytes cultured in L-15 medium supplemented with EGF and DMSO. *In Vitro* Cell Dev Biol Anim 1993; 29A: 714-722.
33. Kost DP and Michalopoulos GK. Effect of 2% dimethylsulfoxide on the mitogenic properties of epidermal growth factor and hepatocyte growth factor in primary hepatocyte culture. J Cell Physiol 1991; 147: 274-280.
34. Chan K, Kost DP and Michalopoulos GK. Multiple sequential periods of DNA synthesis and quiescence in primary hepatocyte cultures maintained on the DMSO-EGF on/off protocol. J Cell Physiol 1989; 141: 584-590.
35. Bour ES, Ward LK, Cornman GA and Isom HC. Tumor necrosis factor-α-induced apoptosis in hepatocytes in long-term culture. Amer J Pathol 1996; 148: 485-495.
36. Michalopoulos GK, Bowen W, Nussler AK, Becich MJ and Howard TA. Comparative analysis of mitogenic and morphogenic effects of HGF/SF and EGF on rat and human hepatocytes maintained in collagen gels. J Cell Physiol 1993; 156: 443-452.
37. Sakata H, Takayama H, Sharp R, Rubin JS, Merlino G and La Rochelle WJ. Hepatocyte growth factor/scatter factor overexpression induces growth, abnormal development and tumor formation in transgenic mouse livers. Cell Growth Differ 1996; 7: 1513-1523.
38. Hansen LK, Mooney DJ, Vacanti JP and Ingber DE. Integrin binding and cell spreading on extracellular matrix act at different points in the cell cycle to promote hepatocyte growth. Mol Biol Cell 1994; 5: 967-975.
39. Kim TH, Bowen WC, Stolz DB, Runge D, Mars WM and Michalopoulos GK. Differential expression and distribution of focal adhesion and cell adhesion molecules in rat hepatocyte differentiation. Exp Cell Res 1998; 244: 93-104.
40. Webber EM, Bruix J, Pierce RH and Fausto N. Tumor necrosis factor primes hepatocytes for DNA replication in the rat. Hepatology 1998; 28: 1226-1234.
41. Watanable Y, Osaki H and Akaike T. TNF-α bifunctionally induces proliferation in primary hepatocytes: role of cell anchorage and spreading. J Immunol 1997; 159: 4840-4847.
42. Rolfe M, James NH and Roberts RA. Tumor necrosis factor alpha (TNF-α) suppresses apoptosis and induces DNA synthesis in rodent hepatocytes: a mediator of the hepatocarcinogenicity of peroxisome proliferators? Carcinogenesis 1997; 18: 2277-2280.
43. Fausto N. Mechanisms of hepatocyte proliferation and apoptosis induced by tumor necrosis factor. 8th Biennial International Congress on Liver Development, Gene Regulation and Disease, Orvietto, Italy 1999: 14.

44. Chevalier S, MacDonald N and Roberts RA. Nafenopin-induced S-phase is independent of both TNF-α priming activity and EGF receptor tyrosine kinase activity. 8th Biennial International Congress on Liver Development, Gene Regulation and Disease, Orvietto, Italy, 1999: 39.
45. Korver W, Schilham MW, Moerer P, van den Hoff MJ, Dam K, Lamers WH, Medema RH and Clevers H. Uncoupling of S-phase and mitosis in cardiomyocytes and hepatocytes lacking the winged-helix transcription factor trident. Current Biol 1998; 8: 1327-1330.
46. Michalopoulos GK, Bowen WC, Zajak VF, Beer-Stolz D, Watkins S, Kostrubsky V and Strom SC. Morphogenetic events in mixed cultures of rat hepatocytes and non-parenchymal cells maintained in biological matrices in the presence of hepatocyte growth factor and epidermal growth factor. Hepatology 1999; 29: 90-100.
47. Mitaka T, Sato F, Mizuguchi T, Yokono T and Mochizuchi Y. Construction of hepatic organoid by rat small hepatocytes and hepatic non-parenchymal cells. Hepatology 1999; 29: 111-125.
48. Takeshita K, Bowen WC and Michalopoulos GK. Three-dimensional culture of hepatocytes in a continuously-flowing medium. *In Vitro* Cell Dev Biol Anim 1998; 34: 482-485.
49. Maher JJ, Bissell DM, Friedman SL and Roll FJ. Collagen measured in primary cultures of normal rat hepatocytes derives from lipocytes within the monolayer. J Clin Invest 1988; 82: 450-459.
50. Simpson GEC and Finckh ES. The pattern of regeneration of rat liver after repeated partial hepatectomies. J Pathol Bacteriol 1963; 86: 361-370.
51. Wu JC, Merlino G, Cveklova K, Mosinger B and Fausto N. Autonomous growth in serum-free medium and production of hepatocellular carcinomas by differentiated hepatocyte lines that over-express transforming growth factor α. Cancer Res 1994; 54: 5964-5973.
52. Hino H, Tateno C, Sato H, Yamasaki C, Katayama S et al. A long-term culture of human hepatocytes which show a high growth potential and express their differentiated phenotypes. Biochem Biophys Res Comm 1999; 256: 184-191.
53. Runge DM, Runge D, Dorko K, Pisarov LA, Leckel K et al. Epidermal growth factor- and hepatocyte growth factor-receptor activity in serum-free cultures of human hepatocytes. J Hepatol 1999; 30: 265-274.
54. Mitaka T, Norioka K, Nakamura T and Mochizuki Y. Effects of mitogens and comitogens on the formation of small-cell colonies in primary cultures of rat hepatocytes. J Cell Physiol 1993; 157: 461-468.
55. Lazaro CA, Rhim JA, Yamada Y and Fausto N. Generation of hepatocytes from oval cell precursors in culture. Cancer Res 1998; 58: 5514-5522.
56. Runge D, Runge DM, Drenning SD, Bowen WC, Grandis JR and Michalopoulos GK. Growth and differentiation of rat hepatocytes: changes in transcription factors HNF-3, HNF-4, STAT-3 and STAT-5. Biochem Biophys Res Commun 1998; 250: 762-768.
57. Runge D, Runge DM, Daskalakis N, Lubecki KA, Bowen WC and Michalopoulos GK. Matrix-mediated changes in the expression of HNF4α isoforms and in DNA-binding activity of ARP-1 in primary cultures of rat hepatocytes. Biochem Biophys Res Commun 1999; 259: 651-655.
58. Michalopoulos GK. Liver regeneration: molecular mechanisms of growth control. FASEB J 1990; 4: 176-187.

59. Cruise JL, Houck KA and Michalopoulos GK. Induction of DNA synthesis in cultured hepatocytes through stimulation of the α-1 adrenoreceptor by norepinephrine. Science 1985; 15: 749-751.
60. Ni N and Yager JD. Comitogenic effects of estrogens on DNA synthesis induced by various growth factors in cultured female rat hepatocytes. Hepatology 1994; 19: 183-192.
61. Yusof YA and Edwards AM. Stimulation of DNA synthesis in primary rat hepatocyte cultures by liver tumor promoters: interactions with other growth factors. Carcinogenesis 1990; 11: 761-770.
62. Fausto N, Laird AD and Webber EM. Liver regeneration 2. Role of growth factors and cytokinase in liver regeneration. FASEB J 1995; 9: 1527-1536.
63. Nishikawa Y, Wang M and Carr BI. Changes in TGF-β receptors of rat hepatocytes during primary culture and liver regeneration: increased expression of TGF-β receptors associated with increased sensitivity to TGF-β-mediated growth inhibition. J Cell Physiol 1998; 176: 612-623.

Chapter 8

Engineering of a miniature 3-D hepatic cell plate culture model

Augustinus Bader[1,2], Erich Knop[3] and Axel Haverich[2]
[1]GBF, Gesellschaft für Biotechnologische Forschung Braunschweig, [2]Leibniz Laboratorien für Biotechnologie und künstliche Organe and [3]Abt. für Zellbiologie und Elektronenmikroskopie, Medizinische Hochschule Hannover (Germany)

Key words: albumin secretion, coculture, collagen gel, cytochrome P450 induction, drug metabolism, electron microscopy, endothelial cells, human hepatocytes, Ito cells, Kupffer cells, matrix geometry, morphology, oxygen supply, sandwich culture, tissue engineering

8.1 Introduction

The development of primary liver cell culture techniques has a long standing history that has contributed significantly to our current knowledge about the relevance of individual aspects of the hepatic microenvironment. These aspects include the composition of the extracellular matrix, matrix geometry, the presence of various cocultured cell types and oxygen supply, to name just a few. In this chapter we have tried to explain the rationale for a rather holistic and integrative approach to the problem of culturing primary liver cells. Our group has developed the concept of reconstructing the three-dimensional (3-D) plate configuration of the hepatic microenvironment to include the adjacent portions of the hepatic sinusoid [1].

The corresponding *in vivo* structure integrates two structural characteristics that function synergistically to provide hepatic function. One is the bipolar enclosure of a hepatocyte monolayer by extracellular matrix, which provides a distinctly better ultrastructural and functional differentiation if it is re-established in culture [2,3] and seems to represent an essential condition for hepatocyte culture. The other characteristic of the hepatic microenvironment is the presence of additional non-parenchymal cell types that populate the extracellular matrix of the Disse space and the sinusoid and are hence also present in a bipolar fashion with respect to the hepatocytes. Together with the parenchymal hepatocytes this too provides a higher order structure, allowing for intercellular communication

M.N. Berry and A.M. Edwards (eds.), The Hepatocyte Review, 97–115.

and functional interaction in our coculture model that represents and rebuilds the physiological characteristics of the liver.

We have investigated the role of these individual components in an attempt to reconstruct this microenvironment at the level of a Petri dish. A second project has looked into upscaling the system to provide a surface area of the order of 10-20 m^2 intended to be used as a bioartificial liver (see Chapter 34). Why should we undertake the effort to reconstruct the hepatic microenvironment in an *in vitro* model? This endeavour has been stimulated by an increased awareness of the importance of following the *in vivo* architecture of the liver in order to avoid certain artefacts regarding the functional performance of the *in vitro* model, particularly in relation to cytochrome P450 (CYP) expression and catalytic activity. Hepatocytes cultured on a single layer of collagen type I lose drug metabolising capabilities rapidly in culture [4]. Use of more complex extracellular matrices, such as Matrigel, prolongs hepatocellular function *in vitro* [5] but this tumor-originating material has the disadvantage of containing as yet not fully identified components (see also Chapter 7). The "sandwich" model [2], in which hepatocytes are cultured between two layers of collagen appears even more effective in this respect, and therefore could be an even more useful system for the study of drug metabolism, especially with respect to CYP induction (see Chapter 25).

In the following, data is presented on key structural characteristics of the hepatic microenvironment, which we sought to reconstruct in an *in vitro* model of the liver. We will particularly address the following features: matrix geometry, 3-D cocultures and oxygen supply.

8.2 Histology of the liver

8.2.1 Distribution of cell types

Hepatocytes *in vivo* are enclosed within the extracellular matrix (ECM) of the space of Disse. This space provides anchorage not only for hepatocytes but also for hepatic sinusoidal cells. It is a key feature that the space of Disse is present on two opposing sinusoidal surfaces enclosing the hepatocyte. Endothelial cell surfaces are located at the border between the space of Disse and the sinusoid, thereby building up a fenestrated lining. Kupffer cells are found overlying the endothelial cells, and the "perisinusoidal" cells or Ito cells can be found inside the space of Disse, with processes extending towards and between the hepatocytes. Pit cells are considered as an additional type of liver cell that are found in the lumen of sinusoids [6,7]. Sinusoids in general are dilated capillary chambers that, inside the liver, carry a mixture of portal venous and arterial blood. Nutrients, toxins and hepatocellular secretory products cross the non-parenchymal cell layer and the space of Disse. The cellular lining of the sinusoids is thought to constitute a coordinated defence system protecting hepatocytes against injury [8]. Kupffer and endothelial cells participate in phagocytosis [8]. Perisinusoidal cells are involved in lipid and vitamin A metabolism [8] Pit cells have been described as large granular lymphocytes, and express natural killer activity.

8.2.2 Cellular biochemical interactions

There is broad evidence of intercellular communication between non-parenchymal cells (NPC) and hepatocytes by means, for example, of prostaglandins [8,9]. Nitric oxide is released in response to bacterial endotoxin in Kupffer cell hepatocyte cocultures [10]. NPC and hepatocytes act as functional units and their communication, as well as collaboration, seems important for the differentiated function of the liver [6,10]. Coculture of NPC with hepatocytes enhances the ability of NPC to express insulin-like growth factor [11]. A proteinaceous substance produced by NPC induces hepatocyte proliferation [12]. Cooperation of NPC with hepatocytes has been suggested for purine metabolism [13]. NPC contribute to the metabolism of xenobiotics and may protect the liver from damage by potentially toxic metabolites [14,15].

8.3 A tissue culture model of the liver in vivo

8.3.1 Effects of matrix geometry on properties of hepatocytes

The importance of the reconstruction of matrix geometry for primary hepatocyte culture was recognized by Dunn *et al* [2]. Sandwiching primary hepatocytes as monolayers within two layers of ECM is aimed at imitating the bilateral presence of ECM with respect to the sinusoidal surfaces of the hepatocytes (space of Disse). Primary rat hepatocytes, when cultured within two layers of ECM in a sandwich configuration, exhibited stable rates of albumin secretion [2]. The adequacy of this model was also apparent for long-term studies of physiological processes such as the acute phase response [16]. We have previously shown the effect of variations in ECM position on biotransformations in primary hepatocytes [4].

8.3.2 Inclusion of non-parenchymal cells

The original sandwich model was monocellular since it included only hepatocytes obtained by means of a Percoll purification process. The addition of non-parenchymal cells, such as endothelial, Kupffer, Pit and Ito cells, offers several potential advantages. These include the opportunity to investigate interactions, to examine means of maintaining differentiation of all cell types, and to study the effects of modification of the microenvironment. Naturally-occurring matrix components of the liver such as collagen type III [17,18] may be obtained by coculturing cells of the hepatic sinusoid. ECM within the space of Disse next to the central vein is predominantly composed of collagen type I [19]. Reid *et al* [20] showed that ECM, composed of collagen type I and proteoglycans, influences hepatocyte growth and transcriptory stability of liver specific functions. Hepatocytes are influenced by variations in ECM constituents [21]. The establishment of a three-dimensional coculture configuration, mimicking more closely the original microenvironment of the liver, was therefore, the next logical step.

8.4 Examination of a three-dimensional model

In order to mimic more closely the microenvironment of the hepatic sinusoid we designed an organotypical three-dimensional coculture model between hepatocytes and NPC by adding liver NPC on top of matrix-overlaid hepatocytes. A gas-permeable membrane used as mechanical support permits oxygen supply to the resulting cell-matrix plate from a second side and not only via the supernatant as in standard culture dishes.

8.4.1 Oxygen supply

In a normal Petri dish oxygen consumption is limited by culture medium thickness and ambient oxygen concentration. Hepatocytes can compensate for hypoxia because of their considerable capacity to satisfy energy requirements by means of anaerobic glycolysis. This, however results in an inefficient utilisation of glucose since the conversion of glucose to lactate leads to the generation of 2 mol ATP/mol glucose compared to 38 mol ATP/mol glucose during oxidative phosphorylation [22]. Wallach and Sherwood [23] have shown by mathematical modelling that pericellular PO_2 levels would approach zero with a medium layer thickness of just 1 mm. Thus, to maintain physiological oxygen supply in cell culture is not a trivial issue, considering the fact that plastic walls and culture medium are efficient barriers to oxygen diffusion [24]. In the 3-D coculture models gas-permeable PTFE films were used to supply oxygen to the liver cells [25]. The use of gas-permeable films permitted an aerobic culture of primary hepatocytes at reduced but physiological ambient PO_2 (10% v/v) while maintaining *in vivo*-like biochemical hepatocellular performance. Conventional hepatocyte cultures rely on ischemic hepatocytes. These findings are also critical for the development of bioartificial liver systems since lactate catabolism and pH regulation is one of the proposed relevant functions of a bioartificial liver [25-27].

8.4.2 Rat and human hepatocyte isolation

Cells were isolated from female Lewis rats weighing 200-250 g. Human hepatocytes were isolated from specimens obtained from patients undergoing resection for hepatic cancer. The methods for both species were similar to those described elsewhere in this volume (see Chapters 1 and 2).

8.4.3 Culture medium

The culture medium, (Williams' E, Biochrom, FRG) was supplemented with 10% fetal bovine serum (Gibco FRG), prednisolone 9.6 ng/ml, glucagon 0.014 µg/ml (Novo, Mainz, FRG) and insulin 0.16 U/ml (Hoechst, Frankfurt, FRG). Penicillin 200 U/ml and streptomycin 200 µg/ml (Gibco) were added. The culture medium was replaced with fresh medium every 24 h. Rat tail collagen was prepared according to the method described by Elsdale and Bard [28].

8.4.4 Sandwich hepatocyte culture

Primary hepatocytes were enclosed within two layers of Type I collagen in a modification of the method described by Hall [29] and Dunn [2]. The first layer of collagen was applied following premixing 1 part of $10 \times$ DMEM concentrate with 9 parts of a rat tail collagen solution (1.11 mg/ml) on ice. 1 ml of the final solution was prepared for a Petri dish with a diameter of 6 cm (Greiner FRG). Thirty minutes after surface coating the Petri dishes with collagen, hepatocytes were seeded with 2×10^6 cells. Four hours following seeding and attachment of the cells, culture medium was removed along with non adherent cells and a second layer of a premixed collagen solution (as described above) was placed on top of the hepatocytes, thus forming a sandwich configuration, or double gel (DG) culture. Thirty minutes after gelation of this second layer of matrix, culture medium was placed on top, as would usually be done in conventional single gel (SG) cultures. A schematic drawing of the culture technique employed is shown in Figure 1. The total volume of the cell culture supernatant was 4 ml.

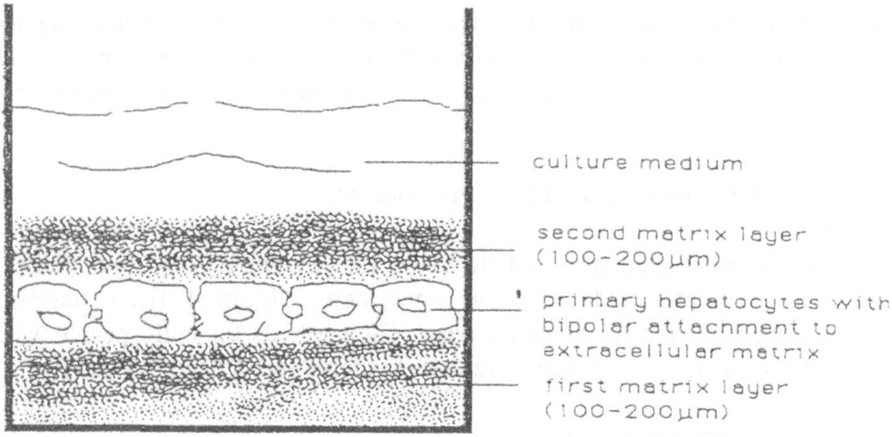

Figure 1. Schematic drawing of the classical sandwich technique for culturing primary hepatocytes by reconstructing matrix geometry *in vitro*

8.4.5 Sandwich hepatocyte and NPC coculture

Freshly isolated hepatocytes were seeded at a density of 3×10^6 cells per dish (28 cm^2) on collagen-coated PTFE membranes. Petriperm^R petri dishes were purchased from Bachofer (Frankfurt, Germany). One hour following seeding, the hepatocytes were overlayed with a second layer of collagen. One hour after gelation of the second collagen layer, NPC were seeded at a density of 4×10^5 cells/ml on top of this second layer of collagen. PC and NPC originating from the same isolation were used for establishing cocultures. Experiments were repeated 4 times using cells from different animals. Four dishes for cocultures of matrix-overlaid-hepatocytes with NPC and four for controls without cocultured NPC were prepared each time. Cells were maintained in culture for 1 week.

8.5 Morphology

8.5.1 Microscopic procedures

For light microscopy, transmission and scanning electron microscopy the cells were fixed on the culture dishes by immersion in glutaraldehyde [1] immediately after removal of the culture medium. Fresh rat whole liver specimens were cut in pieces of 1 × 1 mm and were fixed in the same manner. All specimens were postfixed with 2% OsO_4 and thin specimens of approximately 4 × 10 mm containing cells and the respective collagen layer were cut out of the culture dishes. For sectioning the specimens were embedded in Epoxy resin (Epon™) and cut transversely to the cell layer showing the cells in cross-section. Details of the procedures are given in [1,2].

8.5.2 Morphology of cells in single gel configuration

Hepatocytes cultured in SG configuration appeared to establish polyhedric cell outlines as observed by inverted light microscopy after 2 days in culture, but cross-sections revealed considerable flattening of the cells with flattened nuclei. In transmission electron microscopy (TEM) (Figure 2A) bile canalicular-like structures frequently widened the intercellular space. These were broader than normal and contained few microvilli. At these cell contact zones, the flattened hepatocytes were overlapping. The hepatocytes only rarely possessed microvilli on the outer membrane facing the culture medium. Their surface was relatively smooth on this side, but the inferior surface towards the collagen matrix had numerous microvilli. The cells were connected to each other over long distances until the collagen or medium surface was reached.

After 14 days in culture, cells had lost their original appearance and were extremely flattened and extensively overlapping. Empty spaces had formed between individual clusters in the culture dish. Extensive parts of the cytoplasm underneath the plasma membrane were devoid of cell organelles and filled with a dense filamentous layer. This was more obvious towards the side of the liquid medium and less near the cell surface attached to the collagen. The remaining organelles, especially the mitochondria, were morphologically altered and electron dense. The free cell surface was smooth. Similarly the side facing the matrix had also lost most of its microvilli.

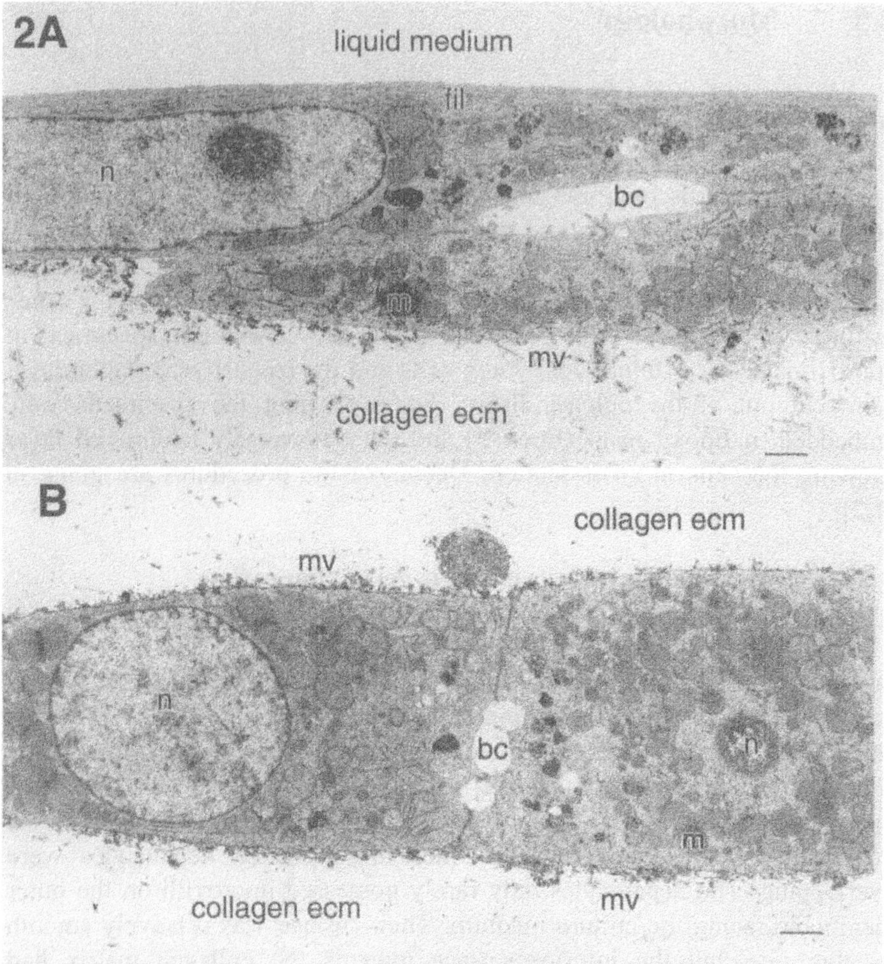

Figure 2. Transmission electron microscopy of hepatocytes in single gel culture (A) and double gel culture (B) after 2 days. The cells in both systems show bile canaliculi (bc) and intact cell organelles (as mitochondria = m). In the single gel the shape of the cell and nucleus (n) is flat, the cells are lying over each other, the cell surface towards the liquid medium is devoid of microvilli (mv) and shows a microfilament layer (fil) underneath (bars = 1 μm).

8.5.3 Double gel configuration

In DG cultures at 2 days the cells were considerably higher. TEM (Figure 2B) revealed that within 2 days the cells had almost re-established a cuboidal cell shape with round nuclei and had formed upright lateral contact zones without major overlapping. No cluster formation occurred as in SG cultures. Bile canalicular-like intercellular spaces showed similar characteristics as in the SG cultures. However, they occurred at very regular intervals between adjoining hepatocytes. They differed from normal bile canaliculi in the wideness of the lumen and a decreased number of microvillous luminal projections. The cell organelles were redistributed inside the cell and displayed an intact morphology. The hepatocytes showed microvilli equally numerous on both sides apposing the collagen layers, as seen *in vivo* for cell membranes bordering the space of Disse. In the DG culture after 14 days the cells had maintained a cuboidal cell shape and an unchanged height of the hepatocyte cords, but showed a tendency to overlap at cell contact zones and to develop enlarged bile canaliculi. The cells, which had round nuclei, were covered on both outer surfaces facing the collagen layers by numerous microvilli- and microplicae-like cytoplasmic projections of various size as known from the sinusoidal faces *in vivo*. The cytoplasmic components were largely unchanged, apart from an increase in mitochondrial size. A comparison of the morphology of hepatocytes in single and double gel culture is given in Figure 2.

8.5.4 Three dimensional cocultures

In the schematic drawing of Figure 3, matrix-overlaid hepatocytes are shown located in a monolayer configuration between two layers of hydrated rat tail collagen. A cocultured non-parenchymal cell (NPC) is depicted on

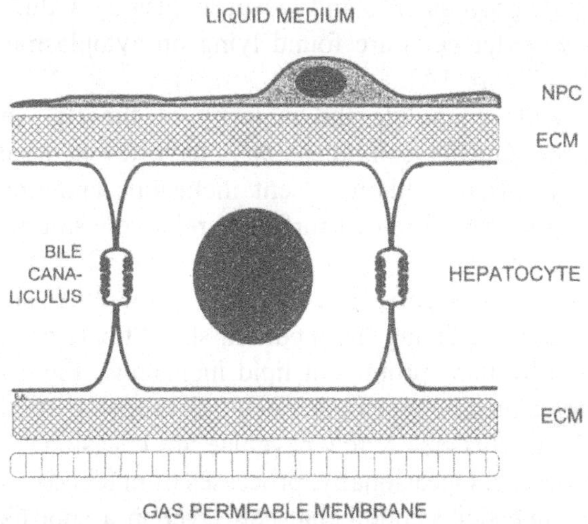

Figure 3. Schematic drawing of the three-dimensional coculture model. Hepatocytes are cultured on a matrix-coated gas-permeable membrane. NPC cover the surface. ECM is present underneath the monolayer of hepatocytes and also between them and a second layer of non-parenchymal cells. Oxygen and nutrients can reach the hepatocytes via either side.

top of the superior ECM layer. Bile canaliculi are shown in between adjacent hepatocytes. Hepatocytes are positioned on the collagen-coated gas-permeable bottom of the Petri dish. Oxygen and nutrients reach the artificial cell plate via two sides. In this model hepatocytes do not show the excessive accumulation of lactate as seen in the classical sandwich model in spite of a 3-D coculture configuration.

In a manner typical for cultures on gas-permeable membranes, attachment resulted in the formation of confluent hepatocellular monolayers. At 7 days in culture, the hepatocytes had maintained their cuboidal cell shape and height. Their nuclei were still roundish and the bile canaliculi were present in regular position with regular frequency between the cells. Sometimes the bile canalicular diameter was slightly increased.

The different cell layers of hepatocytes and NPC, respectively can be visualized by the use of inverted light microscopy in the 3-D coculture model. At the inferior cell layer, the hepatocytes at 2 days constitute a relatively homogeneous population of polyhedric cells with large round nuclei. Each individual hepatocyte is surrounded by a lucent rim which represents the bile canaliculus. A heterogeneous population of NPC can be observed on top of the superior layer of matrix. In TEM an overview cross-section (Figure 4) reveals different NPC in more detail. Round shaped cells with dark cytoplasmic inclusions are identified as Kupffer cells at higher enlargement (Figure 4A, day 1). Large cytoplasmic protrusions like filopodia or lamellopodia are located on their surface and there are numerous cell organelles, phagolysosomes and lucent inclusions within the cytoplasm. Frequently the Kupffer cells are found lying on cytoplasmic extensions of endothelial cells (Figure 4A).

Endothelial cells constitute the majority of the flat shaped cells. In comparison to the Kupffer cells these cells have fewer organelles, a rather homogeneous cytoplasm without lucent inclusions and only a few small dense bodies (Figure 4B). The cell surface is relatively smooth and devoid of microprojections towards the medium but has several coated and uncoated vesicles and caveolae for substance transport at the other side (towards the matrix). On the surface facing the medium, stress fibers are found. Ito cells can be identified by their prominent lipid inclusions (Figure 4C). They are often found intercalated between other cells. Glycogen particles are scattered or found in minute clusters. A few caveolae are occasionally present along the plasma membrane. Occasionally, processes of unidentified cells are lying underneath the superficial endothelial cell layer in a spot-like arrangement (Figure 4 arrowheads). Hepatocytes maintained their typical characteristics as described above for the classical sandwich model and also in the 3-D coculture model in the presence of NPC.

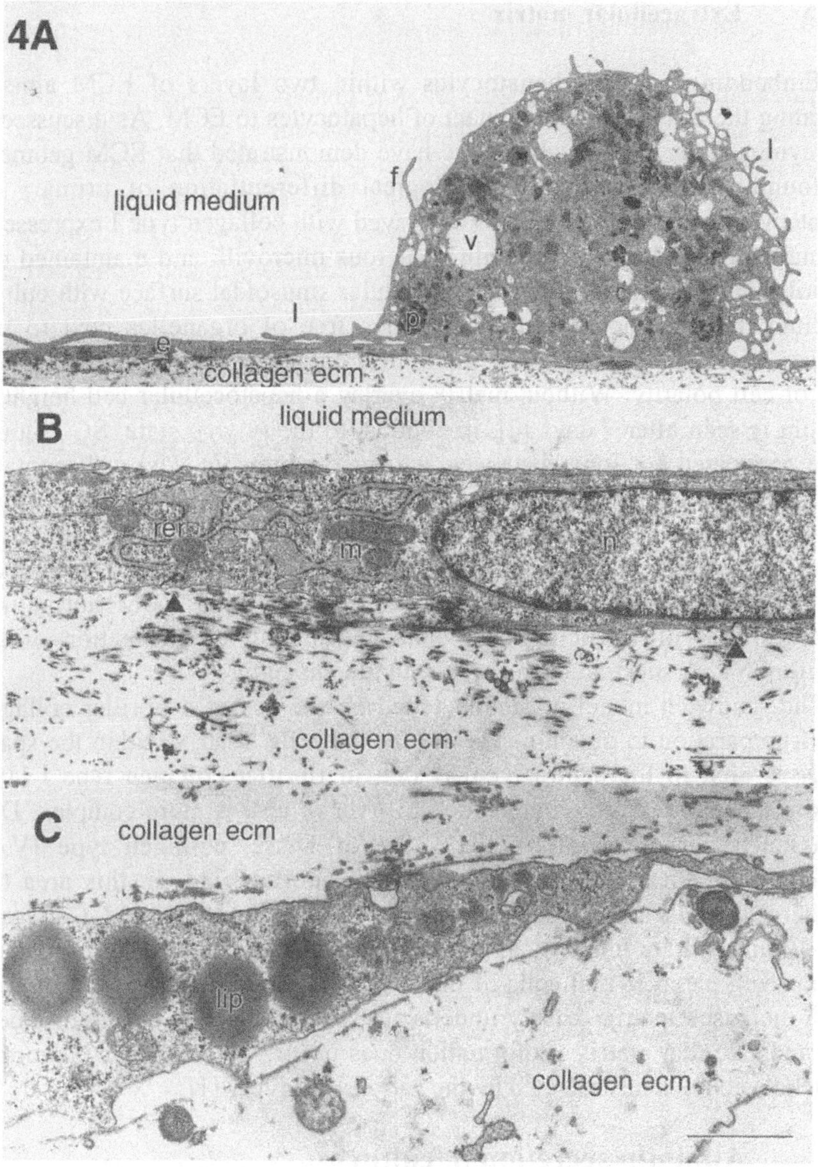

Figure 4. Nonparenchymal cells in the coculture system are shown in transmission electron microscopy (bars = 1µm). The collagen ECM over the hepatocytes is covered by a thin layer of endothelial cells (A: e) with flat nuclei (B: n) showing intact cell organelles (mitochondria = m, and rough endoplasmic reticulum = rer) and formation of vesicles (arrowheads). On top of these are roundish Kupffer cells (A). They have large cytoplasmic protrusions (filopodiae = f, and lamellopodiae = l) and contain dense phagolysosomes (p) and lucent vacuoles (v). Extensions of perisinusoidal Ito cells (C) with lipid droplets (lip) are occasionally seen inside the superficial ECM.

8.5.5 Extracellular matrix

Embedding primary hepatocytes within two layers of ECM aims at imitating the *in vivo* bipolar contact of hepatocytes to ECM. As discussed in a previous study [2] and above, we have demonstrated that ECM geometry profoundly modifies the morphological differentiation of primary rat hepatocytes *in vitro*. Hepatocytes overlayed with collagen type I expressed a normal sinusoidal phenotype with numerous microvilli and maintained cell polarity. Direct contact of the hepatocellular sinusoidal surface with culture medium induced the formation of zones free of organelles next to this surface, the loss of cytoplasmic microdifferentiations and a time dependent loss of cell polarity. Without matrix overlay a hepatocellular cell height of 2-4 μm is seen after 7 days [2]. In contrast to the *in vivo* state, SG cultures were connected for long distances and overlapping. *In vivo*, cell contacts occupy only a small region around the bile canaliculus with the microvillous "sinusoidal" surface reaching deeply between the cells. With increasing culture time bile canaliculi were gradually lost in SG culture. This reflects loss of cell polarisation in SG cultures. Overlay with ECM normalizes hepatocellular ultrastructure and enhances the biotransformation of the specific CYP3A substrate tacrolimus with time in culture.

The sandwich model originally relied on the use of a fibrillar collagen type I preparation to obtain a scaffold for the cells. Matrix within the space of Disse next to the central vein is rich in fibrillar collagen type I [19]. However hepatic matrix composition of liver *in vivo* is more complex. Due to a matrix gradient within the space of Disse, collagen type IV is preferentially located in the vicinity of the portal triad. In this area the smaller stem cells and transitional cells are assumed to be located. Collagens type III and VII are found in an intermediate position possibly corresponding to maturing parenchymal cells of the liver [19]. The ultrastructural findings of an increased matrix density underneath NPC in the 3-D coculture model may indicate that matrix neoformation does occur, which shapes the model to include components secreted by the various cell types [1].

8.6 Albumin secretion in culture

Albumin secretion patterns of DG and SG cultures were highly reproducible. In DG cultures albumin secretion rates continually rose from an initial rate of 1.88 ± 0.09 μg/h per 2×10^6 cells, to reach a secretion rate of 5.28 ± 0.64 μg/h per 2×10^6 cells after 7 days in culture and maintained this rate for a further 7 days. Albumin secretion in SG cultures rapidly declined from an initial secretion rate of 3.93 ± 0.26 µg/h per 2×10^6 cells and the cultures ceased to secrete albumin after 5 days [4].

8.7 Drug metabolism of human hepatocytes culture

8.7.1 Scope and advantages

In drug research, human microsomal preparations have frequently been used. However microsomes deliver an incomplete metabolite profile lacking the phase II metabolites. The required enzymes have a cytosolic location and are lost in the preparation procedure. Microsomes also cannot serve to investigate induction phenomena, which have rightfully gained an increased relevance over the last years (see Chapter 25). Tacrolimus has been shown to be metabolized primarily by CYP3A. Testosterone metabolism is a helpful tool to investigate the maintenance of CYP enzymes *in vitro*, as the various metabolites reflect expression of a large subset of CYP isoenzymes including CYP2C11, CYP2A1, CYP2B1, CYP 3A1 (rat), CYP3A4 (human) [30-33]. The metabolism of urapidil [34] documents the maintenance of monooxygenase activity and can be used to demonstrate metabolic differences between species. Rifampicin can be used to demonstrate CYP induction in human hepatocytes.

8.7.2 Biotransformation of tacrolimus

By means of HPLC/MS analysis it was shown that a consistent quantitative recovery of 8 tacrolimus metabolites could be obtained in prolonged DG cultures, reflecting the sustained activity of CYP3A. The stability of CYP3A expression in culture was previously documented by Western blot analysis [4]. In contrast, hepatocytes without overlay of extracellular matrix showed a pronounced selective decrease of 6 out of 8 metabolites with time and culture. It can be concluded that the extracellular matrix geometry participates in maintaining CYP3A activity of primary rat hepatocytes *in vitro*. Further details are given in [1]. Assessment of the metabolite distribution between the cell-containing collagen phase and the supernatant was performed by HPLC with UV detection. Hepatocytes were maintained in culture for 7 days prior to exposure to tacrolimus (10 µg/ml) for 1 h. As shown in Figure 5, two major metabolite-containing fractions could be identified using this method. Fraction 1 contained the main metabolite, a demethylated compound. Total metabolite formation was significantly higher in DG for both fractions.

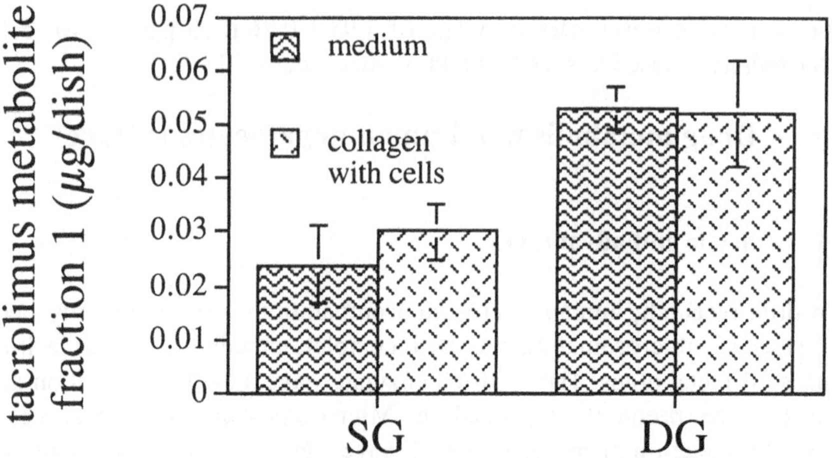

Figure 5. Formation of tacrolimus metabolites in 1 week old primary rat hepatocytes without (SG) and with overlay of ECM (DG). The incubation time was 1 h. Fraction 1 contained demethyl-tacrolimus. Data are shown as the mean ± SD of 60 dishes originating from 5 different isolations for each group. Dishes were seeded with 2×10^6 hepatocytes on day 0.

8.7.3 Induction of drug biotransformation in sandwich culture

Rifampicin shows a threefold to fourfold induction of CYP 3A4 in man *in vivo*, whereas the drug has no inducing effect in the rat *in vivo* [30]. By

Table 1. Testosterone hydroxylation and formation of androstenedione (A) in primary cultures of human hepatocytes with and without rifampicin treatment

Testosterone Oxidation	Control	% change	Rifampicin	% change
2α	0.0 ± 0.0		34.5 ± 27.5	
2β	240.0 ± 182.6	54	624.6 ± 220.2	140
6α	14.2 ± 17.8	40	209.7 ± 313.4	589
6β	1346 ± 890.6	72	5930 ± 904	315
7α	36.7 ± 39.5	61	547.5 ± 217.7	905
11α	17.1 ± 21.5	107	256.6 ± 161.5	1608
15α	18.6 ± 30.3	67	47.5 ± 46.2	171
16α	22.9 ± 21.3	66	60.0 ± 49.0	172
16β	5.2 ± 10.4	62	31.1 ± 26.8	368
A	185.2 ± 109.6	67	289.0 ± 242.5	105

Values are expressed as pmol/min·2×10^6 cells and are means ± SD, n = 5, day 9 in culture. Percentage changes are expressed in terms of the initial value on day 3 in culture (100 %).

using testosterone 6β-hydroxylase activity as an indicator, it was possible to show that rifampicin (RIF) induced CYP3A4 levels in hepatocytes from 5 human donors (Table 1). RIF induction of the enzyme was 5.8-fold that of the control value at day 9 ($p = 0.017$). No effect of RIF (50 μM) on testosterone 6β-hydroxylase activity was found in rat hepatocytes (results not shown). After 7 days of RIF treatment, two testosterone hydroxylase activities, 7α- and 11α-hydroxylase, were significantly induced 9.1- and 16.1-fold at day 9 (Table 1).

8.7.4 Evidence for species-dependent drug biotransformation

Metabolites of the antihypertensive drug, urapadil, in rat hepatocyte sandwich cultures (Figure 6A) were produced in a time dependent manner. M1 concentrations (μg/ml ± SEM) at 1 h following exposure were 0.002 and rose to 0.046 ± 0.016 at 24 h. M2/M3 concentrations at 1 h were 0.02 ± 0.003 and increased to 0.093 ± 0.02 24 h. Figure 6A shows the production of urapidil metabolite M1 and M2/M3 as measured at 1, 2, 5 or 24 h after exposure. M2/M3 was the major metabolite in rat hepatocytes. In contrast, the major metabolite in human hepatocytes is M1 (Figure 6B).

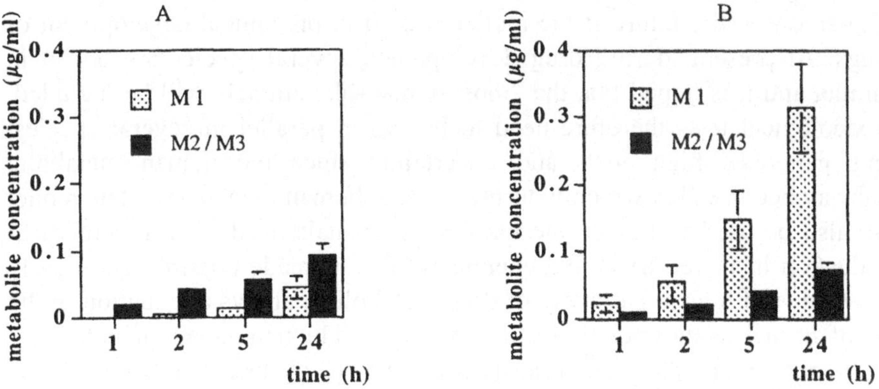

Figure 6. Biotransformation of urapidil (2.5 μg/ml) in sandwich cultures of primary rat hepatocytes (A) and primary human hepatocytes (B). Compounds were analyzed by HPLC, data indicate means ± SEM of 6 dishes for human hepatocytes and means ± SEM of 12 cultures for rat hepatocytes for each time point. M1 and M2/M3 represent the respective metabolites detected.

8.7.5 Metabolism in cocultures

We have not observed any striking differences as yet between sandwich culture and coculture [1]. Thus hepatocyte functions such as albumin secretion as well as phase I and II metabolism of ethoxyresorufin were well maintained and were largely independent of the presence of NPC. Addition of lipopolysaccharide demonstrated the functional interaction between NPC and hepatocytes by depressing the secretion of albumin [1].

8.8 Conclusions

The integration of these characteristic facets of the hepatic microenvironment has provided the basis for establishing a miniature liver cell plate at the level of a Petri dish size. This model can be perfused to obtain a dynamic small scale bioreactor [26]. The model can also be upscaled to obtain a bioreactor with surface areas of 10-30 m^2 [27].

The drug metabolism data presented here shows that the model can be predictive for the *in vivo* situation and reflects the species tested. This has been confirmed in studies that have tested more than 30 compounds (unpublished observations). A major application is found in the area of drug-drug interactions associated with CYP induction. This should advance drug safety in the future if the model is used in preclinical development of drugs. At present, during drug development, several species are tested in parallel and it is hoped that the "most human-like animal" will be included. Toxicological tests therefore need to be run in parallel in several species. This generates high costs and uncertainty since the human metabolic pathway becomes known only during the first human *in vivo* trial. The model can also be used to reduce the number of animals needed in toxicological evaluation of novel drugs and chemicals. This is made possible since early knowledge of human pathways of drug metabolism allows a reduction in the use of animal species not generating the required human metabolic patterns.

In our model glucruonyltransferase activity, gluathatione-S-transferase activity and sulphotransferase activity are equally maintained at high cell-specific levels up to at least 10-14 days in culture [30]. Longer studies have not been undertaken so far and may require improved culture media formulations.

8.9 Summary

The critical relevance of the individual components of the hepatic microenvironment has been raised in the past. Important components include the composition of the extracellular matrix, matrix geometry, the presence of various cocultured cell types and oxygen supply amongst others. We believe that integration of the characteristic facets of the hepatic microenvironment in a comprehensive liver cell culture model could provide the basis for deriving better predictions from the *in vitro* model as to *in vivo* behaviour.

To this purpose we investigated the role of the individual components of the physiologic hepatic microenvironment. Ultrastructural analysis was performed by TEM. Functional performance was studied by HPLC analysis of xenobiotic biotransformation of the antihypertensive drug urapidil, the immunosuppressive agent tacrolimus and the endogenous steroid testosterone. CYP induction was investigated following rifampicin application.

The data clearly indicates that two-sided attachment to ECM induces profound changes of the hepatocellular morphology *in vitro* resulting in the reconstitution of a cuboidal cell shape. This phenomenon is paralleled by an enhanced expression of CYP3A and corresponding metabolic activity. *In vitro* urapidil metabolism by human and rat hepatocytes reflected that occurring *in vivo*. In the 3-D coculture model, differentiated hepatocytes formed a confluent layer below the NPC layer and they regularly expressed bile canaliculi at intercellular contact zones. Both sinusoidal surfaces bore microprojections. Characteristic NPC, including endothelial cells, Kupffer cells and Ito cells completely covered the second layer of extracellular matrix within a week. Kupffer cells were located on top of endothelial cells. Ito cells were intermingled and could be identified by their intracytoplasmic lipid droplets.

This novel three-dimensional coculture model more closely mimics the microenvironment of the hepatic sinusoid by reflecting the plate structure of the liver parenchyma. It may serve as a composite tool to study interactions between hepatocytes and NPC of the liver. The development of hepatic cell culture technology has thus come a long way, starting from the early pioneering work of Berry and Friend published in 1969 [35] in establishing the high-yield isolation technique for these cells. A reliable application has now been developed in the field of drug metabolism.

References

1. Bader A, Knop E, Kern A, Boker K, Frühauf N et al. 3-D coculture of hepatic sinusoidal cells with primary hepatocytes-Design of an organotypical model. Exp Cell Res 1996; 226: 223-233.
2. Knop E, Bader A, Böker K, Pichlmayr R and Sewing KF. Normal ultrastructure of primary hepatocytes depends on bipolar attachment to extracellular matrix. Anatomical Record 1995; 242: 337-349.
3. Bader A, Knop E, Böker KHW, Crome O, Frühauf N et al. Tacrolimus (FK 506) metabolism in primary rat hepatocytes depends on extracellular matrix geometry. NS Arch Pharm 1996; 353: 461-473.
4. Schuetz EG, Li D, Omiecinski CJ, Muller EU, Kleinman HK et al. Regulation of gene expression in adult rat hepatocytes cultured on a basement membrane matrix. J Cell Physiol 1988; 134: 309-323.
5. Dunn JC, Yarmush ML, Koebe HG and Tompkins RG. Hepatocyte function and extracellular matrix geometry: long-term culture in a sandwich configuration. Faseb J 1989; 3: 174-177.
6. Bouwens L, De Bleser P, Vaderkerken K, Geerts B and Wisse E. Liver cell heterogeneity: functions of non-parenchymal cells. Enzyme 1992; 46: 155-168.
7. Bouwens L and Wisse E. Recent insights into hepatic "pit cells" or natural killer cells. Reg Immunol 1988; 2: 137-141.
8. Altin JC and Bygrave FL. Non-parenchymal cells as mediators of physiological responses in the liver. Mol Cell Biochem 1988; 83: 3-14.
9. Casteleijn E, Kuiper J, Van RH, Koster JF and Van BT. Conditioned media of Kupffer and endothelial liver cells influence protein phosphorylation in parenchymal liver cells. Involvement of prostaglandins. Biochem J 1988; 252: 601-605.
10. Billiar TR, Curran RD, Williams DL and Kispert PH. Liver non-parenchymal cells are stimulated to provide interleukin 6 for induction of the hepatic acute-phase response in endotoxemia but not in remote localized inflammation. Arch Surg 1992; 127: 31-36.
11. Villafuerte BC, Koop BL, Pao CI, Birdsong GG and Phillips LS. Coculture of primary rat hepatocytes and non-parenchymal cells permits expression of insulin-like growth factor binding protein-3 *in vitro*. Endocrinology 1994; 134: 2044-2050.
12. Shimaoka S, Nakamura T and Ichihara A. Stimulation of growth of primary cultured adult rat hepatocytes without growth factors by coculture with non-parenchymal liver cells. Exp Cell Res 1987; 172: 228-242.
13. Holstege A, Leser HG, Pausch J and Gerok W. Uridine catabolism in Kupffer cells, endothelial cells, and hepatocytes. Eur J Biochem 1985; 149: 169-173.
14. Oesch F, Lafranconi M and Glatt HR. Role of parenchymal versus non-parenchymal cells in the control of biologically reactive intermediates. Adv Exp Med Biol 1986; 197: 53-61.
15. Steinberg P, Schramm H, Schladt L, Robertson LW, Thomas H and Oesch F. The distribution, induction and isoenzyme profile of glutathione S-transferase and glutathione peroxidase in isolated rat liver parenchymal, Kupffer and endothelial cells. Biochem J 1989; 264: 737-744.
16. Bader A, Rinkes IHB, Closs IE, Ryan CM, Toner M et al. A stable long-term hepatocyte culture system for studies of physiologic processes: Cytokine stimulation of the acute phase response in rat and human hepatocytes. Biotechnol Prog 1992; 8: 219-225.
17. Reid LM and Jefferson DM. Culturing hepatocytes and other differentiated cells. Hepatology 1984; 4: 548-559.

18. Geerts A, Schuppan D, Lazeroms S, De Zanger R and Wisse E. Collagen type I and III occur together in hybrid fibrils in the space of Disse of normal rat liver. Hepatology 1990; : 233-241.
19. Reid LM, Fiorino AS, Sigal SH, Brill S and Holst PA. Extracellular matrix gradients in the space of Disse: relevance to liver biology. Hepatology 1992; 15: 1198-1203.
20. Reid LM, Gaitmaitan Z, Arias I, Ponce P and Rojkind M. Long-term cultures of normal rat hepatocytes on liver biomatrix. Ann N Y Acad Sci 1980; 349: 70-76.
21. Bissell DM, Caron JM, Babiss LE and Friedman JM. Transcriptional regulation of the albumin gene in cultured rat hepatocytes. Role of basement-membrane matrix. Mol Biol Med 1990; 7: 187-197.
22. Jensen MD. Diffusion in tissue culture on gas-permeable and impermeable supports. J Theor Biol 1976; 56: 443-458.
23. Wallach DF and Sherwood P. Diffusion in tissue cultures on gas-permeable and impermeable supports. J Theor Biol 1976; 56: 443-458.
24. Holzer C and Maier P. Maintenance of periportal and pericentral oxygen tensions in primary rat hepatocyte cultures: Influence on cellular DNA and protein content monitored by flow cytometry. J Cell Physiol 1987; 133: 297-304.
25. Bader A, Frühauf N, Tiedge M, Drinkgern M, DeBartolo L et al. Enhanced oxygen delivery reverses anaerobic metabolic states in prolonged sandwich rat hepatocyte culture. Exp Cell Research 1999; 246:221-232.
26. Bader A, Frühauf N, Haverich A and Borlak J. Development of a small scale bioreactor for drug metabolism studies. Xenobiotica 1998; 28: 815-825.
27. Bader A, De Bartolo L and Haverich A. Initial evaluation of the performance of a scaled up flat membrane bioreactor (FMB) with pig liver cells. In: Muraca S, Ed. Bioartificial Liver: The critical issues. Padova: CIC Edizioni Internazionali, 1997: 36-41.
28. Elsdale T and Bard J. Collagen substrata for studies on cell behavior. J Cell Biol 1972; 54: 626-637.
29. Hall HG, Farson DA and Bissell MJ. Lumen formation by epithelial cell lines in response to collagen overlay: a morphogenetic model in culture. Proc Natl Acad Sci USA 1982; 79: 4672-4676.
30. Kern A, Bader A, Pichlmayr R and Sewing K-F. Drug metabolism in hepatocyte sandwich cultures of rats and humans. Biochem Pharmacol 1997; 54: 761-772.
31. Hawksworth GM. Advantages and disadvantages of using human cells for pharmacological and toxicological studies. Human Exp Toxicol 1994; 13: 568-573.
32. Daujat M, Pichard L, Fabre I, Pineau T, Fabre G, Bonfils C and Maurel P. Induction protocols for cytochromes P450IIIA *in vivo* and in primary cultures of animal and human hepatocytes. Meth Enzymol 1991; 206: 345-353.
33. Ged C, Rouillon JM, Pichard L, Combalbert J, Bressot N et al. The increase in urinary excretion of 6b-hydroxycortisol as a marker of human hepatic cytochrome P450IIIA induction. Br J Clin Pharmac 1989; 28: 373-387.
34. Zech K, Eltze M, Kilian U, Sanders KH and Kolassa N. Biotransformation of Urapidil: Metabolites in serum and urine and their biological activity *in vitro* and *in vivo*. Arch Int Pharm Ther 1984; 272: 180-196.
35. Berry MN and Friend DS. High-yield preparation of isolated rat liver parenchymal cells. J Cell Biol 1969; 43: 506-520.

Chapter 9

Freezing and hypothermic storage of porcine hepatocytes

Hans G. Koebe, Sascha A. Pahernik and Friedrich W. Schildberg
Department of Surgery, Klinikum Grosshadern, University of Munich, Marchioninistr. 15, D-81377 Munich, Germany

Key words: albumin secretion, cryopreservation, dimethylsulphoxide, double gel culture, hypothermic storage, porcine hepatocytes

9.1 Storage of hepatocytes

Storage of cells, either in suspension or in primary culture, is highly desirable for *in vitro* investigations on hepatocytes because related experiments can be performed without dependence on the time and logistics of the cell isolation process. Storage techniques in biomedical research include cryopreservation (freezing) at temperatures below 0°C and hypothermic conservation of biological material at 0-4°C. The latter technique, usually performed to preserve organs for transplantation, has recently been introduced as a short term preservation technique for cell suspensions [1]. Since hepatocytes do not proliferate *in vitro*, storage techniques need to be improved in order to reduce cell damage and thereby improve experimental results.

9.2 Cryopreservation of porcine hepatocyte cultures

Two major considerations have stimulated the continuing search for successful freezing techniques for hepatocytes. First, the desire to create a batch-controlled cell bank, and second, the need to supply bioartificial liver (BAL) devices with functional cell material on demand and ready to use.

Current methods of freezing of hepatocytes in suspension, however, usually result in a low overall yield of viable cells and a loss of differentiated cell functions in subsequent cell culture. Even when large numbers of hepatocytes are cryopreserved and, upon thawing, purified by density

M.N. Berry and A.M. Edwards (eds.), The Hepatocyte Review, 117–123.

gradient centrifugation, attachment of cells to culture matrices is impaired, which results in a substantial secondary reduction in cell number [2]. Thus, the principles of mechanisms of successful freezing and thawing of cell material have to be explored and developed into standardized protocols, whatever the design of subsequent investigations on cells or cell cultures may be.

9.2.1 Cryopreservation: problems arising from the freezing of cell suspensions

Although rather contradictory results have been reported in several studies based on the assessment of post-thaw cell viability and function, cryopreservation of cell suspensions until recently could not be generally recommended [3,4]. Hepatocytes with membrane alterations arising from enzymatic digestion during and following the isolation process are unlikely to survive osmotic changes that usually occur during cryoprotection, freezing and thawing. Up to now the one principal advantage of cryopreserving hepatocytes in suspension has been the possibility of clearing dead cells and debris from the suspension after thawing, but this is associated with the disadvantage of a meager cell yield and very low attachment rates. These circumstances have led to initial reports of the successful freezing of hepatocyte cultures [5], and evidence suggests that the presence of an extracellular matrix, the repair of membranes and the cytoskeleton prior to freezing, as well as the monolayer configuration of the cells have added to this improvement in cell survival.

9.2.2 Cryopreservation of hepatocyte cultures

Hepatocytes frozen in culture seem to be more resistant to osmotic changes during the freeze-thaw cycles. For this, cells should be cultured within a gelled collagen matrix, known as the double gel, sandwich or immobilisation culture system [6,7] (see also Chapter 8). Hepatocytes frozen in this culture configuration on day 3 of culture have shown a 70% recovery of cells after thawing, with a persistent protein secretion for at least 2 weeks, as well as cytochrome P450-dependent enzyme activity comparable to non-frozen controls, independent of storage time at −80°C [8,9].

The correct concentration of a cryoprotective agent dimethylsulphoxide (DMSO), as well as the freezing program itself, should prevent cells from harmful intracellular ice formation and from damage due to prolonged exposure to an hyperosmotic environment. To achieve this, the incubation and de-incubation with cryoprotectants should be done in a stepwise fashion and individual protocols must be found for each cell type, culture formation and cryostat employed. The freezing program should allow stabilisation of the cell's internal and external environment prior to cooling towards storage temperature.

9.2.3 Requirements

1. Porcine hepatocytes
2. Programmable cryobiological unit
3. Culture dishes (T-Flask)
4. DMSO as a cryoprotective agent
5. Water bath
6. Cell culture medium

9.2.4 Cryopreservation technique

Hepatocytes should not be cryopreserved before day 3 of culture because cells acquire an increased resistance against osmotic stress during freezing and thawing after a short culture adaptation period. In order to control freezing rates in a programmable unit (Nicool™ ST 20, FPO/TAEMA, Cryobiological Equipment, Sassenage, France), the inflow of liquid N_2 into the freezing chamber is regulated by a platinum resistor that serves as a temperature probe within the culture matrix [9]. Liver cells are cryopreserved in the immobilized culture configuration. Hepatocytes are incubated with 20% (v/v) of DMSO in culture medium (final concentration 10%, v/v) for 15 min at 37°C and for 10 min at 4°C. Afterwards, cultures are placed at defined positions within the freezing chamber and cooled at −1°C/min to a probe temperature of −4°C.

Next, rapid cooling (−7.2°C/min) to −10°C, induces extracellular crystalisation. Linear freezing curves in the probe-culture are achieved by N_2 inflow regulation during the critical phase of extracellular ice formation to compensate for the generated latent heat of fusion. After 20 min of adjustment, cultures are cooled at −1°C/min to −30°C and subsequently at −10°C/min to −80°C, and stored at this temperature for various intervals [9].

9.2.5 Thawing of cultures

After various storage intervals, cultures are transported on dry ice to a laminar flow unit. Instant thawing is achieved by adding 5 ml of 10% DMSO in culture medium (v/v) at 40°C to the frozen culture surface and placement of cultures in a 37°C waterbath. Next, the supernatant is replaced every 5 min by solutions of decreasing DMSO concentrations (7.5%, 5%,2.5%, 0%, v/v) at 4°C and finally, hepatocytes are transferred to the incubator and cultured in culture medium (37°C). Control cultures are held in culture without cryopreservation.

9.2.6 Outcome of cryopreservation

Viability of cryopreserved hepatocytes as assessed by fluoroscopy after fluorescein diacetate and propidium iodide staining of cells is 75 ± 5% (viability after cell isolation: 93 ± 2%). The duration of storage interval does not influence viability results. With respect to their ultrastructural appearance, cells demonstrate well preserved mitochondria, lysosomes, rough endoplasmic reticulum as well as intact intercellular contact sites. Almost no signs of freezing damage to the subcellular components can be observed, except some minor edematous swelling of bile canaliculi and some loss of microvilli therein [9]. With respect to their functional performance, specific liver functions such as albumin secretion of cells are monitored. Cryopreserved hepatocyte cultures show a marked depression of albumin secretion in the early post-thaw period and a recovery of activity to near normal levels after day 3 to 4 of post-thaw cultures. Mean albumin secretion of hepatocytes during the entire culture period (day 1 to 14) is 75 ± 37, 73 ± 29 and 83 ± 41 ng/h/µg DNA for cells cryopreserved for 3 h, 3 days or 30 days, respectively (albumin secretion of non cold-stored hepatocyte cultures: 102 ± 27 ng/h/µg DNA) [10]. These data also indicate that length of storage time does not influence post thaw results.

9.2.7 Establishment of a batch-controlled cell bank

Successful freezing protocols offer a practical means of establishing a cell bank for liver cells. Hepatocytes could be isolated, cultured, frozen and stored long-term for further characterisation and batch-controlled experiments under standardized conditions, independent of the actual availability of fresh donor tissue. Among other applications, this could help to bring about inclusion of primary human hepatocyte cultures in the process of xenobiotic characterisation prior to clinical trials (see Chapter 23).

9.3 Short term cold storage of cell suspensions

If cells are to be stored for hours to days rather than weeks or months, an alternative approach to preserving cells for their subsequent use in primary cultures is the short-term hypothermic storage of hepatocytes. This technique, which employs storage solutions that allow organ preservation for liver transplantation, avoids the inevitable hazards of freezing and thawing cells in suspension. Cells must be incubated with effective protective solutions prior to storage to avoid bleb formation and progressive cell death otherwise encountered for cell suspensions after short periods on ice [11].

9.3.1 Requirements

1. Porcine hepatocytes
2. Temperature controlled refrigerator 4°C
3. Simplified UW solution containing (25 mM potassium phosphate, 5 mM $MgSO_4$, 100 mM lactobionate, 30 mM raffinose and 3 mM gluthatione, pH is adjusted to 7.4
4. Cell culture medium (supplemented DMEM, see Chapter 3)
5. Centrifuge

9.3.2 Method of cold storage

Immediately after isolation, cells are centrifuged at $17 \times g$, for 5 min and finally resuspended in T 75-flasks at a cell concentration of 10×10^6 cells/ml in a simplified UW solution under an air atmosphere at 4°C without stirring. In regard to hypothermic storage of freshly isolated hepatocytes, the simplified UW solution appeared to be equal to the original UW solution and superior to the Leibovitz L-15 solution [12]. The higher the cell concentration in the storage solution (5×10^7 cells/ml), the better the outcome [9]. Cells can be stored up to 48 h under these conditions. At the end of the preservation period, hepatocytes are gently shaken in the storage solution. Cells are then centrifuged at $23 \times g$ for 5 min and resuspended in culture medium before seeding.

9.3.3 Results of cold storage

After hypothermic storage in the simplified UW-solution at 4°C, viability of cells is 82.6 ± 6.8, 69.6 ± 3.9 and 58.7 ± 7.9% for time intervals of 24, 48 and 72 h, respectively (viability after cell isolation: 93 ± 2%). These data are obtained from suspensions after the post-storage washing procedure without any additional density gradient purification. With respect to cell morphology, cells do not show electron microscopically detectable alterations in cell morphology. Hepatocytes exhibit a typical ultrastructure with intact nuclei and nucleoli, as well as a normal lysosomal and mitochondrial profile, without any signs of vesicular transformation.

However, after 72 h of storage, cell viability is disturbed in some specimens, with a loss of nuclei and a commencing lysis of cytoplasmic contents [12]. Hypothermically stored hepatocytes show high albumin secretion during primary culture, comparable to non-stored cells. The rate of albumin secretion between day 1 and 14 of culture averages 121 ± 53, 141 ± 39 and 136 ± 47 ng/h/µg DNA for time intervals of storage of 24, 48 and 72 h, respectively (albumin secretion of non cold-stored hepatocytes: 102 ± 27 ng/h/µg DNA) [12].

9.3.4 Cryopreservation at –80°C versus hypothermic storage at 4°C

Hypothermic conservation of liver cells is a reasonable approach to meeting the requirements for on-demand supply of hepatocytes for BAL-devices. Hepatocytes could be isolated from slaughterhouse organs in qualified laboratories in large scale quantities and then sent to research or clinical centers in a protective solution at 4°C. Here, cells would be washed, purified on density gradients (*e.g.* Percoll^{TM}) and placed as viable hepatocytes in whatever reactor configuration is used. Advantages of this approach are obvious. The technique is easy to apply and should be considered whenever the need exists for short term storage of cell suspensions. The results after 48 h of hypothermia in terms of functional performance and morphological viability are good. The main disadvantage is to be seen in the limited time of storage.

On the other hand, cryopreservation of hepatocyte cultures, best done in Units established for this purpose, is from the technical point of view sophisticated, requires cost-intensive technical equipment, broad cryotechnological knowledge and remains a challenge for further studies. This technique might well support the long-term storage of primary hepatocyte cultures for on-demand supply as ready-to-use functional (biological) units of hybrid artificial liver devices and might help to establish batch-controlled cell banks, especially for human hepatocytes.

In conclusion, both preservation protocols are of considerable value for research and clinical applications of *in vitro* technologies based upon liver cell suspensions and/or hepatocyte cultures. Further investigations have to be directed towards achieving a more predictable and reliable outcome of protocols.

References

1. Poullain MG, Fautrel A, Guyomard C, Chesne C, Grislain L and Guillouzo A. Viability and primary culture of rat hepatocytes after hypothermic preservation: the superiority of the Leibovitz medium over the University of Wisconsin solution for cold storage. Hepatology 1992; 15: 97-106.
2. Innes GK, Fuller BJ and Hobbs KE. Functional testing of hepatocytes following their recovery from cryopreservation. Cryobiology 1988; 25: 23-30.
3. Chesne C and Guillouzo A. Cryopreservation of isolated rat hepatocytes: a critical evaluation of freezing and thawing conditions. Cryobiology 1988; 25: 323-330.
4. De Loecker P, Fuller BJ, Koptelov VA and De Loecker W. Metabolic activity of freshly prepared and cryopreserved hepatocytes in monolayer culture. Cryobiology 1993; 30: 12-18.
5. Koebe HG, Dunn JC, Toner M, Sterling LM, Hubel A et al. A new approach to the cryopreservation of hepatocytes in a sandwich culture configuration. Cryobiology 1990; 27: 576-584.
6. Dunn JCY, Tompkins RG and Yarmush ML. Long-term *in vitro* function of adult hepatocytes in a collagen sandwich configuration. Biotechnology Prog 1991; 7: 237-245.
7. Koebe HG, Pahernik S, Eyer P and Schildberg FW. Collagen gel immobilization: a useful cell culture technique for long-term metabolic studies on human hepatocytes. Xenobiotica. 1994; 24: 95-107.
8. Borel Rinkes IH, Toner M, Sheeha SJ, Tompkins RG and Yarmush ML. Long-term functional recovery of hepatocytes after cryopreservation in a three-dimensional culture configuration. Cell Transplant 1992; 1: 281-292.
9. Koebe HG, Dahnhardt C, Muller-Hocker J, Wagner H and Schildberg FW. Cryopreservation of porcine hepatocyte cultures. Cryobiology 1996; 33: 127-141.
10. Koebe HG, Pahernik SA, Thasler WE and Schildberg FW. Porcine hepatocytes for biohybrid artificial liver devices: a comparison of hypothermic storage techniques. Artif Organs 1996; 20: 1181-1190.
11. Marsh DC, Lindell SL, Fox LE, Belzer FO and Southard JH. Hypothermic preservation of hepatocytes. I. Role of cell swelling. Cryobiology 1989; 26: 524-534.
12. Pahernik SA, Thasler WE, Mueller-Hoecker J, Schildberg FW and Koebe HG. Hypothernic storage of pig hepatocytes: influence of different storage solutions and cell density. Cryobiology 1996; 33: 552-566.

Chapter 10

Storage of isolated hepatocytes

André Guillouzo[1], Claire Guyomard[2], Alain Fautrel[1] and Christophe Chesné[2]
[1]INSERM U 456, Détoxication et Réparation Tissulaire, Faculté de Pharmacie, Université de Rennes 1 and [2]BIOPREDIC, Rennes-Atalante Villejean, RENNES, France

Key words: alginate, cell attachment, cell swelling, cryopreservation, dimethylsulphoxide, dog hepatocytes, drug metabolism, hypothermic preservation, hepatocyte immobilisation, human hepatocytes, polyethylene glycol, storage media

10.1 Introduction

Isolated hepatocytes can be obtained in large numbers from numerous species including man by the classical two-step collagenase perfusion method and are widely used as a model system either in suspension or in primary culture in various research fields [1,2]. They are of particular interest for the study of xenobiotic metabolism and as a bioartificial liver support. Since various hepatic functions exhibit qualitative and quantitative interspecies variations, particularly when comparisons involve experimental animals and human beings, human hepatocytes are highly desirable for answering questions relative to the human situation. However, major problems are encountered with hepatocytes from human beings and some large animals. Although high cell yields can be obtained from a single organ, their sources are limited. Moreover availability of human liver samples is erratic and unpredictable and further reduced by legal and ethical considerations in certain countries [3]. Therefore there is a need for short-term and long-term preservation of isolated hepatocytes and for the establishment of liver parenchymal cell banks. However, after storage the cells must still be viable and capable of expressing their liver-specific functions at levels close to those of their unstored counterparts. Short-term preservation protocols are based on experimental conditions designed for organ storage before transplantation. Numerous cryopreservation protocols have been proposed, mostly for rat hepatocytes: they all involve the use of well-defined freezing conditions in a medium containing a cryoprotectant followed by storage in liquid nitrogen. In a number of studies, cell recovery

M.N. Berry and A.M. Edwards (eds.), The Hepatocyte Review, 125–145.

was low and limited information was provided on maintenance of liver-specific functions after thawing. In this review both the technical aspects for hypothermic preservation and cryopreservation of isolated hepatocytes and their survival and function after storage are discussed.

10.2 Hypothermic preservation

10.2.1 Experimental procedure

The experimental conditions for cell isolation from the livers of various species including man have been described elsewhere [2]. Because of the progress in liver transplantation, human hepatocytes are now mostly prepared from surgical tissue pieces resected from primary and secondary tumors. The protocol used in our laboratory for hypothermic preservation of isolated hepatocytes was designed for the rat [4]. Freshly isolated parenchymal cells are suspended in either the Leibovitz (L-15) culture medium or the University of Wisconsin (UW) cold storage solution. These two storage solutions contain antibiotics (penicillin 10 IU per ml and streptomycin 50 µg per ml). Their pH is adjusted to 7.4 prior to addition of suspended hepatocytes. Isolated parenchymal cells (10^6 per ml) are stored at 0-5°C in sealed 80 cm^2 flasks without stirring for 24 h or more under an atmosphere of air. At the end of the preservation period, hepatocytes are gently suspended in the storage medium and viability is assessed by Trypan blue exclusion. Cells are then centrifuged (100 × *g* for 1 min) and resuspended in culture medium before seeding. The medium consists of 75% Minimum Essential Medium and 25% Medium 199 buffered with 0.22% (w/v) bicarbonate added together with 0.1% bovine serum albumin, 10 µg/ml bovine insulin, antibiotics and 10% fetal calf serum (FCS).

Cocultures are set up by adding rat liver epithelial cells (RLEC) 4 h after the hepatocytes in FCS-supplemented medium [5]. After 24 h RLEC have divided and form a confluent monolayer with hepatocytes. The medium added with 3.5×10^{-6} M hydrocortisone hemisuccinate is renewed every day thereafter.

10.2.2 Critical parameters

Various attempts have been made to preserve viable isolated rat hepatocytes at 0-5°C in either a simple buffer (*e.g.* the Krebs-Henseleit buffer), a culture medium or the UW solution that is used for organ preservation before transplantation. They have shown limited loss of viability and maintenance of morphology, biochemical and metabolic parameters for at least 24 h in the last two media [4,6-8]. Some studies have reported little further damage occurring during an additional 24 h storage [4,9,10]. These conclusions were based on analyses performed immediately after the storage period. However, the UW solution was found to be less effective than the L-15 medium after 24-48 h of storage, particularly when cell attachment rates to plastic were used as endpoints [4] (Figures 1 and 2).

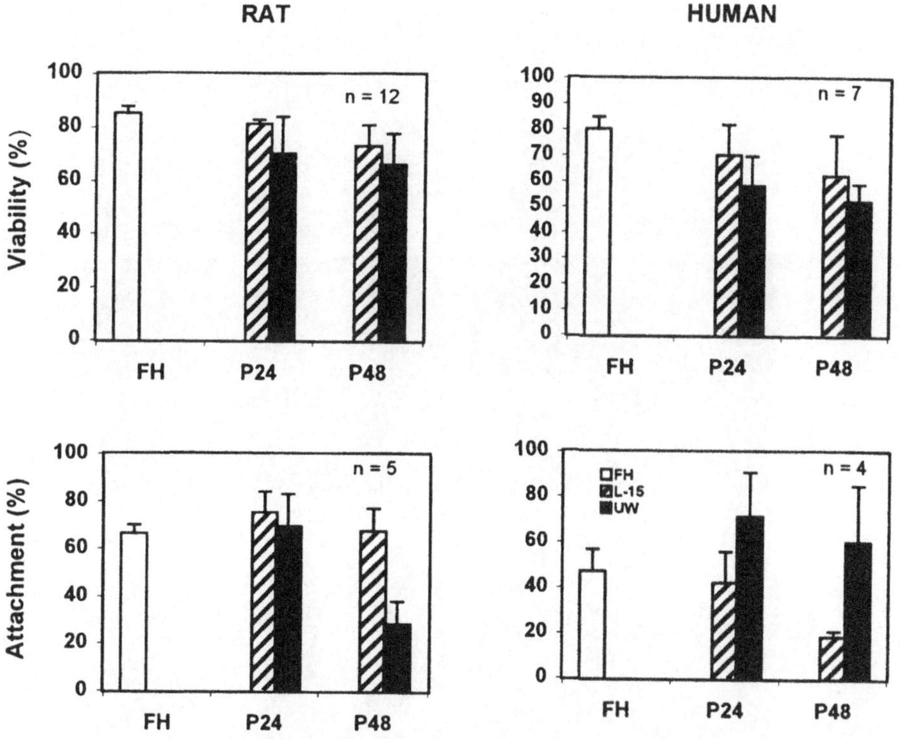

Figure 1. Viability and attachment of rat and human hepatocytes either unpreserved (FH) or hypothermically preserved for 24 h (P24) or 48 h (P48) in L-15 medium (L-15) or UW solution (UW). Taken from [4].

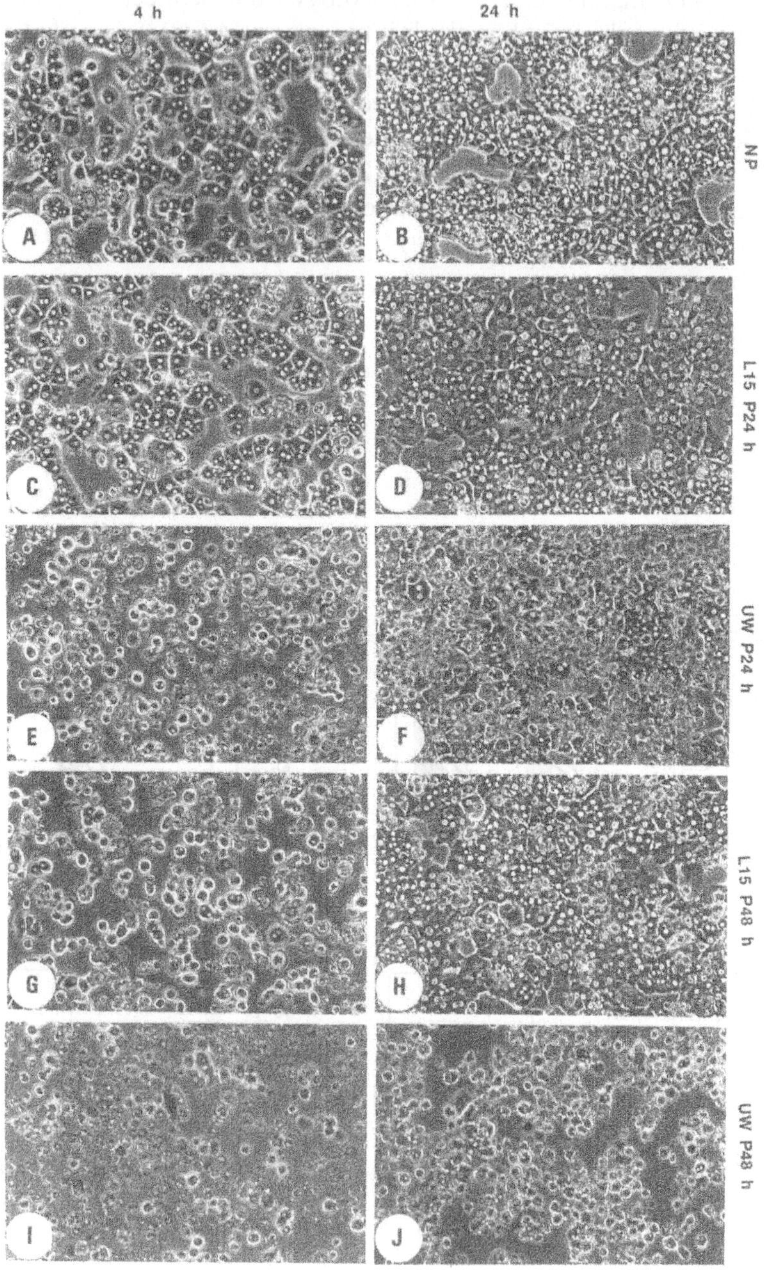

Figure 2. Phase-contrast micrographs of adult rat hepatocytes cultured for 4 h (A, C, E, G, I) or 24 h (B, D, F ,H, J) without previous storage (NP) or after 24 h (P24) or 48 h (P48) of cold storage in either L-15 medium (L15) or UW solution (UW). (180 ×)

Hepatocyte damage during hypothermic preservation has been attributed to various factors, including cell volume changes, disruption of Ca^{2+} homeostasis and the formation of free radicals. Addition of polyethylene glycol (PEG) which is a potent suppressor of ischemia-induced cell swelling, was found to improve preservation of hepatocytes in both L-15 [4] and UW [11] solutions. Its addition at the concentration of 4% to isolated rat hepatocytes suspended in L-15 medium, resulted in a lower decrease in cell viability and a better maintenance of functions but to a limited extent only [4]. The positive effect of PEG was probably reduced because a fraction of cells attached to plastic during hypothermic preservation in L-15 plus PEG medium and were damaged during detachment, which was performed by gentle stirring at the end of the storage period [4]. The effect of PEG was also tested on hypothermic preservation of cultured rat hepatocytes in the UW solution. The cells were stored for various times after 7 days of culture in a collagen sandwich configuration. While alterations were observed in cells stored in PEG-free UW solution after 4 h of cold storage, they appeared only after a 24 h storage in UW solution containing 2.5% PEG [11]. This compound delayed deleterious effects of cold on the cytoskeleton.

The reasons for the lower preservation of rat hepatocytes in UW solution are unclear. After 48 h, certain cells exhibit shrinkage and accumulation of condensed chromatin at the periphery of nuclei that could be indicative of apoptosis. By contrast cells stored in L-15 medium show typical nuclei and nucleoli; however blebbing is observed. It is of interest that storage of rat hepatocytes in a UW/L-15 mixture (vol/vol) results in higher viability, attachment and protein synthesis rates than storage in L-15 or UW solutions alone. After 72 h of storage in such a mixture of media, cell viability and attachment values still represented 60% and 30% of the values obtained with unstored cells, respectively [12]. These observations suggest that neither L-15-medium nor UW solution are ideal media for hypothermic storage of hepatocytes and consequently that a more powerful medium remains to be defined. Some studies suggest that ATP content and energy charge are critical parameters [10] and that amino acids such as glycine can have protective effects, particularly when added during rewarming [6]. The importance of rewarming conditions has also been emphasized for cold preserved organs before grafting: a medium free of oxygen, and addition of antioxidants and fructose have been reported to have beneficial effects. It must be borne in mind that various modified UW solutions have been used successfully for organ preservation.

10.2.3 Functions of hypothermically preserved hepatocytes

Functional activities of cold preserved hepatocytes have been examined either immediately after the end of storage or after plating in culture. ATP content was found to be unchanged after 48 h of storage in UW [10,13]. However, some authors observed a loss of ATP content during storage [14]. We also demonstrated a 29% decrease in ATP in rat hepatocytes isolated from 48 h UW-stored livers [15]. In fact several authors have provided evidence that some functions are impaired after a 48 h storage in the UW solution, especially protein synthesis [4,7,13]. By contrast, membrane transport is well-preserved even after 48 h of storage (Figure 3). Also Hammond and Fry [16] reported that after 1-day storage in UW medium, rat hepatocytes had retained alcohol dehydrogenase and 7-ethoxy 3-4-dimethylcoumarin O-deethylase activities when placed in culture.

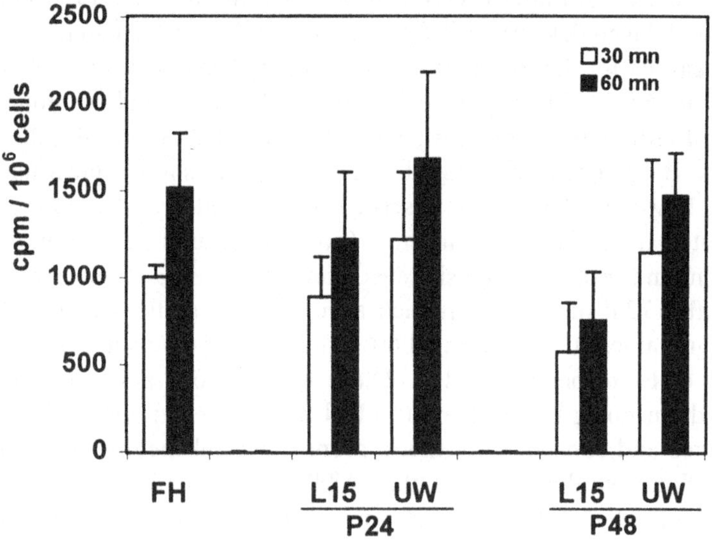

Figure 3. Influence of cold storage on plasma membrane transport in rat hepatocytes. 3 x 10^6 fresh cells (FH) and cells stored for either 24 h (P24) or 48 h (P48) in L-15 medium or UW solution were incubated with 0.1 mCi/ml 1-^{14}C α-aminoisobutyric acid for 30 and 60 min. The values are expressed as cpm/10^6 cells and are means ± SD of 4 experiments.

Parenchymal cells stored in L-15 for 48 h showed only a slight decrease in viability and adherence efficiency and retained their metabolic competence for 72 h in culture. Indeed intracellular GSH content, albumin secretion rate and several phase I and phase II xenobiotic metabolizing enzymes remain very close to the values measured in their unstored counterparts cultured under similar experimental conditions [4].

As shown for fresh cells, survival and function *in vitro* of hypothermically preserved cells are also dependent on culture conditions. While they survive for only a few days in pure culture, rat hepatocytes hypothermically preserved in the L-15 medium for 24 h survive and function for at least two weeks when set up in coculture with rat liver epithelial cells [16].

The question arises as to whether hepatocyte subpopulations are more sensitive to hypothermic preservation. Both rat perivenous and periportal hepatocytes isolated by the digitonin-collagenase technique were preserved either in the L-15 medium or the UW solution for 24 h. While viability and attachment rates of periportal and centrilobular hepatocytes stored in L-15 medium were close to those of corresponding fresh and L-15 hypothermically preserved whole cell populations, periportal cells stored in the UW solution were more fragile, exhibiting much lower viability and attachment rates (Table 1). This could explain some of the differences between L-15- and UW-preserved whole cell populations observed after 24 h of storage.

Table 1. Comparative viability and attachment rates of whole rat hepatocyte populations and perivenous and periportal hepatocyte subpopulations after hypothermic preservation

Condition	**Whole hepatocyte population**	**Perivenous hepatocytes**	**Periportal hepatocytes**
	A – Cell viability		
Freshly isolated	82.4 ± 2.1	77.7 ± 2.8	77.0 ± 0.7
L-15 preserved	75.0 ± 3.5	75.3 ± 2.1	66.8 ± 6.4
UW preserved	66.9 ± 6.4	63.8 ± 8.5	47.0 ± 1.4
	B – Cell attachment		
Freshly isolated	69.0 ± 4.2	59.8 ± 4.9	59.2 ± 12.7
L-15 preserved	63.1 ± 4.9	63.0 ± 11.3	59.6 ± 11.3
UW preserved	57.2 ± 12.0	59.6 ± 5.7	48.7 ± 4.9

A set of cell samples was hypothermically preserved in either L-15 medium or UW solution for 24 h before measurements. All the values are the means ± S.D. of 5 independent experiments. Taken from [16].

Human hepatocytes appear to be better preserved in UW solution: after 24 or 48 h of storage intracellular protein synthesis level was 1.5 to 3 times higher than that measured in corresponding L-15-preserved cells. The latter appear swollen and surrounded with numerous blebs [12]. Similarly membrane transport of α-aminoisobutyric acid was better preserved in UW-hepatocytes than in their L-15 counterparts even after 3 days of storage. Taken together the published observations suggest that the storage medium can be improved and that composition of the optimum medium may vary as a function of the species of origin of the cells.

10.3 Cryopreservation

10.3.1 Experimental procedure

The first freeze-thaw protocol used in our laboratory, was designed for rat hepatocytes [17], and only slight modifications have been introduced. Presently, rat liver cells are cryopreserved as follows. Briefly, to 6×10^6 free hepatocytes in 1 ml of L-15 medium supplemented with 0.2% bovine serum albumin, an equal volume of the same medium containing 32% dimethylsulphoxide (DMSO) and 20% FCS is slowly added. Then the cell suspension is distributed in freezing vials (1.6 ml of the suspension per vial) and kept for 20-30 min at room temperature before transfer to a −20°C freezer. The vials are maintained for 12 min at −20°C, gently stirred every 3-4 min and placed at −70°C. After 2 hours they are transferred to and stored in liquid nitrogen for various periods.

For thawing the vials are placed in a waterbath at 37°C. As soon as ice is no longer visible the cell suspension is collected slowly and diluted in 7-10 ml of L-15 medium containing 0.6 M glucose. An homogeneous cell suspension is obtained by gentle stirring and maintained for 1-2 min at room temperature before centrifugation (1000 rpm). The cell pellet is suspended in the culture medium and viability is estimated by the Trypan blue exclusion test. A similar protocol is used for cell suspensions from other species except that for species other than rats (*e.g.* human and other animals), the optimal DMSO concentration has been found to be 12%.

In thawed free cell suspensions, dead cells can be eliminated by centrifugation on a Percoll cushion. Five ml of the cell suspension are layered onto 15 ml of a 30% Percoll solution (density = 1.124 g/ml) in phosphate buffered saline and centrifuged at 1,000 rpm for 5 min. Viable hepatocytes are pelleted at the bottom and dead cells are left in the upper phase.

Hepatocytes can also be cryopreserved after immobilisation in an alginate gel. For this purpose, freshly isolated parenchymal cells are first suspended in the culture medium (without the above supplements), which contains 2.2% (w/v) sodium alginate (low viscosity, Sigma) at a density of 10^6 and 1.3×10^6 cells per ml for use before and after cryopreservation, respectively. The higher cell density with cryopreserved alginate beads enables the number of viable cells after thawing to be very close to that present in their unfrozen counterparts. Droplets of the suspension, formed by a nitrogen jet droplet generator are gelled in a 1.1% (w/v) solution of $CaCl_2$. Spherical gel beads (0.9-1.1 mm in diameter) containing 400-500 hepatocytes are usually prepared.

After thawing, hepatocytes are seeded at a density of 7×10^5 viable cells per petri dish (35 mm in diameter) in 2 ml of the culture medium supplemented with 10% FCS and incubated at 37°C in a humidified atmosphere composed of 95% air: 5% CO_2. The medium without FCS, and to which 5×10^{-5} M hydrocortisone hemisuccinate is added, is renewed after either 4 h (animal cells) or 15 to 24 h (human cells) and every day thereafter.

Alginate-entrapped hepatocytes are incubated in the culture medium used for suspended cells.

10.3.2 Critical parameters

The most critical freeze-thaw parameters include the choice of the cryoprotectant, composition of the freezing medium and cooling and thawing rates.

When compared to other cryoprotective agents such as glycerol, 1.2 propanediol and dextrans [17,18], DMSO was found to be the most potent cryoprotectant at concentrations ranging between 10 and 20% on the basis of cell viability, cell attachment and metabolic activities after thawing [17,18,20-24]. In an initial interspecies study we observed that the best results were obtained with a 16% DMSO concentration for rat hepatocytes, 14% for other animal parenchymal cells and 10 to 12% for human cells [19]. A variation of 1% to 2% did not result in marked decrease in cell recovery except for dog hepatocytes which appear to be very sensitive to DMSO [19,23]. However, a recent study from Swales and Utesh [25] did not report a higher decrease in cell viability for dog hepatocytes using a final 10% DMSO concentration. It should be noted that these authors incubated dog liver parenchymal cells at a density of 12×10^6 viable cells/ml in a buffer containing 4% DMSO on ice for 5 min before adding an equal volume of buffer containing 16% DMSO and starting the freezing process. We have some evidence that cell density and incubation conditions may markedly improve cell recovery after freezing (Guyomard C., unpublished observations). Starvation of animals before liver disruption should also be considered. It must be borne in mind that the dog is the common non-rodent species used in human risk assessment and therefore it is critical to cryopreserve dog hepatocytes correctly. DMSO delays and reduces ice formation during the freezing process, effects which result in decreased cellular damage. It interacts electrostatically with phospholipid membranes [26]: such effects are thought to be critical for membrane stability during freezing and thawing.

DMSO has been used with various culture media, *e.g.* Leibovitz medium [19]. The most reproducible results appear to be obtained with culture media supplemented with 10 to 20% FCS [18,19,22]. Although some discrepancies exist [20], most authors have come to the conclusion that higher serum concentrations do not improve cell viability and recovery.

A cooling rate of 1 to 5°C/min has been usually selected [17,18,21-24]. In some studies hepatocytes have been frozen in a computer-controlled freezing chamber [22,24,25,27,28]. However, the results have not been shown to be obviously better than those obtained by the use of a –20°C freezer and then a –70°C freezer before plunging the vials into liquid nitrogen.

Cell thawing is another critical step. Quick thawing and immediate dilution of the cryoprotectant give the most satisfactory results. Some authors dilute thawed cells in a solution containing DMSO.

Hepatocytes have also been cryopreserved attached to microcarriers [29], cultured between two layers of collagen [30], microencapsulated [31], entrapped in a gel [32] or after 2 days culture on gelatin [33], using freezing conditions similar [17] or close to those employed for cell suspensions. A higher optimal cooling rate has been found for freezing of hepatocytes cultured in a double collagen gel [34].

10.3.3 Survival and function of cryopreserved free cells

After cryopreservation, hepatocyte viability, based on the Trypan blue exclusion test, is decreased by 10-25% depending on the species (Table 2). The duration of storage in liquid nitrogen has limited, if any, influence on cell viability, even if it lasts several years (Table 3).

Table 2. Survival of hepatocytes after cryopreservation

Species	Viability (%)		Attachment (%)	
	Fresh Cells	Cryopreserved Cells	Fresh Cells	Cryopreserved Cells
Rat	80 ± 4	72 ± 4	68 ± 12	39 ± 9
Man	78 ± 6	75 ± 6	57 ± 13	35 ± 7
Rabbit	82 ± 3	66 ± 10	71 ± 3	37, 38
Monkey	86 ± 5	61 ± 1	60 ± 1	30 ± 6

Cell viability was estimated by the Trypan blue exclusion test prior to separation Percoll [19].

Table 3. Influence of storage duration on viability and drug metabolism capacity of human hepatocytes from one donor

Months of storage	Cell viability (%)	EROD	PROD	Nifedipine oxidation
8	92	2.0	1.3	39
10	96	1.8	ND	ND
12	98	1.6	1.3	22
23	96	1.8	1.8	ND
33	97	ND	ND	29

Viability and drug metabolizing enzyme activities were determined after various periods of deep-freeze storage, using different cell vials from the same human hepatocyte suspension. ND Cell viability was estimated by the trypan blue exclusion test after Percoll separation. Not determined. Ethoxyresorufin O deethylase (EROD) and pentoxyresorufin O-deethylase (PROD) activities expressed as pmol/mn/mg protein; nifedipine oxidation as nmol/h/mg protein.

Most dead cells can be eliminated by centrifugation of the cell suspension over a Percoll solution, so that cell viability reaches 90% or more [18,19,22,25,27]. However cell recovery is decreased: values as low as 40% and 22% of the fresh hepatocyte suspension have been reported for dog [25] and rat [26] cells, respectively. Whether such cell suspensions are representative of the initial heterogeneous cell population may be questioned. This may be the case since we have observed that centrilobular and periportal hepatocyte subpopulations are equally sensitive to the freeze/thaw process as shown by determination of cell viability rates (Table 1). The cell viability parameter, however, does not reflect maintenance of the functional capacity. Cell attachment is a more appropriate criterion. It may be assumed that hepatocytes that attach and spread on the substrate are viable and functional. The percentages of attachment of viable animal parenchymal cells before and after cryopreservation were quite similar for most species and were close to 60-70% and 30-40%, respectively. The values obtained with human hepatocytes were usually close to those observed with animal cells (Table 2). However poor attachment was observed with human liver cell populations showing low viability. It must be emphasized that the quality of freshly isolated hepatocytes is quite important, not only their viability rate but also their energy status. This last parameter has led some authors to incubate isolated rat hepatocytes in a well oxygenated buffer containing 15 mM glucose for 30 min at 37°C before freezing [35].

Thawed, viable, attached cells looked quite similar to their nonfrozen counterparts under phase-contrast microscopy. However usually after 2-3 days, cytoplasmic alterations resulting in appearance of vacuoles and lipid droplets occurred and cell detachment was noticed, indicating that thawed parenchymal cells survived for a shorter period than their fresh counterparts in standard primary culture. By contrast when set up as cocultures, they retained their morphological features and survived for two weeks or more [16]. Electron microscopic examination confirmed light microscopy observations. Major alterations were evidenced in thawed hepatocytes in pure culture, which demonstrated, in addition to the presence of vacuoles and lipid droplets, elongated mitochondria and dilated rough endoplasmic reticulum with less bound ribosomes. By contrast, well-preserved subcellular structures were demonstrated in cocultured thawed hepatocytes. Even 10 days after seeding, vacuoles were scarce, and typical rough endoplasmic cisternae and glycogen particles were seen.

A number of studies have dealt with metabolic activities of cryopreserved hepatocytes. However the majority of these concern only cells of rat origin and a limited number of functions have been measured, mostly within a few hours following thawing. In cryopreserved isolated rat hepatocytes having a high ATP/ADP ratio, metabolism of benzo[a]pyrene was quite similar to that observed in their unfrozen counterparts whether phase 1 or phase 2 metabolites were considered [35]. However, functional alterations are frequently observed even just after thawing, and phenotypic changes occur more rapidly and to a higher extent particularly with respect to phase II xenobiotic metabolizing enzyme activities [22,23,27,36-38]. Cytosolic enzymes appear to be more sensitive: thus sulphoconjugation of biphenyl used as a substrate was much more decreased than glucuronidation in cryopreserved rat hepatocytes [23]. In freshly thawed dog hepatocytes, glucuronidation and sulphation activities were found to be dependent upon the concentration of cofactors, *i.e.* uridine 5 -diphosphoglucuronic acid and 3'-phosphoadenosine 5'-phosphosulphate respectively [25]. However comparative analysis of the published data must take into account that both cell attachment and functional activities are improved after density separation of viable cryopreserved hepatocytes on Percoll [22,23].

Studies on cryopreserved human hepatocytes have been carried out on only a few cell populations [18,19,22,39]. Interindividual functional variations due not only to the donor (genetic polymorphism, liver diseases, premedication, etc.) but also to the conditions of hepatocyte preparation (the time between liver resection and dissociation, duration of warm ischemia etc.) make interpretation of the data more complicated. Traiser *et al.* [39] have reported with one human cell population analysed just after thawing that microsomal epoxide hydrolase and phenol sulphotransferase activities are better preserved than those of cytosolic epoxide hydrolase and glutathione S-transferase (GST). In a study on five different human hepatocyte populations we found that in most cases phenacetin deethylase activity was decreased whereas procainamide N-acetylation and paracetamol glucuronidation and sulphoconjugation were increased in cryopreserved cells compared to their unfrozen counterparts when measured after 44 h of culture [19]. These observations suggest that at least some cytosolic enzymes are not more sensitive to the freeze-thaw process than phase I enzymes. However after 2-3 days alterations are often more extensive in cryopreserved cells [24,40]. Interestingly, human hepatocytes from the same donor, as well as animal hepatocytes, exhibit similar levels of functions when samples are thawed after different times of storage as shown for ethoxyresorufin O-deethylase, pentoxyresorufin O-deethylase and nifedipine oxidation activities in human hepatocytes (Table 3).

Long-term cultures of cryopreserved hepatocytes have been obtained by using sophisticated culture conditions, *e.g.* cell plating on an extracellular matrix [41] or cocultivation with rat liver non-parenchymal cells [16]. In the latter case, various functions were investigated and were found to be either maintained or increased after 7-12 days compared to initial values (Figure 4). A similar decrease was observed in new protein synthesis in cocultures of both unfrozen and cryopreserved rat hepatocytes.

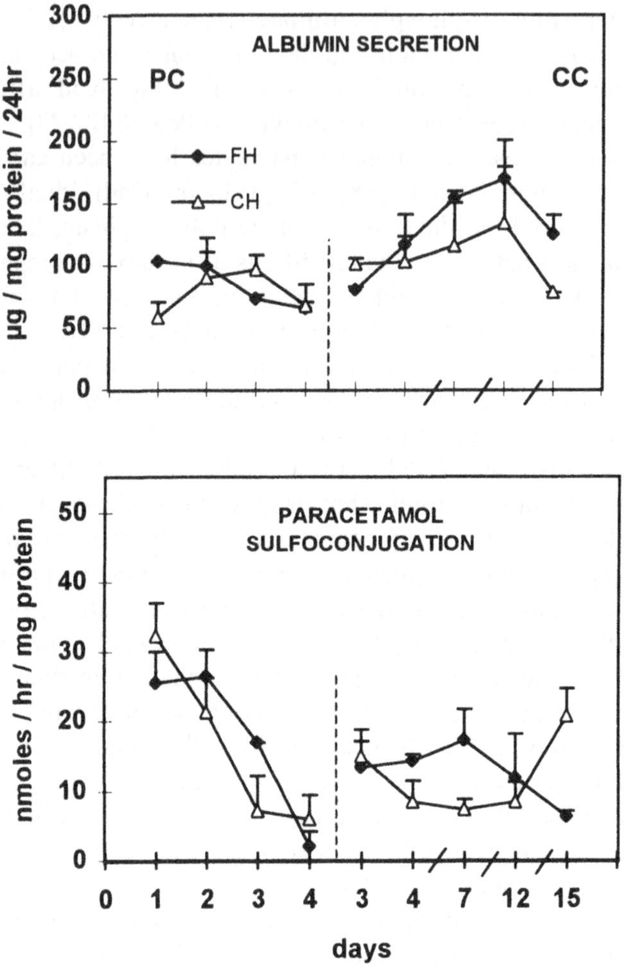

Figure 4. Influence of cryopreservation on albumin secretion and paracetamol sulphoconjugation activity in rat hepatocytes as a function of culture conditions. Rat hepatocytes were set up in pure culture (PC) or in coculture (CC) either fresh (FH) or after cryopreservation (CH). The two functions were measured over 4 days (PC) or 15 days (CC) of culture. Values are means ± SD of two or three independent experiments in duplicate. Taken from [16].

The responsiveness of cryopreserved hepatocytes to inducers is poorly documented. We have shown that dog, monkey and human hepatocytes were still capable of responding to inducers such as phenobarbital, 3-methylcholanthrene and oltipraz after freezing but to a lower extent [42]. Levels of CYP1A and GST-α mRNAs were increased compared to the corresponding fresh cells. Interestingly, as demonstrated *in vivo* and *in vitro* in the rat [43], oltipraz treatment resulted in increased CYP1A transcripts in both fresh and cryopreserved monkey hepatocytes. Enzyme activities are also augmented after exposure to inducers. Thus four, three and five times increases respectively in 6-β-hydroxylation of testosterone were observed after treatment of fresh and cryopreserved rat hepatocytes with pregnenolone 16α-carbonitrile (PCN), a CYP3A inducer, (Table 4).

Table 4. Effects of PCN exposure on 6-β-hydroxylation of testosterone in fresh and thawed rat hepatocytes

Cells	Control	PCN-Treated	Ratio (T/C)
Fresh	3.3 ± 1.7	13.3 ± 5.7	4.0
Thawed	2.0 ± 0.5	7.0 ± 1.0	3.5

10^{-5} M PCN was added to the nutrient medium after 4, 24 and 48 hours and 6-β-hydroxy-testosterone metabolites were measured after 72 h of culture. Values are expressed as nmoles/h/mg protein. They are means ± SD of two experiments in duplicate. T/C : treatment/control.

10.3.4 Survival and function of cryopreserved attached or entrapped hepatocytes

When rat hepatocytes were cryopreserved after immobilisation in calcium-alginate beads, cell loss was quite low, not exceeding 10% [32]; similar values were obtained by labeling the cells with either diacetate fluorescein (40 µg/ml) or propidium diacetate (5 µg/ml). We also investigated various functions, including protein, lipid and urea synthesis, albumin secretion, glycogen content and various phase I and phase II enzyme activities in such cryopreserved immobilized hepatocytes. Most of the functions tested were not altered by the freeze-thaw process (Figure 5) and were expressed at levels close to those measured in unfrozen corresponding hepatocyte monolayers for 48 h or more [32]. Ethoxyresorufin O-deethylase activity, the substrate-specific activity for CYP1A1, was markedly and similarly induced (>tenfold) by 3-methylcholanthrene in fresh and cryopreserved alginate-entrapped rat hepatocytes [32]. Similar observations were recently made with other enzymes, including 6-hydroxylation of testosterone in PCN-treated cells; and GST activity in cells exposed to 10^{-4} M oltipraz (Rialland *et al.*, manuscript in preparation).

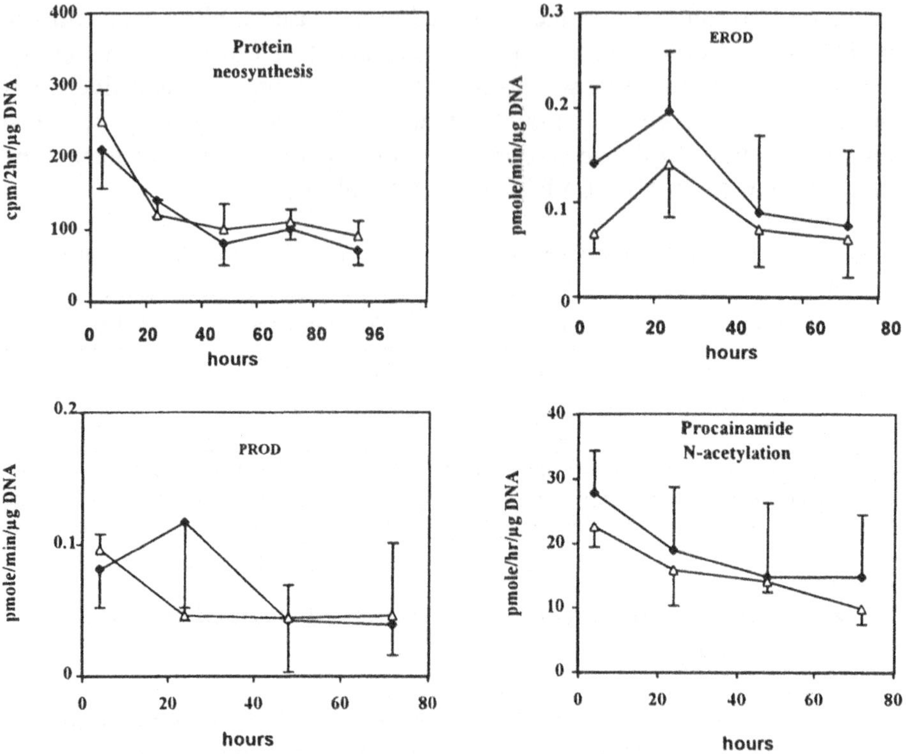

Figure 5. Influence of cryopreservation on protein neosynthesis, ethoxyresorufin O-deethylation (EROD), pentoxyresorufin O-deethylation (PROD) and procainamide N-acetylation in alginate-entrapped rat hepatocytes over a 3- or 4- day incubation in a nutrient medium. The values are means ± SD of three independent experiments in duplicate. In all cases differences were not significant. Taken from [32]

In contrast, viability and functions of hepatocytes cryopreserved either attached to microcarriers or in culture [29-31], were found to be impaired. Thus, hepatocytes cryopreserved, attached to collagen-coated dextran microcarriers, showed gross cytoplasmic damage after thawing; only 30% of the cells survived and produced little albumin [30]. Rat hepatocytes cultured on a collagen film were also cryopreserved. Cells incubated in the freezing medium showed a 20-30% loss of viability and four hours after thawing less that 10% of them excluded Trypan blue. However, a remarkable repair process appeared to occur over 24-48 h post-thaw; indeed 80-90% of thawed rat hepatocytes were viable. CYP content was retained but not GSH content or urea synthesis (possibly reflecting loss of cytosolic enzymes) which decreased to 50% and 25% of the pre-free values respectively [44]. Recently Pang *et al.* [45] found that sulphation and glucuronidation activities were maintained for at least 24 h in thawed rat hepatocytes placed in sandwiches of hydrated collagen gel.

10.4 Conclusions

Isolated hepatocytes can now be stored for either short-term or long-term periods and such preserved cells have been successfully used for xenobiotic metabolism and toxicity studies [46] and for liver support [47]. After storage, hepatocytes retain functional activities *in vitro* for at least a short period including those of phase I and phase II drug metabolizing enzymes, making possible studies on identification of interspecies metabolic pathways of a new drug, drug-drug interactions or mechanisms of toxicity of chemicals. Cryopreserved hepatocytes offer peculiar advantages since they can be stored for long periods and are easily distributed to other laboratories. After thawing, however, cell recovery is generally low and phenotypic alterations occur earlier and to a higher extent than in their corresponding fresh counterparts. Additional studies should be performed on the freeze-thaw protocol in order to reduce cellular damage caused during freezing and rewarming and to define experimental conditions for freezing of large numbers of hepatocytes. The superiority of computer-controlled freezing equipment is not well-established and deserves further investigation. If the loss of some viable cells during the freeze-thaw process cannot be avoided, conditions of incubation and culture of cryopreserved liver parenchymal cells are critical for prolonged maintenance of functional activities. However, the quality of the cells before and after storage should be assessed more carefully. Recent studies have shown that the energy status [35], gel embedding [32], and separation of viable cells on Percoll gradient [23] favor maintenance of expression of liver-specific functions after thawing. Alginate-entrapped hepatocytes are now commercially available under the trade name of Liverbeads™. Cell density, as shown for unfrozen human hepatocytes, should also be taken into account since both cell survival and function appear to be cell density-dependent [48]. Probably in a number of studies, fresh and cryopreserved hepatocytes were investigated at different densities. The observations on cryopreserved hepatocytes published in the literature suggest that some improvements in both the art of cryopreservation and the conditions of incubation after thawing can be expected in the future.

Acknowledgements

We thank Mrs. A. Vannier for typing the manuscript. Personal studies were supported in part by EEC (contract n° 11277-95-10 Field ISP NL and BIOMED contract BMH4 CT96-02-254).

References

1. Berry MN, Edwards AM and Barritt GJ. Isolated hepatocyte preparation; properties and applications. In: Burdon RH and van Knippenberg PH, Eds. Laboratory techniques in biochemistry and molecular biology 21. Amsterdam: Elsevier, 1991: 1-460.
2. Isolated and cultured hepatocytes. In: Guillouzo A and Guguen-Guillouzo C, Eds. Paris: Les Editions INSERM, John Libbey Eurotext, 1986: 1-408.
3. Skett P, Tyson C, Guillouzo A and Maier P. Report on the international workshop on the use of human *in vitro* liver preparations to study drug metabolism in drug development. Biochem Pharmacol 1995; 50: 280-285.
4. Poullain MG, Fautrel A, Guyomard C, Chesné C, Grislain L and Guillouzo A. Viability and primary culture of rat hepatocytes after hypothermic preservation : the superiority of the Leibovitz medium over the UW solution for cold storage. Hepatology 1992; 15: 97-106.
5. Guguen-Guillouzo C, Clement B, Baffet G, Beaumont C, Morel-Chany E, Glaise D and Guillouzo A. Maintenance and reversibility of active albumin secretion by adult rat hepatocytes co-cultured with another liver epithelial cell type. Exp Cell Res 1983; 143: 47-54.
6. Marsh DC, Vreugdenhil PK, Mack VE, Belzer FO and Southard JH. Glycine protects hepatocytes from injury caused by anoxia, cold ischemia and mitochondrial inhibitors, but not injury caused by calcium ionophores or oxidative stress. Hepatology 1993; 17: 91-98.
7. Sandker GW, Slooff MJH and Groothuis GMM. Drug transport, viability and morphology of isolated rat hepatocytes preserved for 24 h in University of Wisconsin solution. Biochem Pharmacol 1992; 43: 1479-1485.
8. Sorrentino D, Van-Ness K, Ribeiro I and Miller CM. Functional and morphological features of isolated hepatocytes preserved in University of Wisconsin solution. Hepatology 1991; 14: 331-339.
9. Sandker GW, Weert B, Merema MT, Kuipers W, Slooff MJH, Meijer DKF and Groothuis GMM. Maintenance of viability and transport function after preservation of rat hepatocytes in various simplified University of Wisconsin solutions. Biochem Pharmacol 1993; 46: 2093-2096.
10. Olinga P, Merema M, Slooff MJH, Meijer DKF and Groothuis GMM. Influence of 48 hours of cold storage in University of Wisconsin organ preservation solution on metabolic capacity of rat hepatocytes. J Hepatol 1997; 27: 738-743.
11. Stefanovich P, Ezzell RM, Sheehan SJ, Tompkins RG, Yarmush ML and Toner M. Effects of hypothermia on the function, membrane integrity and cytoskeletal structure of hepatocytes. Cryobiology 1995; 32: 389-403.

12. Rissel M, Guiguen K, Meunier B, Joly B, Fautrel A and Guillouzo A. Viability and function of rat and human hepatocytes after hypothermic preservation. Hepatology 1993; 18: 321A.
13. Vreugdenhil PK, Marsh DC, Belzer FO and Southard JH. Urea and protein synthesis in cold-preserved isolated rat hepatocytes. Hepatology 1992; 16: 241-246.
14. Svensson G, Soussi B, Wiklund L and Karlberg I. Failure of adenine nucleotide and water content of cold-stored rat livers to serve as graft viability indices for transplantation. Eur Surg Res 1994; 26: 125-132.
15. Guyomard C, Chesné C, Meunier B, Fautrel A, Clerc C et al. Primary culture of adult rat hepatocytes after 48 hours preservation of the liver with cold UW solution. Hepatology 1990; 12: 1329-1336.
16. Fautrel A, Joly B, Guyomard C and Guillouzo A. Long-term maintenance of drug-metabolizing enzyme activities in rat hepatocytes after cryopreservation. Toxicol Appl Pharmacol 1997; 147: 110-114.
17. Chesné C and Guillouzo A. Cryopreservation of isolated rat hepatocytes : a critical evaluation of freezing and thawing conditions. Cryobiology 1988; 25: 323-330.
18. Loretz LJ, Li AP, Flye MW and Wilson AGE. Optimization of cryopreservation procedures for rat and human hepatocytes. Xenobiotica 1989; 19: 489-498.
19. Chesné C, Guyomard C, Fautrel A, Poullain MG, Frémond B et al. Viability and function in primary culture of adult hepatocytes from various animal species and human beings after cryopreservation. Hepatology 1993; 18: 406-414.
20. Gomez-Lechon MJ, Lopez P and Castell JV, Biochemical functionality and recovery of hepatocytes after deep freezing storage. *In Vitro* 1984; 20: 826-832.
21. Rijntjes PJM, Moshage HJ, Van Gemert PJL, de Waal R and Yap SH. Cryopreservation of adult human hepatocytes: the influence of deep freezing storage on the viability, cell seeding, survival, fine structure and albumin synthesis in primary cultures. J Hepatol 1986; 3: 7-18.
22. Lawrence JN and Benford DJ. Development of an optimal method for the cryopreservation of hepatocytes and their subsequent monolayer culture. Toxicol *In Vitro* 1991; 5: 39-51.
23. Powis G, Santone KS, Melder DC, Thomas L, Moore DJ and Wilke TJ. Cryopreservation of rat and dog hepatocytes for studies of xenobiotic metabolism and activation. Drug Metab Disp 1987; 15: 826-832.
24. Dou M, De Sousa G, Lacarelle B, Placidi M, Lechène de la Porte P, Domingo M, Lafont H and Rahmani R. Thawed human hepatocytes in primary culture. Cryobiology 1992; 29: 454-469.
25. Swales NJ and Utesch D. Metabolic activity of fresh and cryopreserved dog hepatocyte suspensions. Xenobiotica 1998; 28: 937-948.
26. Anchordoguy TJ, Cecchini CA, Crowe JH and Crowe LM. Insights into the cryoprotective mechanism of dimethyl sulfoxide for phospholipid bilayers. Cryobiology 1991; 28: 467-473.
27. Utesch D, Diener B, Molitor E, Oesch F and Platt KL. Characterization of cryopreserved rat liver parenchymal cells by metabolism of diagnostic substrates and activities of related enzymes. Biochem Pharmacol 1992; 44: 309-315.
28. Sun E, Aspar DG, Ulrich RG and Melchior GW. Cryopreservation of cynomolgus monkey (Macaca fascicularis) hepatocytes for subsequent culture and protein synthesis studies. *In Vitro* Cell Dev Biol 1990; 25: 147-150.

29. Arnaout WS, Moscioni AD, Barbour RL and Demetriou AA. Development of bioartificial liver: bilirubin conjugation in Gunn rats. J Surg Res 1990; 48: 379-382.
30. Koebe HG, Dunn JC, Tonner M, Sterling LM, Hubel A, Cravalho EG, Yarmush ML and Tompkins RG. A new approach to the cryopreservation of hepatocytes in a sandwich culture configuration. Cryobiology 1990; 27: 576-584.
31. Dixit V, Darvasi R, Arthur M, Lewin K and Gitnick G. Cryopreserved microencapsulated hepatocytes: transplantation studies in Gunn rats. Transplantation 1993; 55: 616-622.
32. Guyomard C, Rialland L, Fremond B, Chesne C and Guillouzo A. Influence of alginate gel entrapment and cryopreservation on survival and xenobiotic metabolism capacity of rat hepatocytes. Toxicol Appl Pharmacol 1996; 141: 349-356.
33. Strom SC, Dorko K, Thompson MT, Pisarov LA and Nussler AK. Large scale isolation and culture of human hepatocytes. In: Franco D, Boudjema K and Varet B, Eds. Ilots de Langerhans et hépatocytes. Paris: Les Editions INSERM, 1998: 196-205.
34. Hubel A, Toner M, Cravalho EG, Yarmush ML and Tompkins RG. Intracellular ice formation during the freezing of hepatocytes cultured in a double collagen gel. Biotechnol Prog 1991; 7: 554-559.
35. Zaleski J, Richburg J and Kauffman FC. Preservation of the rate and profile of xenobiotic metabolism in rat hepatocytes stored in liquid nitrogen. Biochem Pharmacol 1993; 46: 111-116.
36. Adams RM, Wang M, Crane AM, Brown B, Darlington GJ and Ledley FB. Effective cryopreservation and long-term storage of primary human hepatocytes with revovery of viability, differentiation and replicative potential. Cell Tranplant 1995; 4: 579-586.
37. Condouris JA, Grant MH, Engeset J, Pétrie JC and Hawksworth GM. Cryopreservation of human adult hepatocytes for use in drug metabolism and toxicity studies. Xenobiotica 1993; 23: 1399-1409.
38. Diener B, Abdel-Latif H, Arand M and Oesch F. Xenobiotic metabolizing enzyme activities and viability are well preserved in EDTA-isolated rat liver parenchymal cells after cryopreservation. Toxicol Appl Pharmacol 1995; 130: 149-153.
39. Traiser M, Diener B, Fändrich F, Vogel I, Leissner J, Hohenfellner R et al. Isolation and cryopreservation of human hepatocytes. Transplantationsmedizin 1994; 6: 84-90.
40. Chesné C, Gripon P and Guillouzo A. Primary culture of cryopreserved rat hepatocytes. In: Guillouzo A, Ed. Liver Cells and Drugs. Paris: Les Editions INSERM, John Libbey Eurotext, 1988; 164: 343-350.
41. Moshage HJ, Rijntjes PMJ, Hafkenscheid JCM, Roelofs HMJ and Yap SH. Primary culture of cryopreserved adult human hepatocytes on homologous extracellular matrix and the influence of monocytic products on albumin synthesis. J Hepatol 1988; 7: 34-44.
42. De Sousa G, Langouët S , Nicolas F, Placidi M, Rahmani R and Guillouzo A. Induction of cytochrome P-450 1A and glutathione S-transferase transcripts in primary cultures of hepatocytes from dog, monkey and man after cryopreservation. Cell Biol Tox 1996; 12: 351-358.
43. Langouët S, Maheo K, Berthou F, Morel F, Lagadic-Gossmann D et al. Effects of administration of the chemoprotective agent oltipraz on CYP1A and CYP2B in rat liver and rat hepatocytes in culture. Carcinogenesis 1997; 18: 1343-1349.
44. Watts P and Grant MH. Cryopreservation of rat hepatocyte monolayer cultures. Human Exp Toxicol 1996; 15: 30-37.

45. Pang JM, Zaleski J and Kauffman FC. Toxicity of allyl alcohol in primary cultures of freshly isolated and cryopreserved hepatocytes maintained on hydrated collagen gels. Toxicol Appl Pharmacol 1997; 142: 87-94.
46. Guillouzo A. Liver cell models in *in vitro* toxicology. Environ Health Perspect 1998; 106 (suppl 2): 511-532.
47. Demetriou AA, Rozga J, Podesta L, Lepage E, Morsiani E et al. Early clinical experience with a hybrid biopharmatical liver. Scand J Gastroenterol 1995; 30 (suppl 208): 111-117.
48. Ferrini JB, Pichard L, Domergue J and Maurel P. Long-term primary culture of adult human hepatocytes. Chem-Biol Interact 1997; 107: 31-45.

Chapter 11

Hepatocyte swelling: techniques and effects on metabolism

Alfred J. Meijer
Department of Biochemistry, Academic Medical Center, University of Amsterdam, Amsterdam, The Netherlands

Key words: acetylCoA carboxylase, amino acid, autophagy, cell volume, chloride, fatty acid, glycogen, glycogen synthase, hepatocyte, ion transport, lipogenesis, p70S6kinase, phosphatidylinositol 3-kinase, phosphorylase, protein synthesis, proteolysis, S6, signal transduction, swelling,

11.1 Introduction

Cells are frequently exposed to variations in osmotic stress. Some cells, such as intestinal epithelia, kidney medullary cells and red blood cells, are regularly exposed to anisotonic extracellular fluid. In absorbing and secreting epithelia, such as hepatocytes, the large transcellular and solute fluxes also represent an osmotic challenge to the cells. In principle, variations in intracellular osmolarity can occur in any type of cell because during metabolism, changes continuously occur in intracellullar accumulation of nutrients and impermeant metabolites derived from them.

The supply of substrates that reaches the liver via the portal vein varies with the nutritional and/or hormonal conditions. Among the various nutrients entering the liver, amino acids are especially important from an osmotic point of view. Although variations in their portal vein concentrations, *e.g.* during starvation or feeding, result in only minor changes in extracellular osmolarity, this may not be true for the osmolarity within the hepatocyte. Several amino acids are actively transported across the plasma membrane of the hepatocyte by Na^+-dependent amino acid transport systems. Because they are driven by the plasma membrane Na^+ gradient, these amino acids can be highly concentrated in the cells. For this reason and because of the production of impermeant amino acid catabolites within the cells, the intracellular osmolarity of the hepatocyte can undergo considerable changes (Figure 1). The plasma membrane of the hepatocyte, like that of most mammalian cells, is highly permeable to water so that the cell volume varies in response to changes in intracellular osmolarity. Volume changes are buffered by mechanisms that are known as regulatory volume decrease (RV_d) and regulatory volume increase (RV_i), in which the cell attempts to

M.N. Berry and A.M. Edwards (eds.), The Hepatocyte Review, 147–167.

restore its original volume by releasing or gaining inorganic salts and/or organic osmolytes [1,2].

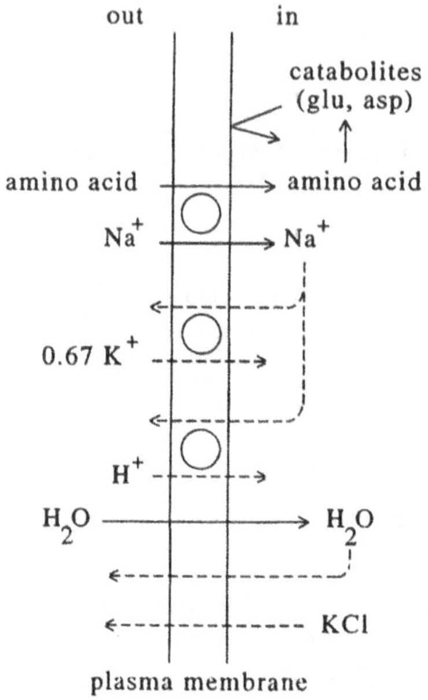

Figure 1. Cell swelling following Na^+ dependent amino acid transport across the plasma membrane; release of KCl (and water) during regulatory volume decrease.

11.1.1 Volume changes in isolated hepatocytes

In rat hepatocytes, the rapid initial cell swelling induced by intracellular amino acid accumulation, and also by hypo-osmotically induced cell swelling (although this is not a physiological condition), is followed by a release of KCl from the cells (and possibly of some $KHCO_3$ although its contribution in hepatocytes is probably minor) [2-4] (Figure 1). Due to the subsequent release of H_2O the cells shrink back to almost, but not quite their original volume [2,3]. In the long-term, hepatocytes may also lose water by vesicular transport, *e.g.* by increasing the rate of exocytosis [5]. Although efflux of KCl contributes most to RV_d in mammalian hepatocytes, this may not be the case in hepatocytes fom other species. In skate hepatocytes, for example, taurine is the most important osmolyte to be released by the cells during RV_d [6] (see also Chapter 5.)

We [7,8] and others [2,9,10] observed that the volume of the hepatocyte not only varies with its metabolic state, but that, conversely, hepatocyte volume can also influence metabolism. For mammalian cells, this is a hitherto unidentified mechanism of metabolic regulation that is not confined to hepatocytes but can also be found in many other cell types [1]. Before the relationship between hepatocyte volume and metabolism is analysed we will first discuss some of the methods that are currently used to measure hepatocyte volume.

11.2 Measurement of hepatocyte volume

Several methods for the estimation of hepatocyte volume are available, depending on the experimental model.

11.2 1 Perfused liver

An elegant method has been described by Lang *et al.* [10] in which the liver of a rat is perfused while mounted on a balance. Any change in liver volume is reflected by a change in liver mass which can be monitored continuously. Another way to determine cell water during perfusion is to compare the wash-out profiles of simultaneously infused [^{3}H]inulin or [^{3}H]raffinose (as extracellular markers) and of [^{14}C]urea (as a marker for both extracellular and intracellular space) [11]. A typical value for intracellular water found this way is 548 μl/g liver wet mass under normosmotic conditions in the absence of amino acids [11].

11.2.2 Hepatocytes in suspension

Measurement of hepatocyte wet mass by centrifugation and weighing of the cell pellet [7].

Typically, a sample of the hepatocyte suspension is placed in a tared Eppendorf tube and subjected to a 1 s pulse of centrifugation in a microcentrifuge. The supernatant is carefully removed and the wall of the tube is blotted dry with a small piece of filter paper, after which the tube is weighed again. If the dry mass of the cells is also known (pellet dried overnight, at 105°C), the water volume per g dry mass can be calculated. For freshly isolated hepatocytes, 1 *g* dry mass equals about 3.8 *g* wet mass [7]. For freshly isolated hepatocytes the water volume is, therefore, $((3.8 - 1)/3.8) \times 100 = 74\%$ of the wet weight [7]. This very simple method also includes extracellular water which is about 40% [7,12]. A refinement of this method involves the inclusion of 0.5 μCi ^3H_2O/ml (in order to measure both the intra-and extracellular water space) and of a labeled inert compound, that is unable to cross the plasma membrane, such as [^{14}C]inulin or [^{14}C]carboxydextran (0.05 μCi/ml). These compounds are added to the hepatocyte suspension shortly (*e.g.* 1 min) before the end of the incubation period [7] in order to avoid incorporation of ^{3}H into metabolites and to prevent endocytosis of the extracellular marker. Dead cells, always present in hepatocyte preparations, contribute to the extracellular water space. A typical value determined this way is 2.69 μl/mg protein which is equivalent to 0.54 ml/g wet mass of cells [13].

Hepatocrit method (in analogy to the hematocrit method of red blood cells) [11].

In our laboratory, we use a pyrex glass tube, consisting of a 65 mm long capillary with a diameter of 1 mm. On top of the capillary, there is a reservoir (30 mm high with an internal diameter of 6.6 mm) which can contain about 0.8 ml cell suspension. The cells are centrifuged at 800 × *g* for 90 s in a tabletop centrifuge with an adaptor that fits the capillaries. An advantage of the hepatocrit method, besides the fact that it is quick and easy, is that information on the presence of damaged cells in the hepatocyte preparation is also obtained: intact hepatocytes are represented by the brownish column in the capillary, whereas damaged cells and DNA are present as a white column on top of the column with intact cells. The capillaries must be uniform in size, and are calibrated with known volumes of water. It is important to stress that the packed cell volume still contains some extracellular water (in between the cells), but this is not more than about 20% of total cell water.

Measurement of hepatocyte volume by electronic sizing.

The volume of hepatocytes in suspension can be measured with a Coulter counter (aperture diameter, 100 μm) coupled to a channel analyser. For the precise settings of the instrument, the reader is referred to Corasanti *et al.* [3]. For each cell a voltage pulse is registered, the height of which is related to the cell volume. The channel analyser divides the spectrum of these pulse heights into a series of channels and plots the frequency of pulses in each channel. The mean channel calculated from the frequency distribution provides a measure of the average cell volume of the cell population [3]. Calibration occurs with latex beads of different sise. Under normosmotic conditions, the median volume is 6168 μm^{3}/cell [3]. The disadvantage of the method is that it does not discriminate between intact and damaged cells.

11.2.3 Hepatocytes in culture

The methods discussed above are not suitable for measuring the volume of hepatocytes in culture because the cells must then be detached from the culture dish by either trypsinisation or by mechanical means and these procedures change the morphology of the cells. A convenient and simple method to measure the intracellular water of cultured hepatocytes is described by Kletzien *et al.* [14]. The technique employs 3-O-methyl[^{14}C]glucose, a non-metabolisable hexose, together with phloretin, an inhibitor of sugar transport. The labeled sugar (1-10 mM; 1 μCi/ml), is taken up by facilitated diffusion and equilibrates across the plasma membrane of the hepatocyte in 10-20 min. Before intracellular radioactivity can be measured, extracellular radioactivity has to be removed. During the washing procedure phloretin is then added to prevent leakage of 3-O-methyl[^{14}C]glucose from the cells. 3-O-methyl[^{14}C]glucose is not metabolized as indicated by a complete back exchange with an excess of unlabeled 3-O-methylglucose following preloading of the cells with 3-O-methyl[^{14}C]glucose and removal of the extracellular fluid (in the absence of phloretin).

Assuming that the intra- and extra-cellular concentrations of the labeled sugar become equal when transport equilibrium has been reached, the volume of the cells can be calculated from the intracellular radioactivity and the radioactivity per litre of extracellular fluid. Typical values are 1.4-1.7 μl/mg cell protein under iso-osmotic culture conditions and this value increases to 2.1 μl/mg upon addition of insulin [15]. Another way to measure cell volume, which requires more work, is to measure the amount of 3-O-methylglucose taken up at transport equilibrium as a function of the extracellular 3-O-methylglucose concentration. The slope of the straight line is equal to the volume of intracellular water. In this way a value of 2.48 μl/mg cell protein was found [14]. This value is slightly higher than that found by Al-Habori *et al.* [15] and may be due to small differences in experimental conditions.

An advantage of the method is that only one label is required and that it can also be applied to cell suspensions. The presence of dead cells contaminating such preparations is not a problem because radioactivity present in dead cells is simply removed by the washing procedure. However, a disadvantage of the method is that it cannot be applied in the presence of glucose.

11.3 Changes in cell volume and hepatocyte metabolism

11.3.1 History

It has been known for a long time that several species of bacteria are able to cope with large changes in extracellular osmolarity by changing the activities of enzymes and/or transport systems, the purpose of which is to adapt the intracellular osmolarity by a change in the intracellular concentration of certain osmolytes (for review, see [16]). Escherichia coli, for example, can survive a hyperosmotic stress as high as 1-2 OsM by inhibition of the synthesis of macromolecules and by increasing the concentration of intracellular glutamate and K^+. The mechanism by which this occurs is an inhibition of glutamate utilisation in the synthesis of N-containing compounds, and a stimulation of K^+ transport across the cell membrane [17].

The notion that changes in either extracellular or intracellular osmolarity can also affect metabolism in mammalian cells originated only about ten years ago. One of the immediate causes that led to this notion was an old observation of Katz and colleagues [18], related to glycogen synthesis in hepatocytes, which had remained unexplained in the literature for almost 15 years. The basic observation was that synthesis of glycogen from dihydroxyacetone or glucose in isolated hepatocytes was dependent on the presence of amino acids in the incubation medium. In collaboration with the group of Hue in Brussels we discovered that an increase in hepatocyte volume caused by amino acid transport and metabolism was largely responsible for the stimulation of glycogen synthesis by amino acids [7]. This will be discussed in the next section. Simultaneously and independently, Häussinger and colleagues, in studying glycogen degradation in the perfused livers of fed rats, observed that an increase in osmolarity of the perfusion medium stimulated the rate of glycogenolysis while hypo-osmolarity resulted in inhibition of the process [10]. These and other observations related to the effects of cell volume on hepatic nitrogen metabolism were communicated during a meeting in Mutters, Austria, in 1990 [19] and have triggered an explosion of the literature on the effect of cell volume on metabolism, not only in hepatocytes but also in other cell types.

11.3.2 Glycogen synthesis

As indicated in the previous section, synthesis of glycogen from glucose in isolated liver cell preparations is a slow process unless amino acids are also present [18]. The amino acids do not serve as a precursor for glucose 6-phosphate synthesis because in the presence of 3-mercaptopicolinate, an inhibitor of phosphoenolpyruvate carboxykinase, and thus of gluconeogenesis, they still stimulate glycogen synthesis from glucose [7,20]. A clue was provided by the observation that certain non-metabolisable amino acid analogues were also able to stimulate glycogen synthesis [7,21], although their effect was not as potent as that observed with metabolisable amino acids such as alanine, glutamine and proline. Apparently, metabolism of amino acids was not essential for stimulation of glycogen production. Stimulation of the production of glycogen by amino acids was the consequence of their transport across the plasma membrane of the hepatocytes. Moreover, only those amino acids that were transported via Na^+-dependent transport systems were effective. These amino acids are concentrated inside the cells [12,22] which results in an increase in intracellular osmolarity and thus in cell swelling. It is the increase in cell volume that triggers the stimulation of glycogen synthesis [7]. Amino acids that are rapidly metabolized (*e.g.* proline, glutamine, alanine, asparagine) can further increase intracellular osmolarity, and thus cell volume, because they also produce non-permeant catabolites such as glutamate and aspartate [7,23] (*cf.* Figure 1). Hepatocyte swelling induced by exposure of the cells to a hypo-osmotic environment in the absence of amino acids also results in stimulation of glycogen synthesis from glucose [7,24]. An example of such data is given in Figure 2A. In this experiment, hepatocytes were incubated in media, the osmolarity of which was varied by changes in the concentration of NaCl. Decreasing the osmolarity of the medium resulted in stimulation of glycogen synthesis. Addition of a complete mixture of all amino acids at the various NaCl concentrations also increased glycogen synthesis and the effects of hypo-osmolarity and of amino acid addition were almost additive. A similar relationship was observed when cell volume was plotted as a function of the osmolarity of the medium in the absence or presence of amino acids (Figure 2B): both hypo-osmolarity and amino acid addition resulted in cell swelling and the effects were additive. A plot of glycogen synthesis against cell volume revealed an almost linear relationship (Figure 2C). This indicates that cell swelling, whether induced by hypo-osmolarity of the extracellular milieu or by an increase in intracellular osmolarity (amino acid addition), had stimulated glycogen synthesis. A similar relationship between hepatocyte volume and glycogen synthesis was also observed under other conditions [7].

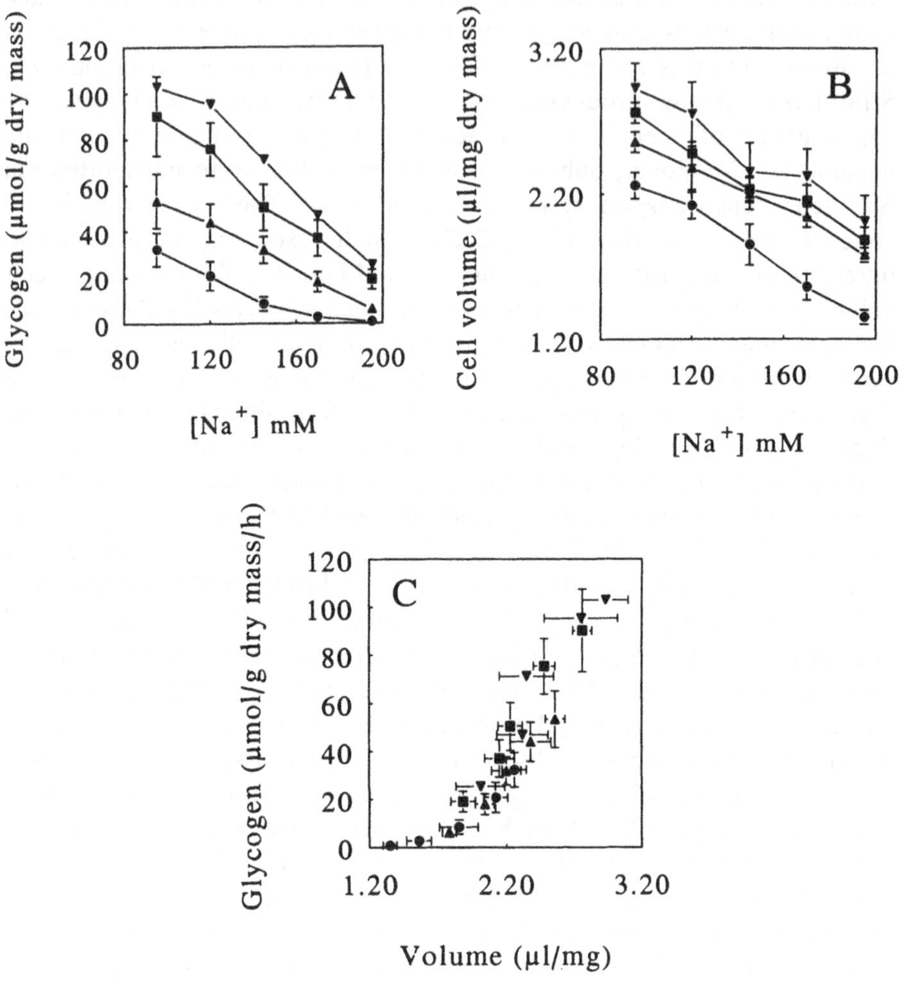

Figure 2. Cell volume changes and glycogen synthesis in isolated rat hepatocytes. Hepatocytes isolated from 20 h-starved male Wistar rats were incubated in Krebs-Henseleit bicarbonate medium for 60 min with 20 mM glucose at 37°C in the absence (●) or presence of a complete mixture of amino acids, in which each of the amino acids is present at 1× (▲), 2× (■) and 4× (▼) its concentration in the portal vein of a starved rat [28]. The osmolarity of the medium was manipulated by changing the NaCl concentration of the medium which was 70, 95, 120 (iso-osmotic), 145 and 170 mM, respectively. Because the medium also contained 25 mM $NaHCO_3$, the total Na^+ concentration ranged from 95 to 195 mM. A, glycogen synthesis as a function of the Na^+ concentration of the medium; B, cell volume as a function of the Na^+ concentration of the medium; C, glycogen synthesis as a function of cell volume. Data are the means (± S.E.) of experiments carried out with 4 different hepatocyte preparations.

The next problem to be solved was the mechanism behind the stimulation of glycogen synthesis by cell swelling. It appeared that an increase in hepatocyte volume was always accompanied by an increase in the activity of glycogen synthase [7]. The increased activity of the enzyme could not be ascribed to changes in intracellular H^+ or Ca^{2+} [25] nor was it explained by an increase in the intracellular concentration of glucose 6-phosphate which activates glycogen synthase [26]. However, we discovered that glycogen synthase phosphatase, the enzyme responsible for activation of glycogen synthase, is inhibited by chloride ions at concentrations that occur in the hepatocyte, and the fall in intracellular chloride during RV_d may be sufficient to deinhibit the enzyme [8]. While chloride is an inhibitor of glycogen synthase phosphatase, glutamate is an activator of this enzyme [8]. Because glutamate is an important intermediate in the degradation of many amino acids, this further contributes to the activation of glycogen synthase by metabolisable amino acids.

Recently, yet another mechanism has been suggested by Krause *et al.* [27]. It was found that activation of glycogen synthase by hypo-osmotic or amino acid-induced cell swelling was accompanied by activation of phosphatidylinositol 3-kinase, and was prevented, albeit partly, by the phosphatidylinositol 3-kinase inhibitors wortmannin and LY294002. It was proposed that the activity of phosphatidylinositol 3-kinase, which is an important component of amino acid and insulin-dependent signal transduction [28,29,30], is also involved in the regulation of glycogen synthesis by cell volume [27].

Stimulation of glycogen production by hypo-osmotically and amino acid-induced cell swelling is not a unique property of liver cells. In muscle cells, too, synthesis of glycogen from glucose is stimulated by hypo-osmotically and glutamine-induced cell swelling [31]. The effect was not at the level of glucose transport across the plasma membrane so that presumably it was caused by activation of glycogen synthase, as in hepatocytes.

11.3.3 Glycogenolysis

In contrast to glycogen synthesis, degradation of glycogen is inhibited by cell swelling and accelerated by cell shrinkage. This was demonstrated in experiments in which livers from fed rats were perfused with solutions of variable osmolarities [10]. Although this has not been studied, there is no doubt that activation of glycogen phosphorylase b by cell shrinkage must have been the underlying mechanism under these conditions.

11.3.4 Protein degradation

A linear relationship between cell volume and the rate of degradation of intracellular protein under a variety of conditions was observed in liver perfusion experiments by Häussinger *et al.* [2,32]. Cell swelling inhibited proteolysis whereas cell shrinkage caused activation. Degradation of intracellular protein in the liver predominantly occurs by the lysosomal pathway, *i.e.* by autophagy [29]. Using isolated hepatocytes we have been able to confirm that cell swelling inhibits proteolysis [33] although the details of our results were slightly different. While in the perfused liver and in cultured hepatocytes cell swelling alone is sufficient to inhibit protein degradation [2,32], this is not the case in freshly isolated hepatocytes [33,34] (*cf.* [35], and the answer of vom Dahl and Häussinger in the same reference, for a discussion of these differences in results obtained with the two experimental systems). Instead, in this cell preparation, proteolysis is inhibited by cell swelling only in combination with specific amino acids (leucine, phenylalanine, tyrosine) which by themselves do not affect cell volume but are strong inhibitors of autophagic sequestration (see further). This is in line with observations showing that proteolysis in either perifused hepatocytes [36] or in the perfused liver [37] is very sensitive to inhibition by low concentrations of a combination of leucine, which specifically inhibits autophagic sequestration, and alanine, the amino acid causing cell swelling.

What is the mechanism of inhibition of proteolysis by cell swelling? As indicated above, hepatic proteolysis predominantly occurs via the autophagolysosomal pathway. Leucine, tyrosine and phenylalanine are potent inhibitors of autophagic sequestration [38]. Sequestration is the surrounding of cytoplasmic material by endoplasmic reticulum-derived membranes (*cf.* [29] for review) to form autophagosomes which subsequently fuse with lysosomes to allow the degradation of the sequestered material. It is likely that the effect of cell swelling, like the effect of the three amino acids mentioned, is exerted at the level of autophagic sequestration. However, this effect may not be direct but may be mediated by changes in signal transduction (*cf.* Section 11.3.12).

An increase in intralysosomal pH, as originally proposed [39], does not explain the inhibition of proteolysis by cell swelling [40]. Rather, it is the pH of prelysosomal, presumably endocytic, vesicles that is affected by cell volume changes [1].

11.3.5 Protein synthesis

While hepatic protein degradation is inhibited by cell swelling, protein synthesis at low, but not at high, concentrations of amino acids, is stimulated by hypo-osmotically induced cell swelling [33,41]. An increase in cell volume caused by alanine accounts for the ability of this amino acid to stimulate protein synthesis [42]. The mechanism responsible for stimulation of protein synthesis by cell swelling has not yet been studied.

11.3.6 Amino acid metabolism and urea synthesis

Hypo-osmotically induced cell swelling stimulates urea synthesis from some amino acids (glutamine, glycine) because the mitochondrial enzymes participating in their degradation are activated when the mitochondrial volume increases and also because plasma membrane transport of amino acids (*e.g.* glutamine, alanine) is increased [1,43,44]. The increase in the rate of amino acid transport is presumably related to the increase in plasma membrane potential in swollen hepatocytes [2,45], and the cytosolic concentration of those amino acids that are co-transported with Na^+ increases with the plasma membrane potential [46,47]. Urea synthesis from physiological concentrations of ammonia alone was inhibited by hepatocyte swelling but this was due to the particular experimental conditions used because addition of lactate and pyruvate prevented this inhibition. It was concluded that cell swelling inhibited the supply of oxaloacetate for aspartate synthesis, presumably by inhibition of malate transport into the mitochondria [2,44]. An alternative explanation is that the antiproteolytic effect of cell swelling inhibited the endogenous supply of amino acids which, in the course of their degradation, can produce oxaloacetate which is only catalytically active in urea synthesis from ammonia.

The effect that cell volume changes have on amino acid transport is confusing because in some cell types other than hepatocytes, neutral amino acid transport is increased by hyperosmolarity [1]. In these cases, this adaptation may be related to a function of amino acids as osmolytes to counteract cell shrinkage [1].

11.3.7 Fatty acid synthesis

In liver cells, both hypo-osmotically and amino acid-induced cell swelling strongly stimulate the activity of acetylCoA carboxylase, the rate controlling enzyme of fatty acid synthesis, and also stimulate the synthesis of fat from glucose [9]. Interestingly, under various experimental conditions the kinetics of the activation of acetylCoA carboxylase and glycogen synthase by cell swelling were almost identical [9,48]. This suggests that the mechanisms responsible for the activation of these two enzymes are similar. Gaussin *et al.* [49] have recently purified a protein phosphatase-2A from liver that is activated by glutamate and Mg^{2+} and inhibited by chloride. However, activation by glutamate was stronger than was observed for glycogen synthase phosphatase, while inhibition of the new protein by chloride was less (*cf.* [48]). Krause *et al.* [27] have proposed that acetylCoA carboxylase may also be activated by the same phosphatidylinositol 3-kinase-containing signal transduction pathway that activates glycogen synthase (see Section 11.3.2). The increase in lipogenesis by hepatocyte swelling is consistent with stimulation of the pentose phosphate pathway by an increase in cell volume [1].

11.3.8 Fatty acid oxidation

Hepatocytic carnitine palmitoyltransferase is inhibited by cell swelling [50]. This enzyme is part of the machinery responsible for the import of long-chain fatty acids into the mitochondria. Cell swelling, indeed, inhibited fatty acid oxidation [50]. Interestingly, activity of carnitine palmitoyltransferase in digitonin-permeabilized hepatocytes was inhibited by glutamate so that also in this case (as for glycogen synthase and acetylCoA carboxylase) intracellular glutamate may be, at least in part, the mediator of the effects of amino acid-induced cell swelling [50].

11.3.9 Glycolysis and gluconeogenesis

Short-term effects of hepatocyte swelling on glycolysis are not very clear. According to Lang *et al.* [10], in the fed liver hypo-osmotically induced swelling resulted in inhibition of the production of glucose and lactate. However, this does not mean that glycolysis is inhibited (*cf.* [1]) because under these conditions glycogen is the precursor of lactate formation and all one can say is that glycogenolysis is inhibited by cell swelling. In experiments with isolated hepatocytes incubated with glucose, there was little effect on glycolytic flux [24,33], except at low extracellular CO_2 concentration when swelling actually slightly stimulated glycolysis [24]. Likewise, there is very little effect of swelling on the short-term regulation of gluconeogenesis [24]. It is likely however that in the long-term, gluconeogenic flux will be decreased because of a decrease in transcription of phosphoenolpyruvate carboxykinase (PEPCK) and fructose 1,6-bisphosphatase genes (see Section 11.3.12).

11.3.10 Energy metabolism

The effect that cell swelling has on mitochondrial volume depends on whether hepatocyte swelling is brought about by hypo-osmolarity of the extracellular medium or by amino acid accumulation. Thus, hypo-osmotically induced cell swelling increases both the volume of the cytosolic and of the mitochondrial compartment [51], whereas amino acid-induced cell swelling results in an increase in cytosolic volume with no change in mitochondrial volume [46]. Hypo-osmotically induced cell swelling did not affect the rate of respiration but increased the overall thermodynamic force over the mitochondrial electron transport chain, which points to an increase in kinetic control at the level of the respiratory chain [51]. Amino acid-induced hepatocyte swelling *per se*, *i.e.* under conditions where amino acid metabolism was prevented, neither affected the rate of respiration nor the mitochondrial proton motive force and the mitochondrial and cytosolic phosphate potentials [46].

11.3.11 Membrane flow

As might be expected, several processes that are dependent on membrane flow, are affected by hepatocyte volume. Examples are autophagy, already discussed in Section 11.3.4, exocytosis (stimulated by hepatocyte swelling [5,52]) and receptor-mediated [53,54] (inhibited by hyperosmolarity), but not fluid phase endocytosis [54]. Another example is bile flow, a process also dependent on vesicular transport, which is stimulated by swelling [1,52].

11.3.12 Gene transcription

Whereas the effects of cell volume on hepatocyte function discussed in the previous sections are short-term effects, long-term effects related to changes in gene expression have also been described. Examples of proteins that are induced by hepatocyte swelling are ornithine decarboxylase [55], c-*jun* [1], β-actin [1,56], tubulin [1], tyrosine aminotransferase [1], argininosuccinate synthase [57] and α2-macroglobulin [58]. By contrast, hepatocyte swelling decreases the transcription of the gluconeogenic enzymes PEPCK [59,60] and fructose 1,6-bisphosphatase [59]. Changes in mRNA stability may also contribute to the effects observed [59].

Although the data clearly show that transcription of several genes can be affected by changes in hepatocyte volume, the underlying mechanisms are largely unknown. Possibly, changes in signal transduction are involved [1].

11.3.13 Signal transduction

Remarkably, the same amino acids (leucine, phenylalanine, tyrosine) that inhibit autophagic protein degradation in hepatocytes (Section 11.3.4) stimulate a signal transduction pathway that is similar, but unlikely to be identical, to that of insulin. This pathway probably starts with an amino acid receptor in the plasma membrane [29] and has p70S6 kinase, ribosomal protein S6 and perhaps phosphatidylinositol 3-kinase as some of its components [29,38]. The phosphorylation of ribosomal protein S6 at low concentrations of leucine, phenylalanine and tyrosine is potentiated by cell swelling [38]. The same holds true for the phosphorylation of p70S6 kinase, as measured by band shift assay (D.A. van Sluijters and A.J. Meijer, unpublished observation). These and other observations have led us to propose that autophagic proteolysis is controlled by this signal transduction pathway [28,29,38].

Activation of the p70S6 kinase pathway by amino acids has recently been confirmed in hepatoma and muscle cells [61]. The specificity of the amino acid stimulation of p70S6 kinase activity was identical to that found in hepatocytes in that the combination of leucine, phenylalanine and tyrosine was particularly effective. Interestingly, a complete mixture of all amino acids was relatively ineffective unless leucine was also present [61]. It is likely that in these cell types, too, as in hepatocytes, a combination of leucine and amino acid-induced cell swelling results in activation of signal transduction.

Because in hepatocytes cell swelling potentiates the inhibition of protein degradation, as well as the stimulation of protein synthesis and the phosphorylation of ribosomal protein S6 by amino acids, it may be argued that these metabolic changes are due to a swelling-induced increase in amino acid transport across the plasma membrane. However, we consider this unlikely because under the conditions that cell swelling affects these various processes no such effect on amino acid transport was observed [33]. Moreover, for leucine to exert its inhibitory effect on proteolysis, and perhaps also on signal transduction [29], this amino acid does not have to permeate the plasma membrane [62].

In various cell types, cell swelling also appears to activate the MAP kinase signalling pathway [1]. In hepatocytes, however, activation of this pathway does not seem to be related to the effect cell swelling has on glycogen synthase, acetylCoA carboxylase, autophagic proteolysis and S6 phosphorylation, since these effects were observed under conditions where p42MAP kinase and p44MAP kinase were not activated [27,29].

11.3.14 Hormones and changes in cell volume

Because many of the metabolic effects of changes in hepatocyte volume resemble those of insulin and glucagon and because insulin and glucagon by themselves are able to cause hepatocyte swelling and shrinkage, respectively, it has been proposed that these (and possibly other) hormones exert their metabolic effects in part via changes in cell volume [1,63]. In this view, cell volume is considered as another second messenger. Although this is an interesting idea, it is presumably an oversimplification because hormone action and changes in cell volume do not always follow the prediction [13,64]. For example, addition of glucagon to hepatocytes from fed rats resulted in increases in cAMP content, phosphorylase *a* activation and glycogen degradation that preceded any change in cell volume [13]. Likewise, inhibition of phosphorylase activity by insulin in the presence of glucagon, under the same conditions, inhibited phosphorylase activity in the absence of any change in hepatocyte volume [13]. Other examples are the short-term stimulation of hepatocyte plasma membrane amino acid transport by glucagon [47] even though glucagon causes hepatocyte shrinkage [63], and the stimulation of glycogenolysis by phenylephrine despite the fact that this agent causes hepatocyte swelling [63].

11.4 Mechanisms involved in the regulation of metabolism by cell volume

Although changes in cell volume can affect the flux through metabolic pathways, it is not at all clear how these volume changes can be sensed by cells and translated into flux changes. Several mechanisms have been proposed in the literature. The mediating event may be a change in intracellular ion concentration, *e.g.* following the opening of stretch-activated ion channels [65,66]. An example was the decrease in chloride, which is released during regulatory volume decrease, as was discussed earlier in relation to the regulation of glycogen synthesis. Another possibility is that cell volume changes affect macromolecular crowding, *i.e.* the way intracellular proteins interact with each other because of their high concentrations, and thus affect their kinetic properties [67]. It is also possible that the cytoskeleton is involved in sensing changes in cell volume [1]. This is an intriguing possibility because many enzyme sytems are associated with the cytoskeleton. For example, glycogen synthase [68] and acetylCoA carboxylase [69] are both connected with the cytoskeleton. Finally, membrane stretch may affect membrane receptors and in this way can influence signal transduction. Integrins may also participate in these mechanisms [70].

11.4.1 Changes of cell volume and metabolism *in vivo*

From all these *in vitro* data it is clear that hepatocyte volume may be considered as an important parameter in the coordination of fluxes through anabolic and catabolic pathways. In this respect, an increase in cell volume has similar effects to those of insulin. The fact that anabolic processes are stimulated by cell swelling and, vice versa, that catabolism is activated by cell shrinkage, can be seen as an ability of the hepatocyte to decrease, or increase, its own osmolarity. This is similar to the situation encountered in bacteria (*cf.* Section 11.3.1).

What are the indications that changes in volume of hepatocytes, and of other cell types, do affect metabolism *in vivo*? The increase in amino acid concentration in the portal vein during feeding after a period of starvation is sufficient to cause an increase in liver volume. Indeed, the volume of the liver after feeding is bigger than during starvation and this cannot be accounted for by the increase in glycogen and protein content of the hepatocytes alone [71]. Thus, there is reason to believe that amino acid-induced hepatocyte swelling after feeding can help in stimulating the anabolic pathways in this tissue.

Indications that cell swelling (or the cellular hydration state) may be an important factor in the control of muscle protein *in vivo* were provided by studies from Häussinger *et al.* [72] and Finn *et al.* [73]. In these studies, a clear relationship was observed between muscle cell volume and the whole-body nitrogen balance in a large number of critically ill patients, and the loss of muscle cell water in these patients exceeded the loss of protein [73]. Although these results provide no proof, they strongly suggest that protein metabolism *in vivo* may be influenced by changes in cell volume.

A third example of the possible influence cell swelling has on tissue function can be found in astrocytes. In these cells large amounts of glutamine are formed by glutamine synthase when plasma ammonia concentrations are high. The astrocytes increase their volume because glutamine efflux is too slow. This swelling of the astrocytes may account for the increase in intracranial pressure in hepatic encephalopathy [74]. Most significantly, large amounts of glycogen are present in the astrocytes under these conditions [75].

Addendum

Häussinger *et al.* [76] have recently provided evidence that p38MAP kinase, but not p44MAP kinase and p42MAP kinase (also known as ERK-1 and ERK-2, respectively), also contributes to the regulation of hepatic autophagic proteolysis by changes in liver cell volume. This pathway does not require phosphatidylinositol 3-kinase [76] and is different from the p70S6 kinase signalling pathway discussed in Section 11.3.13.

References

1. Häussinger D. The role of cellular hydration in the regulation of cell function. Biochem J 1996; 313: 697-710.
2. Häussinger D and Lang F. Cell volume in the regulation of hepatic function: a mechanism for metabolic control. Biochim Biophys Acta 1991; 1071: 331-350.
3. Corasanti JG, Gleeson D and Boyer JL. Effects of osmotic stress on isolated rat hepatocytes. I. Ionic mechanisms of cell volume regulation. Am J Physiol 1990; 258: G290-G298.
4. Haddad P, Beck JS, Boyer JL and Graf J. Role of chloride ions in liver cell volume regulation. Am J Physiol 1991; 261: G340-G348.
5. Russo MA, Morgante E, Mariani MF, Yeh H-I, Farber JL and van Rossum GDV. Effects of ouabain and chloride-free medium in isoosmotic volume control and ultrastructure of hepatocytes in primary culture. Eur J Cell Biol 1994; 64: 229-242.
6. Ballatori N and Boyer JL. Taurine transport in skate hepatocytes. I. Uptake and efflux. Am J Physiol 1992; 262: G445-G450.
7. Baquet A, Hue L, Meijer AJ, van Woerkom GM and Plomp PJAM. Swelling of hepatocytes stimulates glycogen synthesis. J Biol Chem 1990; 265: 955-959.
8. Meijer AJ, Baquet A, Gustafson, L, van Woerkom GM and Hue L. Mechanism of activation of liver glycogen synthase by swelling. J Biol Chem 1992; 267: 5823-5828.
9. Baquet A, Maisin L and Hue L. Swelling of rat hepatocytes activates acetyl-CoA carboxylase in parallel to glycogen synthase. Biochem J 1991; 278: 887-909.
10. Lang F, Stehle T and Häussinger D. Water, K^+, H^+, lactate and glucose fluxes during cell volume regulation in perfused rat liver. Pflugers Arch 1989; 413: 209-216.
11. vom Dahl S, Hallbrucker C, Lang F, Gerok W and Häussinger D. A non-invasive technique for cell volume determination in perfused rat liver. Biol Chem Hoppe-Seyler 1991; 372: 411-418.
12. Bakker-Grunwald T. Potassium permeability and volume control in isolated rat hepatocytes. Biochim Biophys Acta 1983; 731: 239-242.
13. Gaussin V, Baquet A and Hue L. Cell shrinkage follows, rather than mediates, the short-term effects of glucagon on carbohydrate metabolism. Biochem J 1992; 287: 17-20.
14. Kletzien RF, Pariza MW, Becker JE and Potter VR. A method using 3-O-methyl-D-glucose and phloretin for the determination of intracellular water space of cells in monolayer culture. Anal Biochem 1975; 68: 537-544.
15. Al-Habori M, Peak M, Thomas TH and Agius L. The role of cell swelling in the stimulation of glycogen synthesis by insulin. Biochem J 1992; 282: 789-796.

16. Csonka LN. Physiological and genetic responses of bacteria to osmotic stress. Microbiol Revs 1989; 53: 121-147.
17. McLaggan D, Naprstek J, Buurman ET and Epstein W. Interdependence of K^+ and glutamate accumulation during osmotic adaptation of *Escherichia coli*. J Biol Chem 1994; 269: 1911-1917.
18. Katz J, Golden S and Wals PA. Stimulation of hepatic glycogen synthesis by amino acids. Proc Natl Acad Sci USA 1976; 73: 3433-3437.
19. Häussinger D and Lang F. Interaction of cell volume and cell function. Abstracts of a symposium in Mutters, Austria, February 1990. Renal Physiol Biochem 1990; 13: 162-179.
20. Okajima F and Katz J. Effect of mercaptopicolinic acid and of transaminase inhibitors on glycogen synthesis in hepatocytes. Biochem Biophys Res Commun 1979; 87: 155-162.
21. Rognstad R. Effects of amino acid analogues and amino acids mixtures on glycogen synthesis in rat hepatocytes. Biochem Arch 1986; 2: 185-190.
22. Kristensen LO. Association between transports of alanine and cations across cell membrane in rat hepatocytes. Am J Physiol 1986; 251: G575-G584.
23. Plomp PJAM, Boon L, Caro LHP, van Woerkom LHP and Meijer AJ. Stimulation of glycogen synthesis in hepatocytes by added amino acids is related to the total intracellular content of amino acids. Eur J Biochem 1990; 191: 237-243.
24. Peak M, Al-Habori M and Agius L. Regulation of glycogen synthesis and glycolysis by insulin, pH and cell volume. Biochem J 1992; 282: 797-805.
25. Baquet A, Meijer AJ and Hue L. Hepatocyte swelling increases inositol 1,4,5-trisphosphate, calcium and cyclic AMP concentration but antagonizes phosphorylase activation by Ca^{++} dependent hormones. FEBS Lett 1991; 278: 103-106.
26. Lavoinne A, Baquet A and Hue L. Stimulation of glycogen synthesis and lipogenesis by glutamine in isolated rat hepatocytes. Biochem J 1987; 248: 429-437.
27. Krause U, Rider MH and Hue L. Protein kinase signaling pathway triggered by cell swelling and involved in the activation of glycogen synthase and acetylCoA carboxylase in isolated rat hepatocytes. J Biol Chem 1996; 271: 16668-16673.
28. Blommaart EFC, Krause U, Schellens JPM, Vreeling-Sindelàrovà H and Meijer AJ. The phosphatidylinositol 3-kinase inhibitors wortmannin and LY294002 inhibit autophagy in isolated rat hepatocytes. Eur J Biochem 1997; 243: 240-246.
29. Blommaart EFC, Luiken JJFP and Meijer AJ. Autophagic proteolysis: control and specificity. Histochem J 1997; 29: 365-385.
30. Moule SK and Denton RM. Multiple signaling pathways involved in the metabolic effects of insulin. Am J Cardiol 1997; 80: 41A-49A.
31. Low SY, Rennie MJ and Taylor PM. Modulation of glycogen synthesis in rat skeletal muscle by changes in cell volume. J Physiol 1996; 495: 299-303.
32. Hallbrucker C, vom Dahl S, Lang F and Häussinger D. Control of hepatic proteolysis by amino acids. The role of cell volume. Eur J Biochem 1991; 197: 717-724.
33. Meijer AJ, Gustafson LA, Luiken JJFP, Blommaart PJE, Caro LHP, van Woerkom GM, Spronk C and Boon L. Cell swelling and the sensitivity of autophagic proteolysis to inhibition by amino acids in isolated rat hepatocytes. Eur J Biochem 1993; 215: 449-454.
34. Fosse M, Berg TO, O'Reilly DSJ and Seglen PO. Vanadate inhibition of hepatocytic autophagy. Calcium-modulated and osmolality-modulated antagonism by asparagine. Eur J Biochem 1995; 230: 17-24.

35. Meijer AJ. Inhibition of autophagic proteolysis by cell swelling in hepatocytes. Biochem J 1995; 312: 987-988.
36. Leverve XM, Caro LHP, Plomp PJAM and Meijer AJ. Control of proteolysis in perifused hepatocytes. FEBS Lett 1987; 219: 455-458.
37. Mortimore GE, Khurana KK and Miotto G. Amino acid control of proteolysis in perfused livers of synchronously fed rats. Mechanism and specificity of alanine coregulation. J Biol Chem 1991; 266: 1021-1028.
38. Blommaart EFC, Luiken JJFP, Blommaart PJE, van Woerkom GM and Meijer AJ. Phosphorylation of ribosomal protein S6 is inhibitory for autophagy in isolated rat hepatocytes. J Biol Chem 1995; 270: 2320-2326.
39. Busch GL, Schreiber R, Dartsch PC, Völkl H, vom Dahl S, Häussinger D and Lang F. Involvement of microtubules in the link between cell volume and pH of acidic cellular compartments in rat and human hepatocytes. Proc Natl Acad Sci USA 1994; 91: 9165-9169.
40. Luiken JJFP, Aerts JMFG and Meijer AJ. The role of the intralysosomal pH in the control of autophagic proteolytic flux in rat hepatocytes. Eur J Biochem 1996; 235: 564-573.
41. Stoll B, Gerok W, Lang F and Häussinger D. Liver cell volume and protein synthesis. Biochem J 1992; 287: 217-222.
42. Rivas T, Urcelay E, Gonzalez-Manchon C, Parrilla R and Ayuso MS. Role of amino acid-induced changes in ion fluxes in the regulation of hepatic protein synthesis. J Cell Physiol 1995; 163: 277-284.
43. Bode BP and Kilberg MS. Amino acid-dependent increase in hepatic system N activity is linked to cell swelling. J Biol Chem 1991; 266: 7376-7381.
44. Häussinger D, Lang F, Bauers K and Gerok W. Control of hepatic nitrogen metabolism and glutathione release by cell volume regulatory mechanisms. Eur J Biochem 1990; 193: 891-898.
45. Wang K and Wondergem R. Mouse hepatocyte membrane potential and chloride activity during osmotic stress. Am J Physiol 1992; 263: G566-G572.
46. Espié P, Devin A, Guérin B and Rigoulet M. Energetics of isolated hepatocyte swelling induced by sodium co-transported amino acids. J Bioenerg Biomembr 1997; 29: 591-601.
47. Moule SK, Bradford NM and McGivan JD. Short-term stimulation of Na^+-dependent amino acid transport by dibutyryl cyclic AMP in hepatocytes. Biochem J 1987; 241: 737-743.
48. Baquet A, Gaussin V, Bollen M, Stalmans W and Hue L. Mechanism of activation of liver acetyl-CoA carboxylase by cell swelling. Eur J Biochem 1993; 217: 1083-1089.
49. Gaussin V, Hue L, Stalmans W and Bollen M. Activation of hepatic acetylCoA carboxylase by glutamate and Mg^{++} is mediated by protein phosphatase-2A. Biochem J 1996; 316: 217-224.
50. Guzmán M, Velasco G, Castro J and Zammit VA. Inhibition of carnitine palmitoyltransferase I by hepatocyte swelling. FEBS Lett 1994; 344: 239-241.
51. Espié P, Guérin B and Rigoulet M. On isolated hepatocytes mitochondrial swelling induced in hypoosmotic medium does not affect the respiratory rate. Biochim Biophys Acta 1995; 1230: 139-146.
52. Bruck R, Haddad P, Graf J and Boyer JL. Regulatory volume decrease stimulates bile flow, bile acid excretion, and exocytosis in isolated perfused rat liver. Am J Physiol 1992; 262: G806-G812.

53. Krämer-Guth A, Busch GL, Kaba NB, Schwedler S, Wanner C and Lang F. Effect of osmolarity on LDL binding and internalization in hepatocytes. Am J Physiol 1997; 273: C1409-C1415.
54. Oka JA, Christensen MD and Weigel PH. Hyperosmolarity inhibits galactosyl receptor-mediated but not fluid endocytosis in isolated rat hepatocytes. J Biol Chem 1989; 264: 12016-12024.
55. Tohyama Y, Kameji T and Hayashi S. Mechanisms of dramatic fluctuations of ornithine decarboxylase activity upon tonicity changes in primary cultured rat hepatocytes. Eur J Biochem 1991; 202: 1327-1331.
56. Husson A, Quillard M, Fairand A, Chedeville A and Lavoinne A. Hypoosmolarity and glutamine increased the β-actin gene transcription in isolated rat hepatocytes. FEBS Lett 1996; 394: 353-355.
57. Quillard M, Husson A and Lavoinne A. Glutamine increases argininosuccinate synthetase mRNA levels in rat hepatocytes. The involvement of cell swelling. Eur J Biochem 1996; 236: 56-59.
58. Meisse D, Renouf S, Husson A and Lavoinne A. Cell swelling increased the α2-macroglobulin gene expression in cultured rat hepatocytes. FEBS Lett 1998; 422: 346-348.
59. Kaiser S. Cell volume regulates liver phosphoenolpyruvate carboxykinase and fructose-1,6-bisphosphatase genes. Am J Physiol 1998; 274: G509-G517.
60. Newsome WP, Warskulat U, Noe B, Wettstein M, Stoll B, Gerok W and Häussinger D. Modulation of phosphoenolpyruvate carboxykinase mRNA levels by the hepatocellular hydration state. Biochem J 1994; 304: 555-560.
61. Patti ME, Brambilla E, Luzi L, Landaker EJ and Kahn CR. Bidirectional modulation of insulin action by amino acids. J Clin Invest 1998; 101: 1519-1529.
62. Miotto G, Venerando R, Marin O, Siliprandi N and Mortimore GE. Inhibition of macroautophagy and proteolysis in the isolated rat hepatocyte by a non-transportable derivative of the multiple antigen peptide leu$_8$-lys$_4$-lys$_2$-lys-βala. J Biol Chem 1994; 269: 25348-25353.
63. Häussinger D. Cell volume and hormone action. Trends Pharmacol Sci 1992; 13: 371-373.
64. Zhande R and Brownsey RW. Cell volume and the metabolic actions of insulin. Biochem Cell Biol 1996; 74: 513-522.
65. Sarkadi B and Parker JC. Activation of ion transport pathways by changes in cell volume. Biochim Biophys Acta 1991; 1071: 407-427.
66. Watson PA. Function follows form: generation of intracellular signals by cell deformation. FASEB J 1991; 5: 2013-2019.
67. Garner MM and Burg MB. Macromolecular crowding and confinement in cells exposed to hypertonicity. Am J Physiol 1994; 266: C877-C892.
68. Fernández-Novell JM, Bellido D, Vilaró S and Guinovart JJ. Glucose induces the translocation of glycogen synthase to the cell cortex in rat hepatocytes. Biochem J 1997; 321: 227-231.
69. Geelen MJH, Bijleveld C, Velasco G, Wanders RJA and Guzmán M. Studies on the intracelllular localization of acetyl-CoA carboxylase. Biochem Biophys Res Commun 1997; 233: 253-257.
70. Low SY, Rennie MJ and Taylor PM. Involvement of integrins and the cytoskeleton in modulation of skeletal muscle glycogen synthesis by changes in cell volume. FEBS Lett 1997; 417: 101-103.

71. Belloni AS, Rebuffat P, Gottardo G, Meneghelli V, Coi A, Mazzocchi G and Nussdorfer GG. A morphometric study of the effects of short-term starvation on rat hepatocytes. J Submicrosc Cytol Pathol 1988; 20: 751-757.
72. Häussinger D, Roth E, Lang F and Gerok W. Cellular hydration state: an important determinant of protein catabolism in health and disease. Lancet 1993; 341: 1330-1332.
73. Finn PJ, Plank LD, Clark MA, Connolly AB and Hill GL. Progressive cellular dehydration and proteolysis in critically ill patients. Lancet 1996; 347: 654-656.
74. Willard-Mack CL, Koehler RC, Hirata T, Cork LC, Takahashi H, Traystman RJ and Brusilow SW. Inhibition of glutamine synthetase reduces ammonia-induced astrocyte swelling in rat. Neuroscience 1996; 71: 589-599.
75. Conn HO and Lieberthal MM. Pathology of portal-systemic encephalopathy. In: The hepatic coma syndrome and lactulose. Baltimore: Williams & Wilkins, 1979: 37-45.
76. Häussinger D, Schliess F, Domdrowski F and vom Dahl S. Involvement of $p38^{MAPK}$ in the regulation of proteolysis by liver cell hydration. Gastroenterology 1999; 116: 921-935.

Chapter 12

Permeabilisation of hepatocytes with α-toxin

Peter E. Declercq and Myriam I. Baes
Laboratory of Clinical Chemistry, Faculteit Farmaceutische Wetenschappen, Katholieke Universiteit Leuven, Leuven, Belgium

Key words: α-toxin preparation, cell integrity, channel-forming toxins, permeabilisation methods, plasma membrane permeability

12.1 The advantages of cell permeabilisation

In large part, current concepts of eukaryotic cell metabolism are based on results obtained with disrupted cells and subcellular fractions. These preparations have been preferred to isolated cell suspensions and cell cultures for analysing enzymatic steps in different pathways, since the plasma membrane makes the interior of intact cells inaccessible to various substrates and modulators. However, apart from disrupting the normal spatial relationship between cellular organelles, homogenisation may also disintegrate functional enzyme units in the cell, to the detriment of complex reaction sequences and control systems. As a consequence, the coupling of individual steps in metabolic pathways, *e.g.* "channelling" of intermediates between enzymes, is obscured [1]. Evidence for such "topodynamic regulation" of intermediary metabolism has accumulated over the past years and has led to a dramatic increase in our knowledge of metabolic regulation [2-4]. Another important, and sometimes overlooked, drawback of homogenates and subcellular fractions is the risk for proteolytic or other alteration of the proteins, enzymes or other macromolecules, that are the subject of investigation.

Since extrapolation of results obtained with broken cell preparations to the situation in the intact cell may thus not always be justified, the study of intracellular processes requires the development of experimental systems to study cellular processes *in situ*. "Selective permeabilisation" of the plasma membrane to low molecular weight compounds, whilst keeping the

M.N. Berry and A.M. Edwards (eds.), The Hepatocyte Review, 169–180.

intracellular membranes more or less intact, is an attractive technique for this purpose because it combines the advantages of intact cells, *i.e.* preserved intracellular architecture and, presumably, unchanged interaction between enzymes and organelles, with the ease of substrate manipulation that is typical of broken cell preparations [5-8]. Most permeabilisation techniques also offer the advantage that the interior of the cells is made accessible in a very short time span, without lengthy pretreatment of the cells. This significantly reduces the possibility for artefactual changes in the system under study. Of course, a permeabilized cell is an artefact in itself, by the very fact that small molecules are exchanged between the cytosol and the incubation medium. Therefore, results obtained with such a preparation merely present a different approach that must be compared with other results. The other major drawback of permeabilized cells is their limited stability (see below).

It is worth mentioning that permeabilisation is also applied to organelles, *e.g.* some interesting studies have been done with α-toxin permeabilized microsomes from rat liver [9].

12.2 Overview of cell permeabilisation techniques

The optimal permeabilisation of cells is limited to the cell membrane and does not affect intracellular organelles or cell viability. It results in free passage of low molecular weight substances across the membrane without release of cytosolic macromolecules. Reproducibility, easy manipulation and a reasonable cost are other important considerations.

The permeabilising agents used so far can be divided into different groups:

1. physical methods such as osmotic shock, brief freeze thawing and electropermeabilisation [7,8]
2. organic solvents such as toluene [10]
3. detergents or cholesterol binding agents such as lysolecithin, saponin, digitonin and filipin [11-13]
4. dextran sulphate [14]
5. channel forming or cytolytic bacterial toxins such as α-toxin from Staphylococcus aureus and streptolysin O from β-hemolytic streptococci [5,15-18]

Of the physical methods available only electropermeabilisation is sufficiently specific for the plasma membrane. It creates small pores (1-2 nm) but these are not stable. Moreover, a rather expensive instrument is required.

The lesions reported for dextran sulphate are so large that even globular proteins of approximately 400 kDa can cross the cell membrane [19]. The effect of solvents, detergents and polyene antibiotics is difficult to control and they are not, or only partially, specific for the plasma membrane [20].

Furthermore, the lesions they create are sufficiently large to allow cytosolic proteins with a molecular weight up to at least 200 kDa to leave the cell [6]; although this might be desirable for certain applications. Nonetheless, much valuable work has been done with this class of membrane permeabilising agents, *e.g.* on the short term regulation of hepatic lipogenesis [21,22], fatty acid β-oxidation [23,24], glucokinase activity [25] and cholesterol synthesis [22,26] in digitonin-permeabilized hepatocytes.

Membrane permeabilisation by channel-forming toxins is very appropriate for metabolic studies since it is technically simple and reproducible (see below). However, not all channel formers are equally suited: streptolysin O pores were reported to have an approximately tenfold larger diameter (30-35 nm) than α-toxin channels [5], large enough not only for the free passage of immunoglobulins but also for the passage of the permeabilising molecules themselves [6].

This chapter will focus mainly on the use of α-toxin with isolated rat hepatocytes. Nevertheless, several issues (*e.g.* the viability of permeabilized cells) go beyond this particular permeabilisation technique. The reader is referred to the above mentioned literature and [27] for more extensive information on other permeabilisation techniques.

12.3 α-Toxin permeabilized hepatocytes

α-Toxin (α-hemolysin) is an exotoxin produced by most strains of *Staphylococcus aureus* and is considered a major virulence factor of this microorganism. The substance is cytotoxic, hemolytic and lethal to most laboratory animals [28]. It is secreted as a water-soluble 293 amino acid monomeric protein with a molecular weight of 33-34 kDa. Upon binding to a target membrane, and through a number of intermediate stages, the monomers self-associate to an amphiphilic heptamer, at least in sensitive cells [29-31]. In some cell types (*e.g.* rabbit erythrocytes) α-toxin appears to bind to ill-defined receptors, which makes the particular cells very sensitive to the lytic effect of the toxin, while in other cell types and in artificial membrane systems a non-specific association with the lipid bilayer occurs [28,32]. Electron micrographs of α-toxin-treated erythrocytes show annular structures, approximately 10 nm in diameter, harbouring central pits, 1 to 2 nm in diameter [33-35]. These appear to be channels that allow the passive flux of ions and other small molecules up to a exclusion limit of approximately 4 kDa [5,6,35]. Since these channels are too small to allow the passage of cytoplasmic proteins, an ionic imbalance is created over the cell membrane, resulting in colloid osmotic swelling and lysis. This biological effect of α-toxin shows pertinent analogies with the activity of the complement system. As mentioned above, different cell types show different sensitivities towards α-toxin. The optimal α-toxin concentration must therefor be established for each cell type. The technique to do this basically depends on the type of experiment and not on the type of cell.

12.4 Preparation and purification of α-toxin

The commercial availability of α-toxin is somewhat problematic. Gibco recently stopped selling the product; it is still available from Calbiochem but the price is prohibitive when planning to perform extensive experiments. The protein can be prepared in any lab that has the basic facilities for growing bacteria and for performing column chromatography. There are several published methods to prepare and purify the protein [30,31,36,37]. The procedure given below is based on the work of Harshman *et al.* [36], and has been used successfully in the authors' laboratory for several years. It is lengthy, taking 10 days to obtain the final preparation, but fortunately the hands-on time is much shorter and, in the experience of the authors, almost pure α-toxin is obtained at a high and reproducible yield.

12.4.1 Materials

A stock preparation of *Staphyloccus aureus* (strain Wood 46) was kindly supplied by Dr. S. Harshman, Vanderbilt University, Nashville, TN, and kept at –80°C. The microorganism is cultivated in a medium containing, per litre, 25 g Yeast extract (Difco), 20 g Casamino acid (Difco), 2.5 g glucose, 0.125 mg thiamine (Sigma), added as a diluted solution, 1.16 mg nicotinic acid (Sigma), 11.52 g K_2HPO_4 and 4.61 g KH_2PO_4 (final pH 7.1). We usually prepare a total of 3.64 litres of this solution and divide it into four 900 ml aliquots and two 20 ml aliquots, all of which are sterilized in the autoclave.

The solution of phenylmethylsulfonyl fluoride (PMSF; Sigma) is 100 mM in 2-propanol and is kept at room temperature.

Filtration of the bacterial culture is done with a pressurized Millipore system (filter type GUWP P142-50).

Glass permeation chromatography is performed on controlled pore-size glass beads (74-125 μm; CPS-GB; CPG Inc., New Jersey, USA), in a column with an approximately 400 ml bed-size. This is washed with 0.1 M potassium phosphate pH 6.8 and eluted with 1 M potassium phosphate pH 7.5, of which we prepare 2.5 litres each.

For dialysis we require 9 to 12 litres of 10 mM potassium phosphate pH 6.0 and use Spectra/Por No 3 membranes (MW cut-off 3500).

Cation exchange chromatography is done on CM-Sepharose fast-flow (Pharmacia), column bed size approximately 200 ml, with buffer A: 10 mM potassium phosphate pH 6 (2 litres is prepared) and buffer B: 10 mM potassium phosphate pH 6 + 0.5 M NaCl (1 litre).

Gel filtration is performed with Pharmacia PD-10 columns and 20 mM HEPES buffer pH 7.4.

12.4.2 Methods

All procedures are performed at low temperature (<10°C), unless stated otherwise.

Day 1: Prepare and autoclave the cultivation medium. Inoculate two 20 ml aliquots of cultivation medium ("starters") with the stock preparation of *Staphyloccus aureus* and incubate at 37°C for 24 h, shaking at approximately 200 rpm with aeration. Prepare buffers for glass bead chromatography (GBC).

Day 2: With 3/4 of the starter cultures, inoculate the 900 ml aliquots of cultivation medium, and incubate for 19 h (same conditions as above). The remaining starter culture can be kept at –80°C for future use; it should be stabilized by adding an equal volume of sterile glycerol.

Prepare the filtration system (put in new filters, test for leakage). Prepare the GBC column following the instructions of the supplier; repack this column for each preparation and avoid air bubbles. Equilibrate the column with washing buffer.

Day 3: Centrifuge the contents of the culture flasks at 10000 × *g* for 35 min. Pool and filter the supernatants; collect in 1 ml of PMSF solution per 0.5 litres of filtrate. Adjust the pH of the filtrate to 6.8.

Load the filtrate onto the GBC column and pump washing buffer until the eluate is clear (approximately 1.5 litres, flow rate 1 ml/min). Prepare dialysis buffer and put it in the cold room. Elute the GBC column overnight with eluting buffer (1 ml/min); monitor absorbance at 280 nm and collect fractions of 10 to 15 ml.

Day 4: Pool the fractions containing the protein (absorbance >0.6) and measure the volume (approximately 300 ml; this volume depends on the growth of the bacteria and can be up to 600 ml) Dialyse for 24 h against a tenfold volume (routinely 3 litres) of dialysis buffer.

Regenerate the GBC column as follows: Wash with 1.5 litres of distilled water, followed by 1 litres of 1% (w/v) SDS, followed by 1.5 litres of distilled water, followed by 0.5 litres of washing buffer containing 0.02% (w/v) sodium azide. Store in the cold room.

Day 5: Change the dialysis buffer and continue the dialysis for another 24 h. Prepare buffers A and B for ion exchange chromatography and put them in the cold room.

Day 6: Collect the dialysed solution and verify that its conductivity is approximately the same as that of unused dialysis buffer; if not the dialysis should be continued for another 24 h. Centrifuge at 700 × *g* for 10 min; collect and pool the supernatants.

Load the preparation onto the CM-Sepharose column (flow rate 1 ml/min) and wash with 200 ml of buffer A. Elute overnight using a linear gradient of 800 ml buffer A and 800 ml buffer B (flow rate 1 ml/min). Monitor the eluate and collect fractions as described for the GBC chromatography step.

Day 7: Pool the fractions containing the protein (absorbance >0.6) and measure the volume (approximately 300 ml). Precipitate the protein by adding 60.9 g ammonium sulphate per 100 ml solution. The salt must be added very slowly (*e.g.* 30 min for 60.9 g) while mixing slowly at 4°C. Transfer to 50 ml conical glass tubes and leave in the cold room for up to 3 days.

Regenerate the CM-Sepharose column as follows: Wash with 300 ml 10 mM potassium phosphate solution pH 6 containing 1 M NaCl, followed by 250 ml 10 mM potassium phosphate pH 6 containing 0.02% (w/v) sodium azide. Keep in the cold room.

Day 10: Centrifuge the conical tubes at 1100 × *g* for 30 min at 4°C. Prepare the PD10 columns (one per conical tube): rinse 5 times with HEPES buffer.

Dissolve each protein pellet in as little HEPES buffer as possible (1-1.5 ml) and measure the volume. Apply each fraction onto a separate column and add HEPES buffer to a total volume of exactly 2.5 ml (*e.g.* if 1.5 ml of sample is used, add 1 ml of HEPES). Discard the effluent. Apply exactly 2 ml of HEPES buffer and collect the effluent containing α-toxin.

Take an aliquot for the protein assay, which may yield 1 to 5 mg/ml, and freeze the preparation at –80°C, in fractions of appropriate volume. Under those conditions α-toxin keeps its original activity for at least a year. SDS-PAGE of the solution, and staining with Coomassie blue, routinely shows only one band. In a hemolytic assay [31] the specific hemolytic activity should be 25,000-30,000 IU per mg of protein (see also Section 12.6).

12.5 Conditions for permeabilisation

In our experience [38,39], successful studies with permeabilized cells rely heavily on the choice of an appropriate incubation buffer, which obviously also depends on the aim of the experiment. To some extent optimisation of the incubation conditions remains a matter of trial and error, because too little is known about the composition of the cytosol, which one tries to mimic. Our standard conditions for permeabilisation of freshly isolated rat hepatocytes are based on those described by McEwen and Arion [18]. It proved to be essential to add a high-molecular-mass compound, such as BSA or dextran (MW. 55-73 kDa) at a final concentration of 5% (w/v), in order to compensate for the colloid osmotic pressure of intracellular proteins. Intact cells can deal with this oncotic pressure by redistribution of electrolytes across the plasma membrane, but permeabilized cells have lost this ability. The addition of glutathione (5 mM or more) and Mg-ATP (4 mM or more) improves the stability of the permeabilized cells. Needless to say, the stability of permeabilized hepatocytes also depends on the quality of the original cell preparation. We routinely isolate rat hepatocytes using the method of Zahlten and Stratman [40], modified by the EGTA-preperfusion technique described by Seglen [41]. α-Toxin can be added to the freshly prepared incubation medium before addition of the cells. Effective permeabilisation, upon addition of the hepatocytes, occurs within seconds.

Although there is probably room for improvement and further optimisation of the experimental conditions, we were satisfied to find that intact and permeabilized rat hepatocytes are equally stable up to 30 min of incubation at 37°C, as determined by lactate dehydrogenase (LDH) leakage (see below). After 45 min less than 75% of the cells retained LDH.

12.6 Assessing the effectiveness of permeabilisation

Setting up a permeabilisation experiment implies finding the appropriate dosage window to allow effective permeabilisation while keeping the cells stable long enough to do the experiment. The α-toxin concentration needed for optimal permeabilisation depends on the cell type and also varies somewhat from one preparation to the other. We have not routinely standardized the activity of our α-toxin on the basis of an erythrocyte lysis assay [31] because we favour a direct calibration in the experimental system under study. The concentrations mentioned below therefore refer to the protein measured in the α-toxin preparation. There are different ways to approach the α-toxin dosage adjustment, which thus needs to be done with each new preparation of the agent.

Under our experimental conditions Trypan blue staining appeared unreliable as an indicator of plasma membrane permeability, presumably because the pigment is adsorbed to BSA, of which at least 2% (w/v) is present in our incubation medium (together with 3% dextran) [38].

It is rather easy to set up a general permeability assay, in which one compares the cellular uptake or distribution volume of three inert radiolabeled compounds as a function of the α-toxin concentration in the medium. We use: 1) dextran (MW. 70 kDa), which should penetrate neither permeabilized nor intact cells; 2) sucrose, that does not penetrate intact cells but should have free access to the interior of permeabilized cells; 3) 3-methylglucose, that can enter hepatocytes almost unhindered under all conditions [35,42]. Experimental details and results can be found elsewhere [38] but in the final picture the distribution volumes ("spaces") of 3-methylglucose and dextran should not change significantly with increasing α-toxin concentrations, while the net sucrose space (*i.e.* corrected for the dextran space) should rise from approximately zero to the level of the net 3-methylglucose space.

A more direct and often more interesting approach, is to focus right away on the experimental question at hand. Since we have been interested in the activity of glycerolipid synthesis, we measured the incorporation of radiolabeled glycerol 3-phosphate into triacylglycerol and selected the α-toxin concentration that produced maximal incorporation without significant cell lysis (see below). We routinely arrived at a concentration of 100 to 300 μg per 10^8 hepatocytes. It is important to note that very different results can be obtained with other types of cells, *e.g.* for C6 rat glioma cells in culture approximately three times as much α-toxin is needed as for rat hepatocytes in culture (L. M. G. van Golde and M. J. H. Geelen, personal communication).

12.7 Determining the viability of permeabilized cells

As mentioned before, permeabilized cells are not stable, and disintegrate or revert to the intact, *i.e.* non-permeabilized state. In the case of electro-permeabilisation, the pores reseal when the cells are incubated at 37°C for less than 15 min [43]. In most of the other permeabilisation techniques, including the use of α-toxin, the cells eventually break up completely, although this depends on the concentration of the permeabilising agent (see below). It is therefore important to validate the experimental system with respect to the viability of the cells. At the same time, and depending on the permeabilising agent and the purpose of the experiments, it may be important to assess the intactness of intracellular organelles.

A very simple and sensitive method is the measurement of the leakage of lactate dehydrogenase (LDH). After the hepatocytes have been centrifuged, the LDH activity in the supernatant should be low in comparison to the activity in the pellet. Under optimal conditions the soluble fraction, expressed as a percentage of total LDH activity (pellet + supernatant), will be less than 20% and not significantly different from what is obtained with control cells that have not been permeabilized.

Analogous to the description by Jorgensen *et al.* [44], our α-toxin treated hepatocytes routinely appear normal in size, round and nucleated; little cellular debris is seen.

We confirmed the intactness of subcellular organelles by substantiating the latency of appropriate marker enzymes (results not shown). The ratio of mannose 6-phosphatase to glucose 6-phosphatase can be used as a quantitative index for intactness of the endoplasmic reticulum [18,45]. The latency of catalase can be measured to evaluate the integrity of the peroxisomal membrane [46]. Comparable rates of fatty acid β-oxidation, and L-carnitine dependence thereof, in intact and permeabilized hepatocytes are good indicators for mitochondrial integrity of the α-toxin-treated cells [38].

12.8 Conclusion

Although there are important shortcomings, permeabilisation of hepatocytes, and other cell types, offers unique advantages for the study of regulatory mechanisms in biochemistry and cell biology. The choice of the appropriate permeabilisation technique depends on the purpose of the experiment. The application of bacterial toxins, such as α-toxin, is not difficult and should be considered whenever the permeability of a biological membrane limits the experimental options of the investigator.

References

1. Srere PA. Complexes of sequential metabolic enzymes. Annu Rev Biochem 1987; 56: 89-124.
2. Kent C. Eukaryotic phospholipid biosynthesis. Annu Rev Biochem 1995; 64: 315-343.
3. Cheung C-W, Cohen NS and Raijman L. Channeling of urea cycle intermediates in situ in permeabilized hepatocytes. J Biol Chem 1989; 264: 4038-4044.

4. George TP, Cook HW, Byers DM, Palmer FB and Spence MW. Channeling of intermediates in the CDP-choline pathway of phosphatidylcholine biosynthesis in cultured glioma cells is dependent on intracellular Ca^{2+}. J Biol Chem 1991; 266: 12419-12423.
5. Bhakdi S, Weller U, Walev I, Martin E, Jonas D et al. A guide to the use of pore-forming toxins for controlled permeabilization of cell membranes. Med Microbiol Immunol 1993; 182: 167-175.
6. Schulz I. Permeabilizing cells: some methods and applications for the study of intracellular processes. Meth Enzymol 1990; 192: 280-299.
7. Felix H. Permeabilized and immobilized cells. Meth Enzymol 1988; 137: 637-641.
8. Lepers A, Cacan R and Verbert A. Permeabilized cells as a way of gaining access to intracellular organelles: an approach to glycosylation reactions. Biochim 1990; 72: 1-5.
9. Bossuyt X and Blanckaert N. Topology of nucleotide-sugar: dolichyl phosphate glycosyltransferases involved in the dolichol pathway for protein glycosylation in native rat liver microsomes. Biochem J 1993; 296: 627-632.
10. Rutter GA and Denton RM. Regulation of NAD^+-linked isocitrate dehydrogenase and 2-oxoglutarate dehydrogenase by Ca^{2+} ions within toluene-permeabilized rat heart mitochondria. Interactions with regulation by adenine nucleotides and $NADH/NAD^+$ ratios. Biochem J 1988; 252: 181-189.
11. Mayorek N and Bar-Tana J. Lipid synthesis in permeabilized cultured rat hepatocytes. J Biol Chem 1989; 264: 4450-4455.
12. Kaminskas E and Li JC. DNA fragmentation in permeabilized cells and nuclei. Biochem J 1989; 261: 17-21.
13. Stephens TW and Harris RA. Effect of starvation and diabetes on the sensitivity of carnitine palmitoyltransferase I to inhibition by 4-hydroxyphenylglyoxylate. Biochem J 1987; 243: 405-412.
14. Jackson SA, Thomson MJ and Clegg JS. Glycolysis compared in intact, permeabilized and sonicated L-929 cells. FEBS Lett 1990; 262: 212-214.
15. Bauldry SA and Wooten RE. Induction of cytosolic phospholipase A2 activity by phosphatidic acid and diglycerides in permeabilized human neutrophils: interrelationship between phospholipases D and A2. Biochem J 1997; 322: 353-363.
16. De Groote K, Naesens L, Balzarini J, Baes MI and Declercq PE. Effects of 2',3'-dideoxycytidine and 2',3'-dideoxycytidine 5'-triphosphate on phospholipid metabolism in permeabilized rat hepatocytes. Biochem Pharmacol 1997; 54: 713-719.
17. Maechler P, Kennedy ED, Pozzan T and Wollheim CB. Mitochondrial activation directly triggers the exocytosis of insulin in permeabilized pancreatic β-cells. EMBO J 1997; 16: 3833-3841.
18. McEwen BF and Arion WJ. Permeabilization of rat hepatocytes with *Staphylococcus aureus* α-toxin. J Cell Biol 1985; 100: 1922-1929.
19. Clegg JS and Jackson SA. Glycolysis in permeabilized L-929 cells. Biochem J 1988; 255: 335-344.
20. Gankema HS, Laanen E, Groen AK et al. Characterization of isolated rat-liver cells made permeable with filipin. Eur J Biochem 1981; 119: 409-414.
21. Geelen MJH. Medium-chain fatty acids as short-term regulators of hepatic lipogenesis. Biochem J 1994; 302: 141-146.

22. Gaussin V, Skarlas P, Ching YP, Hardie DG and Hue L. Distinct type 2A protein phosphatases activate HMG-CoA reductase and acetyl-CoA carboxylase in liver. FEBS Lett 1997; 413: 115-118.
23. Guzmán M and Geelen MJH. Short-term inhibition of carnitine palmitoyltransferase I activity in rat hepatocytes incubated with ethanol. Biochem Biophys Res Commun 1988; 154: 682-687.
24. Skrede S and Bremer J. Acylcarnitine formation and fatty acid oxidation in hepatocytes from rats treated with tetradecylthioacetic acid (a 3-thia fatty acid). Biochim Biophys Acta 1993; 1167: 189-196.
25. Agius L. Control of glucokinase translocation in rat hepatocytes by sorbitol and the cytosolic redox state. Biochem J 1994; 298: 237-243.
26. Geelen MJH, Papiez JS, Girgis K and Gibson DM. In situ measurement of HMG-CoA reductase activity in digitonin-permeabilized hepatocytes. Biochem Biophys Res Commun 1991; 180: 525-530.
27. Berry MN, Edwards AM and Barritt GJ. Isolated hepatocyte preparation; properties and applications. In: Burdon RH and van Knippenberg PH, Eds. Laboratory techniques in biochemistry and molecular biology 21. Amsterdam: Elsevier, 1991: 1-460.
28. Bhakdi S and Tranum-Jensen J. α-toxin of *Staphylococcus aureus*. Microbiol Rev 1991; 55: 733-751.
29. Valeva A, Walev I, Pinkernell M, Walker B, Bayley H et al. Transmembrane β-barrel of staphylococcal α-toxin forms in sensitive but not in resistant cells. Proc Natl Acad Sci USA 1997; 94: 11607-11611.
30. Valeva A, Palmer M and Bhakdi S. Staphylococcal α-toxin: formation of the heptameric pore is partially cooperative and proceeds through multiple intermediate stages. Biochem 1997; 36: 13298-13304.
31. Bortoleto RK, de Oliveira AHC, Ruller R, Arni RK and Ward RJ. Tertiary structural changes of the α-hemolysin from *Staphylococcus aureus* on association with liposome membranes. Arch Biochem Biophys 1998; 351: 47-52.
32. Hildebrand A, Pohl M and Bhakdi S. *Staphylococcus aureus* α-toxin. J Biol Chem 1991; 266: 17195-17200.
33. Belmonte G, Cescatti L, Ferrari B, Nicolussi T, Ropele M et al. Pore formation by *Staphylococcus aureus* α-toxin in lipid bilayers. Dependence upon temperature and toxin concentration. Eur Biophys J 1987; 14: 349-358.
34. Thelestam M. Assay of pore-forming toxins in cultured cells using radioisotopes. Methods Enzymol 1988; 165: 278-285.
35. Füssle R, Bhakdi S, Sziegoleit A, Tranum-Jensen J, Kranz T et al. On the mechanism of membrane damage by *Staphylococcus aureus* α-toxin. J Cell Biol 1981; 91: 83-94.
36. Harshman S, Sugg N and Cassidy P. Preparation and purification of staphylococcal α toxin. Methods Enzymol 1988; 165: 3-7.
37. Lind I, Ahnert-Hilger G, Fuchs G and Gratzl M. Purification of α-toxin from *Staphylococcus aureus* and application to cell permeabilization. Anal Biochem 1987; 164: 84-89.
38. Stals HK, Mannaerts GP and Declercq PE. Factors influencing triacylglycerol synthesis in permeabilized rat hepatocytes. Biochem J 1992; 283: 719-725.
39. Stals HK, Top W and Declercq PE. Regulation of triacylglycerol synthesis in permeabilized rat hepatocytes. Role of fatty acid concentration and diacylglycerol acyltransferase. FEBS Lett 1994; 343: 99-102.

40. Zahlten RN and Stratman FW. The isolation of hormone-sensitive rat hepatocytes by a modified enzymatic technique. Arch Biochem Biophys 1974; 163: 600-608.
41. Seglen O. Preparation of isolated rat liver cells. Methods Cell Biol 1976; 13: 29-83.
42. Craik JD and Elliott KRF. Kinetics of 3-*O*-methyl-D-glucose transport in isolated rat hepatocytes. Biochem J 1979; 182: 503-508.
43. Gordon PB and Seglen PO. Use of electrical methods in the study of hepatocytic autophagy. Biomed Biochim Acta 1986; 45: 1635-1645.
44. Jorgenson RA and Nordlie RC. Multifunctional glucose-6-phosphatase studied in permeable isolated hepatocytes. J Biol Chem 1980; 255: 5907-5915.
45. Vanstapel F, Pua K and Blanckaert N. Assay of mannose-6-phosphatase in untreated and detergent-disrupted rat-liver microsomes for assessment of integrity of microsomal preparations. Eur J Biochem 1986; 156: 73-77.
46. Van Veldhoven PP, Just WW and Mannaerts GP. Permeability of the peroxisomal membrane to cofactors of β-oxidation. J Biol Chem 1987; 262: 4310-4318.

Chapter 13

Perifused immobilized hepatocytes for metabolic studies

Hassan Farghali
Institute of Pharmacology, 1st Faculty of Medicine Charles University, Czech Republic

Key words: agarose gel threads, bioreactor, drug metabolism, hepatocyte immobilisation, nuclear magnetic resonance, perifusion

13.1 Introduction

Freshly isolated cells in suspension, primary cell cultures, and clonal cell lines are examples of well established cellular models from various laboratories that have contributed significantly to the understanding of many aspects of mammalian cell physiology and molecular biology. Nevertheless, such cells may be metabolically less active than the corresponding cells *in vivo*, due to inefficient oxygenation and build-up of waste products [1]. Therefore the development of a perifusion method for isolated cells [1,2] that improves the process of oxygenation and helps in end-product removal, is likely to be of considerable value in improving cell maintenance. The perifusion of mammalian cells is not straightforward because they tend to flow with the perifusate and hence block the filters that are used to restrain them. Hence, it is desirable that the cells be immobilized by using an appropriate carrier which enables cellular perifusion whilst permitting normal cellular functions. The immobilisation of cells involves the fixing of cells on or in an inert support. The combination of cells and support is frequently termed a "bioreactor".

Hepatocytes, isolated by the collagenase method [3], are now being very extensively used in biological and biomedical studies. Thus, the availability of an effective hepatocyte perifusion system would be of considerable value. More than 20 years ago, there were a few preliminary trials of hepatocyte perifusion [4,5] followed by several reports on various methods of cell immobilisation for perifusion studies [1,6,7], using either animal or plant

M.N. Berry and A.M. Edwards (eds.), The Hepatocyte Review, 181–193.

cells and. This chapter concentrates on the concepts of cellular immobilisation perifusion in general with special emphasis on hepatocytes. A description of hepatocyte immobilisation for perifusion purposes in a small or research-scale laboratory bioreactor is presented.

13.1.1 Advantages and shortcomings of cellular immobilisation

Advantages:

1. Perifusion of immobilized cells facilitates the oxygenation process, and assists end-product removal, thus contributing to cell survival.
2. Cells can be used for continuous monitoring of various metabolic and pathophysiological processes.
3. Uncontaminated cellular products can be recovered readily for further processing.
4. The flexibility of bioreactor design allows new approaches to the solution of special problems in biomedical research.
5. The availability of a bioreactor reduces the need for whole organ studies.

Shortcomings:

The limited use of immobilized hepatocytes to date may reflect:

1. The lack of agreed optimal procedures for cell immobilisation.
2. Mechanical problems in achieving efficient and uniform perifusion.
3. The relative instability of the immobilized hepatocyte systems so far developed.

13.1.2 Methods of cell immobilisation

Currently cell immobilisation may be summarized under 5 processes, namely, adsorption, covalent binding, encapsulation, entrapment and cross-linking [8]. Since there are large numbers of possible permutations involving methods of immobilisation and support material, it is easier to simplify the methods of cell immobilisation into four major categories [1,2,9]:

1. The thread technique: cells are mixed with a liquid gel (agarose gel, calcium alginate or matrigel) that is allowed to solidify, entrapping the cells in threads.
2. Microcarrier beads: cells are cultured on the surface of solid or porous charged plastic beads.

3. Hollow fiber systems: Cells are grown using commercially available hollow fiber systems which are suitable for cellular proliferation.
4. Perifusion of cells in culture dishes.

13.2 Hepatocyte immobilisation in agarose gel threads

Although the hollow fiber method is perhaps more physiological than other methods, and is used in bioartificial liver support systems, the thread method will be described here because of its simplicity and the ease with which it can be performed after short practice.

13.2.1 Isolation of rat hepatocytes

Male rats weighing 200-250 *g* are used for the isolation of hepatocytes by the standard two-phase perfusion method where collagenase is included in the second phase [3]. The hepatocytes are counted and examined for viability. Hepatocyte suspensions in which Trypan blue is excluded by more than 90% of the cells should be used.

13.2.2 Hepatocyte immobilisation

The method as outlined below is from reference [10]

1. A 1.8% solution of agarose that gels at low temperature (sea plaque agarose, FMC, Rockland, ME) should be used. It is dissolved in Krebs-Henseleit buffer (KHB) at 70°C and cooled to 37°C.
2. Equal volumes of the agarose solution at 37°C and the isolated hepatocyte suspension (4-5 $\times 10^7$ cells/ml) are mixed (*e.g.* 4 ml each depending on the requirement of the experiment).
3. The liquid agarose-hepatocyte mixture is introduced into a two-necked water-jacketed thermostatted bottle (37°C) that contains a magnetic stirrer. The cells are immobilized in agarose threads by extruding the agarose-hepatocyte mixture through ice-bath-cooled Chemfluor TFE tubing (I.D. 0.5 mM, Berghof/America, Concord, CA) into an appropriate bioreactor tube. This contains a medium that will depend on the purpose and duration of the experiment *e.g.* KHB medium or Williams' E medium). The process of extrusion, known simply as threading, is performed by means of the release of a smooth stream of carbogen into the well sealed two-necked glass bottle that contains the hepatocyte-agarose mixture. This results in the outflow of the agarose/hepatocytes slurry and its passage through the pre-cooled thin-wall TFE tubing [10]. Alternatively, the extrusion of the hepatocyte/agarose mixture can be achieved manually by using a syringe with a 24-guage needle in place of the two-necked bottle.

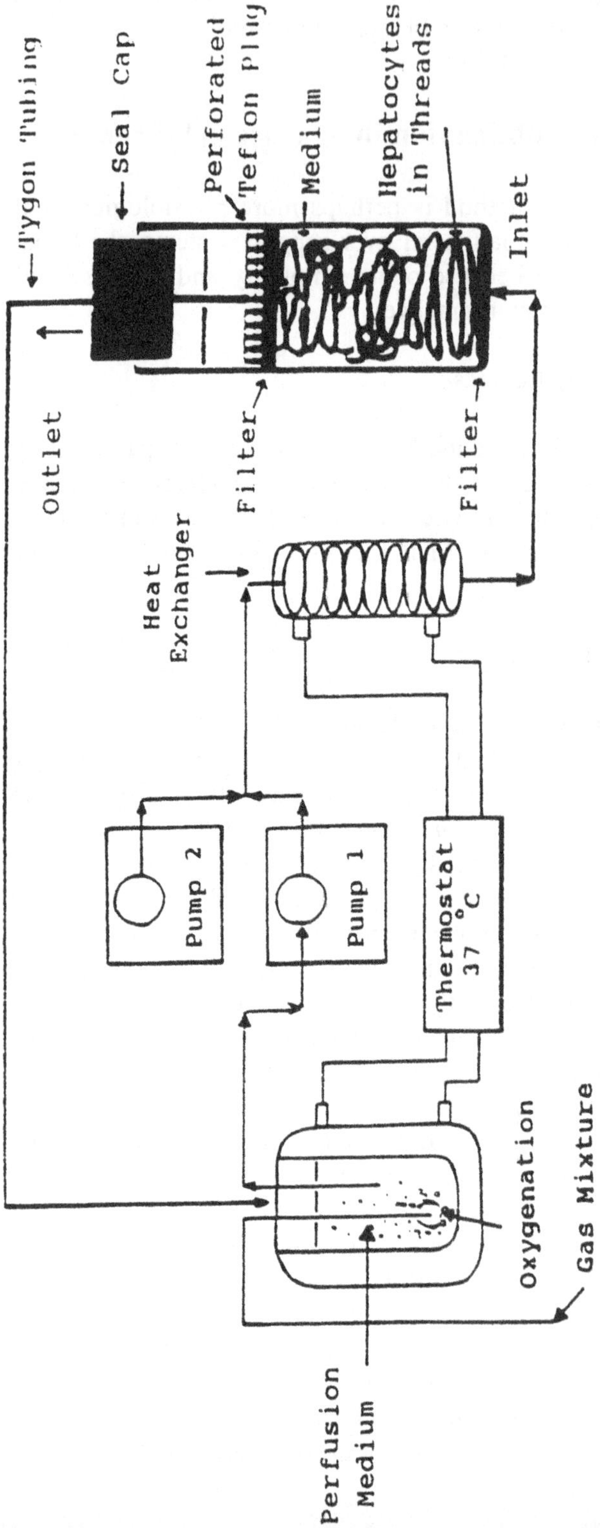

Figure 1. Schematic illustration of the perifusion system of immobilized hepatocytes in agrose gel threads. On the left side is a thermoregulated chamber through which oxygenated perifusion fluid is pumped at a rate of 10 ml/min. Pump 2 is connected, if necessary, to the perifusion system by a three-way stopcock. On the right side is a heat exchanger, which warms the inflow perfusate before it enters the threading tube. (Courtesy of Humana Press Inc., ref. 10 with permission.)

The tube used to collect the entrapped hepatocytes should permit cell perifusion and be customized according to the specific goal of the experiment. The tube described here was used for NMR as well as for conventional metabolic studies (Figure 1). This tube which allows medium inflow and outflow from the lower and upper ends respectively has 2 filters. One is fixed below the extruded hepatocyte-agarose mixture after it solidifies into threads, while the second is placed over the threads to keep them in position. The lower filter is a PE 20 μm filter (Kontes Scientific Glassware/Instruments, Vineland, New Jersey), which allows the passage of water and solutes in the perifusate while trapping the threads or any unattached cells or cellular debris that were washed out during the first non–recirculating phase of perifusion. The upper filter is made of several layers of medical gauze which enables the loose cells and debris to be washed out until the perifusate is clear, whilst restraining the cell-gel threads which are kept in position by a perforated Teflon plug (Figure 1). The perforated Teflon plug, which fits the tube tightly, is introduced and removed from the tube by a rod that fits into the plug.

13.2.3 Perifusion of the immobilized hepatocytes

The agarose threads containing hepatocytes are gently compressed into the perifusion tube and densely packed according to the delivered volume of the agarose/hepatocyte slurry. For instance, 8 ml of agarose/hepatocyte may be packed to a final volume of approximately 9 ml in a perifusion tube of 20 ml. In this way immobilized hepatocytes are not squeezed out of the agarose matrix and an efficient perifusion of the cells is guaranteed. The inlet of the tube is connected to a roller pump by tygon tube through a three-way stopcock. This stopcock is used to eliminate any air bubbles that may accidentally reach the cells by aspiration, and to rapidly introduce any compound as a bolus by means of a syringe. It can also be used to connect another roller pump for perifusion of a desired compound. The upper end of the cell perifusion tube is sealed with a rubber seal cap which contains an air-tight fixed tygon tube. This serves as an outlet for perifusion medium, keeps the Teflon plug and filter in position and minimizes the dead volume of the perifusion medium.

A perifusion scheme for immobilized hepatocytes is shown in Figure 1. The flow rate of the oxygenated medium is kept at an adequate rate for oxygenation (*e.g.* at 10 ml/min per 2.0×10^8 cells). The perifusion is started in a non-recirculating manner to remove unattached hepatocytes and cell debris (30 min). Then perifusion continues in a recirculating or non-recirculating system (depending on the goal of the experiment). The oxygenated medium is circulated for 1 h to stabilize the cells before carrying out the subsequent study (this is based on previous studies using ^{31}P NMR in which stable maximum ATP levels and oxygen consumption rates were achieved). The total volume of the perifusion medium is adjusted to fulfill the requirement of the planned experiment.

13.2.4 Assessing hepatocyte functionality and viability in the agarose gel matrix

Most of the methods used are similar to those used for isolated hepatocytes. Trypan blue exclusion, albumin synthesis, ATP content, the response to various hormones, drug metabolising capacity, maintenance of transmembrane Na^+ and K^+ ionic gradients, alanine aminotransferase and aspartate aminotransferase leakage, gluconeogenesis, lactate dehydrogenase leakage, oxygen consumption, and ureogenesis are examples among many others. ATP monitoring by ^{31}P-NMR, xenobiotic metabolic capacity, intracellular ion concentrations that are not routine methods will be mentioned under the section on applications.

13.3 Some applications of immobilized mammalian cells

An attractive application of bioreactors is the study of nuclear magnetic resonance spectroscopy (NMR) of cells, which allows for non-invasive and on-line analyses of many biochemical events in cells (real time measurements). By utilising recent advances in small laboratory bioreactor systems developed by NMR spectroscopists [2,11], it is possible to measure the rates at which various reactions of a metabolic pathway take place in a cell population. For instance, ^{31}P-NMR can be employed to gain information about the intracellular pH and the energetic status of cells and some phosphorus-containing xenobiotics can be used for drug metabolism studies. ^{13}C- and ^{1}H-NMR spectroscopy are also potential powerful tools for cellular metabolic studies.

Further applications of immobilized cells include: (a) *In vitro* biochemical, physiological, pharmacokinetic, pharmacodynamic and toxicity studies. (b) Immobilized and perifused cells as bioartificial organs (see Chapters 30-34). (c) Monoclonal antibody production and study of recombinant proteins. The feasibility of producing high amounts of specific monoclonal antibodies was assessed and it was found that cells in a hollow fiber bioreactor can provide significant amounts [12,13]. (d) Viral vectors.

The use of mammalian cell bioreactor technology has enabled scientists to successfully implement bioreactor technology for the engineering of viral vectors [14].

1.3.1 On-line, nondestructive nuclear magnetic resonance

NMR can be employed for accurate real-time metabolic studies of hepatocytes which investigate some metabolic functions of the cells by on-line monitoring of intracellular ATP and other phosphorus compounds [15,16]. Proton and ^{13}C-labeled compound can also be applied to perifused hepatocytes for NMR studies as for other cell types [17-20].

For NMR studies, a spectrum with a good signal-to-noise ratio and spectral resolution can be obtained by using a densely packed cellular population that is achieved by the fabrication of a convenient hepatocyte bioreactor as in Figure 1. Immobilized hepatocytes were perifused with KHB medium at a rate of 10 ml/min in a non-recirculating system. The threads were equilibrated until a constant ^{31}P-NMR base line spectrum and oxygen consumption were observed (approximately one hour). ^{31}P-NMR spectra were obtained at 81 MHz with a 3 cm solenoid coil at 37°C. Spectra were

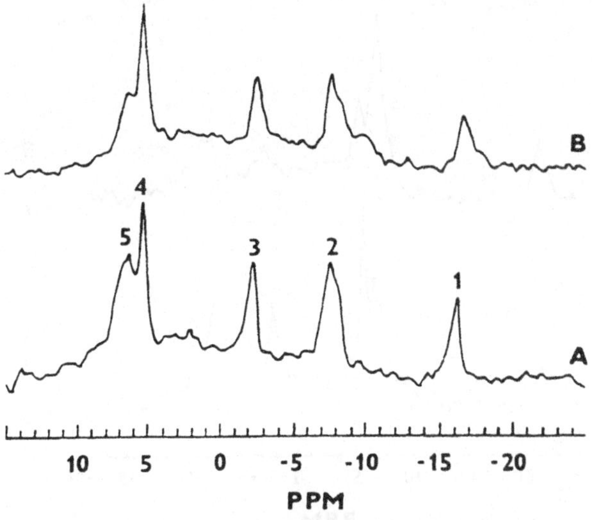

Figure 2. ^{31}P-NMR spectra of immobilized perifused hepatocytes over a 6-h period. A and B are spectra obtained after perifusion of 0 (after stabilisation) and 6 h, respectively. Peak assignments: (1) β-phosphate of ATP; (2) α-phosphate of ATP; (3) γ-phosphate of ATP; (4) inorganic phosphate (P_i); and (5) phosphomonoesters (PME). (Courtesy of Biochimica Biophysica Acta, ref. 15 with permission.)

continuously recorded in blocks of 256 scans accumulated at a repetition rate of 2 seconds using a Bruker Biospec II spectrometer operating at 4.7 Tesla and equipped with a 40 cm bore-sized, horizontal, superconducting solenoid. Figure 2 demonstrates the ability of perifused hepatocytes within the gel-threads to reach and maintain a stable cellular energy status for several hours which is evident from the stable peaks of ATP. On the other hand, oxygenated hepatocyte suspensions that were not perifused did not exhibit visible ^{31}P-NMR signals for ATP. It should be noted that hepatocytes in suspension generally contain less ATP than perifused hepatocytes as measured enzymatically by the freeze clamp method.

The addition of a fructose load at 1.5 μmol/ml/min for 1 min (Figure 3) results in a sharp reduction in Pi signal and a concomitant reduction in the ATP signal and a large increase in the phosphomonoesters due to the

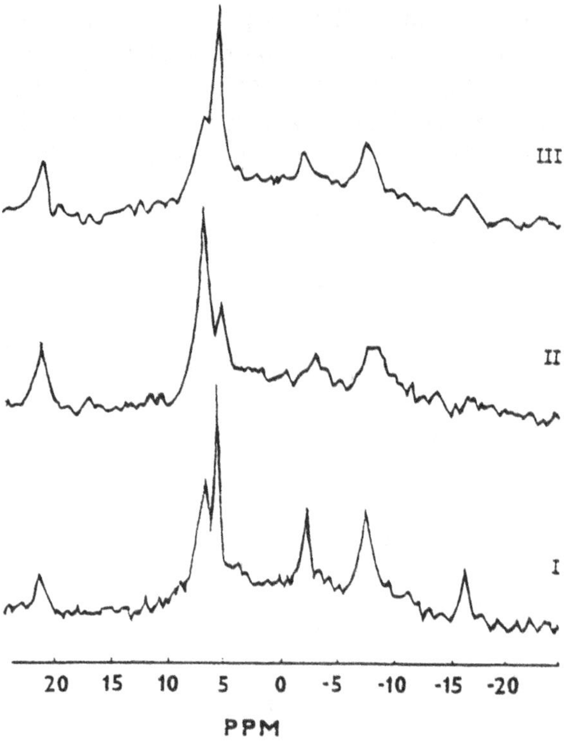

Figure 3. ^{31}P-NMR spectra of immobilized perifused hepatocytes exposed to fructose. After stabilization, fructose was perifused at a rate of 1.5 μmol/ml/min in KHB solution for 1 min. Spectra I-III were obtained before, and 14 and 32 min, respectively after fructose perifusion. The immediate appearance of fructose-1-phosphate peak (peak 5) and its gradual disappearance over time is evident. The peak assignments are the same as in Figure 2. (Courtesy of Biochimica Biophysica Acta, ref. 15 with permission.)

accumulation of fructose 1-phosphate. The spectra demonstrate that 32 min after fructose challenge, ATP and Pi peaks are recovered and the phosphomonoester signal has disappeared.

13.3.2 Measurement of intracellular ions and pH

We have successfully used isolated hepatocytes immobilized in agarose threads to measure intracellular free Na^+, Ca^{2+}, and pH [21] in hepatocytes during anoxic injury. Cytosolic free Ca^{2+} was determined, using aequorin-loaded cells, before immobilisation by gravity loading [22].

The concentration of Mg^{2+} was measured from the ^{31}P-NMR during normoxia, anoxia and reoxygenation. This measurement is based on the chemical shift of the ATP β-phosphate relative to the ATP α-phosphate on the basis of an effective K_d of the MgATP complex of 86 μM [23]. Na^+ was measured using Na^+-binding benzofuranisophthalate acetoxymethylester (SBFI AM). 2'7'bis(carboxyethyl)-5,6-carboxyfluorescein acetoxymethyl-ester (BCECF) was used to measure pH_i after hepatocytes were incubated for 45 min at room temperature with BCECF (5 μg/ml) followed by washing and immobilisation in a cuvette. Details of the techniques used and the results are described in [22].

13.3.3 Study of drug pharmacokinetics

This immobilized hepatocyte system can be used for the study of phase I biotransformation reactions [24]. Isolated hepatocytes from rats treated with sodium phenobarbital (80 mg/kg/day) for 4 successive days, or saline treated rats, were immobilized and perifused. After stabilisation, the cells were perifused in a recirculating system with a total volume of 100 ml perifusion medium. Salicylamide (140 μM) was added to the perifusion medium for an additional 20 min followed by 7-ethoxycoumarin (7-EC, 20 μM). Salicylamide inhibit the conjugation steps subsequent to 7-EC deethylation. Thus, the de-ethylation product, umbelliferone (UM), is accumulated and can be quantified in the medium. Serial samples were collected from the reservoir to determine the time course of formation of 7-EC deethylation product, UM, by UV spectrofluorometry. Experiments were repeated by using hepatocytes from phenobarbital- or saline-treated rats. UM concentration in the perfusate was calculated from a calibration curve (excitation 366 nm and emission 454 nm). Results of such a study are shown in Figure 4.

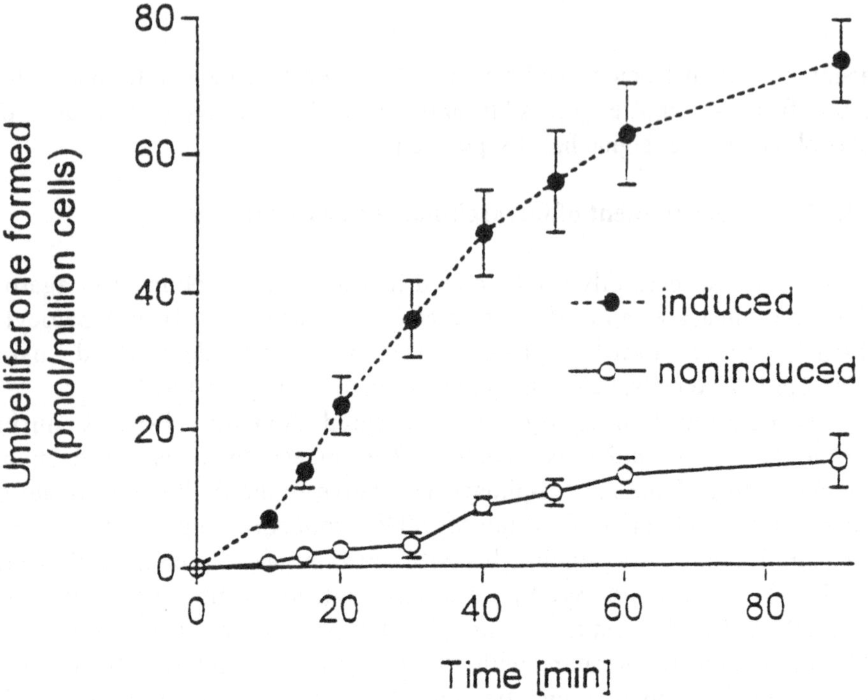

Figure 4. The time course of the accumulation of the product 7-EC deethylation (umbelliferone) (mean ± SEM, n=5) in the perfusate of induced and noninduced hepatocytes. (Courtesy of Humana Press Inc., ref. 10 with permission.)

This cellular model is sufficiently sensitive for detection of minimal enzyme metabolising activity whereas it was reported that rat liver microsomes display, practically no O-dealkylating activity toward 7-ethoxycoumarin without induction [25]. Figure 4 demonstrates that even a minimum deethylase activity in rat liver could be detected in a hepatocyte bioreactor.

13.4 Further studies with immobilized isolated hepatocytes

Isolated hepatocytes immobilized in agarose gel threads can also be used for studying oxidative stress injury and other hepatic insults due to chemicals such as ethyl alcohol, allyl alcohol or acetaminophen [10]. For instance, it is known that fructose is a glycolytic substrate that can protect the cells during mitochondrial dysfunction and the decrease of ATP supply that may occur during intoxication, hypoxic or anoxic cell injury [26]. Moreover, it has been shown that fructose 1,6-biphosphate (F16P) protects rat liver against D-galactosamine toxicity, and possesses protective effects in animal and human shock and ischemia and during hepatotoxic injury due to carbon tetrachloride [27]. Our data demonstrate that pretreatment with a low concentration of F16P protects immobilized hepatocytes from oxidative stress injury as produced by the prooxidant cumene hydroperoxide, which is added to the perifusion medium. Generally hepatoprotective and hepatotoxic agents can be readily studied using this hepatocyte model.

In the liver, the multifunctional messenger nitric oxide (NO) was found to play an important but paradoxical role in injuries produced by toxins or sepsis [28,29]. We demonstrated that endogenous NO can be produced by immobilized hepatocytes after lipopolysaccharide (LPS) administration *in vivo* and that the degree of hepatocyte injury as measured by lipid peroxide formation is LPS concentration-dependent [30]. Hepatocytes that were isolated 5-6 h after LPS treatment *in vivo*, expressed NOS activity and NO_2^- + NO_3^- were measurable in the medium a few hours after perifusion. However, there was an inverse correlation between NO_2^- levels in the perifusion medium and lipid peroxide production in hepatocytes of LPS treated rats.

13.5 Concluding remarks

The application of the concept of cellular immobilisation and perifusion to hepatocytes for various biomedical applications is still in its infancy. At present an application of considerable potential is its use as a bioartificial liver support system which perhaps will represent the first bioartifical organ to be prepared. On the other hand, applications in metabolic research studies are hampered by the frustrations and apparent complications found during the process of immobilisation. Further progress is required in relation to the simplification of methods of immobilisation, the commercial availability of ready made bioreactors for hepatocyte immobilisation, and better understanding and standardisation of the culture conditions for immobilized and perifused hepatocytes. If this can be achieved the cellular model described here may become popular due, not only to the better performance of immobilized hepatocytes as compared to conventional hepatocyte culture, but also for economic and ethical reasons.

Acknowledgements

This work was supported by the research grants from UK 95/96, GACR 307/97/0474 and IGA MZ 4180-3.

References

1. Gillies RJ, Chresand TJ, Drury PD and Dale BE. Design and application of bioreactors for analyses of mammalian cells by NMR. Reviews of Magnetic Resonance in Medicine 1986; 1: 155-179.
2. Gillies RJ, Galons JP, McGovern KA, Scherer PG, Lien YH, Job C, Ratcliff R, Chapa F, Cerdan S and Dale BE. Design and application of NMR-compatible bioreactor circuits for extended perifusion of high-density mammalian cell cultures. NMR in Biomedicine 1993; 8: 95-104.
3. Berry MN, Edwards AM and Barritt GJ. Isolated hepatocyte preparation, properties and applications. In Burdon RH and van Knippenberg PH, Eds. Laboratory techniques in biochemistry and molecular biology 21. Amsterdam: Elsevier, 1991: 1-460.
4. Cohen SM, Ogawa S, Rottenberg H, Glynn P, Yamane T, Brown TR and Shulman RG. ^{31}P nuclear magnetic resonance studies of isolated rat liver cells. Nature 1978; 273: 554-556.
5. Van der Meer R and Tager JM. A simple method for the perifusion of isolated liver cells. FEBS Letters 1976; 67: 36-40.
6. Brodelins P, Lilja H and Mosbach K. Overview of immobilization techniques for cells/organelles. In: Mosbach K, Ed. Meth Enzymol. Orlando, FL: Academic Press, 1987; 135B: 173-175.
7. Foxall DL, Cohen JS and Mitchell JB. Continuous perifusion of mammalian cells embedded in agarose gel threads. Experimental Cell Research 1984; 154: 521-529.
8. Bickerstaff GF. Immobilization of enzymes and cells, some practical considerations. In: Bickerstaff GF, Ed. Immobilization of enzymes and cells. Methods in Biotechnology. Totowa, New Jersey: Humana Press, 1997: 1-11.
9. Dich J and Grunnet N. A perifusion system for cultured hepatocytes. Anal Biochem 1992; 206: 68-72.
10. Farghali H and Hynie S. Hepatocytes immobilization in agars and functional integrity testing. In: Bickerstaff GF, Ed. Immobilization of Enzymes and cells. Methods in Biotechnology. Totowa, New Jersey: Humana Press, 1997: 175-184.
11. Daly PF, Lyon RC, Straka EJ and Cohen JS. 31P-NMR spectroscopy of human cancer cells proliferating in a basement membrane gel. FASEB J 1988; 2: 2596-2604.
12. Goodall M. A simple hollow-fiber bioreactor for the "in-house" production of monoclonal antibodies. In: Immunochemical protocols, 2nd ed. Totowa, New Jersey: Humana Press, 1998: 80: 39-56.
13. Goter A, van de Griend RJ, van Endenburg JDH, Haasnoot WHB and Fleuren GJ. Production of bi-specific monoclonal antibodies in a hollow-fibre bioreactor. J Immun Meth 1993; 161: 145-150.
14. Shankar R, Whitely CB, Pan D, Burger S, McCullough J and Stroncek D. Retroviral transduction of peripheral blood leukocytes in a hollow-fiber bioreactor. Transfusion 1997; 37: 685-690.
15. Farghali H, Rossaro L, Gavaler JS, Van thiel DH, Dowd SR, Williams DS and Ho C. 31P-NMR Spectroscopy of perifused rat hepatocytes immobilized in agarose threads: application to chemical-induced hepatotoxicity. Biochim Biophys Acta 1992; 1139: 105-114.
16. Macdonald JM, Grillo M, Schmidlin O, Tajiri DT and James TL. NMR spectroscopy and MRI investigation of a potential bioartificial liver. NMR in Biomedicine 1998; 11: 55-66.

17. Donoghue C, Brideau M, Newcomer P, Pangrle B, DiBiasio D, Walsh E and Moore S. Use of magnetic resonance imaging to analyze the performance of hollow-fiber bioreactors. Ann NY Acad Sci 1992; 665: 285-300.
18. Kaplan O, Van Zijl PCM and Cohen JS. NMR studies of metabolism of cells and perfused organs. In: Seeling J and Rudin M, Eds. NMR Basic Principles and Progress. Berlin: Springer-Verlag, 1992: 3-52.
19. Mancuso A, Sharfstein ST, Tucker SN, Clark DS and Blanch HW. Examination of primary metabolic pathways in a murine hybridoma with carbon-13 nuclear magnetic resonance spectroscopy. Biotech Bioeng 1994; 44: 563-585.
20. Szwergold BS. NMR spectroscopy of cells. Annu Rev Physiol, 1992; 54: 775-789.
21. Gasbarrini A, Borle AB, Farghali H, Bender C, Francavilla A and Van Thiel DH. Effect of anoxia on intracellular ATP, Na^{+}_{i}, Ca^{2+}_{i} and cytotoxicity in rat hepatocytes. J Biol Chem 1992; 267: 6654-6663.
22. Borle AB and Snowdowne KW. Measurement of intracellular ionized calcium with aequorin. In: Conn PM, Ed. Hormone action. Meth Enzymol 1986; 124: 90-116.
23. Malloy CR, Cunnigham CC and Radda GK. The metabolic state of the rat liver *in vivo* measured by ^{31}P-NMR spectroscopy. Biochem Biophys Acta 1986; 885: 1-11.
24. Hynie S, Kamenikova L and Farghali H. Immobilized hepatocytes in xenobiotic biotransformation studies. In: Bickerstaff GF, Ed. Immobilization of Enzymes and cells. Methods in Biotechnology, Totowa, New Jersey: Humana Press, 1997: 185-192.
25. Ullrich V, Rommer U and Weber P. Characterisation of cytochrome P-450 species in rat liver microsomes, I. Differences in the O-dealkylation of 7-ethoxycoumarin after pretreatment with phenobarbital and 3-methylcholanthrene. Hoppe-Seyler Z Physiol Chem 1973; 354: 514-520.
26. Gasbarrini A, Borle AB, Farghali H, Francavilla A and Van Thiel D. Fructose protects rat hepatocytes from anoxic injury. J Biol Chem 1992; 267: 5745-7552.
27. Roig T, De Oliveira JR, Bartrons R and Bermudez J. Fructose 1,6-biphosphate protects against D-galactosamine toxicity in isolated rat hepatocytes. Am J Physiol 1994; 266: C1722-C1728.
28. Freeswick PD, Geller DA, Lancaster J R Jr and Billiar TR. Nitric oxide and the liver. In: Arias IM, Boyer L, Fausto N, Jakoby WB, Schachter DA and Shafriz DA, Eds. The Liver: Biology and Pathobiology, New York: Raven Press Ltd, 1994: 1031-1045.
29. Pastor CM, Morris SM and Billiar TR. Sources of arginine for induced nitric oxide synthesis in the isolated perfused liver. Am J Physiol 1995; 269: G861-G866.
30. Farghali H, Hynie S, Vohnikova Z and Masek K. Possible dual role of nitric oxide in oxidative stress injury: a study in perfused hepatocytes. Int J Immunopharmacol 1997; 19: 599-605.

Chapter 14

Small hepatocytes in primary cultures

Toshihiro Mitaka and Fumihiko Sato
Department of Pathology, Cancer Research Institute, Sapporo Medical University School of Medicine, Japan

Key words: extracellular matrix, hepatic non-parenchymal cells, maturation, nicotinamide, progenitor cells, proliferation, small hepatocytes

14.1 Introduction

Recently, we and other investigators have made significant advances toward the development of culture conditions that promote proliferation of primary hepatocytes isolated from normal adult rats [1-6]. Inoue *et al.* [1] first showed that the number of hepatic nuclei increased more than twofold in a medium supplemented with 5% calf serum, 10 mM nicotinamide, and 10 ng/ml epidermal growth factor (EGF)*. The cultured hepatocytes maintained a high level of albumin mRNA expression and had the potential to express tryptophan 2,3-dioxygenase (TO) mRNA, which is thought to be a differentiated function of mature hepatocytes. Thereafter, we reported that primary rat hepatocytes cultured in a serum-free chemically defined medium could repeatedly replicate their DNA [2]. This was the first report proving that many hepatocytes that maintained hepatic differentiated functions and structures can complete the cell cycle more than twice in a serum-free medium. In this culture we found a remarkable increase of small mononucleate cells amongst mature hepatocytes. Although the size of these cells is about 1/2 to 1/3 that of typical hepatocytes in culture, the phenotypic appearance is close to that of mature hepatocytes [7-9]. Therefore, we call the cells small hepatocytes. In this review article we show that the small hepatocytes appear in primary culture and discuss the role of the cells in liver growth and hepatic diseases.

M.N. Berry and A.M. Edwards (eds.), The Hepatocyte Review, 195–208.

14.2 Preparation of small hepatocytes

14.2.1 General considerations

We normally use adult male Sprague-Dawley rats for isolating hepatocytes though we have also confirmed that small hepatocytes appear in primary cultured cells isolated from Fischer 344, Wistar, Long-Evans Cinnamon and analbuminemic (Nagase) rats regardless of sex (unpublished results). Although we and others have been trying to isolate a pure population of small hepatocytes, it is, to our knowledge, so difficult that nobody has yet succeeded in obtaining a single cell type. Therefore, we are now observing the proliferation of small hepatocytes either together with parenchymal cells (P-population) or among the mixed populations of hepatic non-parenchymal cells (NPC) (Mixed-populations). As the isolation method for parenchymal cells is the standard procedure, in this review we describe the method for Mixed-populations.

14.2.2 Preparation of cell suspensions

A liver of an adult rat is perfused *in situ* through the portal vein at a flow rate of 40 ml/min with 150 ml of Ca^{2+}, Mg^{2+}-free Hanks' balanced salt solution supplemented with 0.5 mM EGTA, 0.5 µg/ml insulin and antibiotics. After the initial brief perfusion, the liver is perfused at a flow rate of 20 ml/min with 200 ml of Hanks' balanced salt solution containing 40 mg of collagenase (Yakult Co., Tokyo, Japan). After perfusion, cells are dispersed and suspended in L-15 medium with 0.2% BSA, 20 mM HEPES, 1 mg/ml galactose, 30 µg/ml proline, 0.5 µg/ml insulin, 10^{-7} M dexamethasone, and antibiotics. The cell suspension is filtered through 250 µm and 70 µm nylon meshes and then centrifuged at $50 \times g$ for 1 min. The pellet is used for isolation of parenchymal cells (P-population). We use the supernatant for the isolation of small hepatocytes.

14.2.3 Separation from liver epithelial cells

The size of small hepatocytes results in the dilemma that this fraction is also rich in NPC, especially liver epithelial cells (LEC). In primary cultures of rat hepatocytes the major contaminants of NPC are the epithelial cells. The cells have been studied for a long time and are known to clonally proliferate in primary culture. In addition, the cells are considered candidates for hepatic stem cells because they have the potential to differentiate into hepatocytes and/or bile duct epithelial cells when they are transplanted into appropriate sites *in vivo*. LEC have been thought to be derived from hepatic parenchymal cells but hepatocytes in collagenase-dispersed cell suspensions are now not considered to be the obligate precursors of LEC because identifiable hepatocytes have never proliferated to form epithelial colonies under conventional culture conditions. Therefore, until the colonies of small hepatocytes were discovered, epithelial cell colonies had been derived only from LEC. As shown in Figure 1C, small hepatocytes are morphologically distinct from LEC, which have a flattened shapes, possess a translucent cytoplasm, and a variable size. Whereas the small hepatocytes have a round nucleus and a dense cytoplasm, and the size and shape of most cells are similar.

To observe the proliferation of a single small hepatocyte in the culture, the method for the isolation of LEC was used with some modifications [10]. The supernatant obtained after centrifugation of the isolated cell suspension at $50 \times g$ for 1 min is recentrifuged at $50 \times g$ for 5 min and the pellet is suspended in L15 medium. The following centrifugation steps are then employed. 1) $50 \times g$ for 5 min; 2) $50 \times g$ for 5 min; 3) $150 \times g$ for 5 min; 4) $150 \times g$ for 5 min; 5) $50 \times g$ for 5 min. The pellet is finally suspended in an appropriate volume of the medium and then the viability of the cells is examined by the Trypan blue exclusion test. The number of viable cells is counted without regard to cell size. Most large hepatocytes are dead after the repeated centrifugation at $150 \times g$. Although many large hepatocytes incorporate the dye into their cytoplasm, the dead cells have no effect on the proliferation of small hepatocytes.

14.2.4 Culture procedure

Then 1.5×10^5 viable cells are plated on 35 mm culture dishes coated with rat tail collagen and placed in a 100% air incubator at 37°C. Two to 3 h after plating, the medium is changed to DMEM with 20 mM HEPES, 25 mM $NaHCO_3$, 30 µg/ml L-proline, 0.5 µg/ml insulin, 10^{-7} M dexamethasone, 10% fetal bovine serum (FCS), 10 mM nicotinamide, 1 mM ascorbic acid 2-phosphate, 10 ng/ml EGF, and antibiotics. The cells are then placed in a humidified, 5% CO_2/95% air incubator at 37°C. The medium is replaced with fresh medium three times a week. After 4 days of culture, 1% dimethyl sulphoxide (DMSO), which concentration suppresses the growth of NPC but not small hepatocytes, is added to the culture medium [7].

14.3 Culture medium

14.3.1 Role of nicotinamide in small hepatocyte proliferation

The growth of rodent hepatocytes cultured in medium supplemented with nicotinamide has been reported by other laboratories [1,4,5,11,12]. In spite of the fact that they used a variety of different media (Table 1), all contained 5 to 10 mM nicotinamide. As we reported [2], although an increase of the concentration of nicotinamide in the medium delayed the first DNA synthesis of primary hepatocytes, the multiplication and survival of the cells were promoted. A concentration of 10 mM nicotinamide was the most effective and more than 20 mM was toxic to primary hepatocytes [1,2].

Table 1. Media for the proliferation of small hepatocytes

Component	**Mitaka et al.**	**Mitaka et al.**	**Mitaka et al.**	**Wu et al.**	**Tateno et al.**	**Block et al.**	**Cable et al.**
Rat/Mouse Strain	male SD rat	male SD rat	male SD rat	male TGF-α transgenic mice	male F344 rat	male F344 rat	male F344 rat
Basal medium	DMEM: Ham's F12	DMEM	L-15	DMEM: Ham's F12	DMEM	DMEM	RPMI (M6a)
tryptophan (mg/l)	9	16	20	9	16	16	5
Nicotinamide (μM)	16.5	32.8	8.2	16.5	32.8	32.8	8.2
Additives							
Serum (FCS)	(-)	10%	(-)	5%	10%	(-)	(-)
proline (mg/l)	30	30	30	(-)	30	30	(-)
insulin (mg/l)	5	0.5	0.5	5	0.5	5	0.06
Dexamethasone (M)	10^{-7}	10^{-7}	10^{-7}	10^{-7}	10^{-7}	10^{-7}	10^{-6}
Glucagon (mg/l)	(-)	(-)	(-)	(-)	(-)	(-)	0.04
Nicotinamide (mM)	**10**	**10**	**(-)**	**10**	**10**	**5**	**(-)**
Transferrin (mg/l)	(-)	(-)	5	(-)	(-)	5	100
Ethanolamine (mg/l)	(-)	(-)	(-)	(-)	(-)	(-)	4.1
linolenic acid (mg/l)	(-)	(-)	(-)	(-)	(-)	(-)	2
L-ascorbic acid 2-phosphate (mM)	(-)	1	(-)	(-)	0.2	(-)	(-)
Cu (μg/l)	(-)	(-)	2.2	(-)	(-)	(-)	2.5
Fe (μg/l)	(-)	(-)	0.33	(-)	(-)	(-)	834
Zn (μg/l)	(-)	(-)	0.01	(-)	(-)	(-)	1465
Se (μg/l)	2	(-)	2	5	(-)	5	17
EGF (μg/l)	10	10	10	10	10	20	25
HGF (μg/l)	(-)	(-)	(-)	(-)	(-)	40	(-)
DMSO	(-)	1%	(-)	(-)	1%	(-)	2%
Dish coating	Type I collagen	Rat tail collagen	Rat tail collagen	Rat tail collagen	(-)	Type I collagen	Rat tail collagen
Appearance of small hepatocytes	(+)	(+)	(+)	(+)	(+)	(+)	(+)
References	[2]	[22]	[3]	[12]	[11]	[5]	[6]

Nicotinamide, which is an aqueous vitamin, is converted in the cell to NAD and NADP, both of which are the most abundant cellular coenzymes necessary for a number of biochemical reactions. Consequently, it is reasonable to assume that a decrease in intracellular NAD and NADP may influence many functions of hepatocytes. In primary culture of rat hepatocytes it is known that the cells rapidly lose their intracellular NAD with time in culture [1,13]. This loss can be prevented by supplementation with nicotinamide [1].

Nicotinamide is known to inhibit poly(ADP-ribose) polymerase (PARP), an enzyme that catalyzes polymerisation of the ADP-ribosyl moiety of NAD. A concentration of 10 mM in the medium can completely inhibit PARP activity [14]. A sharp increase of PARP activity was reported after isolated hepatocytes were plated onto dishes [15]. Furthermore, rapidly proliferating cells exhibit relatively higher activity of poly (ADP-ribosyl)ation than quiescent cells, as seen in the precancerous state of hamster livers [16] and SV-40 transformed cells [17]. Recently, it was reported that PARP could bind p53 protein and that pharmacological inhibitors of PARP could extend the replicative life span of cultured fibroblasts [18]. On the other hand, a decrease in poly (ADP-ribosyl)ation during differentiation has been observed in many other types of cells. The addition of nicotinamide to cultured hepatocytes has been reported to prevent the decreases of cytochrome P450 activity [13] and mRNA expression of albumin and TO [1]. Although treatment of primary hepatocytes with 3-acetylpyridine, a weak inhibitor of PARP, exhausted the cellular NAD, mRNA expression of albumin, TO and Cx32 was well maintained [19]. These results suggest that maintenance of the NAD level in the cells may be more important for the proliferation of primary hepatocytes than for the maintenance of the differentiated functions. Besides supplying NAD and suppressing the consumption of NAD by inhibition of PARP activity, nicotinamide may play a separate role in primary hepatocytes, namely the induction of the growth of small hepatocytes.

Although nicotinamide could enhance the appearance and proliferation of small mononucleate hepatocytes, the appearance of the cells was not limited to cultures containing a high concentration of nicotinamide. Table 1 shows the major contents of culture media in which the appearance of small hepatocytes has been reported. A high concentration of nicotinamide was added to most media. However, Cable and Isom [6] and we [3,20] reported the appearance of small hepatocytes in media without a high concentration of nicotinamide. They observed that when primary hepatocytes were cultured in chemically defined medium, supplemented with 2% DMSO and EGF, the addition of trace elements could trigger the proliferation of hepatocytes after about 10 days of culture [6]. In their culture conditions, large clusters of small hepatocytes appeared and grew with time in culture. They concluded that the addition of Fe^{2+}, Cu^{2+} and Zn^{2+} to the medium induced the DNA synthesis of hepatocytes and cell proliferation during the first two months without loss of hepatic differentiation. On the other hand, we observed the appearance of small hepatocytes in L-15 medium supplemented with EGF in a CO_2-incubator [3,20], although the frequency of the appearance of small cell colonies was less than that of cells cultured in nicotinamide-supplemented medium. Although the two media mentioned above do not include a high concentration of nicotinamide, as shown in Table 1, their basal media contain small amounts of nicotinamide and tryptophan. In mammals nicotinamide can be biosynthesized from tryptophan. These results suggest that nicotinamide is essential for the appearance of small hepatocytes and that a small amount of it may be enough for their appearance although 10 mM nicotinamide in the medium is better.

14.4 Characteristics of small hepatocytes

14.4.1 Phenotypic expression

Table 2 shows the characteristics of small hepatocytes that appeared in the P-population and Mixed-population. Except for their size, the cells show the morphologic characteristics of mature hepatocytes. Immunocytochemically, the small hepatocytes surrounded by mature ones are positive for albumin, transferrin, α_1-antitrypsin, and ceruloplasmin [7,8]. CK8 and CK18, intermediate filaments expressed in hepatocytes, are also positive in small hepatocytes. However, CK7 and CK19, which are expressed in bile duct cells, desmin in stellate cells, smooth muscle actin and SE1 [21] in sinusoidal endothelial cells, and ED 1 and ED2 in Kupffer cells are never stained in the small hepatocytes that appeared at day 6. Nor are Cx32 and Cx26, which are usually expressed in differentiated hepatocytes, detected in the cells at day 6. Furthermore, ultrastructurally, small hepatocytes possess mitochondria and diaminobenzidine-positive peroxisomes in their cytoplasm and bile-canalicular structures between the cells [7,8], although these are fewer in number than in mature hepatocytes. When 2% DMSO, which has the ability to maintain hepatic differentiated functions in primary culture, is added to the culture medium from day 4 or 5, small hepatocytes can continue to proliferate, although the DNA synthesis of relatively large hepatocytes is inhibited [7]. However, several days later, the small hepatocytes increase their volume of cytoplasm and the colonies are difficult to identify. This result is one piece of evidence that small hepatocytes can differentiate into mature ones.

Table 2. Characteristics of small hepatocytes isolated from adult rat livers

	Parenchymal cells[1]	Coculture	
	Day 6	**Day 10**	**<Day20**
Characteristics of hepatocytes			
Albumin	(++)	(++)	(++)
α-Fetoprotein	(±)[2]	(±)	some cells
Transferrin	(++)	(++)	(++)
Cytokeratin 8	(++)	(++)	(++)
Cytokeratin 18	(++)	(++)	(++)
Connexin 32	(NT)[3]	(±)	(+)
Connexin 26	(NT)[3]	(-)	Induced[4]
Peroxisome	small[5]	small	large with nucleoid
Glycogen	(+)	(±)	(++)
Characteristics of bile duct cells			
GGT	(-)	(-)	bile canaliculus
GST-P	(-)	(±)	some cells
Cytokeratin 7	(-)	(±)	some cells
Cytokeratin 19	(-)	(±)	some cells
Characteristics of other types of cells			
Vimentin	(+/-)[6]	(-)	some cells
Desmin	(-)	(-)	(-)
Smooth muscle actin	(-)	(-)	(-)
ED 1/2	(-)	(-)	(-)
SE 1	(-)	(-)	(-)
Connexin 43	(-)	(-)	some cells

[1]Small hepatocytes appear from the fraction after a collagenase-digested cell suspension is centrifuged at $50 \times g$ for 1 min. [2]Some cells in colonies are positively stained. [3]Not tested. [4]Some cells in a focal colony are stained. [5]Small hepatocytes possess many small peroxisomes. [6]Some colonies show positive staining. Modified from Mitaka T, et al. [26] with permission.

14.4.2 Appearance of small hepatocyte colonies

A small hepatocyte in a mixed-population proliferates and forms a round colony in the early culture period (Figure 1). In these colonies all cells are equally stained positively for albumin, transferrin, CK8 and CK18, and negatively for AFP, CK7 and CK19 [22,23]. However, with time in culture, some cells in a colony, although not all, show phenotypic alterations in that they express AFP, CK7 and/or CK19. There is no relationship between the

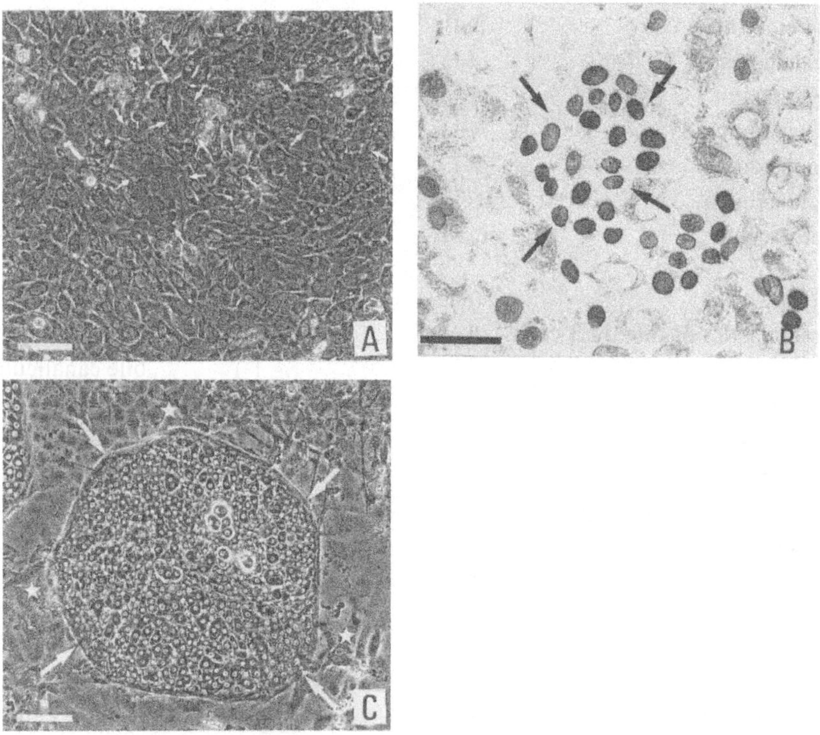

Figure 1. (A) Phase-contrast micrograph at day 5 of primary rat hepatocytes cultured in serum-free Dulbecco's modified Eagle's medium (DMEM) supplemented with 10 mM nicotinamide and 10 ng/ml epidermal growth factor (EGF). Arrows show small hepatocyte colonies. Bar, 100 μm. (B) Double immunocytochemistry for albumin and BrdU of the small hepatocytes in focal colonies. 40 μM Bromodexoyuridine (BrdU) was added to the medium between 96 and 144 h after plating. The cells were fixed with cold absolute ethanol at 144 h. Darkened nuclei are those stained with BrdU. Cells with darkened cytoplasm are those stained for albumin. Small mononuclear cells in the colony surrounded by arrows are stained with both albumin and BrdU. Bar, 50 μm. (C) Phase-contrast photographs of a small hepatocyte colony cultured in DMEM supplemented with 10% fetal bovine serum, 10 mM nicotinamide, 1 mM ascorbic acid 2-phosphate, and 10 ng/ml EGF at day 30. 1% DMSO was added to the medium from day 4. Arrows show the small hepatocyte colony surrounded by hepatic non-parenchymal cells (stars). Bar, 100 μm.

size of cells and AFP-positivity, but CK7- and CK19-positive cells have a tendency to locate at the edge of the colony. About 2 weeks after plating, many colonies are surrounded by NPC, especially LEC and stellate cells. Thereafter, we often observe that some large hepatocytes, which are sometimes binucleate, appear within the colonies. On the other hand, piled-up cells in a colony are seen, and they can proliferate to show plate-like structures. They are similar to the trabeculae *in vivo* in the liver acinus, and bile-canalicular structures are seen between the cells, which form two- to three-cell-thick plates. Corresponding to the structure, GGT activity is strongly shown along the bile-canalicular structures. Albumin secretion into culture medium also increases with time in culture [22,23].

14.5 Growth capacity

14.5.1 Effects of growth factors

The first cell division of small hepatocytes is observed from day 2 to day 3, and thereafter rapid growth and colony formation takes place. The average number of cells in a colony reaches about 27 at day 10 [22]. Tateno & Yoshizato [11] reported colonies that consisted of more than 100 cells at day 10. As we reported [22], the colonies can develop without EGF. However, EGF stimulates their formation. Therefore, we examined the effects on the formation of the colonies of other factors such as HGF, TGF-α, and FGFs, which are well known to induce the proliferation of mature hepatocytes [9]. HGF and TGF-α had the potential to cause as many colonies to develop as EGF. However, the number of cells in colonies was greatest when TGF-α was in the medium (unpublished data).

We have observed that the proliferation of small hepatocytes is promoted remarkably when the growth of NPC is suppressed (manuscript in preparation). This result suggests that NPC may produce growth inhibitors such as TGF-β in the medium, in keeping with our finding that TGF-β can inhibit the DNA synthesis of small hepatocytes [24]. These results indicate that, after the initial division of small hepatocytes, their growth rate may be determined by the balance of the concentrations of growth factors in the medium. At present, we do not have exact data concerning whether or not small hepatocytes can divide in an unlimited manner. Tateno & Yoshizato [11] reported that a single small cell divided to produce 1400 cells. This number equals about 10 cell doublings. It seems that small hepatocytes may have the potential to divide without limit if the cells are maintained in ideal culture conditions.

14.5.2 Small hepatocytes in liver growth

We have reported that small hepatocytes were mixed among mature ones and that the small hepatocytes appeared to form colonies with time in culture. The frequency of the colony formation varied and decreased with the age of the rat: about 6% of hepatocytes isolated from 4-week old rats, about 2.5% of those from 8-week old rats, and about 1.5 to 2.0% of those from adult rats were small hepatocytes [25]. In aged rats (more than 1.5 years old) the frequency was less than 0.5%. Furthermore, the cells isolated from rats younger than 3 weeks old were very small, and more than 90% of the cells were mononucleate. Taking these results into consideration, we have proposed that hepatocytes may be classified into three types of cell with respect to their ability to divide [26,27]:

1. Cells that have a high potential to proliferate and form colonies and, therefore, continue to supply daughter cells (Type I cells, small hepatocytes);
2. Cells for which the number of possible cell divisions is limited; these cells normally express fully differentiated functions (Type II cells); and
3. Cells that lose the ability to divide and reach the final differentiated state (Type III cells).

As Type II and Type III cells are functionally sufficiently differentiated, they are considered to be mature hepatocytes. On the basis of this classification of hepatocytes, we can interpret postnatal liver growth and the proliferation of parenchymal cells [27].

Soon after birth, liver plates may consist of small hepatocytes, although periportal hepatocytes may have bipotential progenitor capacity. We make this assumption because the periportal cells are positively stained by antibodies against calf keratin and human CK, which react exclusively with bile duct cells in adult liver [28-30]. During weaning, some small hepatocytes in liver plates differentiate into Type II cells. At present we do not know what possible mechanisms induce the differentiation of small hepatocytes. However, we have a clue indicating that hepatic NPC play an important role in the differentiation of small hepatocytes, because as mentioned above most small hepatocytes can continue to proliferate without differentiation when the growth of NPC is inhibited (unpublished data). In addition, extracellular matrix (ECM) produced by NPC, especially stellate cells, may promote their maturation [23]. Thus, direct effects of the products of NPC such as cytokines and/or ECM may regulate the proliferation and differentiation of small hepatocytes into mature ones.

With growth, many small hepatocytes become Type II cells, thus decreasing the number of small hepatocytes. In this period both small hepatocytes and Type II cells divide, and this activity contributes to the rapid increase in the total number of parenchymal cells. Consequently, small hepatocytes randomly distribute in the liver acinus. With increasing age, in rats, some Type II cells complete their role of proliferation and become Type III cells. After completing their role as differentiated hepatocytes, Type III cells will die. Thus, in the normal adult rat liver, to compensate for cells eliminated by death or for other reasons, Type II cells may produce new cells contiguous to the eliminated ones. In addition, during liver regeneration after two-thirds partial hepatectomy, division of Type II cells may be sufficient to restore the original mass of the liver, although small hepatocytes possibly also participate in the event.

14.6 Perspectives

The development of small cell colonies was shown in *in vitro* experiments using rat hepatocytes. If that is the case, we need to ask does this occur in human liver? Preliminary experiments have shown that colonies of small hepatocytes appear among human hepatocytes isolated from small pieces of liver obtained at operation, although the frequency of appearance is quite low. In addition, it has been reported that small hepatocytes appear in some human liver diseases. Liver biopsy specimens from patients who suffered acute viral hepatitis and recovered from fulminant hepatitis, showed relatively small polygonal hepatocytes with small nuclei arranged in clusters in periportal areas [31,32]. In addition, we pathologists often observe small cell clusters in liver specimens from patients who are suffering from advanced chronic hepatitis and liver cirrhosis. These cells have been understood as the regeneration of mature hepatocytes, but we now regard them as a new liver cell population, namely small hepatocytes. Therefore, we may interpret certain liver diseases from a new point of view. For example, in the livers of patients who are suffering from chronic hepatitis, mature hepatocytes frequently die. Initially, the neighbouring hepatocytes can compensate for the loss of cells. Thereafter, frequent growth stimuli due to the loss of neighbouring cells cause the exhaustion of the replicative capacity of Type II cells. To fill up the space formed by dead cells, hepatic NPCs such as stellate cells may invade into the space and produce ECM. Simultaneously small hepatocytes may also proliferate and fill up the free space. Therefore, large clusters of small hepatocytes may be often observed in the late stage of chronic hepatitis and liver cirrhosis.

In this review we described the proliferation and maturation of small hepatocytes. Recent advances in culture conditions and the discovery of small hepatocytes are great steps forward to the dream that we can freely regulate the proliferation and differentiation of hepatocytes. As Strain discussed in an Editorial [33] relating to our work [23] and that of Michalopoulos [34], coculture of small hepatocytes and hepatic NPCs instead of mature hepatocytes may be exploited to improve the bioartificial liver. For this purpose, we have to investigate the role of individual NPCs on small hepatocytes. Purification of each cell population is an urgent priority. In addition, although we have failed to identify small hepatocytes in the normal liver acinus, it is necessary to seek a specific marker to clarify whether or not they are hepatic stem cells.

Acknowledgements

We thank Drs. Takashi Kojima, Toru Mizuguchi, and Nagato Higaki for their contributions to this work and Dr. Yohichi Mochizuki for his encouragement. We also thank Ms. M. Kuwano, Ms. Y. Takahashi, and H. Itoh for their skillful technical assistance. We are grateful to Mr. K. Barrymore for help with the manuscript.

References

1. Inoue C, Yamamoto H, Nakamura T, Ichihara A and Okamoto H. Nicotinamide prolongs survival of primary cultured hepatocytes without involving loss of hepatocyte-specific functions. J Biol Chem 1989; 264; 4747-4750.
2. Mitaka T, Sattler CA, Sattler GL, Sargent LM and Pitot HC. Multiple cell cycles occur in rat hepatocytes cultured in the presence of nicotinamide and epidermal growth factor. Hepatology 1991; 13: 21-30.
3. Mitaka T, Sattler GL and Pitot HC. Amino acid-rich medium (Leibovitz L-15) enhances and prolongs proliferation of primary cultured rat hepatocytes in the absence of serum. J Cell Physiol 1991; 147: 495-504.
4. Tateno C and Yoshizato K. Long-term cultivation of adult rat hepatocytes that undergo multiple cell divisions and express normal parenchymal phenotypes. Am J Pathol 1996; 148: 383-392.
5. Block GD, Locker J, Bowen WC, Peterson BE, Katyal S, Strom SC, Riley T, Howard TA and Michalopoulos GK. Population expansion, clonal growth, and specific differentiation patterns in primary cultures of hepatocytes induced by HGF/SF, EGF, and TGFα in a chemically defined (HGM) medium. J Cell Biol 1996; 132: 1133-1149.
6. Cable EE and Isom HC. Exposure of primary rat hepatocytes in long-term DMSO culture to selected transition metals induces hepatocyte proliferation and formation of duct-like structures. Hepatology 1997; 26: 1444-1457.

7. Mitaka T, Mikami M, Sattler GL, Pitot HC and Mochizuki Y. Small cell colonies appear in the primary culture of adult rat hepatocytes in the presence of nicotinamide and epidermal growth factor. Hepatology 1992; 16: 440-447.
8. Mitaka T, Sattler GL, Pitot HC and Mochizuki Y. Characteristics of small cell colonies developing in primary cultures of adult rat hepatocytes. Virchows Archiv B Cell Pathol 1992; 62: 329-335.
9. Mitaka T, Norioka K, Nakamura T and Mochizuki Y. Effects of mitogens and co-mitogens on the formation of small-cell colonies in primary cultures of rat hepatocytes. J Cell Physiol 1993; 157: 461-468.
10. Furukawa K, Shimada T, England P, Mochizuki Y and Williams GM. Enrichment and characterization of clonogenic epithelial cells from adult rat liver and initiation of epithelial cell strains. *In Vitro* Cell Dev Biol 1987; 23: 339-348.
11. Tateno C and Yoshizato K. Growth and differentiation in culture of clonogenic hepatocytes that express both phenotypes of hepatocytes and biliary epithelial cells. Am J Pathol 1996; 149: 1593-1605.
12. Wu JC, Merlino G, Cveklova K, Mosinger B Jr and Fausto N. Autonomous growth in serum-free medium and production of hepatocellular carcinomas by differentiated hepatocyte lines that overexpress transforming growth factor α. Cancer Res 1994; 54, 5964-5973.
13. Paine AJ, Hockin LJ and Legg RF. Relationship between the ability of nicotinamide to maintain nicotinamide adenine dinucleotide in rat liver cell culture and its effect on cytochrome P-450. Biochem J 1979; 184: 461-463.
14. Uchigata Y, Yamamoto H, Kawamura A and Okamoto H. Protection by superoxide dismutase, catalase, and poly(ADP-ribose) synthetase inhibitors against alloxan- and streptozotocin-induced islet DNA strand breaks and against the inhibition of proinsulin synthesis. J Biol Chem 1982; 257: 6084-6088.
15. Althaus FR, Lawrence SD, Sattler GL and Pitot HC. ADP-ribosyltransferase activity in cultured hepatocytes. J Biol Chem 1982; 257: 5528-5535.
16. Romaschin AD, Kirsten E, Jackowski G and Kun E. Quantitative isolation of oligo- and poly adenosine-diphosphoribosylated proteins by affinity chromatography from livers of normal and dimethylnitrosamine-treated Syrian hamsters. J Biol Chem 1981; 256: 7800-7805.
17. Miwa M, Oda K, Segawa K, Tanaka M, Irie S, Yamaguchi N, Kuchino T, Shiroki K, Shimojo H, Sakura H, Matsushima T and Sugimura T. Cell density-dependent increase in chromatin-associated ADP-ribosyltransferase activity in simian virus 40-transformed cells. Arch Biochem Biophys 1977; 181: 313-321.
18. Vaziri H, West MD, Allsopp RC, Davison TS, Wu Y-S, Arrowsmith CH, Poirier GG and Benchimol S. ATM-dependent telomere loss in aging human diploid fibroblasts and DNA damage lead to the post-translational activation of p53 protein involving poly(ADP-ribose) polymerase. EMBO J 1997; 16: 6018-6033.
19. Sato F, Mitaka T, Mizuguchi T, Mochizuki Y and Hirata K. Effects of nicotinamide-related agents on the growth of primary rat hepatocytes and formation of small hepatocyte colonies. Liver 1999; 19:481-488.
20. Mitaka T, Norioka K and Mochizuki Y. Redifferentiation of proliferated rat hepatocytes cultured in L15 medium supplemented with EGF and DMSO. *In Vitro* Cell Dev Biol 1993; 27A: 714-722.

21. Ohmura T, Enomoto K, Satoh H, Sawada N and Mori M. Establishment of a novel monoclonal antibody, SE-1, which specifically reacts with rat hepatic sinusoidal endothelial cells. J Histochem Cytochem 1993; 41: 1253-1257.
22. Mitaka T, Kojima T, Mizuguchi T and Mochizuki Y. Growth and maturation of small hepatocytes isolated from adult rat liver. Biochem Biophys Res Commun 1995; 214: 310-317.
23. Mitaka T, Sato F, Mizuguchi T, Yokono T and Mochizuki Y. Reconstruction of hepatic organoid by rat small hepatocytes and hepatic non-parenchymal cells. Hepatology 1999; 29: 111-125.
24. Mitaka T, Kojima T, Norioka K and Mochizuki Y. TGF-β completely blocks the formation of small-cell colonies: effects of mito-inhibitory factors on the proliferation of primary cultured rat hepatocytes. Cell Struct Funct 1995; 20: 167-176.
25. Mitaka T, Norioka K, Sattler GL, Pitot HC and Mochizuki Y. Effect of age on the formation of small-cell colonies in cultures of primary rat hepatocytes. Cancer Res 1993; 53: 3145-3148.
26. Mitaka T, Mizuguchi T, Sato F, Mochizuki C and Mochizuki Y. Growth and maturation of small hepatocytes. J Gastroenterol Hepatol 1998; 13: S70-S77.
27. Mitaka T. The current status of primary hepatocyte culture. Int J Exp Path 1998; 79: 393-409.
28. Shiojiri N, Lemire JM and Fausto N. Cell lineages and oval cell progenitors in rat liver development. Cancer Res 1991; 51: 2611-2620.
29. Shiojiri N. Transient expression of bile-duct-specific cytokeratin in fetal mouse hepatocytes. Cell Tissue Res 1994; 278: 117-123.
30. Van Eyken P, Sciot R, Callea F, Van der Steen K, Moerman P and Desmet V. The development of the intrahepatic bile ducts in man: a keratin-immunohistochemical study. Hepatology 1988; 8: 1586-1595.
31. Karvountzis GG, Redeker AG and Peters RL. Long term follow-up studies of patients surviving fulminant viral hepatitis. Gastroenterology 1974; 67: 870-877.
32. Uchida T, Kronborg I and Peters RL. Acute viral hepatitis: Morphologic and functional correlations in human livers. Hum Pathol 1984; 15: 267-277.
33. Strain AJ. *Ex vivo* liver cell morphogenesis: one step nearer to the bioartificial liver? Hepatology 1999; 29: 288-290.
34. Michalopoulos GK, Bowen WC, Zajac VF, Beer-Stolz D, Watkins S, Kostrubsky V and Strom SC. Morphogenetic events in mixed cultures of rat hepatocytes and non-parenchymal cells maintained in biological matrices in the presence of hepatic growth factor and epidermal growth factor. Hepatology 1999; 29: 90-100.

Chapter 15

Mechanisms of hepatocyte growth regulation by hormones and growth factors

Thoralf Christoffersen, G. Hege Thoresen, Olav F. Dajani, Øyvind Melien, Tormod Guren, Magne Refsnes and Dagny Sandnes
Department of Pharmacology, Faculty of Medicine, University of Oslo, P.O. Box 1057 Blindern, N-0316 Oslo, Norway

Key words: Ca^{2+}, catecholamines, cell cycle, comitogens, cyclic AMP, cytokine, DNA synthesis, EGF, glucagon, glucocorticoids, G-proteins, HGF, human hepatocytes, inositol 1,4,5-trisphosphate, insulin, MAP kinase, mitogens, prostaglandins, protein kinase C, protooncogene expression, restriction point, signal transduction, TGF-α, TGF-β, transcription factors, tyrosine kinase

15.1 Introduction

The striking ability of mature, quiescent liver cells to rapidly reenter the cell cycle and proliferate to compensate for loss or injury of hepatic tissue has long fascinated scientists [1-3]. Evidence from *in vivo* experiments, dating back from early, pioneering work [4-9], strongly suggested that hepatic growth is under humoral control. The development of techniques for hepatocyte isolation [10,11] has permitted the study of hepatocyte growth in culture. This has contributed greatly to identifying agents and mechanisms that initiate, sustain, and terminate liver cell proliferation.

The extensive research on liver and hepatocyte growth *in vivo* and *in vitro* has been thoroughly reviewed, with emphasis on the basic nature of hepatic growth and regeneration [12-15], the identification and effects of humoral liver growth regulators [13,16-18], and the clinical implications of the regenerative potential of the liver [16,19,20]. The present review focuses primarily on hepatocytes studied in culture and discusses some of the mechanisms by which hormones and growth factors control hepatocyte proliferation.

M.N. Berry and A.M. Edwards (eds.), The Hepatocyte Review, 209–246.

15.2 A variety of hormonal factors influence hepatocyte growth

Numerous biological substances, including recognized endocrine hormones, polypeptide growth factors, cytokines, other agents that exert local (paracrine) effects, and neurotransmitters are now known to affect hepatocyte growth. Much of the work on hormonal growth regulation in hepatocytes *in vitro* has been motivated by the experience from studies on liver regeneration and the search for hepatotrophic factors. However, questions arising in general cell biology have also been approached in the hepatocyte model, with fruitful exchange of insights between work on liver *in vivo* and hepatocytes *in vitro*.

A brief summary of hormonal factors found to have effects on hepatocyte growth is given below. More detailed description of the work that has led to the identification of the many hepatoregulatory substances is beyond the scope of this article. Excellent reviews have treated this literature comprehensively [13,17,18].

15.2.1 Insulin and glucagon

Early studies suggested that factors in the portal blood have a role in maintaining the viability and proliferative capacity of the liver [4,8,21]. That insulin promotes liver growth is supported by massive evidence. Experiments where partial hepatectomy has been combined with various kinds of surgical or chemical endocrine deprival followed by controlled hormonal replenishment have indicated that insulin is essential for an adequate regenerative response [8,16]. Insulin increases DNA synthesis in primary cultures of rat hepatocytes [22-26], typically in synergism with other hormonal factors, although a significant effect can be obtained by insulin alone. While it has often been speculated that insulin stimulates growth through IGF-1 receptors, there is now strong evidence that the effect is mediated largely via genuine insulin receptor signaling [27]. Insulin is commonly included in hepatocyte culture media.

Glucagon has also been strongly implicated in the control of hepatocyte proliferation, but it has been hard to define its role precisely, because seemingly conflicting results have been reported, suggesting stimulatory as well as inhibitory effects (see [18] for review). In primary hepatocyte monolayers, glucagon, likewise, has been found to either enhance [23,24] or inhibit [28] DNA synthesis. It now seems clear that glucagon (through cAMP) is actually involved in both positive and negative regulation of hepatocyte proliferation [29,30].

15.2.2 Other hormones and neurotransmitters

Much interest has focused on the role of catecholamines. *In vivo* experiments, using denervation and various pharmacological strategies, have indicated that hepatocyte proliferation is subject to adrenergic control, through sympathetic nerves and circulating catecholamines. However, these data have been partly contradictory and difficult to interpret (see [13,18,31,32] for review of this literature). Experiments in primary hepatocyte monolayer cultures using various adrenergic agonists and antagonists [24,33-37] have demonstrated that the cells are responsive to growth modulation by catecholamines. There is most extensive documentation for growth-promoting effects of α_1-adrenergic stimulation [34,35], but as will be discussed below, the studies have also uncovered more complex regulations, with inhibitory as well as stimulatory adrenergic effects and involvement of both α_1- and β_2-adrenoceptors [37].

Various other hormones may augment hepatic growth [18]. In cultured human, rat or mouse hepatocytes, vasopressin, angiotensin II, serotonin, and several gastrointestinal hormones such as gastrin, neurotensin, and vasoactive intestinal polypeptide, enhance DNA synthesis [18,38-43]. Hormones that act via nuclear receptors also affect hepatocyte growth. Estrogens enhance the response of cultured hepatocytes to various mitogens [44]. Glucocorticoids have profound and complex effects on hepatocyte growth. While largely inhibitory, the glucocorticoid effects are highly conditional, depending on the interaction with other agents such as insulin and cAMP-elevating agonists [45,46]. Thyroid hormones seem to be required for an hepatic proliferative response *in vivo*, but this has so far not been demonstrated in primary hepatocyte cultures [47,48].

15.2.3 Eicosanoids

Many studies have suggested that prostaglandins are somehow involved in promoting liver regeneration [49-51], and they also appear to have a cytoprotective effect in the liver [52], being produced largely in non-parenchymal cells and acting on hepatocytes. Although their physiological and pathophysiological role in hepatic growth is not fully clarified, the ability of several prostaglandins or other eicosanoids to promote DNA synthesis in neonatal or adult rat hepatocytes in culture is well documented [53-58]. More generally, there is an increasing awareness of the possibility that prostaglandins and other eicosanoids are important in cell growth and even tumor development [59]. Hepatocytes are a model for exploration of underlying mechanisms.

15.2.4 The epidermal growth factor family

Epidermal growth factor (EGF) stimulates hepatic growth [17,18,60] and also regulates several other liver functions [61]. Other ligands that activate the EGF receptor include transforming growth factor alpha (TGF-α), heparin-binding EGF (HB-EGF) and amphiregulin. It is possible that while EGF provides a more continuous trophic stimulus, TGF-α has a role in acute growth responses. TGF-α is not normally expressed in hepatocytes, but both its mRNA and the protein appear upon growth stimulation, suggesting an autocrine role [62,63]. The fact that EGF can induce TGF-α expression in hepatocytes [63] might reflect a kind of relay and amplification mechanism for EGF receptor ligands. In transgenic rats that overexpress TGF-α, the most consistent finding is hyperplasia and tumour formation in the liver (and pancreas and mammary gland) [64]. There is abundant evidence from work on cultured hepatocytes that the EGF receptors are capable of conveying strong growth stimulation in these cells. Hepatocytes have a high number of EGF receptors compared to most other normal cells. Approximately 200,000 surface EGF receptors per cell were found in freshly isolated rat hepatocytes, about 5% of which are in a high-affinity state [65]. Stimulation of DNA synthesis can be induced in hepatocytes by EGF [22,23,25,26,66,67] as well as by TGF-α [62,68-71], HB-EGF [72], and amphiregulin (unpublished results, this laboratory).

15.2.5 Hepatocyte growth factor

This polypeptide factor deserves particular consideration here, not only because of its interesting actions, but also because its discovery resulted from the use of hepatocytes *in vitro* as an experimental model in the search for hepatotrophic factors. Different lines of investigation had demonstrated that a hepatocyte growth-stimulatory activity that could not be accounted for by any known hormone or growth factor was present in rat serum [73,74] and platelets [75-78], and also in plasma from patients with acute fulminant hepatic failure [79]. Using hepatocytes in primary culture as a bioassay system, factors from different sources were purified and cloned [80-82] and were found to be identical, encoded by a gene that is highly conserved in different species. With its 728 amino acids, hepatocyte growth factor (HGF) is a quite large molecule compared to other growth factors. It shows structural similarity to plasminogen and related proteases and requires proteolytic cleavage to become active. When HGF was in the process of being isolated, it was believed to be a liver-specific stimulator of proliferation, and no homology was found with any previously known growth factor. It has later become clear, however, that HGF acts on many cell types, and also that several other independently discovered proteins, including the so-called scatter factor (SF), are identical with HGF and encoded by the same gene [83,84]. Its receptor is the *c-met* proto-oncogene product, a tyrosine kinase-type receptor [83,85]. HGF (or HGF/SF) is now known to be a pluripotent factor, typically produced in mesenchymal cells, acting on epithelia and several other cells, and its actions include stimulation of cell proliferation, cell movement and morphogenesis, and even metastasis [86-88].

While thus far from liver-specific, HGF clearly is a strong stimulator of hepatocyte proliferation. It appears that this ubiquitous regulator plays a decisive role in the mechanisms of liver regeneration, although it is does not drive the process alone and apparently does not initiate it [15]. In contrast to EGF, the HGF level rises many-fold in the blood in response to partial hepatectomy or toxic or infectious liver damage both in humans and rats [79,89]. Studies on rat [68] and human [90,91] hepatocytes in primary culture have demonstrated that HGF stimulates DNA synthesis, acting with extremely high potency.

15.2.6 Other growth factors

In addition to the EGF family and HGF, certain other receptor tyrosine kinase-activating polypeptide growth factors, particularly members of the fibroblast growth factor (FGF) family, including keratinocyte growth factor, have been found to stimulate DNA synthesis in rat hepatocytes in primary culture [92-95], but not in human hepatocytes [94]. The effects of FGF-1 in rat hepatocytes seem to be determined by the balance between different FGF receptors and occur in a complex interplay with serum and EGF [95]. FGFs may be important in the regulation of hepatocyte proliferation, but their exact role is not clear.

A more detailed account of the effect of HGF and other growth factors is given in Chapter 16.

15.2.7 Cytokines

The hepatocytes are targets of various cytokines, incuding several interleukins, interferons, tumor necrosis factor alpha (TNF-α), and transforming growth factor beta (TGF-β), many of which are produced in non-parenchymal liver cells [96]. Several observations implicate cytokines in initiation of liver growth. In hepatocytes *in vitro*, most studies have shown inhibition of DNA synthesis and/or induction of apoptosis in response to

IL-6 or IFN-γ [97-99], while TNF-α has been found to stimulate DNA synthesis in cultured mouse [99,100] and rat [101] hepatocytes.

TGF-β, which exists in three isoforms (β1, β2, β3) in mammalian tissues and exerts many effects in a broad range of cells [102], is a strong inhibitor of hepatocyte proliferation *in vitro* [46,103-106] and *in vivo* [107,108]. This is consistent with the experience that while the TGF-βs generally are growth-stimulatory in cells of mesenchymal origin, they inhibit proliferation in most epithelia and a variety of other cell types [102] and probably are the most potent growth-inhibitory hormonal factors so far identified. The inhibition of DNA synthesis by TGF-β1 in cultured hepatocytes typically occurs with ID_{50} as low as 1-20 pM [46]. Activin, which is structurally related to TGF-β, also inhibits hepatocyte growth and may have an autocrine role [109].

15.3 Hepatocytes in short term primary culture as an experimental model for studies of cell proliferation

Several features of cultured hepatocytes make them a valuable model for study of cellular growth control. These cells have receptors and intact signal transduction mechanisms that enable them to respond to a wide range of hormones. Although mammalian cells *in vivo* are normally exposed to a multitude of humoral growth-regulating influences, most of the widely used experimental *in vitro* models are cell lines with a restricted repertoire of hormone responses, often transfected to overexpress one or a few components of the signal pathways to be studied. Hepatocytes provide a useful supplement to these models. Additional advantages are the possibility of comparing *in vitro* results with the *in vivo* liver regeneration model, and the wealth of information available in the literature on general signal transduction mechanisms in hepatocytes, much of which is of potential relevance to growth regulation.

It has been argued that a problem with the interpretation of growth regulation studied in hepatocytes is that, although these cells are highly differentiated when isolated, they undergo phenotypic changes during culturing that affect the expression of receptors and other components of the signal transduction pathways. It is clear that such changes do occur [37,110-113], and often reflect dedifferentiation of the cells, but they commonly are part of biological prereplicative phenomena and may also have their equivalent in alterations that take place during liver regeneration and other conditions *in vivo* [33,114-116]. Hepatocytes in primary culture also show a pattern of expression of protooncogenes [117,118] that in principle resembles that observed in regenerating liver [119,120], although generally developing more slowly in culture.

One limitation in the use of primary cultures of hepatocytes in studies of growth-regulating mechanisms has been that these cells are not as readily amenable to targeted genetic manipulations as many other experimental models. However, the hepatocytes are increasingly being used in experimental strategies where genetic modifications are achieved by various kinds of transfection, antisense nucleotide-mediated inhibition of gene expression, or cell isolation from transgenic animals (see for example [27,121-123]). Further exploitation of such methods will be important in future work in this field.

Culture conditions influence the response of the hepatocytes to growth-regulatory hormones [124] so that comparison of different studies may be difficult. However, there are certain points that may be helpful in the interpretation of the results obtained in different experimental settings. First, many of the modulating effects of constituents of the culture media may reflect physiological regulation by nutrients, in terms of integration of metabolic and hormonal control. For example, the enhancement by pyruvate [125] or oxaloacetate [126], and the inhibition by glucose [127], of hormone- or growth factor-induced DNA synthesis in cultured hepatocytes is compatible with a link between metabolism and the rate of proliferation in order to adapt liver mass to the metabolic demand [18]. Second, the great variation of culture conditions used in studies involving hepatocyte growth is to a large extent dictated by the different purposes of the investigations. Thus, in some cases the prime goal is to keep the cells as differentiated as possible, or control specific gene expression and study its relationship to cell proliferation. This has been achieved with the use of extracellular matrix or various substrata [128-131] and agents such as DMSO or butyrate [132,133]. On the other hand, in studies attempting to explore the molecular mechanisms of growth control it is often desirable to use defined and relatively simple culture conditions, although this may remove the hepatocytes further from the physiological situation. Hepatocytes can be stimulated by hormones and growth factors to proceed to S phase and further to mitosis in chemically defined media [23,26,66,128,134]. Discussion of the influence of various culture conditions is beyond the scope of this review. However, in general, various matrix components, high cell density and cell-cell contact, high glucocorticoid concentration, and differentiating chemical agents, are factors that tend to attenuate or delay proliferative responses [29,45,129,132,135,136].

Most of the studies on mechanisms of hepatocyte growth regulation, and hence most of those cited here, have used rat hepatocytes, but cells from other species have also been employed. On the whole, despite differences in modes of culturing, the main findings seem relatively consistent when comparing hepatocytes from different species, including human cells.

15.4 Hepatocyte growth responses to hormonal agents: temporal aspects and integration of multiple synergistic and antagonistic stimuli

The fact that hepatocytes can adjust their growth in response to a variety of hormonal factors indicates that they have mechanisms enabling them to coordinate and integrate multiple growth-regulatory signals, which they may receive concomitantly or sequentially.

15.4.1 Mitogenic and comitogenic effects

It is common to distinguish between those factors that are directly mitogenic in hepatocytes and those that appear to prime the cells and/or enhance their responsiveness to the mitogens [17,63]. Members of the first category, the mitogens, are often termed "complete mitogens" since they, by definition, are capable of eliciting all the signals necessary to induce cell proliferation. Notable examples are the EGF family of growth factors and HGF. Agents that belong to the second category are commonly called comitogens or "incomplete mitogens" as they are considered not to induce cell growth alone but promote the effect of the mitogens. Examples include insulin, several agonists of heptahelical membrane receptors, estrogens, etc. However, while the distinction between mitogens and comitogens is convenient and seems largely to reflect real and biologically important properties, it should be remembered that this difference is not absolute. For example, by treating rat hepatocytes with insulin in primary culture in the absence of other added growth factors, under completely defined serum-free conditions, it may be possible to induce S phase entry in a substantial proportion (often as many as 25-30%) of the cells [26].

15.4.2 Differential hormone responsiveness in distinct parts of the G_1 phase

Studies of growth control in various models have indicated that there are a number of control points in the mammalian cell cycle, where the cells need external stimuli to be able to progress further, or where inhibitory control is exerted [137]. These control points are not located similarly in all cells. Cell cycle control is not discussed in depth here (see Chapters 16 and 17), but we shall summarize data on how different hormones and growth factors exert their growth-regulatory effects in different parts of the prereplicative period (G_1) (Figure 1).

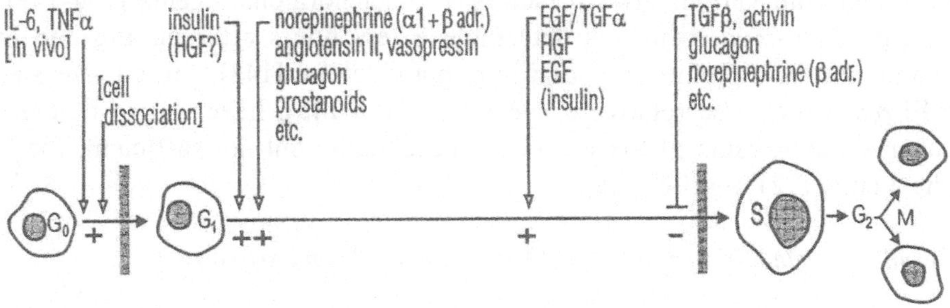

Figure 1. Time points in the G_1 phase of the cell cycle where hepatocytes *in vitro* seem to be most responsive to different growth-regulatory hormonal agents. (The initiation, *i.e.* G_0 to G_1 transition, appears usually to have occurred already at the time when the hepatocytes are isolated and plated.) For further explanation, see the text.

Trigger mechanisms for the G_0 to G_1 transition

There is increasing evidence that to proliferate, hepatocytes need some kind of triggering stimulus that makes them competent to respond to subsequent mitogenic activation [63,138,139]. Many agents and mechanisms have been proposed (and need not be mutually exclusive). Recently, several findings have suggested that inflammatory cytokines, particularly TNF-α and IL-6, play a critical role in priming the hepatocytes [63,120,140]. Their effects are believed to be mediated by activation of the preformed transcription factors NFκB and the STAT protein family. Thus, NF-κB and stat3 are activated in the liver early after partial hepatectomy [120]. Liver regeneration is impaired in mice genetically deficient in IL-6 [141]. Furthermore, infusion of TNF-α in rats in vivo activates both NFκB (which may induce IL-6) and stat3 (possibly via IL-6) and also increases the mitogenic response of the intact liver to HGF and TGF-α [142].

A notable difference between intact liver and cultured hepatocytes is the response to IL-6 and TNF-α. Thus, while IL-6 inhibits DNA synthesis [97-99], and TNF-α has been reported to exert a comitogenic effect [99-101], none of these agents is required to prime the hepatocytes *in vitro*. A likely explanation, supported by experimental evidence, is that the isolation (dissociation) of the hepatocytes is sufficient to trigger the cells. Thus, c-*fos* and c-*jun* are rapidly expressed upon collagenase perfusion, apparently due to disruption of cell/cell contact [118,143]. Therefore, it appears that the hepatocytes are already competent to respond to further proliferogenic stimuli (*i.e.* have moved from G_0 to G_1) when they are plated).

Upon plating, the hepatocytes are then able to proceed further through part of G_1, as indicated by the sequential appearance of transcripts of several protooncogenes [118,143-146]. This occurs in the absence of any mitogenic stimulus, although the expression of some of the genes may be enhanced by treatment with various growth factors or other hormonal agents [144-147]. Studies on c-*myc*, which is detectable a few hours after plating and is maintained for a prolonged period in primary culture [143], and c-H-*ras* and c-Ki-*ras*, which rise relatively late in G_1 [118,145], have shown that the enhanced expression of these genes is neccessary, but not sufficient, for S phase entry [121,144,145,148].

Permissive effect of insulin requires its presence from early G_1

The ability of insulin to enhance hepatocyte DNA synthesis *in vitro*, and its synergism with EGF, was observed from the earliest studies of hormonal induction of DNA synthesis in hepatocytes [22] and is well established [23,25,26]. The hepatocytes seem to require insulin from the early part of the prereplicative period [26]. Thus, the growth-promoting effect of insulin, in contrast to EGF, declined dramatically with delay in the time of addition, and hepatocytes that did not receive insulin during the first 48 h of culturing became virtually non-responsive to EGF [26]. Insulin facilitates mitogenic signaling at least partly through Ras [149]. It is not clear whether the ability of insulin to increase the expression of c-Ki-*ras* in hepatocytes [150] is relevant for the permissive effect.

Heptahelical membrane receptors mediate comitogenic effects early in G_1

Growth-promoting agonists that bind to heptahelical receptors (*i.e.* G protein-coupled receptors with seven membrane-spanning domains) generally seem to act as comitogens in hepatocytes. When examined in primary hepatocyte cultures, there are certain phenomenological characteristics of their effects that apply to all these agonists as a group (although their molecular mechanisms of action may be different). First, the hepatocytes in primary culture usually are most responsive to their growth-promoting action early in G_1 (*e.g.* 0-10 h after plating) [30,35,151]. Second, their comitogenic effects do not require simultaneous administration of the mitogen. For example, full effect was obtained when comitogens of this kind were given 24 h prior to EGF [40]. Furthermore, when more than one comitogen is used, their effects commonly are additive, while their interaction with EGF typically is multiplicative (synergistic) [56]. Finally, they usually exert their effect by accelerating the traverse of the hepatocyte through G_1 [37,40], thereby increasing the pool of cells that can react to a mitogenic stimulus. Taken together, the data suggest that this group of comitogenic hormones induce biochemical changes that facilitate a mitogen-responsive state.

Glucagon is one example of these hormones. It exerts dual, bidirectional, regulation, with either stimulatory or inhibitory effects on DNA synthesis in the hepatocyte model, depending on the dose, time of application, etc [29,30,152]. The stimulatory effect, as seen in the enhancement of the action of a mitogen such as EGF, predominates if glucagon (usually 10 pM - 10 nM) is administered early after plating in low doses. It should be emphasized, however, that the growth stimulation by glucagon is variable and depends greatly on other factors such as interacting hormonal stimuli and nutritional status, and is not clearly evident under all culture conditions.

Another example of this is seen in the effect of catecholamines. It was found in primary cultures of hepatocytes from regenerating [33] and normal liver [24] that the β-adrenergic agonist isoproterenol augmented the growth response to EGF and insulin. Later it was demonstrated that α_1-adrenergic activation produces a marked comitogenic effect [17,32,34-36]. Further studies attempting to sort out the relative contributions of α_1- and β-adrenergic mechanisms, using various agonists and antagonists in adult rat hepatocytes in primary serum-free culture [37], showed that both α_1-adrenoceptors and β-adrenoceptors may be involved in enhancing hepatocyte growth. The α_1-component was most prominent when a maximal response was observed, but the β-component contributed significantly and was greater than the α_1-mediated effect at low concentrations of norepinephrine [37]. The effect of norepinephrine given early in G_1 was essentially to advance the response to the subsequent addition of EGF, left-shifting the time curve by about 10 h [37,40]. However, in addition to the early G_1 effect augmenting the proliferative response, adrenergic agonists may also strongly inhibit S phase entry in hepatocytes, by an effect late in G_1. In contrast to the stimulation, this inhibitory effect seems to be mediated entirely by β-adrenoceptors, as it was completely blocked by timolol and was unaffected by prazosin [37].

Hepatocytes are responsive to mitogenic growth factors in mid/late G_1

Much evidence suggests that the major growth-stimulatory action of EGF in cultured hepatocytes is exerted in the middle or relatively late part of G_1. Studies of the concerted actions of EGF and insulin, showed that the addition of EGF, unlike insulin, can be postponed as much as 40-50 h and still give a maximal response [26,67]. Thus, if the cells received insulin at the time of plating, a delay in the addition of EGF (to 48 h instead of at plating) reduced the time from the exposure of the cells to the start of the DNA synthesis, increased the sensitivity to EGF, and markedly enhanced the rate of S phase entry [26]. The sensitivity to TGF-α also increases with time in culture [70]. It is of interest that since TGF-α is expressed at a late stage in growth-stimulated cultured hepatocytes (and in regenerating liver) [62,63], the appearance of the growth factor, and the development of an optimal cellular responsiveness to it, roughly coincide and seem to be relatively late events in the prereplicative process.

These results suggest that, provided the cells receive insulin from an early stage, activation of the EGF receptor is not required until relatively late in G_1. Other studies have led to similar conclusions and have also indicated that while cultured hepatocytes may traverse the early part of G_1 independently of growth factor stimulation, as evidenced by the expression of early genes (see above), they reach a restriction point in mid/late G_1 which they cannot pass in the absence of a mitogen [118]. When exposed to mitogenic stimulation (EGF plus pyruvate) the hepatocytes progress through the last part of G_1, as indicated by activation of cell cycle genes, with expression of cyclin D1 being the earliest and largest mitogen-associated response [118]. The expression of cyclin A late in G_1 is also necessary for S phase entry, as shown by experiments using antisense oligonucleotides in rat hepatocyte cultures [122].

Inhibitory control near the transition from G_1 to S phase

Negative regulation of hepatocyte recruitment to S phase and mitosis probably may take place in different parts of G_1 [46,153]. However, several agonists, such as glucagon [28,30,46,152], β-adrenergic agonists [37], TGF-β [46], and apparently activin [109], seem to exert their main inhibitory effect at a point late in G_1. The inhibition can be almost complete, but is usually reversible [45,46,152]. Studies that have monitored the kinetics in detail have located the inhibitory point for glucagon (through cAMP) and TGF-β at 1-2 h before S phase entry [30,45,46,152], and the points of action for glucagon/cAMP and TGF-β could not be distinguished temporally [46]. Thus, it seems that there is a major regulatory point responsive to negative control by several converging hormonal signals immediately before the G_1/S border in hepatocytes.

It also seems that the hepatocytes may increase their sensitivity to inhibitory stimuli late in the G_1 phase. After rat hepatocytes are plated and proceed through G_1, their level of β-adrenoceptors and β-adrenergic responsiveness increase progressively [110,154-156], accompanied by a reciprocal decrease in α_1-adrenoceptors [37,111]. It may be noted that similar changes in β-adrenergic receptors and responsiveness have been found in hepatocytes isolated from regenerating liver [33,115] and early during hepatocarcinogenesis [116,157]. There is also increased expression of TGF-β receptors in rat hepatocytes during culturing and liver regeneration [158].

15.4.3 Differentiation versus proliferation

Since hepatocytes have the unique characteristic of being highly differentiated and yet capable of very quickly entering the cell cycle, they need mechanisms that can integrate and balance these functions at the transcriptional level. The expression of the genes that determine the differentiated characteristics of the mature hepatocytes as well as those required for proliferation is controlled by a number of transcription factors, some of which are ubiquitously expressed while others are relatively liver-specific (or liver-enriched) [120,159-161]. Two examples are the coordinating functions of the C/EBP family of transcription factors and the glucocorticoid receptor.

The C/EBPs are known to activate the expression of various liver-typical genes [160] and seem to also be involved in control of hepatocyte growth. C/EBPα is regulated inversely to hepatocyte proliferation and may act as a growth inhibitor [162,163]. Several lines of evidence indicate that the interplay between different C/EBP subtypes may be important in coordinating hepatic growth and differentiation [161].

Glucocorticoids control a variety of hepatic genes, generally in a way that promotes liver-specific functions. In addition, they may depress hepatocyte proliferation *in vivo* [164] and *in vitro* [45], but the exact cellular response depends on many interacting hormonal stimuli and other regulatory mechanisms. In rat hepatocytes in primary culture, a strong inhibitory effect of dexamethasone on DNA synthesis [28,45,136,165], reflecting decreased rate of S phase entry [45], is exerted in synergism with glucagon (or cAMP) [28,136], and is most pronounced at high cell densities [45], but is strongly counteracted by insulin [136]. Furthermore, when investigated under various experimental conditions, dexamethasone is found to regulate inversely DNA synthesis and the expression of a liver-specific gene product [136].

15.5 Intracellular signalling mechanisms involved in growth control in hepatocytes

Growth factors and hormones that regulate cell growth do so by altering gene expression in a way that eventually activates, or inhibits, the cell cycle machinery [166]. The remainder of this review deals with mechanisms of signal transduction in hepatocytes that link stimulation of cell surface receptors to the nuclear events.

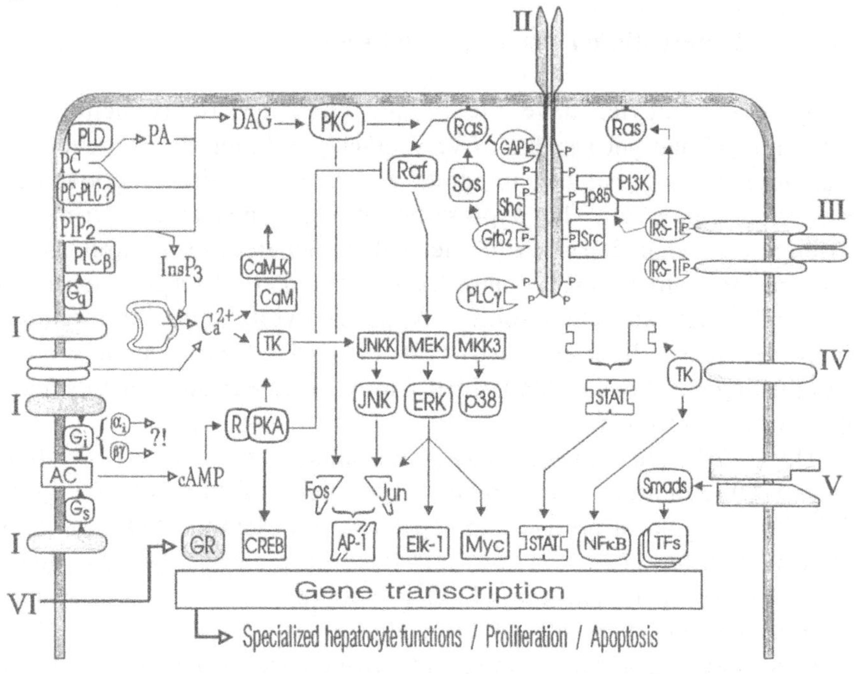

Figure 2. Some of the pathways that convey hormonal signals from various receptors to transcriptional regulation. Many of these mechanisms may be involved in hepatocyte growth regulation, and some of them (not all) are discussed in the text. The figure (although rather crowded) is simplified and primarily is meant as an overview to support the text. Notable areas of complexity not shown in the figure are for example: different isoforms of many of the key proteins involved (*e.g.* PKC), subunits with differential regulatory functions (*e.g.* α vs. βγ subunits of heterotrimeric G proteins), many transactivating mechanisms between distinct pathways, formation of multiprotein functional complexes (*e.g.* many of the transcription factors), and several regulated intracellular translocations. The receptors indicated are: I, the group of G protein-coupled (heptahelical) receptors; II, growth factor type of receptor tyrosine kinases; III, insulin receptor; IV, cytokine receptors; V, receptor for ligands like TGF-β or activin; VI, steroid receptors (GR = glucocorticoid receptor). Abbreviations: AA, arachidonic acid; AC, adenylyl cyclase; TF, various unspecified transcription factors; TK, various unspecified tyrosine kinases. Other abbreviations are given in the text or are standard abbreviations.

15.5.1 Brief overview

Some major signaling pathways are shown schematically in Figure 2. Various heptahelical receptors couple via different heterotrimeric G proteins to effectors such as adenylyl cyclase, initiating the cAMP/PKA pathway, or phosphoinositide-specific phospholipase C (PI-PLC), generating inositol 1,4,5-trisphosphate ($InsP_3$) and diacylglycerol (DAG) with resultant signaling through Ca^{2+} and protein kinase C (PKC), respectively [167-169]. Receptor tyrosine kinases convey their effects by recruiting various signal-transducing proteins, including Shc/Grb2, (Ras)GAP, Src, PI 3-kinase, PLC-γ, and members of the STAT family [170,171]. There is a vast amount of data on many of these and other pathways in hepatocytes, due to their extensive use as experimental models in different contexts, and much of this is relevant for mitogenic mechanisms. However, because studies are progressively revealing more complex signalling pathways, with multiple signals, crosstalk, and promiscuous couplings, it is not straightforward to decide exactly which mechanisms mediate the growth-regulatory effects.

Many questions regarding the initial signal steps are unresolved. For example, which G proteins (and which α and βγ subunits) are involved in mitogenic signalling by heptahelical receptors in hepatocytes? What are the principal signal-transducing substrates of the receptor tyrosine kinases in these cells? Which are the main lipid-generated messengers involved in growth regulation in hepatocytes, and by which mechanisms are they generated? And, while free cytosolic Ca^{2+} is controlled by many mechanisms (extensively studied in hepatocytes), and Ca^{2+} is essential at several critical points in the cell cycle [167], what is the role of Ca^{2+} in mediating mitogenic stimuli in hepatocytes?

Downstream mechanisms that convey receptor-induced messages to the appropriate transcriptional regulation are partly understood [172]. Some of the signaling pathways to the target transcription factors are indicated in Figure 2. Of great importance is the identification of the phosphorylation cascades involving protein kinases of the MAP kinase family that transduce various extracellular stimuli into biological responses [173]. The most extensively studied of the MAP kinases are the two closely related p44 and p42 MAP kinases, termed ERK1 and ERK2, which seem to have a crucial role in mitogenic signaling. However, other similar cascades, the p38 and JNK pathways also operate in parallel with ERK [173]. In cultured hepatocytes, many growth factors and growth-promoting hormones, *e.g.* EGF, HGF, insulin, and several G protein-coupled receptor agonists, activate both the ERK pathway [71,174-178] and p38 and JNK [179-182]. All three pathways have been implicated in growth regulation in hepatocytes [71,179,183] but their specific roles are not clarified.

We shall consider some of the mechanisms in more detail. Rather than trying to provide a comprehensive review of growth-regulatory signal transduction in hepatocytes, we have concentrated the discussion on some of the questions that have been much debated and studied in these cells.

15.5.2 Mitogenic signaling from the EGF receptor

Targets and downstream pathways for the EGF receptor in hepatocytes

There are many potential targets for the EGF receptor [170]. It has been realized that not only can different kinds of receptor tyrosine kinases vary in their affinity and preference for signal-mediating substrates but also that one type of receptor, such as the EGF receptor, may show cell-specific differences in their use of downstream pathways. Therefore it is important to try to determine which pathways are involved in the signaling from the EGF receptor in hepatocytes.

1. *ERK1/2* are rapidly activated upon exposure of hepatocytes to EGF [71,174,178,180,184]. The activation is maximal within 5 min, and is sustained for several hours [71]. TGF-α causes a more prolonged activation of ERK than EGF [71]. While it has been reported that inhibition of ERK (using the MEK inhibitor PD98059) does not substantially reduce basal DNA synthesis (or cause apoptosis) in primary hepatocyte cultures [180], other published data [185] and our results (Thoresen *et al.*, unpublished) indicate that PD98059 decreases both basal and EGF-induced DNA synthesis. This suggests a role for ERK in mitogenic signaling in these cells. However, other members of the MAP kinase superfamily are also activated upon stimulation of the EGF receptor. Thus, increased activity of both JNK and p38 is found after treatment of hepatocytes with EGF [180] (and findings in this laboratory), and the p38 inhibitor SB203580 reduces the DNA-synthesis response to EGF [182]. Thus, it is possible that all the members of the MAP kinase family, ERK, JNK, and p38, are involved in mitogenic pathways that are used, directly or indirectly, by the EGF receptor in hepatocytes. Further work is needed to clarify these mechanisms in more detail.
2. *PLC-γ* is a target of several receptor tyrosine kinases, including the EGF receptor [170], but it is debated whether this pathway is obligatory in EGF-mediated mitogenic signaling. In hepatocytes, PLC-γ is present in substantial amounts [186] and is tyrosine-phosphorylated in response to

EGF [186-188], although very slightly when compared to two cell lines (A431 and Rat1hER) that overexpress the EGF receptor [186]. EGF-induced increase in $InsP_3$ in hepatocytes has been observed in some [187,188] but not in other [189,190] studies, or has been very small, requiring high, supraphysiological concentrations of the growth factor [40,191,192] (and unpublished data, this laboratory). A rise in $[Ca^{2+}]_i$ has been a more consistent finding [186,187,189-191,193,194] also at relatively low concentrations (0.1-1 nM) of EGF [187,189,194] and seems to originate from internal stores [187,189,191]. However, evidence has been given to suggest that although Ca^{2+} is likely to be a permissive factor for growth-promoting effects in hepatocytes, it apparently is not a major second messenger for EGF [195]. An intriguing finding is that in hepatocytes the EGF-induced $[Ca^{2+}]_i$ rise [187,190,191,196] is sensitive to pertussis toxin, suggesting involvement of a G_i protein. Data have also been provided in support of the interesting, but not yet proved, hypothesis that the EGF receptor induces a translocation of PLC-γ to the cytoskeleton where it interacts with $G\alpha_i$ [197]. There is disagreement regarding the diacylglycerol (DAG) response to EGF in hepatocytes [188,192]. We have found that EGF, at concentrations that stimulate DNA synthesis, does not detectably elevate DAG in cultured rat hepatocytes, and the effects of EGF on ERK or DNA synthesis are not affected by the PKC blocker GF109203X [151]. Thus, although the EGF receptor in rat hepatocytes elicits a modest rise in $[Ca^{2+}]_i$, possibly via PLC-γ and in a $G\alpha_i$-dependent way, the activation of PLC-γ seems to be too weak to induce a signal via DAG/PKC that significantly contributes to the growth-stimulatory effect of EGF. (Instead, EGF acts in synergism with other hormones that can provide the DAG formation needed for an optimal mitogenic effect, see below.)

3. *STAT proteins* have recently been found to be potential targets for the EGF receptor. Thus, in some transformed cells overexpressing EGF receptors growth stimulation, as well as antiproliferative or pro-apoptotic signals from the EGF receptor, seem to require STAT1 and STAT3, respectively [198,199]. After intraperitoneal injection of EGF into mice, activation of STAT1, 3 and 5 has been demonstrated in preparations from whole liver [200-202]. However, the role of STAT-dependent mechanisms in EGF-mediated effects in hepatocytes is not known. We have recently investigated the activation of several STAT proteins in cultured rat hepatocytes after exposure to EGF, IFN-γ, or IL-6. In hepatocytes that were highly responsive to the stimulatory effect of EGF on the DNA synthesis, the EGF-induced

activation of STAT1, STAT3 and STAT5b was completely downregulated, while ERK1/2, JNK, and p38 were stimulated (Guren *et al.*, submitted). In contrast, under the same conditions INF-γ and IL-6 strongly activated STAT1 and STAT3, respectively.

Relationship between receptor binding and mitogenic response

A well known observation that is not fully understood, is the consistent inverse relationship between the EGF receptor level and the ability of the hepatocytes to respond to the mitogenic stimulus of EGF. The progressive increase in EGF responsiveness that occurs when the cells are plated and proceed through G_1 [26,67] is accompanied by a decrease in the number of surface EGF receptors and an almost complete loss of high-affinity receptors [67,132]. Growth-promoting agents such as insulin, norepinephrine, and TPA enhance this downregulation, while glucocorticoids and differentiating/growth-inhibitory agents like butyrate and DMSO counteract it and preserve the high-affinity receptor population [113,132,203]. Further complexity is seen when the binding and effects of different EGF receptor ligands are examined. Thus, while EGF binds with about twentyfold higher affinity than TGF-α to the receptor, TGF-α elicits a more sustained MAP kinase activation and more effective stimulation of DNA synthesis [70,71]. The hepatocyte EGF receptors are internalized and recycled [204], and there is evidence that endocytosed receptors may remain active [205]. If so, it is possible that there are dissimilarities in EGF receptor signaling in different compartments and also differential binding of EGF and TGF-α.

15.5.3 HGF and c-Met-mediated signalling

The HGF receptor, c-Met, is also a tyrosine kinase-type receptor. It mediates a mitogenic effect of HGF, with efficacy roughly similar to EGF (and somewhat lower than TGF-α), but a potency that is about tenfold higher than that of EGF. Half maximal stimulation of DNA synthesis by HGF in human and rat hepatocytes occurs in the range 10-50 pM [68,90,91]. HGF elicits responses in many cells that are qualitatively different from EGF. Certain differences between the two growth factors have also been observed in hepatocytes, *e.g.* a greater ability of HGF to induce motility [206]. It is of interest to know how the signal transduction mechanisms for HGF compare with EGF. Unfortunately, few studies have made direct and systematic comparisons of the main signaling pathways for these two growth factors in hepatocytes, so in most cases we have to extrapolate from one investigation to another.

1. *ERK* is activated in response to HGF, as with EGF [207]. This activation is sensitive to genistein [177], consistent with tyrosine kinase dependency, and is also depressed by treatment with pertussis toxin [208], suggesting that G_i plays a role in the signaling.
2. HGF stimulates phosphorylation of *PLC-γ* in hepatocytes [209] and causes elevation of $InsP_3$, but most studies have found only small increases [193,209-211], which were of the same magnitude as the response to EGF [193,210]. HGF also raises $[Ca^{2+}]_i$ in a manner comparable with EGF [193,210,212,213]. The effects of EGF and HGF on $InsP_3$ and $[Ca^{2+}]_i$ levels are partly additive [193,210]. Stimulation of hepatocytes with HGF results in DAG accumulation, which at least in part is generated through the activation of phospholipase D (PLD) [214].
3. *PI 3-kinase* mediates a variety of effects in many cells and seems to be involved in HGF signaling [215]. Stimulation of hepatocytes with HGF results in association of the PI 3-kinase regulatory subunit with c-Met, and wortmannin (although at relatively high concentrations) reduces HGF-stimulated DNA synthesis, suggesting a role of PI 3-kinase in the growth-stimulatory effect of HGF [216].

In conclusion, the existing data suggests that HGF and EGF largely activate the same signalling pathways in hepatocytes, and induce similar responses when directly compared. Thus, alhough c-Met and the EGF receptor mediate different effects in many cells, it is conceivable that the range of signalling targets available for them to interact with are quite similar in hepatocytes. However, the qualitative differences in cellular responses that have been observed also in these cells [206] suggest that there are distinct signaling pathways. The possibility that PI 3-kinase might be involved in the effect of HGF, but not EGF, deserves further study.

15.5.4 The role of phospholipids in hormonal regulation of hepatocyte proliferation

Several hormones that exert comitogenic effects in hepatocytes activate PI-PLC and raise $[Ca^{2+}]_i$ [40,57,217-220]. A permissive function of Ca^{2+} for the proliferative response is likely. However, the role of PI-PLC in the growth-stimulation by these agents is not clear [40,221]. The generation of $InsP_3$ does not correlate well with the magnitude of the subsequent stimulation of DNA synthesis by vasopressin, angiotensin II, norepinephrine, and $PGF_{2\alpha}$ in primary cultures of rat hepatocytes [40]. Although hormone-stimulated PI-PLC (PLC-β) activity rapidly declines upon culturing of hepatocytes [112,222,223], norepinephrine and $PGF_{2\alpha}$ still enhance DNA synthesis under conditions where the $InsP_3$ response to these agonists is completely down-regulated [223]. Together the findings suggest that activation of PI-PLC is not sufficient to explain the growth-augmenting effect of these agents in hepatocytes.

Activation of DAG/PKC-mediated pathways is believed to be important for mitogenic responses [224]. Several observations suggest that DAG and PKC are involved in hepatic growth [225-227]. We have found that vasopressin, angiotensin II and norepinephrine produce a sustained increase in the DAG level in cultured hepatocytes, and that the protein kinase C inhibitor, GF109203X, almost abolishes the enhancement by these agents of EGF-stimulated DNA synthesis [151]. While the DAG/PKC pathway was not detectably activated by EGF (see above), an optimal effect of EGF was dependent on the permissive effect of DAG. Thus, synthesis of DAG with activation of PKC appears to be one of the mechanisms underlying the comitogenic effect of several agents that activate the cells early in G_1.

The fact that norepinephrine and $PGF_{2\alpha}$ are still able to stimulate DAG accumulation and enhance DNA synthesis under conditions where the $InsP_3$ response to these agonists is completely down-regulated, suggests that PIP_2 is unlikely to be the source of the sustained DAG accumulation, although presently we cannot exclude the possibility that there is some breakdown of PIP_2, with a resultant long-term accumulation of DAG, in the absence of measurable increases in $InsP_3$ mass. Previous studies in freshly isolated hepatocytes have suggested that this may occur [228]. However, sustained elevation of DAG is believed to result largely from phosphatidylcholine (PC) breakdown [229]. The best characterized route of DAG generation from PC involves PLD, forming phosphatidic acid which can then be converted to DAG by phosphatidate phosphohydrolase [230]. Whereas all the hormones that elevate DAG also stimulate PLD activity, it seems unlikely that PLD activation produces the sustained elevation of DAG, since the PLD activity is rapidly desensitized [151,228,231]. DAG might also be formed directly from PC by PC-specific PLC (although the existence of PC-PLC in mammalian cells is disputed) or by sphingomyelin synthase activity [232].

Other mechanisms are likely to be involved, and it is also possible that vasopressin, angiotensin II, and norepinephrine may act through partly distinct mechanisms, in accordance with the finding that the growth-stimulatory effects of these agonists are largely additive [40,56,57]. Previous work has suggested that in hepatocytes the receptors of various Ca^{2+}-mobilizing agents differ qualitatively in their interaction with G proteins [233] and that they mediate dissimilar effects on growth-related gene expression [147].

15.5.5 The adenylyl cyclase/cAMP pathway mediates bidirectional growth modulation in hepatocytes

Since both glucagon receptors [234,235] and β-adrenoceptors [236] activate adenylyl cyclase in hepatocytes, the dual effects on DNA synthesis by glucagon and β-adrenergic stimulation clearly implicate cAMP as a bidirectional regulator of hepatocyte growth. This has been substantiated by the use of various agents that stimulate adenylyl cyclase, or of cAMP analogs that activate protein kinase A [29,30,152]. The cAMP effects modulate the response to mitogens such as EGF and typically are growth-promoting if the cAMP pathway is activated in early G_1 and strongly inhibitory if this occurs late in G_1.

The stimulatory effect is usually elicited at quite small elevations of intracellular cAMP or low concentrations of exogenous cAMP analogs [29,30], but under certain conditions higher levels are required. For example, the depression of DNA synthesis caused by elevated glucose in the medium is reversed by cAMP, and in this case larger increases in the intracellular concentration of the nucleotide are optimal [127]. The growth-promoting effect of cAMP in cultured hepatocytes is enhanced by insulin [136,152,237] and is more prominent at low cell density [29]. The mechanisms by which cAMP augments growth responses are not clear. Unlike the phospholipid-mediated pathways discussed above, cAMP does not activate ERK in hepatocytes [238]. The growth-augmenting effect of cAMP might reflect more general trophic actions. It is of interest, for example, that cAMP may decrease autophagic protein degradation in hepatocytes [239,240].

The cAMP-mediated inhibition of S phase entry exerted late in G_1 is dose dependent at physiological levels of adenylyl cyclase-stimulating agonists or corresponding concentrations of activating cAMP analogs [28,30,46,136,152]. The specificity of the effect has been further documented by showing that the inhibition is mimicked by microinjection of free catalytic (CIα) subunit of PKA and that this is counteracted by coinjection of stoichiometrically matching amounts of regulatory (RIα) subunit [152]. The inhibitory effect of cAMP on the DNA synthesis is synergistically enhanced by pretreatment with glucocorticoid [28,136,152] and favored by high cell density [28,29]. The inhibition is reversible [28,30,46,152], at least partly due to downregulation of PKA [152]. The mechanism of the inhibition by cAMP is not clear. Dibutyryl cAMP reduces retinoblastoma protein phosphorylation in hepatocytes late in G_1 [241].

15.5.6 The role of prostaglandins

Two quite different questions concerning the function of prostaglandins in hepatocyte growth have received particular attention. Do prostaglandins serve as mediators for other agonists? And, by which mechanisms do they act to exert their growth-promoting or comitogenic effects?

Do prostanoids mediate growth stimulation by other agonists?

It is well known that in the intact liver various eicosanoids are produced in non-parenchymal cells and may be released in response to nervous stimuli or inflammatory cytokines; and thus may stimulate hepatocytes as part of complex paracrine regulations in the liver. However, it has also been proposed that prostaglandins may be involved in autocrine mechanisms, by participating in signaling pathways of other agonists that act on the hepatocytes. In particular, prostaglandins have been suggested to mediate effects of mitogenic growth factors [54,55,177,207,242]. This idea is based on studies showing that EGF and TGF-α [54,55,242] or HGF [177,207] cause release of prostaglandins from hepatocytes and that cyclooxygenase inhibitors prevent the stimulation by these growth factors of DNA synthesis and c-*myc* expression. On the other hand, combination experiments have shown that exogenous PGE_2, $PGF_{2\alpha}$ and several other prostaglandins amplify the effect of EGF [56], suggesting that the prostaglandins facilitate and synergize with, rather than mediate, the action of EGF. However, the two models are not necessarily incompatible. Thus, it is conceivable that one of the mechanisms of EGF/TGF-α and HGF is to generate eicosanoids which reinforce the effects of the growth factor, thereby making an amplifying loop. A mechanism for enhanced eicosanoid synthesis exists through activation of PLA_2 by the MAP kinases, both ERK and p38 [173].

Mechanism of growth-promoting actions of prostaglandins

Several prostanoid receptors are expressed in liver, including EP_3 and FP [243,244]. Since these may be coupled to different heterotrimeric G proteins, linked in turn to various effectors, several mechanisms theoretically may mediate the growth effects.

One possibility is stimulation via $G\alpha_s$ and adenylyl cyclase and thus involvement of the cAMP/PKA-dependent mechanisms. cAMP has often been implicated in trophic or cytoprotective effects of prostaglandins, particularly PGE_2 and PGI_2. However, our results have not supported a role of cAMP in the growth-promoting effects of PGE_2 or other prostaglandins.

No significant cAMP rise is elicited in rat hepatocytes in response to PGE_2 or any other prostaglandin (even PGI_2 and PGD_2, which elevate cAMP in many other cells) unless special precautions are taken, including the blocking of G_i by pertussis toxin [245]. Under conditions where several prostaglandins enhance DNA synthesis, no cAMP elevation was detected [57]. Furthermore, the comitogenic stimulation exerted by PGE_2 and $PGF_{2\alpha}$ was markedly greater than stimulation by a series of agents that elevate or mimic cAMP, and the effects of various prostaglandins on the DNA synthesis were fully additive with those of cAMP [57]. A second possibility is coupling through $G\alpha_q$ to PI-PLCβ, with activation of the $InsP_3/Ca^{2+}$ and DAG/PKC pathways. Several prostaglandins moderately increase $InsP_3$ and single cell measurement shows elevated $[Ca^{2+}]_i$ in 100% of the hepatocytes [57]. However, when compared for example with vasopressin and angiotensin II, the growth-enhancing efficacy of PGE_2 and $PGF_{2\alpha}$ is markedly larger while the $InsP_3$ responses they elicit are much smaller. Therefore, G_q/PLCβ activation alone probably does not explain the comitogenic effect of prostaglandins. Finally, G_i is strongly activated by several prostaglandins in hepatocytes [245] and pertussis toxin inhibits prostaglandin-induced activation. It is conceivable that G_i (through α_i and/or $\beta\gamma$) is involved in the growth-augmenting effects of prostaglandins. It has been reported that the stimulation of MAP kinase by PGE_2 is mediated by receptors of the EP_3 subtype [246], which couple with G_i. The stimulation of MAP kinase by $PGF_{2\alpha}$ is partly blocked by pertussis toxin, also implicating G_i, although this appears to reflect a more general role of G_i that also relates to several other agonists [178].

15.6 Conclusions

Hepatocytes are responsive to a broad range of hormones and growth factors that stimulate or inhibit their proliferation and this is revealed in the studies of these cells in primary culture. There is extensive synergism, or in some cases antagonism, between different hormonal agents. The pattern of responsiveness is consistent with the mechanisms of action of separate hormones and the changes observed in gene expression during the prereplicative period. A simplified working model is given in Figure 3.

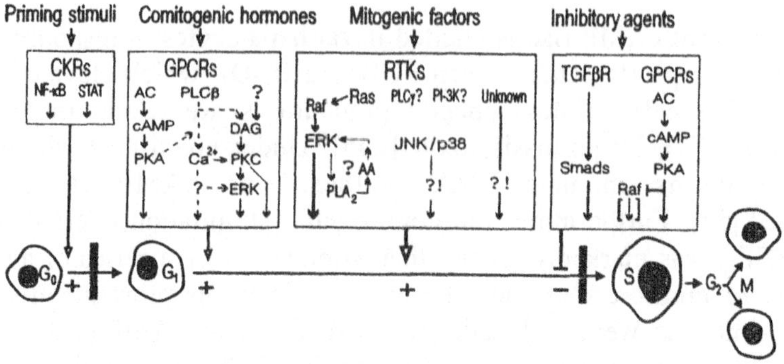

Figure 3. Tentative model of some of the mechanisms that may be involved in hormonal regulation at distinct stages of the hepatocyte traverse through the prereplicative period from G_0 to S phase *in vitro*. Specific points: Early in G_1 the cells are sensitive to various factors, including a number of hormones that act via heptahelical (G protein-coupled) receptors (GPCRs), which are permissive or facilitating (comitogenic) for subsequent mitogen stimulation by factors such as EGF/TGF-α and HGF that act via receptor tyrosine kinases (RTKs). The GPCRs-activating hormones stimulate pathways that eventually produce gene activation which accelerate cell cycle progression and increase the responsiveness to the mitogens. The mitogenic growth factors seem not to be required until mid/late G_1 in the cultured hepatocytes. One aspect of the synergism between various hormones and EGF/TGF-α is that the EGF receptor in hepatocytes apparently is more or less uncoupled from some of its potential signal mechanisms, such as PLCγ/PKC. Thus, activation through the EGF receptor is dependent on preexposure to additional hormones to yield maximal effect. This creates a situation with a flexible and complex interdependence of several hormones and growth factors that are involved in hepatocyte growth regulation. For further details see the text.

Work on hepatocytes has in the past led to the discovery of several novel mechanisms in cell regulation and continues to raise important questions for further research. This, combined with many novel experimental strategies, including improved methods for genetic manipulation, makes it probable that hepatocytes will continue to provide a valuable model for the study of hepatic growth and general biology of cell proliferation.

Acknowledgements

Work in the authors' laboratory referred to here has been supported by grants from The Norwgian Cancer Society, The Research Council of Norway, and Novo Nordisk Foundation. We thank Eva Østby for skilful hepatocyte isolation and culturing over many years.

Magne Refsnes' present address is: National Institute of Public Health, N-0403, Oslo, Norway.

References

1. Higgins GM and Anderson RM. Experimental pathology of the liver. I: Restoration of the liver of the white rat following partial surgical removal. Arch Pathol 1931; 12: 186-202.
2. Brues AM, Drury DR and Brues MC. A quantitative study of cell growth in regenerating liver. Arch Pathol 1936; 22: 658-673.
3. Pack GT, Islami AH, Hubbard JC and Brasfield RD. Regeneration of human liver after major hepatectomy. Surgery 1962; 52: 617-623.
4. Rous P and Larimore LD. Relation of the portal blood to liver maintenance. J Exp Med 1920; 31: 609-632.
5. Bucher NLR and Scott J. Regeneration of liver in parabiotic rats. Cancer Res 1951; 11: 457-465.
6. Saetren H. A principle of autoregulation of growth: Production of organ-specific mitosis inhibitors in kidney and liver. Exp Cell Res 1956; 11: 229-232.
7. Younger LR, King J and Steiner DF. Hepatic proliferative response to insulin in severe alloxan diabetes. Cancer Res 1966; 26: 1408-1414.
8. Moolten FL and Bucher NLR. Regeneration of rat liver: Transfer of humoral agent by cross circulation. Science 1967; 158: 272-274.
9. Simek J, Chmelar VL, Melka J, Pazderka J and Charvat Z. Influence of protracted infusion of glucose and insulin on the composition and regeneration activity of liver after partial hepatectomy in rats. Nature 1967; 213: 910-911.
10. Berry MN and Friend DS. High-yield preparation of isolated rat liver parenchymal cells. J Cell Biol 1969; 43: 506-520.
11. Seglen PO. Preparation of isolated rat liver cells. Methods Cell Biol 1976; 13: 29-83.
12. Bucher NLR and Malt RA. Regeneration of the Liver and Kidney. Boston: Little Brown, 1971.
13. Alison MR. Regulation of hepatic growth. Physiol Rev 1986; 66: 499-541.
14. Steer CJ. Liver regeneration. FASEB J 1995; 9: 1396-1400.
15. Michalopoulos GK and Defrances MC. Liver regeneration. Science 1997; 276: 60-66.
16. Starzl TE and Terblanche J. Hepatotrophic substances. Progr Liver Dis 1979; 6: 135-151.
17. Michalopoulos GK. Liver regeneration: Molecular mechanisms of growth control. FASEB J 1990; 4: 176-187.
18. Bucher NLR and Strain AJ. Regulatory mechanisms in hepatic regeneration. In: Milward-Sadler GH, Wright R and Arthur MJP, Eds. Wright's liver and biliary disease. London: WB Saunders, 1992: 258-259.
19. Strain AJ. Isolated hepatocytes: Use in experimental and clinical hepatology. Gut 1994; 35: 433-436.
20. Kay MA and Fausto N. Liver regeneration: prospects for therapy based on new technologies. Mol Med Today 1997; 3: 108-115.
21. Grisham JW, Leong GF, Albright ML and Emerson JD. Effect of exchange transfusion on labeling of nuclei with thymidine-3H and on mitosis in hepatocytes of normal and regenerating rat liver. Cancer Res 1966; 26: 1476-1485.
22. Richman RA, Claus TH, Pilkis SJ and Friedman DL. Hormonal stimulation of DNA synthesis in primary cultures of adult rat hepatocytes. Proc Natl Acad Sci USA 1976; 73: 3589-3593.
23. McGowan JA, Strain AJ and Bucher NLR. DNA synthesis in primary cultures of adult rat hepatocytes in a defined medium: Effects of epidermal growth factor, insulin, glucagon, and cyclic AMP. J Cell Physiol 1981; 108: 353-363.
24. Friedman DL, Claus TH, Pilkis SJ and Pine GE. Hormonal regulation of DNA synthesis in primary cultures of adult rat hepatocytes. Exp Cell Res 1981; 135: 283-290.

25. Sand TE, Brønstad G, Digernes V, Killi A, Amara W, Refsnes M and Christoffersen T. Quantitative aspects of the effects of insulin, epidermal growth factor, and dexamethasone on DNA synthesis in cultured adult rat hepatocytes. Acta Endocrinol 1985; 109: 369-377.
26. Sand TE and Christoffersen T. Temporal recquirement for epidermal growth factor and insulin in the stimulation of hepatocyte DNA synthesis. J Cell Physiol 1987; 131: 141-148.
27. Rother KI, Imai Y, Caruso M, Beguinot F, Formisano P and Accili D. Evidence that IRS-2 phosphorylation is required for insulin action in hepatocytes. J Biol Chem 1998; 273: 17491-17497.
28. Vintermyr OK, Mellgren G, Bøe R and Døskeland SO. Cyclic adenosine monophosphate acts synergistically with dexamethasone to inhibit the entrance of cultured adult rat hepatocytes into S-phase: With a note on the use of nucleolar and extranucleolar [^{3}H]-thymidine labelling patterns to determine rapid changes in the rate of onset of DNA replication. J Cell Physiol 1989; 141: 371-382.
29. Brønstad GO, Sand TE and Christoffersen T. Bidirectional concentration-dependent effects of glucagon and dibutyryl cyclic AMP on DNA synthesis in cultured adult rat hepatocytes. Biochim Biophys Acta 1983; 763: 58-63.
30. Thoresen GH, Sand TE, Refsnes M, Dajani OF, Guren TK et al. Dual effects of glucagon and cyclic AMP on DNA synthesis in cultured rat hepatocytes: Stimulatory regulation in early G1 and inhibition shortly before the S phase entry. J Cell Physiol 1990; 144: 523-530.
31. Morley CGD and Royse VL. Adrenergic agents as possible regulators of liver regeneration. Int J Biochem 1981; 13: 969-973.
32. Cruise JL, Knechtle SJ, Bollinger RR, Kuhn C and Michalopoulos GK. α1-Adrenergic effects and liver regeneration. Hepatology 1987; 7: 1189-1194.
33. Brønstad G and Christoffersen T. Increased effect of adrenaline on cyclic AMP formation and positive β-adrenergic modulation of DNA synthesis in regenerating hepatocytes. FEBS Lett 1980; 120: 89-93.
34. Cruise JL, Houck KA and Michalopoulos GK. Induction of DNA synthesis in cultured rat hepatocytes through stimulation of α1 adrenoceptors by norepinephrine. Science 1985; 227: 749-751.
35. Cruise JL and Michalopoulos GK. Norepinephrine and epidermal growth factor: Dynamics of their interaction in the stimulation of hepatocyte DNA synthesis. J Cell Physiol 1985; 125: 45-50.
36. Takai S, Nakamura T, Komi N and Ichihara A. Mechanism of stimulation of DNA synthesis induced by epinephrine in primary culture of adult rat hepatocytes. J Biochem 1988; 103: 848-852.
37. Refsnes M, Thoresen GH, Sandnes D, Dajani OF, Dajani L and Christoffersen T. Stimulatory and inhibitory effects of catecholamines on DNA synthesis in primary rat hepatocyte cultures: Role of α1- and β-adrenergic mechanisms. J Cell Physiol 1992; 151: 164-171.
38. Hasegawa K and Carr BI. Neurotensin-amplification of DNA synthesis stimulated by EGF or TGF-α in primary cultures of adult rat hepatocytes. Cell Struct Funct 1993; 18: 105-110.
39. Bhora FY, Kothary PC, Imanishi H, Eckhauser FE and Raper SE. Vasopressin stimulates DNA synthesis in cultured rat hepatocytes. J Surg Res 1994; 57: 706-710.
40. Dajani OF, Rottingen JA, Sandnes D, Horn RS, Refsnes M et al. Growth-promoting effects of Ca^{2+}-mobilizing agents in hepatocytes: Lack of correlation between the acute activation of phosphoinositide-specific phospholipase C and the stimulation of DNA synthesis by angiotensin II, vasopressin, norepinephrine, and prostaglandin $F_{2\alpha}$. J Cell Physiol 1996; 168: 608-617.

41. Kar S, Hasegawa K and Carr BI. Comitogenic effects of vasoactive intestinal polypeptide on rat hepatocytes. J Cell Physiol 1996; 168: 141-146.
42. Metcalfe AMJ, Phillips P, Dixon RM and Radda GK. Vasopressin synergistically stimulates DNA synthesis in normal and regenerating rat liver cell cultures in the presence of hepatocyte growth factor. J Mol Endocrinol 1997; 18: 161-166.
43. Balasubramanian S and Paulose CS. Induction of DNA synthesis in primary cultures of rat hepatocytes by serotonin: possible involvement of serotonin S2 receptor. Hepatology 1998; 27: 62-66.
44. Ni N and Yager JD. Comitogenic effects of estrogens on DNA synthesis induced by various growth factors in cultured female rat hepatocytes. Hepatology 1994; 19: 183-192.
45. Vintermyr OK and Døskeland SO. Characterization of the inhibitory effect of glucocorticoids on the DNA replication of adult rat hepatocytes growing at various cell densities. J Cell Physiol 1989; 138: 29-37.
46. Thoresen GH, Refsnes M and Christoffersen T. Inhibition of hepatocyte DNA synthesis by transforming growth factor β 1 and cyclic AMP: effect immediately before the G1/S border. Cancer Res 1992; 52: 3598-3603.
47. Short J, Klein K, Kibert L and Ove P. Involvement of iodothyronines in liver and hepatoma cell proliferation in the rat. Cancer Res 1980; 40: 2417-2422.
48. Francavilla A, Carr BI, Azzarone A, Polimeno L, Wang Z et al. Hepatocyte proliferation and gene expression induced by triiodothyronine *in vivo* and *in vitro*. Hepatology 1994; 20: 1237-1241.
49. MacManus JP and Braceland BM. A connection between the production of prostaglandins during liver regeneration and the DNA synthetic response. Prostaglandins 1976; 11: 609-620.
50. Miura Y and Fukui N. Prostaglandins as possible triggers for liver regeneration after partial hepatectomy. Cell Molec Biol 1979; 25: 179-184.
51. Rixon RH and Whitfield JF. An early mitosis-determining event in regenerating liver and its possible mediation by prostaglandins or thromboxane. J Cell Physiol 1982; 113: 281-288.
52. Quiroga J and Prieto J. Liver cytoprotection by prostaglandins. Pharmac Ther 1993; 58: 67-92.
53. Andreis PG, Whitfield JF and Armato U. Stimulation of DNA synthesis and mitosis of hepatocytes in primary cultures of neonatal rat liver by arachidonic acid and prostaglandins. Exp Cell Res 1981; 134: 265-272.
54. Skouteris GG, Ord MG and Stocken LA. Regulation of the proliferation of primary rat hepatocytes by eicosanoids. J Cell Physiol 1988; 135: 516-520.
55. Skouteris GG and McMenamin M. Transforming growth factor-α-induced DNA synthesis and c-myc expression in primary rat hepatocyte cultures is modulated by indomethacin. Biochem J 1992; 281: 729-733.
56. Refsnes M, Thoresen GH, Dajani OF and Christoffersen T. Stimulation of hepatocyte DNA synthesis by prostaglandin E2 and prostaglandin F2α: Additivity with the effect of norepinephrine, and synergism with epidermal growth factor. J Cell Physiol 1994; 159: 35-40.
57. Refsnes M, Dajani OF, Sandnes D, Thoresen GH, Rottingen JA, Iversen JG and Christoffersen T. On the mechanisms of the growth-promoting effect of prostaglandins in hepatocytes: the relationship between stimulation of DNA synthesis and signaling mediated by adenylyl cyclase and phosphoinositide-specific phospholipase C. J Cell Physiol 1995; 164: 465-473.
58. Hong JT and Glauert HP. Comitogenicity of eicosanoids and the peroxisome proliferator ciprofibrate in cultured rat hepatocytes. J Cell Physiol 1996; 169: 309-319.

59. DuBois RN, Abramson SB, Crofford L, Gupta RA, Simon LS et al. Cyclooxygenase in biology and disease. FASEB J 1998; 12: 1063-1073.
60. Skov Olsen P, Boesby S, Kirkegaard P, Therkelsen K, Almdal T et al. Influence of epidermal growth factor on liver regeneration after partial hepatectomy in rats. Hepatology 1988; 8: 992-996.
61. Marti U, Burwen SJ and Jones AL. Biological effects of epidermal growth factor, with emphasis on the gastrointestinal tract and liver: an update. Hepatology 1989; 9: 126-138.
62. Mead JE and Fausto N. Transforming growth factor α may be a physiological regulator of liver regeneration by means of an autocrine mechanism. Proc Natl Acad Sci 1989; 86: 1558-1562.
63. Fausto N, Laird AD and Webber EM. Liver regeneration. 2 Role of growth factors and cytokines in hepatic regeneration. FASEB J 1995; 9: 1527-1536.
64. Jhappan C, Stahle C, Harkins RN, Fausto N, Smith GH and Merlino GT. TGF-α overexpression in transgenic mice induces liver neoplasia and abnormal development of the mammary gland and pancreas. Cell 1990; 61: 1137-1146.
65. Gladhaug IP and Christoffersen T. Kinetics of epidermal growth factor binding and processing in isolated intact rat hepatocytes. Dynamic externalization of receptors during ligand internalization. Eur J Biochem 1987; 164: 267-275.
66. Sand TE and Christoffersen T. A simple medium for the study of hepatocyte growth in culture under defined conditions. *In Vitro* Cell Dev Biol 1988; 24: 981-984.
67. Wollenberg GK, Harris L, Farber E and Hayes MA. Inverse relationship between epidermal growth factor induced proliferation and expression of high affinity surface epidermal growth factor receptors in rat hepatocytes. Lab Invest 1989; 60: 254-259.
68. Gohda E, Yamasaki T, Tsubouchi H, Kurobe M, Sakiyama O et al. Biological and immunological properties of human hepatocyte growth factor from plasma of patients with fulminant hepatic failure. Biochim Biophys Acta 1990; 1053: 21-26.
69. Webber EM, FitzGerald MJ, Brown PI, Bartlett MH and Fausto N. Transforming growth factor-α expression during liver regeneration after partial hepatectomy and toxic injury, and potential interactions between transforming growth factor-α and hepatocyte growth factor. Hepatology 1993; 18: 1422-1431.
70. Guren TK, Thoresen GH, Dajani OF, Taraldsrud E, Moberg ER and Christoffersen T. Epidermal growth factor behaves as a partial agonist in hepatocytes: Effects on DNA synthesis in primary culture and competition with transforming growth factor-α. Growth Factors 1996; 13: 171-179.
71. Thoresen GH, Guren TK, Sandnes D, Peak M, Agius L and Christoffersen T. Response to transforming growth factor α (TGF-α) and epidermal growth factor (EGF) in hepatocytes: Lower EGF receptor affinity of TGF-α is associated with more sustained activation of p42/p44 mitogen-activated protein kinase and greater efficacy in stimulation of DNA synthesis. J Cell Physiol 1998; 175: 10-18.
72. Ito N, Kawata S, Tamura S, Kiso S, Tsushima H, Damm D et al. Heparin-binding EGF-like growth factor is a potent mitogen for rat hepatocytes. Biochem Biophys Res Commun 1994; 198: 25-31.
73. Michalopoulos G, Cianciulli HD, Novotny AR, Kligerman AD, Strom SJ and Jirtle RL. Liver regeneration studies with rat hepatocytes in primary culture. Cancer Res 1982; 42: 4673-4682.
74. Michalopoulos G, Houck KA, Dolan ML and Luetteke NC. Control of hepatocyte replication by two serum factors. Cancer Res 1984; 44: 4414-4419.
75. Strain AJ, McGowan JA and Bucher NLR. Stimulation of DNA synthesis in primary cultures of adult rat hepatocytes by rat platelet-associated substances. *In Vitro* 1982; 18: 108-116.
76. Russel WE, McGowan JA and Bucher NLR. Partial characterization of a hepatocyte growth factor from rat platelets. J Cell Physiol 1984; 119: 183-192.

77. Nakamura T, Teramoto H, Tomita Y and Ichihara A. Proline is an essential amino acid for hepatocyte growth in culture. Biochem Biophys Res Commun 1984; 122: 884-891.
78. Paul D and Piasecki D. Rat platelets contain growth factor(s) distinct from PDGF which stimulate DNA synthesis in primary adult rat hepatocyte cultures. Exp Cell Res 1984; 154: 95-100.
79. Gohda E, Tsubouchi H, Nakayama H, Hirono S, Takashi K et al. Human hepatocyte growth factor in plasma from patients with fulminant hepatic failure. Exp Cell Res 1986; 166: 138-150.
80. Miyazawa K, Tsubouchi H, Naka D, Takahashi K, Okigaki M, Arakaki N, Nakayama H, Hirono S, Sakiyama O, Gohda E et al. Molecular cloning and sequence analysis of cDNA for human hepatocyte growth factor. Biochem Biophys Res Commun 1989; 163: 967-973.
81. Nakamura T, Nishizawa T, Hagiya M, Seki T, Shimonishi M et al. Molecular cloning and expression of human hepatocyte growth factor. Nature 1989; 342: 440-443.
82. Zarnegar R and Michalopoulos G. Purification and biological characterization of human hepatopoietin A, a polypeptide growth factor for hepatocytes. Cancer Res 1989; 49: 3314-3320.
83. Naldini L, Weidner KM, Vigna E, Gaudino G, Bardelli A et al. Scatter factor and hepatocyte growth factor are indistinguishable ligands for the MET receptor. EMBO J 1991; 10: 2867-2878.
84. Weidner KM, Arakaki N, Hartmann G, Vandekerckhove J, Weingart S et al. Evidence for the identity of human scatter factor and human hepatocyte growth factor. Proc Natl Acad Sci USA 1991; 88: 7001-7005.
85. Bottaro DP, Rubin JS, Faletto DL, Chan AML, Kmiecik TE et al. Identification of the hepatocyte growth factor receptor as the c-Met proto-oncogene product. Science 1991; 251: 802-804.
86. Rubin JS, Bottaro DP and Aaronson SA. Hepatocyte growth factor/scatter factor and its receptor, the c- met proto-oncogene product. Biochim Biophys Acta 1993; 1155: 357-371.
87. Strain AJ. Hepatocyte growth factor: Another ubiquitous cytokine. J Endocrinol 1993; 137: 1-5.
88. Zarnegar R and Michalopoulos GK. The many faces of hepatocyte growth factor: From hepatopoiesis to hematopoiesis. J Cell Biol 1995; 129: 1177-1180.
89. Lindroos PM, Zarnegar R and Michalopoulos GK. Hepatocyte growth factor (hepatopoietin-A) rapidly increases in plasma before DNA synthesis and liver regeneration stimulated by partial hepatectomy and carbon tetrachloride administration. Hepatology 1991; 13: 743-750.
90. Strain AJ, Ismail T, Tsubouchi H, Arakaki N, Hishida T et al. Native and recombinant human hepatocyte growth factors are highly potent promoters of DNA synthesis in both human and rat hepatocytes. J Clin Invest 1991; 87: 1853-1857.
91. Gómez-Lechón MJ, Castelli J, Guillén I, O'Connor E, Nakamura T et al. Effects of hepatocyte growth factor on the growth and metabolism of human hepatocytes in primary culture. Hepatology 1995; 21: 1248-1254.
92. Kan M, Huang J, Mansson PE, Yasumitsu H, Carr B and McKeehan WL. Heparin-binding growth factor type 1 (acid fibroblast growth factor): A potential biphasic autocrine and paracrine regulator of hepatocyte regeneration. Proc Natl Acad Sci USA 1989; 86: 7432-7436.
93. Houck KA, Zarnegar R, Muga SJ and Michalopoulos GK. Acidic fibroblast growth factor (HBGF-1) stimulates DNA synthesis in primary rat hepatocyte cultures. J Cell Physiol 1990; 143: 129-132.

94. Strain AJ, McGuinness G, Rubin JS and Aaronson SA. Keratinocyte growth factor and fibroblast growth factor action on DNA synthesis in rat and human hepatocytes: Modulation by heparin. Exp Cell Res 1994; 210: 253-259.
95. Tanahashi T, Suzuki M, Imamura T and Mitsui Y. Identification of a 79-kDa heparin-binding fibroblast growth factor (FGF) receptor in rat hepatocytes and its correlation with the different growth responses to FGF-1 between hepatocyte subpopulations. J Biol Chem 1996; 271: 8221-8227.
96. Andus T, Bauer J and Gerok W. Effects of cytokines on the liver. Hepatology 1991; 13: 364-375.
97. Nakamura T, Arakaki R and Ichihara A. Interleukin-1 β is a potent growth inhibitor of adult rat hepatocytes in primary culture. Exp Cell Res 1988; 179: 488-497.
98. Huggett AC, Ford CP and Thorgeirsson SS. Effects of interleukin-6 on the growth of normal and transformed rat liver cells in culture. Growth Factors 1989; 2: 83-89.
99. Satoh M and Yamazaki M. Tumor necrosis factor stimulates DNA synthesis of mouse hepatocytes in primary culture and is suppressed by transforming growth factor β and interleukin-6. J Cell Physiol 1992; 150: 134-139.
100. Watanabe Y, Osaki H and Akaike T. TNF-α bifunctionally induces proliferation in primary hepatocytes: role of cell anchorage and spreading. J Immunol 1997; 159: 4840-4847.
101. Beyer HS. Tumor necrosis factor-α is a direct hepatocyte mitogen in the rat. Biochem Int 1993; 29: 1-4.
102. Lawrence DA. Transforming growth factor-β: a general review. Eur Cytokine Netw 1996; 7: 363-374.
103. Nakamura T, Tomita Y, Hirai R, Yamaoka K, Kaji K and Ichihara A. Inhibitory effect of transforming growth factor-β on DNA synthesis of adult rat hepatocytes in primary culture. Biochem Biophys Res Commun 1985; 133: 1042-1050.
104. Carr BI, Hayashi B, Branum EL and Moses HL. Inhibition of DNA synthesis in rat hepatocytes by platelet-derived type β transforming growth factor. Cancer Res 1986; 46: 2230-2334.
105. Wollenberg GK, Semple E, Quinn BA and Hayes MA. Inhibition of proliferation of normal, preneoplastic, and neoplastic rat hepatocytes by transforming growth factor-β. Cancer Res 1987; 47: 6595-6599.
106. Wollenberg GK, Lamarre J, Semple E, Farber E, Gauldie J and Hayes MA. Counteracting effects of dexamethasone and α2-macroglobulin on inhibition of proliferation of normal and neoplastic rat hepatocytes by transforming growth factors-β type-1 and type-2. Int J Cancer 1991; 47: 311-316.
107. Russel WE, Coffey JR, Ouellette AJ and Moses HL. Type β transforming growth factor reversibly inhibits the early proliferative response to partial hepatectomy in the rat. Proc Natl Acad Sci USA 1988; 85: 5126-5130.
108. Fausto N, Mead JE, Gruppuso PA, Castilla A and Jakowlew SB. Effects of TGF-βs in the liver: cell proliferation and fibrogenesis. Ciba Found Symp 1991; 157: 165-177.
109. Yasuda H, Mine T, Shibata H, Eto Y, Hasegawa Y, Takeuchi T et al. Activin A: An autocrine inhibitor of initiation of DNA synthesis in rat hepatocytes. J Clin Invest 1993; 92: 1491-1496.
110. Christoffersen T, Refsnes M, Brønstad GO, Østby E, Huse J et al. Changes in hormone responsiveness and cyclic AMP metabolism in rat hepatocytes during primary culture and effects of supplementing the medium with insulin and dexamethasone. Eur J Biochem 1984; 138: 217-226.
111. Schwarz KR, Lanier SM, Carter EA, Homcy CJ and Graham RM. Rapid reciprocal changes in adrenergic receptors in intact isolated hepatocytes during primary cell culture. Mol Pharmacol 1985; 27: 200-209.

112. Bouscarel B, Augert G, Taylor SJ and Exton JH. Alterations in vasopressin and angiotensin-II receptors and responses during culture of rat liver cells. Biochim Biophys Acta 1990; 1055: 265-272.
113. Gladhaug IP, Refsnes M and Christoffersen T. Regulation of surface expression of high-affinity receptors for epidermal growth factor (EGF) in hepatocytes by hormones, differentiating agents, and phorbol ester. Dig Dis Sci 1992; 37: 233-239.
114. Earp HS and O'Keefe EJ. Epidermal growth factor receptor number decreases during rat liver regeneration. J Clin Invest 1981; 67: 1580-1583.
115. Sandnes D, Sand TE, Sager G, Brønstad GO, Refsnes MR et al. Elevated level of β-adrenergic receptors in hepatocytes from regenerating rat liver: Time study of [^{125}I]iodocyanopindolol binding following partial hepatectomy and its relationship to catecholamine-sensitive adenylate cyclase. Exp Cell Res 1986; 165: 117-126.
116. Refsnes M, Sager G, Sandnes D, Sand TE, Jacobsen S and Christoffersen T. Increased number of β-adrenoceptors in hepatocytes from rats treated with 2-acetylaminofluorene. Cancer Res 1986; 46: 2285-2288.
117. Kruijer W, Skelly H, Botteri F, van der Putten H, Barber JR, Verma IM and Leffert HL. Proto-oncogene expression in regenerating liver is simulated in cultures of primary adult rat hepatocytes. J Biol Chem 1986; 261: 7929-7933.
118. Loyer P, Cariou S, Glaise D, Bilodeau M, Baffet G and Guguen-Guillouzo C. Growth factor dependence of progression through G1 and S phases of adult rat hepatocytes *in vitro* - Evidence of a mitogen restriction point in mid-late G1. J Biol Chem 1996; 271: 11484-11492.
119. Sobczak J, Mechti N, Tournier MF, Blanchard JM and Duguet M. c-myc and c-fos gene regulation during mouse liver regeneration. Oncogene 1989; 4: 1503-1508.
120. Taub R. Liver regeneration 4: Transcriptional control of liver regeneration. FASEB J 1996; 10: 413-427.
121. Skouteris GG and Kaser MR. Expression of exogenous c-myc oncogene does not initiate DNA synthesis in primary rat hepatocyte cultures. J Cell Physiol 1992; 150: 353-359.
122. Zindy F, Lamas E, Chenivesse X, Sobczak J, Wang J, Fesquet D, Henglein B and Brechot C. Cyclin A is required in S phase in normal epithelial cells. Biochem Biophys Res Commun 1992; 182: 1144-1154.
123. Kao CY, Factor VM and Thorgeirsson SS. Reduced growth capacity of hepatocytes from c-myc and c-myc/TGF-α transgenic mice in primary culture. Biochem Biophys Res Commun 1996; 222: 64-70.
124. McGowan JA. Hepatocyte proliferation in culture. In: Guillouzo A and Guguen-Guillouzo C, Eds. Research in isolated and cultured hepatocytes. London: John Libbey Eurotex Ltd./INSERM, 1986: 13-38.
125. McGowan JA and Bucher NLR. Pyruvate promotion of DNA synthesis in serum-free primary cultures of adult rat hepatocytes. *In Vitro* 1983; 19: 159-166.
126. Li Y, Sattler GL and Pitot HC. Oxaloacetate induces DNA synthesis and mitosis in primary cultured rat hepatocytes in the absence of EGF. Biochem Biophys Res Commun 1993; 193: 1339-1346.
127. Dajani OF, Refsnes M, Guren TK, Horn RS, Thoresen GH and Christoffersen T. Elevated glucose concentrations inhibit DNA synthesis and expression of c-myc in cultured hepatocytes. Biochem Biophys Res Commun 1994; 202: 1476-1482.
128. Enat R, Jefferson DM, Ruiz-Opazo N, Gatmaitan Z, Leinwand LA and Reid LM. Hepatocyte proliferation *in vitro*: its dependence on the use of serum-free hormonally defined medium and substrata of extracellular matrix. Proc Natl Acad Sci USA 1984; 81: 1411-1415.
129. Sawada N, Tomomura A, Sattler CA, Kleinman HK and Pitot HC. Extracellular matrix components influence DNA synthesis of rat hepatocytes in primary culture. Exp Cell Res 1986; 167: 458-470.

130. Rana B, Mischoulon D, Xie Y, Bucher NL and Farmer SR. Cell-extracellular matrix interactions can regulate the switch between growth and differentiation in rat hepatocytes: reciprocal expression of C/EBP α and immediate-early growth response transcription factors. Mol Cell Biol 1994; 14: 5858-5869.
131. Block GD, Locker J, Bowen WC, Petersen BE, Katyal S et al. Population expansion, clonal growth, and specific differentiation patterns in primary cultures of hepatocytes induced by HGF/SF, EGF and TGF-α in a chemically defined (HGM) medium. J Cell Biol 1996; 132: 1133-1149.
132. Gladhaug IP, Refsnes M, Sand TE and Christoffersen T. Effects of butyrate on epidermal growth factor receptor binding, morphology, and DNA synthesis in cultured rat hepatocytes. Cancer Res 1988; 48: 6560-6564.
133. Chan K, Kost DP and Michalopoulos G. Multiple sequential periods of DNA synthesis and quiescence in primary hepatocyte cultures maintained on the DMSO-EGF on/off protocol. J Cell Physiol 1989; 141: 584-590.
134. Vintermyr OK and Døskeland SO. Cell cycle parameters of adult rat hepatocytes in a defined medium. A note on the timing of nucleolar DNA replication. J Cell Physiol 1987; 132: 12-21.
135. Nakamura T, Yoshimoto K, Nakayama Y, Tomita Y and Ichihara A. Reciprocal modulation of growth and differentiated functions of mature rat hepatocytes in primary culture by cell-cell contact and cell membranes. Proc Natl Acad Sci USA 1983; 80: 7229-7233.
136. Thoresen GH, Refsnes M, Dajani OF, Johansen EJ and Christoffersen T. Dexamethasone inversely regulates DNA synthesis and phosphoenolpyruvate carboxykinase mRNA levels in cultured rat hepatocytes: interactions with insulin, glucagon, and transforming growth factor β1. Pharmacol Toxicol 1995; 76: 163-170.
137. Norbury C and Nurse P. Animal cell cycles and their control. Annu Rev Biochem 1992; 61: 441-470.
138. Mead JE, Braun L, Martin DA and Fausto N. Induction of replicative competence ("priming") in normal liver. Cancer Res 1990; 50: 7023-7030.
139. Webber EM, Godowski PJ and Fausto N. *In vivo* response of hepatocytes to growth factors requires an initial priming stimulus. Hepatology 1994; 19: 489-497.
140. Diehl AM and Rai RM. Liver regeneration 3: Regulation of signal transduction during liver regeneration. FASEB J 1996; 10: 215-227.
141. Cressman DE, Greenbaum LE, DeAngelis RA, Ciliberto G, Furth EE et al. Liver failure and defective hepatocyte regeneration in interleukin-6-deficient mice. Science 1996; 274: 1379-1383.
142. Webber EM, Bruix J, Pierce RH and Fausto N. Tumor necrosis factor primes hepatocytes for DNA replication in the rat. Hepatology 1998; 28: 1226-1234.
143. Etienne PL, Baffet G, Desvergne B, Boisnard-Rissel M, Glaise D and Guguen-Guillouzo C. Transient expression of c-fos and constant expression of c-myc in freshly isolated and cultured normal adult rat hepatocytes. Oncogene Res 1988; 3: 255-262.
144. Sawada N. Hepatocytes from old rats retain responsiveness of c-myc expression to EGF in primary culture but do not enter S phase. Exp Cell Res 1989; 181: 584-588.
145. Ikeda T, Sawada N, Fujinaga K, Minase T and Mori M. H-ras gene is expressed at the G1 phase in primary cultures of hepatocytes. Exp Cell Res 1989; 185: 292-296.
146. Weir E, Chen Q, Defrances MC, Bell A, Taub R and Zarnegar R. Rapid induction of mRNAs for liver regeneration factor and insulin-like growth factor binding protein-1 in primary cultures of rat hepatocytes by hepatocyte growth factor and epidermal growth factor. Hepatology 1994; 20: 955-960.
147. Gonzalez-Espinosa C and Garcia-Sainz JA. Angiotensin II and active phorbol esters induce proto-oncogene expression in isolated rat hepatocytes. Biochim Biophys Acta 1992; 1136: 309-314.

148. Skouteris GG and Schröder CH. C-myc is required for the G0/G1-S transition of primary hepatocytes stimulated with a deleted form of hepatocyte growth factor. Biochem J 1996; 316: 879-886.
149. Goalstone ML and Draznin B. What does insulin do to Ras? Cell Signal 1998; 10: 297-301.
150. Chan SO, Wong SSC and Yeung DCY. Insulin induction of c-Ki-ras in rat liver and in cultured normal rat hepatocytes. Comp Biochem Physiol 1993; 104B: 341-347.
151. Dajani OF, Sandnes D, Melien O, Rezvani F, Nilssen LS et al. The role of diacylglycerol in hormonal induction of S phase in hepatocytes: The diacylglycerol-dependent protein kinase C pathway is not activated by epidermal growth factor (EGF), but is involved in the enhancement of responsiveness to EGF by vasopressin, angiotensin II, and norepinephrine. J Cell Physiol 1999; 180: 203-214.
152. Vintermyr OK, Boe R, Bruland T, Houge G and Døskeland SO. Elevated cAMP gives short-term inhibition and long-term stimulation of hepatocyte DNA replication: Roles of the cAMP- dependent protein kinase subunits. J Cell Physiol 1993; 156: 160-170.
153. Mellgren G, Vintermyr OK and Doskeland SO. Okadaic acid, cAMP, and selected nutrients inhibit hepatocyte proliferation at different stages in G1: modulation of the cAMP effect by phosphatase inhibitors and nutrients. J Cell Physiol 1995; 163: 232-240.
154. Okajima F and Ui M. Conversion of adrenergic regulation of glycogen phosphorylase and synthase from an α to a β type during primary culture of rat hepatocytes. Arch Biochem Biophys 1982; 213: 658-668.
155. Nakamura T, Tomomura A, Noda C, Shimoji M and Ichihara A. Acquisition of a β-adrenergic response by adult rat hepatocytes during primary culture. J Biol Chem 1983; 258: 9283-9289.
156. Refsnes M, Sandnes D, Melien O, Sand TE, Jacobsen S and Christoffersen T. Mechanisms for the emergence of catecholamine-sensitive adenylate cyclase and β-adrenergic receptors in cultured hepatocytes. Dependence on protein and RNA synthesis and suppression by isoproterenol. FEBS Lett 1983; 164: 291-298.
157. Christoffersen T and Berg T. Altered hormone control of cyclic AMP formation in isolated parenchymal liver cells from rats treated with 2-acetylaminofluorene. Biochim Biophys Acta 1975; 381: 72-77.
158. Nishikawa Y, Wang MF and Carr BI. Changes in TGF-β receptors of rat hepatocytes during primary culture and liver regeneration: Increased expression of TGF-β receptors associated with increased sensitivity to TGF-β-mediated growth inhibition. J Cell Physiol 1998; 176: 612-623.
159. Noda C and Ichihara A. Regulation of liver-specific gene expression. Cell Struct Funct 1993; 18: 189-194.
160. Xanthopoulos KG and Mirkovitch J. Gene regulation in rodent hepatocytes during development, differentiation and disease. Eur J Biochem 1993; 216: 353-360.
161. Diehl AM. Roles of CCAAT/enhancer-binding proteins in regulation of liver regenerative growth. J Biol Chem 1998; 273: 30843-30846.
162. Mischoulon D, Rana B, Bucher NL and Farmer SR. Growth-dependent inhibition of CCAAT enhancer-binding protein (C/EBP α) gene expression during hepatocyte proliferation in the regenerating liver and in culture. Mol Cell Biol 1992; 12: 2553-2560.
163. Rana B, Xie Y, Mischoulon D, Bucher NLR and Farmer SR. The DNA binding activity of C/EBP transcription factors is regulated in the G_1 phase of the hepatocyte cell cycle. J Biol Chem 1995; 270: 18123-18132.
164. Nagy P, Kiss A, Schnur J and Thorgeirsson SS. Dexamethasone inhibits the proliferation of hepatocytes and oval cells but not bile duct cells in rat liver. Hepatology 1998; 28: 423-429.

165. Baribault H and Marceau N. Dexamethasone and dimethylsulfoxide as distinct regulators of growth and differentiation of cultured sucling rat hepatocytes. J Cell Physiol 1986; 129: 77-84.
166. Roussel MF. Key effectors of signal transduction and G1 progression. Adv Cancer Res 1998; 74: 1-24.
167. Berridge MJ. Calcium signalling and cell proliferation. Bioessays 1995; 17: 491-500.
168. Post GR and Brown JH. G protein-coupled receptors and signaling pathways regulating growth responses. FASEB J 1996; 10: 741-749.
169. Gutkind JS. The pathways connecting G protein-coupled receptors to the nucleus through divergent mitogen-activated protein kinase cascades. J Biol Chem 1998; 273: 1839-1842.
170. Van der Geer P, Hunter T and Lindberg RA. Receptor protein-tyrosine kinases and their signal transduction pathways. Annu Rev Cell Biol 1994; 10: 251-337.
171. Leaman DW, Leung S, Li X and Stark GR. Regulation of STAT-dependent pathways by growth factors and cytokines. FASEB J 1996; 10: 1578-1588.
172. Hill CS and Treisman R. Transcriptional regulation by extracellular signals: Mechanisms and specificity. Cell 1995; 80: 199-211.
173. Lewis TS, Shapiro PS and Ahn NG. Signal transduction through MAP kinase cascades. Adv Cancer Res 1998; 74: 49-139.
174. Peak M, Yeaman SJ and Agius L. Epidermal growth factor and insulin stimulate MAP kinase activity in cultured hepatocytes. Biochem Soc Trans 1993; 21: S494.
175. Boylan JM and Gruppuso PA. *In vitro* and *in vivo* regulation of hepatic mitogen-activated protein kinases in fetal rats. Am J Physiol 1994; 267: G1078-G1086.
176. Gines P, Li X, Brown SES, Nakamura T, Guzelian PS, Heasley LE et al. Inhibitory actions of cyclic adenosine monophosphate and pertussis toxin define two distinct epidermal growth factor-regulated pathways leading to activation of mitogen-activated protein kinase in rat hepatocytes. Hepatology 1996; 23: 1167-1173.
177. Adachi T, Nakashima S, Saji S, Nakamura T and Nozawa Y. Mitogen-activated protein kinase activation in hepatocyte growth factor-stimulated rat hepatocytes: Involvement of protein tyrosine kinase and protein kinase C. Hepatology 1996; 23: 1244-1253.
178. Melien O, Thoresen GH, Sandnes D, Ostby E and Christoffersen T. Activation of p42/p44 mitogen-activated protein kinase by angiotensin II, vasopressin, norepinephrine, and prostaglandin F2α in hepatocytes is sustained, and like the effect of epidermal growth factor, mediated through pertussis toxin-sensitive mechanisms. J Cell Physiol 1998; 175: 348-358.
179. Spector MS, Auer KL, Jarvis WD, Ishac EJN, Gao B, Kunos G and Dent P. Differential regulation of the mitogen-activated protein and stress-activated protein kinase cascades by adrenergic agonists in quiescent and regenerating adult rat hepatocytes. Mol Cell Biol 1997; 17: 3556-3565.
180. Jarvis WD, Auer KL, Spector M, Kunos G, Grant S et al. Positive and negative regulation of JNK1 by protein kinase C and p42MAP kinase in adult rat hepatocytes. FEBS Lett 1997; 412: 9-14.
181. Auer KL, Contessa J, Brenz-Verca S, Pirola L, Rusconi S et al. The Ras/Rac1/Cdc42/SEK/JNK/c-Jun cascade is a key pathway by which agonists stimulate DNA synthesis in primary cultures of rat hepatocytes. Mol Biol Cell 1998; 9: 561-573.
182. Chen J, Ishac EJ, Dent P, Kunos G and Gao B. Effects of ethanol on mitogen-activated protein kinase and stress-activated protein kinase cascades in normal and regenerating liver. Biochem J 1998; 334: 669-676.
183. Kitano S, Bohr VA, Reed TD, Haggerty CM, May A and Roth GS. Effect of aging on EGF-stimulated replication of specific genes in rat hepatocytes. J Cell Physiol 1998; 176: 32-39.

184. Liu Y, Guyton KZ, Gorospe M, Xu Q, Kokkonen GC, Mock YD, Roth GS and Holbrook NJ. Age-related decline in mitogen-activated protein kinase activity in epidermal growth factor-stimulated rat hepatocytes. J Biol Chem 1996; 271: 3604-3607.
185. Kitano S, Fawcett TW, Yo Y and Roth GS. Molecular mechanisms of impaired stimulation of DNA synthesis in cultured hepatocytes of aged rats. Am J Physiol 1998; 275: C146-C154
186. Liang M and Garrison JC. Epidermal growth factor activates phospholipase C in rat hepatocytes via a different mechanism from that in A431 or rat1hER cells. Mol Pharmacol 1992; 42: 743-752.
187. Yang LJ, Baffy G, Rhee SG, Manning D, Hansen CA and Williamson JR. Pertussis toxin-sensitive Gi protein involvement in epidermal growth factor-induced activation of phospholipase C-γ in rat hepatocytes. J Biol Chem 1991; 266: 22451-22458.
188. Yang L, Camoratto AM, Baffy G, Raj S, Manning DR and Williamson JR. Epidermal growth factor-mediated signaling of Gi-protein to activation of phospholipases in rat-cultured hepatocytes. J Biol Chem 1993; 268: 3739-3746.
189. Bosch F, Bouscarel B, Slaton J, Blackmore PF and Exton JH. Epidermal growth factor mimics insulin effects in rat hepatocytes. Biochem J 1986; 239: 523-530.
190. Hughes BP, Crofts JN, Auld AM, Read LC and Barritt GJ. Evidence that a pertussis-toxin-sensitive substrate is involved in the stimulation by epidermal growth factor and vasopressin of plasma-membrane Ca^{2+} inflow in hepatocytes. Biochem J 1987; 248: 911-918.
191. Johnson RM and Garrison JC. Epidermal growth factor and angiotensin II stimulate formation of inositol 1,4,5- and inositol 1,3,4-trisphosphate in hepatocytes. Differential inhibition by pertussis toxin and phorbol 12-myristate 13-acetate. J Biol Chem 1987; 262: 17285-17293.
192. Cerpovicz PF and Ochs RS. Effects of EGF on the mass of inositol 1,4,5-trisphosphate and sn (1,2)- diacylglycerol in freshly isolated rat hepatocytes: comparison with vasopressin. Biochem Biophys Res Commun 1992; 187: 1055-1062.
193. Mine T, Kojima I, Ogata E and Nakamura T. Comparison of effects of HGF and EGF on cellular calcium in rat hepatocytes. Biochem Biophys Res Commun 1991; 181: 1173-1180.
194. Tanaka Y, Hayashi N, Kaneko A, Ito T, Miyoshi E et al. Epidermal growth factor induces dose-dependent calcium oscillations in single fura-2-loaded hepatocytes. Hepatology 1992; 16: 479-486.
195. Petronijevic T and Edwards AM. Role of Ca^{2+} in stimulation of DNA synthesis by epidermal growth factor and tumor promoters in cultured rat hepatocytes. J Cell Physiol 1993; 155: 44-53.
196. Liang M and Garrison JC. The epidermal growth factor receptor is coupled to a pertussis toxin-sensitive guanine nucleotide regulatory protein in rat hepatocytes. J Biol Chem 1991; 266: 13342-13349.
197. Yang LJ, Rhee SG and Williamson JR. Epidermal growth factor-induced activation and translocation of phospholipase C-γ1 to the cytoskeleton in rat hepatocytes. J Biol Chem 1994; 269: 7156-7162.
198. Chin YE, Kitagawa M, Su WC, You ZH, Iwamoto Y and Fu XY. Cell growth arrest and induction of cyclin-dependent kinase inhibitor p21 WAF1/CIP1 mediated by STAT1. Science 1996; 272: 719-722.
199. Grandis JR, Drenning SD, Chakraborty A, Zhou MY, Zeng Q et al. Requirement of STAT3 but not STAT1 activation for epidermal growth factor receptor-mediated cell growth *in vitro*. J Clin Invest 1998; 102: 1385-1392.
200. Ruff-Jamison S, Chen K and Cohen S. Induction by EGF and interferon-γ of tyrosine phosphorylated DNA binding proteins in mouse liver nuclei. Science 1993; 261: 1733-1736.

201. Ruff-Jamison S, Zhong Z, Wen Z, Chen K, Darnell JE Jr and Cohen S. Epidermal growth factor and lipopolysaccharide activate STAT3 transcription factor in mouse liver. J Biol Chem 1994; 269: 21933-21935.
202. Ruff-Jamison S, Chen K and Cohen S. Epidermal growth factor induces the tyrosine phosphorylation and nuclear translocation of STAT 5 in mouse liver. Proc Natl Acad Sci USA 1995; 92: 4215-4218.
203. Cruise JL, Cotecchia S and Michalopoulos G. Norepinephrine decreases EGF binding in primary rat hepatocyte cultures. J Cell Physiol 1986; 127: 39-44.
204. Gladhaug IP and Christoffersen T. Rapid constitutive internalization and externalization of epidermal growth factor receptors in isolated rat hepatocytes: Monensin inhibits receptor externalization and reduces the capacity for continued endocytosis of epidermal growth factor. J Biol Chem 1988; 263: 12199-12203.
205. Decker SJ. Epidermal growth factor and transforming growth factor-α induce differential processing of the epidermal growth factor receptor. Biochem Biophys Res Commun 1990; 166: 615-621.
206. Stolz DB and Michalopoulos GK. Comparative effects of hepatocyte growth factor and epidermal growth factor on motility, morphology, mitogenesis, and signal transduction of primary rat hepatocytes. J Cell Biochem 1994; 55: 445-464.
207. Adachi T, Nakashima S, Saji S, Nakamura T and Nozawa Y. Roles of prostaglandin production and mitogen-activated protein kinase activation in hepatocyte growth factor-mediated rat hepatocyte proliferation. Hepatology 1995; 21: 1668-1674.
208. Adachi T, Nakashima S, Saji S, Nakamura T and Nozawa Y. Possible involvement of pertussis toxin-sensitive G protein in hepatocyte growth factor-induced signal transduction in cultured rat hepatocytes: Pertussis toxin treatment inhibits activation of phospholipid signaling, calcium oscillation, and mitogen-activated protein kinase. Hepatology 1997; 26: 295-300.
209. Okano Y, Mizuno K, Osada S, Nakamura T and Nozawa Y. Tyrosine phosphorylation of phospholipase Cγ in c-met/HGF receptor-stimulated hepatocytes: Comparison with HepG2 hepatocarcinoma cells. Biochem Biophys Res Commun 1993; 190: 842-848.
210. Baffy G, Yang L, Michalopoulos GK and Williamson JR. Hepatocyte growth factor induces calcium mobilization and inositol phosphate production in rat hepatocytes. J Cell Physiol 1992; 153: 332-339.
211. Osada S, Nakashima S, Saji S, Nakamura T and Nozawa Y. Hepatocyte growth factor (HGF) mediates the sustained formation of 1,2-diacylglycerol via phosphatidylcholine-phospholipase C in cultured rat hepatocytes. FEBS Lett 1992; 297: 271-274.
212. Kaneko A, Hayashi N, Tsubouchi H, Tanaka Y, Ito T et al. Intracellular calcium as a second messenger for human hepatocyte growth factor in hepatocytes. Hepatology 1992; 15: 1173-1178.
213. Osada S, Saji S, Nakamura T and Nozawa Y. Cytosolic calcium oscillations induced by hepatocyte growth factor (HGF) in single fura-2-loaded cultured hepatocytes: effects of extracellular calcium and protein kinase C. Biochim Biophys Acta 1992; 1135: 229-232.
214. Adachi T, Nakashima S, Saji S, Nakamura T and Nozawa Y. Phospholipase D activation in hepatocyte growth factor-stimulated rat hepatocytes mediates the expressions of c-*jun* and c-*fos*: Involvement of protein tyrosine kinase, protein kinase C, and Ca^{2+}. Hepatology 1996; 24: 1274-1281.
215. Ponzetto C, Bardelli A, Maina F, Longati P, Panayotou G et al. A novel recognition motif for phosphatidylinositol 3-kinase binding mediates its association with the hepatocyte growth factor/scatter factor receptor. Mol Cell Biol 1993; 13: 4600-4608.
216. Skouteris GG and Georgakopoulos E. Hepatocyte growth factor-induced proliferation of primary hepatocytes is mediated by activation of phosphatidylinositol 3- kinase. Biochem Biophys Res Commun 1996; 218: 229-233.

217. Lynch CJ, Blackmore PF, Charest R and Exton JH. Relationship between receptor binding capacity for norepinephrine, angiotensin II, and vasopressin and release of inositol trisphosphate, Ca^{2+} mobilization, and phosphorylase activation in rat liver. Mol Pharmacol 1985; 28: 93-99.
218. Athari A and Jungermann K. Direct activation by prostaglandin F2α but not thromboxane A2 of glycogenolysis via an increase in inositol 1,4,5-trisphosphate in rat hepatocytes. Biochem Biophys Res Commun 1989; 163: 1235-1242.
219. Mine T, Kojima I and Ogata E. Mechanism of prostaglandin E2-induced glucose production in rat hepatocytes. Endocrinology 1990; 126: 2831-2836.
220. Püschel GP, Kirchner C, Schroder A and Jungermann K. Glycogenolytic and antiglycogenolytic prostaglandin E2 actions in rat hepatocytes are mediated via different signalling pathways. Eur J Biochem 1993; 218: 1083-1089.
221. Cruise JL, Muga SJ, Lee YS and Michalopoulos GK. Regulation of hepatocyte growth: α-1 adrenergic receptor and ras p21 changes in liver regeneration. J Cell Physiol 1989; 140: 195-201.
222. Kajiyama Y and Ui M. Switching from α_1- to β-subtypes in adrenergic response during primary culture of adult-rat hepatocytes as affected by the cell-to-cell interaction through plasma membranes. Biochem J 1994; 303: 313-321.
223. Sandnes D, Dajani OF, Bjørneby A and Christoffersen T. Relationship between activation of phosphoinositide-apecific phospholipase C and growth stimulation by Ca^{2+}-mobilizing hormones in hepatocytes. Pharmacol Toxicol 1999; 84: 234-240.
224. Nishizuka Y. Intracellular signaling by hydrolysis of phospholipids and activation of protein kinase C. Science 1992; 258: 607-614.
225. Bocckino SB, Wilson PB and Exton JH. An early elevation of diacylglycerol and phosphatidate in regenerating liver. Biochem Biophys Res Commun 1989; 164: 290-294.
226. Banfic H, Zizak M, Divecha N and Irvine RF. Nuclear diacylglycerol is increased during cell proliferation *in vivo*. Biochem J 1993; 290: 633-636.
227. Tessitore L, Perletti GP, Sesca E, Pani P, Piccinini F and Dianzani MU. Protein kinase C isozymes during liver regeneration. Biochem Biophys Res Commun 1995; 214: 354-360.
228. Allan CJ and Exton JH. Quantification of inositol phospholipid breakdown in isolated rat hepatocytes. Biochem J 1993; 290: 865-872.
229. Exton JH. Phosphatidylcholine breakdown and signal transduction. Biochim Biophys Acta 1994; 1212: 26-42.
230. Exton JH. New developments in phospholipase D. J Biol Chem 1997; 272: 15579-15582.
231. Moehren G, Gustavsson L and Hoek JB. Activation and desensitization of phospholipase D in intact rat hepatocytes. J Biol Chem 1994; 269: 838-848.
232. Luberto C and Hannun YA. Sphingomyelin synthase, a potential regulator of intracellular levels of ceramide and diacylglycerol during SV40 transformation. Does sphingomyelin synthase account for the putative phosphatidylcholine-specific phospholipase C? J Biol Chem 1998; 273: 14550-14559.
233. Dasso LLT and Taylor CW. Interactions between Ca^{2+}-mobilizing receptors and their G proteins in hepatocytes. J Biol Chem 1994; 269: 8647-8652.
234. Christoffersen T, Mørland J, Osnes JB, Berg T, Boman D and Seglen PO. Enrichment of glucagon sensitive adenyl cyclase upon preparation of parenchymal cells from rat liver. Arch Biochem Biophys 1972; 150: 807-809.
235. Sonne O, Berg T and Christoffersen T. Binding of 125I-labeled glucagon and glucagon-stimulated accumulation of adenosine 3':5'-monophosphate in isolated intact rat hepatocytes. Evidence for receptor heterogeneity. J Biol Chem 1978; 253: 3203-3210.
236. Refsnes M, Sandnes D and Christoffersen T. The relationship between β-adrenoceptor regulation and β-adrenergic responsiveness in hepatocytes. Studies on acquisition,

desensitization and resensitization of isoproterenol-sensitive adenylate cyclase in primary culture. Eur J Biochem 1987; 163: 457-466.

237. Ito Y, Uchijima Y, Ariga M, Seki T, Takenaka A, Hakuno F, Takahashi S, Ariga T and Noguchi T. Interaction between cAMP-dependent and insulin-dependent signal pathways in tyrosine phosphorylation in primary cultures of rat hepatocytes. Biochem J 1997; 324: 379-388.

238. Thoresen GH, Johansen EJ and Christoffersen T. Effects of cAMP on ERK mitogen-activated protein kinase activity in hepatocytes do not parallel the bidirectional regulation of DNA synthesis. Cell Biol Int 1999; in press.

239. Holen I, Gordon PB, Stromhaug PE and Seglen PO. Role of cAMP in the regulation of hepatocytic autophagy. Eur J Biochem 1996; 236: 163-170.

240. Seglen PO. DNA ploidy and autophagic protein degradation as determinants of hepatocellular growth and survival. Cell Biol Toxicol 1997; 13: 301-315.

241. Okamoto Y, Kikuchi E, Matsumoto M and Nakano H. Dibutyryl cyclic AMP enhances the down-regulation of RB protein during G1 phase in the proliferating primary rat hepatocytes, but inhibits their entries into S phase and RB's phosphorylation. Biochem Biophys Res Commun 1993; 195: 84-89.

242. Skouteris GG and Kaser MRT. Prostaglandins-E2 and -F2α mediate the increase in c-myc expression induced by EGF in primary rat hepatocyte cultures. Biochem Biophys Res Commun 1991; 178: 1240-1246.

243. Neuschäfer-Rube F, Püschel GP and Jungermann K. Characterization of prostaglandin-F2a-binding sites on rat hepatocyte plasma membranes. Eur J Biochem 1993; 211: 163-169.

244. Neuschäfer-Rube F, DeVries C, Hänecke K, Jungermann K and Püschel GP. Molecular cloning and expression of a prostaglandin E_2 receptor of the $EP_{3\beta}$ subtype from rat hepatocytes. FEBS Lett 1994; 351: 119-122.

245. Melien O, Winsnes R, Refsnes M, Gladhaug IP and Christoffersen T. Pertussis toxin abolishes the inhibitory effects of prostaglandins E1, E2, I2 and F2 α on hormone-induced cAMP accumulation in cultured hepatocytes. Eur J Biochem 1988; 172: 293-297.

246. Hashimoto N, Watanabe T, Ikeda Y, Yamada H, Taniguchi S et al. Prostaglandins induce proliferation of rat hepatocytes through a prostaglandin E_2 receptor EP_3 subtype. Am J Physiol 1997; 272: G597-G604.

Chapter 16

Proliferative response and metabolic effects of growth factors in human hepatocytes

María José Gómez-Lechón[1], Isabel Guillén[1], M.J. Teresa Donato[1], Xavier Ponsoda[2], Ramiro Jover[1] and José V. Castell[1]

[1]Unidad de Hepatología Experimental, Centro de Investigación, Hospital Universitario La Fe. Avda. Ampanar 21, 46009-Valencia. Spain and [2]Departament de Parasitologia i Biologia Cel.lular, Facultat de Ciències Biològiques, Universistat de València. Avda Dr. Moliner 50, 46100 Burjassot, Spain

Key words: acute-phase proteins, cell cycle, collagenase, c-fos, c-jun, c-myc, CYP, cyclin A, cyclin B, cyclin D, cyclin E, dexamethasone, DNA synthesis, EGF, human hepatocyte, hepatocyte proliferation, HGF, insulin, S-phase, TGF-β

16.1 Introduction

Liver regeneration is the outcome of several pathologies, including not only partial hepatectomy but also liver transplantation, viral or toxic hepatitis, cirrhosis, hepatocellular carcinoma, and fulminant hepatic failure [1,2]. Thus, understanding the nature and mechanisms of action of the factors involved in the control of liver cell proliferation is important for establishing the basis for improved therapies for patients with chronic or acute hepatic injury. Human hepatocytes in primary culture retain most of their basic metabolic hepatic functions if adequate chemically-defined culture media and substrata are used. Indeed, adult human hepatocytes can be stimulated to synthesize DNA when they are cultured at very low cell density, under defined culture conditions (supplementation of culture medium with certain nutrients and hormones) and in the presence of specific hepatotrophic growth factors [3-6]. Therefore, primary cultured human hepatocytes have become a useful tool for investigating human liver cell metabolic functionality, drug biotransformation processes, gene expression and the proliferative response to growth factors.

Hepatocyte growth factor (HGF) is a disulphide-linked heterodimeric polypeptide synthesized by non-parenchymal liver cells and other extrahepatic cells of mesodermal origin [7,8]. Increased expression of HGF mRNA has been reported in the course of liver regeneration and inflammation, and the blood levels of this factor increase rapidly after partial hepatectomy [9], liver pathologies involving tissue damage (fulminant

M.N. Berry and A.M. Edwards (eds.), The Hepatocyte Review, 247–261.

hepatitis, liver cirrhosis, chronic hepatitis; [10]) and toxic liver necrosis [11,12]. Moreover, HGF has been found to be one of the most potent signals for stimulation of DNA synthesis in cultured hepatocytes in the absence of other growth factors [3,4,7,13], which suggests that it may play an important role in controlling the growth process *in vivo*.

In this review we examine the basic mechanisms underlying the proliferative response of human hepatocytes to growth factors, in particular to HGF. Effects of HGF on hepatocyte metabolism and the expression of other hepatic genes are also discussed.

16.2 Proliferative response of human hepatocytes to growth factors and hormones

16.2.1 Analytical methods

Protein synthesis (of cyclins, oncoproteins and plasma proteins) was investigated in hepatocytes labelled for 3-12 hours with 75 μCi/ml of L-[^{35}S]-Methionine in methionine-free RPMI-1640 culture medium. Proteins were immunoprecipitated with specific antibodies, electrophoresed and quantified as described [14]. DNA synthesis was determined by measuring the amount of ^{3}H-thymidine incorporated into the TCA precipitable fraction of the cell homogenate. Twenty-four hours before the assay, 5 μCi/ml of ^{3}H-thymidine was added to the wells concomitantly with the change of medium, and the cultures were processed as described [4]. Cell cycle status and cell proliferation of hepatocytes were analysed in isolated nuclei, stained with propidium iodide, on a EPICS profile II Flow Cytometer [4]. Glycogen was extracted and quantified from cells incubated up to 8 hours either with 10 ng/ml HGF or 10^{-7} M glucagon, as described [4]. Intracellular Ca^{2+} levels were measured as described previously [15]. Inositol 1,4,5-IP_3 production was assessed after a 2-hour pulse-labelling with 10 μCi/ml of ^{3}H-inositol and inositol phosphates were separated as described [15]. Cyclic AMP in supernatants of cell homogenates was measured using an enzyme immunoassay. Activity of CYP isozymes and conjugating enzymes was determined as described in [16]. The starting point and duration of the treatments with HGF are indicated in the figure for each experiment, and HGF was renewed every 24 hours when the medium was changed.

16.2.2 Analysis of the proliferative response of human hepatocytes to stimulation by HGF

The kinetics of DNA synthesis in response to HGF was investigated in cultures of human hepatocytes stimulated 20 h after cell plating. As shown in Figure 1, hepatocytes displayed a characteristic lag period before initiating DNA synthesis: the fourfold to sixfold increase over controls was similar to that reported by other authors [3,13,17]. HGF showed a concentration-dependent effect on DNA synthesis. The first significant stimulatory effect was observed at 0.625 ng/ml HGF and the maximal effect was found between 5 and 10 ng/ml. Beyond this concentration, HGF inhibited DNA synthesis. Similar results have been reported for human hepatocytes stimulated by HGF and other growth factors (EGF, TGF-α) [13,17,18].

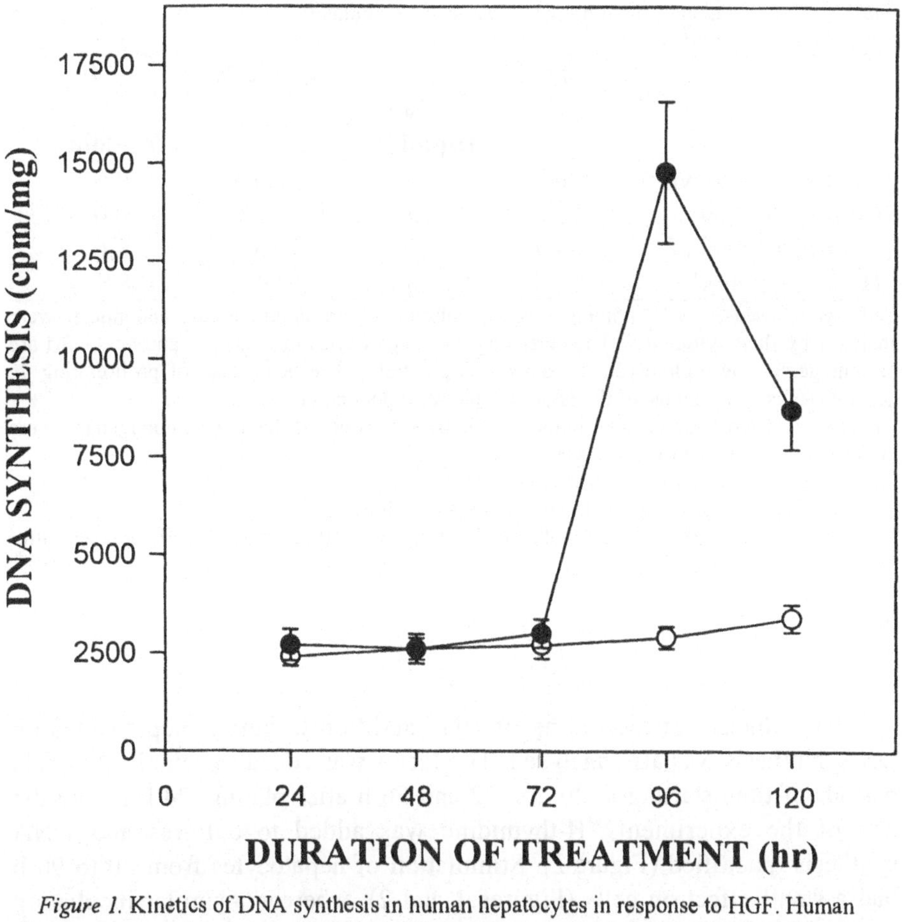

Figure 1. Kinetics of DNA synthesis in human hepatocytes in response to HGF. Human hepatocytes were incubated with 10 ng/ml HGF (filled symbols) from 20 h after cell plating up to 48, 72, 96, 120 and 144 hours of culture and analyzed for DNA synthesis in comparison with non-stimulated cultures.

Cell cycle analysis by flow cytometry of hepatocytes stimulated with HGF showed a progressive displacement of the population to a higher DNA content upon incubation with the factor (Table 1). By 120 hours of incubation with HGF, the number of hepatocytes leaving the G0/G1 state increased, as did the proportion of human hepatocytes entering the S- and G2+M-phases (Table 1). Most of the mitogenic activity, in absolute terms, seems to be associated with 2c nuclei, with a minor contribution on the part of 4c cells. It appears that human hepatocytes are more refractory to initiating DNA synthesis than rat hepatocytes [5,19,20]: however, the delay in the appearance of G2/M+4c cells is not due to a longer S-phase. These results fit well with what is known in human liver regeneration, where the process starts later (2-3 days) and takes longer (2 weeks) than does rat liver regeneration [21].

Table 1. Cell cycle analysis of human hepatocytes stimulated for 96 hours with HGF

	G0/G1 %	**S % Diploid c.**	**4c+G2/M %**	**≥ 4c % Tetraploid c.**
Isolated hepatocytes[a]	86.7	1.3	12.0	0.0
Control 20 hours[b]	83.2	3.6	13.2	0.0
Control 120 hours[c]	84.7	2.8	12.5	8.6
HGF 120 hours[d]	52.5	30.1	17.4	8.9

Cells were lysed in an hypotonic staining solution of propidium iodide and nuclei were analyzed by flow cytometry. The percentages of hepatocytes in G0/G1, S and 4c+G2/M are percentages of the diploid (2c) hepatocyte population. The percentage of proliferating 4c hepatocytes is a percentage of the total of hepatocyte population.

[a] Freshly isolated hepatocytes obtained after perfusion of the liver with collagenase were immediately processed for cell cycle analysis.

[b] Hepatocytes analysed 20 hours after cell plating

[c] Control hepatocytes harvested after 120 hours of culture.

[d] Hepatocytes stimulated 20 hours after cell plating with 10 ng/ml HGF and harvested after 96 hours of incubation with HGF.

The influence of the timing of HGF addition to human hepatocytes on DNA synthesis was also analysed. The factor was added to cells for variable periods of time starting at 20, 48, 72 and 96 h after plating: 24 h before the end of the experiment, ^{3}H-thymidine was added to cultures and DNA synthesis quantified (Figure 2). Stimulation of hepatocytes from 20 to 96 h had a small effect on cells (Figures 1 and 2), whereas a similar incubation period with the factor, but from 48 to 120 h, increased the synthesis of DNA fourfold. Moreover, stimulation from 20 to 72 h had almost no effect, while incubation of cells with the factor for the same period of time but from 72 to 120 h, resulted in maximal DNA synthesis (eightfold to tenfold increase versus controls). These results suggest that not only the duration but, in particular, the timing of HGF stimulation determine the extent of DNA synthesis and presumably the entry of hepatocytes into S phase.

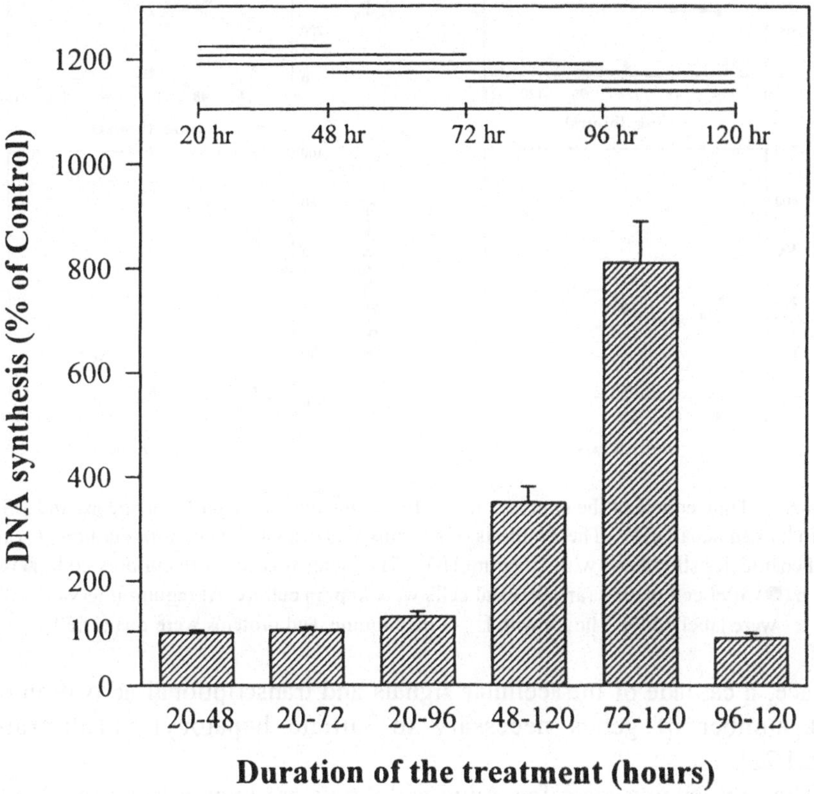

Figure 2. DNA synthesis in human hepatocytes in response to HGF. Human hepatocytes were stimulated for variable periods of time with 10 ng/ml HGF, starting at 20, 48, 72 or 96 hours after cell plating. Data are expressed as percentage of nonstimulated control cultures.

16.2.3 Cyclins, protooncogene expression and signal transduction during the proliferative response of hepatocytes to HGF

Upon stimulation of quiescent cells with growth factors, a series of rapid events occur, including interaction with specific receptors on the cell

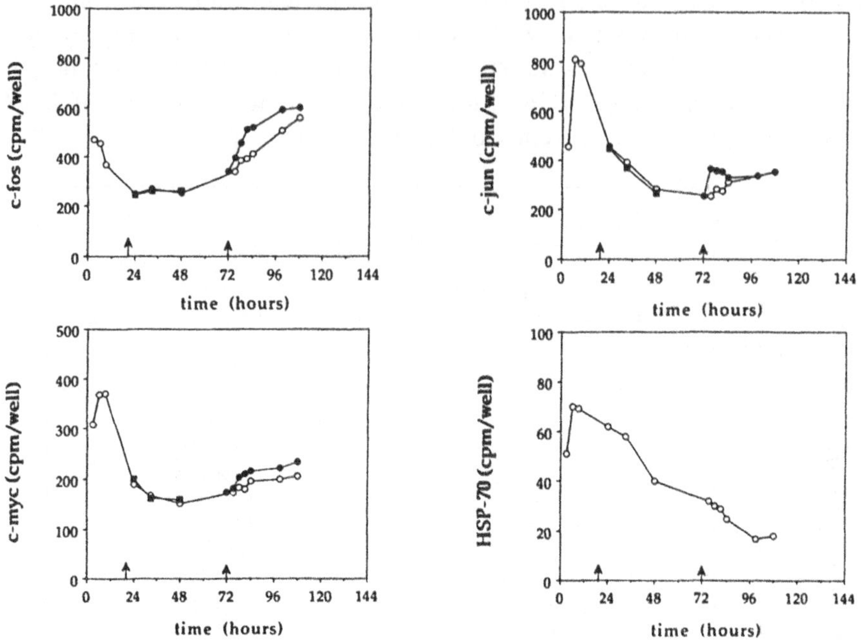

Figure 3. Time course of the expression of c-fos, c-jun and c-myc protooncogenes and HSP-70 in human hepatocytes. The synthesis of proteins was evaluated in control cultures (O) and in hepatocytes stimulated with 10 ng/ml HGF. The factor was added to cultures 20 h (■) or 72 h (●) after cell plating (arrows) and cells were kept in culture. At regular intervals, cells were labelled for 3 hours with L-^{35}S methionine, and proteins were quantified.

surface, a cascade of intracellular signals and transcriptional activation of a vast number of genes necessary to initiate hepatocyte proliferation [15,21,22].

The protooncogenes c-fos, c-jun and c-myc are known to be involved in cell activation upon mitogenic stimulation [23-25]. In the liver, a transient sequential increase in these protooncoproteins occurs shortly after partial hepatectomy, as components of a primary gene response during liver regeneration [26-29]. Stimulation of human hepatocytes with HGF 20 h after cell plating did not produce differences in protooncoprotein expression as compared with the controls (Figure 3). However, if the stimulation with the factor commenced at 72 h of culture (a timing that results in the maximal

DNA synthesis), when protooncoprotein synthesis had decreased to the minimal levels, HGF induced a prompt increase (3 to 6 h after addition of HGF) in c-fos, c-jun and c-myc protein synthesis before the DNA synthesis peak (Figure 3).

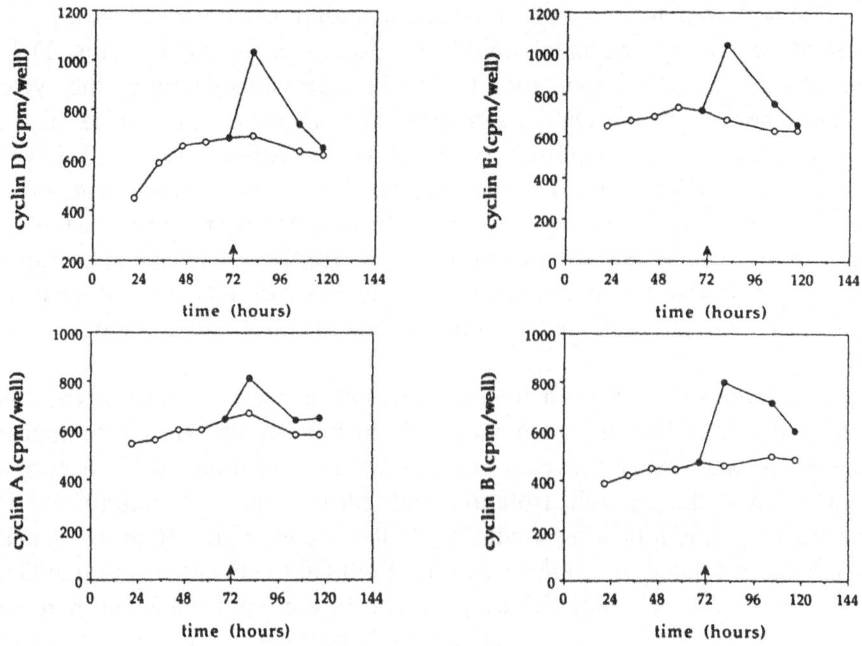

Figure 4. Time course of the expression of cyclin D, A, B and E in human hepatocytes. The synthesis of cyclins was evaluated in nonstimulated cultures (O) and in hepatocytes stimulated with 10 ng/ml HGF. The factor was added to cultures 72 hours (●) after cell plating (arrow) and cells were kept in culture. At regular intervals, cells were labelled for 12 hours with L-^{35}S methionine and proteins were quantified.

When hepatocytes were stimulated with HGF, by 72 h after cell plating an increased synthesis of cyclin A, B, D and E was observed. Cyclins A, B and E were expressed in control hepatocytes at almost constant levels during the first days in culture, whereas cyclin D showed a significant increase during the first 48 hours of culture (Figure 4). The increase in cyclins A and B is consistent with the fact that there is a major population of hepatocytes in progression through S and G2 phases to mitosis, as shown by flow cytometry (Table 1). This observation was also made in the course of liver regeneration where there is an increase in cyclin A in the S phase and transient expression of cyclin B from the late S, until the M, phase [30]. Cyclin A is responsible for activating cdk2 kinase through the S, G2 and early mitotic phases [37,38], whereas cyclin B is a typical mitotic marker in higher eukaryotic cells [33]. The simultaneous appearance of these two cyclins in HGF-stimulated hepatocytes could reveal the existence of postreplicative premitotic cells in unstimulated culture, a phenomenon that has been reported both in plant and animal tissues, including hepatocytes [34,35]. The peak of regulatory cyclin E could be indicative of the ongoing DNA replication of a cell subpopulation that had not previously entered the cycle. Finally, before peak DNA synthesis, an increase in cyclin D, a protooncogene that is operative in the G0-G1 transition and G1 [36], was observed in HGF-stimulated human hepatocytes. This contrasts with other reports [30], but is in agreement with recent observations made in regenerating mouse, rat and human livers [37,38]. However, the appearance of cyclin D close to that of cyclins E and A is a further indication of a partial asynchrony in the cell cycle progression of HGF stimulated human hepatocytes.

Most hepatocytes in adult liver are arrested in the G0 phase of the cell cycle [26], and although hepatocytes isolated from adult liver might appear quiescent, our study reveals that G0/G1 transition had occurred spontaneously during cell isolation and early stages of culture, as is supported by the following findings: a) the increase in cyclin D, a cell protein that is related to cellular transition from G0 to G1 and/or through G1, during the first 48 hours of culture [36]; and b) the rapid, transient increase in the early expression of the G1 protooncogenes c-fos, c-jun and c-myc (Figure 4). This transition occurred in human hepatocytes cultured in serum-free medium, that is, in total absence of any other trophic factor. A possible explanation, applicable to a certain extent to rat hepatocytes [39,40], is that the transition is triggered by the stress caused by tissue disruption, loss of cell-cell interactions and metabolic adaptation of hepatocytes to the new environment [39,40]. The early increase in HSP-70 in hepatocytes (Figure 4) is another indication that the cells are suffering a stress reaction during the very early stages of culture. In fact, it has been reported in other cellular systems that stress can stimulate c-fos, c-jun and c-myc expression in association with HSP-70 [41,42].

It seems that to enter the cell cycle, quiescent cells pass through a "priming" (induction to replicative competence) and "progression" phase, each one mediated by its own signals [22,26,43]. Priming is a frequent and reversible event, easily inducible in culture by serum, growth factors, stress and extracellular matrix components [26,42]. However, cell cycle progression is limited by a restriction point at the G1 phase, in which important mechanisms controlling cell proliferation exist [25,43]. The events observed during the "priming" and "progression" phases, leading to the restriction point, should determine the entrance of hepatocytes into the S phase and are consistent with the lag period displayed by hepatocytes before initiating DNA synthesis (Figures 1, 2 and 3).

The expression of early protooncogenes and cyclin D in control cultures provides evidence that hepatocytes are primed to pass through G0/G1, regardless of HGF stimulation, up to the restriction point located in the G1 phase [25,44]. In the absence of HGF stimulation, hepatocytes remain blocked at this point and do not replicate DNA. Thus, the progressive accumulation of hepatocytes at this restriction point would have the effect of partially synchronising cells. At this stage, the consequence of stimulation with HGF (at 72 h) would be that a larger number of cells simultaneously enter the S phase, resulting in greater DNA synthesis than when cultures are stimulated at an early time of culture.

The biological actions of HGF must be mediated by binding to its cell surface receptor which is a product of the c-met protooncogene. Activation of receptor tyrosine kinase [45] leads to changes in the concentration of one or more intracellular messengers generated by different transduction pathways. It has been suggested that autophosphorylation of the receptor is a key event in the signal transduction of HGF in rat hepatocytes [46]. Stimulated tyrosine kinase receptors are known to activate phospholipase C, which causes an increase in inositol 1,4,5-IP_3 that triggers an increase in intracellular Ca^{2+} [46]. We have found evidence indicating that tyrosine phosphorylation is involved in the stimulatory effect of HGF. Genistein, a tyrosine kinase inhibitor, caused a dose-dependent inhibition of DNA synthesis in HGF-stimulated cells [15]. An early transient dose-dependent increase in 1,4,5-IP_3 was detected shortly after addition of HGF, followed by an increase in intracellular-free calcium, but there were no changes in cAMP [15,46,47]. Finally, the effects of HGF on human hepatocytes were negatively regulated by protein kinase C. The inhibition of this activity by H7, a specific inhibitor of protein kinase C, increased in a dose-dependent manner the stimulatory effect of HGF on DNA synthesis. This contrasts with the effects observed after inhibiting tyrosine kinase activity. However, both findings are consistent with previous reports showing that kinase activity of the c-met protein is positively regulated by autophosphorylation on tyrosine but negatively regulated by serine phosphorylation, which is mediated by protein kinase C [46,47].

16.2.4 Combined effect of hormones and growth factors on HGF-induced DNA synthesis

Many biological signals are believed to be involved in the control of liver regeneration. Among the factors known to influence hepatic growth, hormones like glucagon and insulin have long been recognized to be positive regulators and, in fact, their plasma levels change greatly in the early course of liver regeneration [15,19]. The stimulation of DNA synthesis in human hepatocytes by growth factors like EGF [3] has been demonstrated previously, but a direct involvement in liver regeneration has not been clearly shown [48]. We compared the ability of hormones (glucagon and insulin) and other growth regulatory factors (EGF, TGF-β) to promote DNA synthesis in human hepatocytes with that of HGF. Increasing concentrations of each factor were added to culture medium alone or in combination with an optimal concentration of HGF (10 ng/ml). The analysis of results showed that in molar terms HGF is the most potent mitogen for human hepatocytes (Figure 5). Glucagon, increased the effect of the hepatic growth factor whereas, the effect of HGF was gradually inhibited by TGF-β, a factor claimed to act as a hepatic chalone (hepatocyte DNA synthesis inhibitor) in rat liver regeneration [18,49,50]

16.3 Effects of HGF on the expression of hepatic functions by human hepatocytes

In addition to its probable effects on genes controlling the cell cycle of human hepatocytes, HGF also influenced the expression of other hepatic genes not related to growth. This factor modulates the expression of acute-phase plasma proteins, whose levels dramatically change in the course of inflammation. HGF increased the synthesis of albumin, a protein that decreases during an acute-phase reaction, and decreased the synthesis of α_1-antichymotrypsin (a protein that increases in inflammatory states). These effects are virtually opposite to the actions of inflammatory cytokines (*i.e.* IL-6) in human liver [51,52]. The effects of HGF were counteracted by the simultaneous presence of IL-6 in the incubation medium in the cases of albumin, transferrin, fibronectin, α_1-antichymotrypsin and haptoglobin, but not for C-reactive protein and α_2-macroglobulin. A clear additive effect was observed only in the case of fibrinogen. The results indicate that HGF not only has a mitogenic activity, but also regulates hepatic specific genes and might, in conjunction with IL-6, influence the expression of acute-phase plasma proteins in man.

During the early course of liver regeneration, liver glycogen diminishes

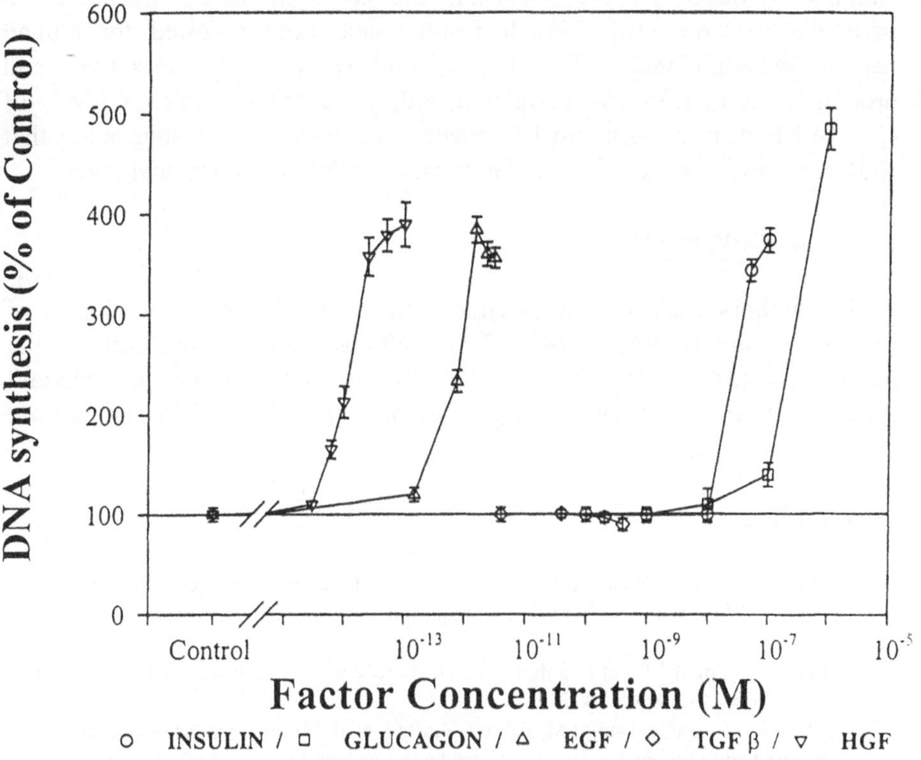

Figure 5. Combined effect of HGF and growth factors on DNA synthesis. Human hepatocytes were stimulated from 20 to 120 hours with increasing concentration of HGF, EGF, TGF-β, insulin or glucagon, and analyzed for DNA synthesis. Data are expressed as percentage of ^{3}H-thymidine incorporation of the unstimulated hepatocyte control.

rapidly [19-21]. This is correlated with a dramatic increase of glucagon in serum [17,53]. The HGF that also increases during the first stages of liver regeneration [11] has very faint effects on glycogenolysis, whereas glucagon has no mitotic activity. The results point to glucagon acting in conjunction with HGF, potentiating the mitogenic stimuli of HGF and mobilising hepatic glycogen stores.

A decrease in CYP activities in regenerating rat liver has been detected after partial hepatectomy [54]. Therefore, it was of interest to examine the effects of HGF on the CYP system and conjugating reactions in cultured human hepatocytes. The results show that HGF produced a general down-regulation of the activity of all the CYP isozymes studied (CYP1A2, CYP2A6, CYP2B6, CYP2E1 and CYP3A4), but had no effects on the activity of phase II enzymes (UGT and GST) and GSH levels [16]. Moreover, the effect of HGF on the CYP system was observed both in basal and induced activities (3-methylcholanthrene-induced for CYP1A2 and rifampicin-induced for CYP3A4) and did not seem to be specific for a particular isozyme [16]. Similar results have been reported for human hepatocytes stimulated by EGF [53]. The observed changes in activity and protein levels of CYP3A4 correlated with a reduction in mRNA levels of CYP3A4 both in control and RIF-treated hepatocytes, thus suggesting that HGF can down-regulate CYP3A4 expression at the transcriptional level.

Acknowledgements

The authors wish to acknowledge with thanks the financial support of European Union (projects AIR2-CT93-0860 and BIO4-CT96-0052) and the ALIVE Foundation. Dr. T. Nakamura for providing human recombinant HGF and Dr. E. O'Connor for the flow cytometry analysis of the samples are also acknowledged.

References

1. Higgins GM and Anderson AM. Experimental pathology of the liver. I Restoration of the liver of the white rat liver following partial surgical removal. Arca Pathol 1931; 12: 186-192.
2. Lewan L, Ynguer T and Engelrecht C. The biochemistry of the regenerating liver. Int J Biochem 1977; 8: 457-477.
3. Blanc P, Etienne H, Daujat M, Fabre I, Zindy F et al. Mitotic responsiveness of cultured adult human hepatocytes to epidermal growth factor, transforming growth factor a, and human serum. Gastroenterology 1992; 102: 1340-1350.
4. Gómez-Lechón MJ, Castell JV, Guillen I, O'Connor E, Nakamura T et al. Effects of hepatocyte growth factor in the growth and metabolism of human hepatocytes in primary culture. Hepatology 1995; 21: 1248-1254.
5. Michalopoulos G, Cianciulli HD, Novotny AR, Kligerman AD, Strom SC and Jirtle RL. Liver regeneration studies with rat hepatocytes in primary culture. Cancer Res 1982; 42: 4673-4682.
6. Nakamura T, Tomita Y and Ichiara A. Density-dependent growth control of adult rat hepatocytes in primary culture. J Biochem 1983; 94: 1020-1035.
7. Matsumoto K and Nakamura T. Hepatocyte growth factor: molecular structure, roles in liver regeneration and other biological functions. Crit Rev Oncogenesis 1992; 3: 27-54.
8. Mizuno K and Nakamura T. Molecular characteristics of HGF and the gene, and its biological aspects. In: Goldberg ID and Rosen EM, Eds. Hepatocyte Growth Factor-Scatter (HGF-SF) and the C-Met receptor. Basel: Birkghauser Verlag, 1993: 1-29.
9. Nishizaki T, Takenake K, Yoshizumi T, Yanaga K, Soejinma Y, Shirabe K and Sugimachi K. Alteration in levels of human hepatocyte growth factor following hepatectomy. J Am Coll Surg 1995; 18: 6-10.
10. Shimizu I, Ichiara A and Nakamura T. Hepatocyte growth factor in ascites from patients with cirrhosis. J Biochem 1991; 109: 14-18.

11. Bezerra JA, Laney DW and Friezner Degan SJ. Increased expression of mRNA for hepatocyte growth factor-like protein enduring liver regeneration and inflammation. Biochem Biophys Res Commun 1994; 303: 666-673.
12. Lindros PM, Zarnegar R and Michalopoulos GK. Hepatocyte growth factor (hepatopoietin A) rapidly increases in plasma before DNA synthesis and liver regeneration stimulated by partial hepatectomy and carbon tetrachloride administration. Hepatology 1991; 13: 743-749.
13. Strain AJ, Ismail T, Tsubouchi H, Arakaki N, Hishida T, Kitamura H, Daikuhara Y and McMaster P. Native and recombinant human hepatocyte growth factors are highly potent promoters of DNA synthesis in both human and rat hepatocytes. Clin Invest 1991; 87: 1853-1857.
14. Guillén MI, Gómez-Lechón MJ, Nakamura T and Castell JV. The hepatocyte growth factor regulates the synthesis of acute-phase proteins in human hepatocytes. Divergent effects on IL-6 synthesis. Hepatology 1996; 23: 1345-1352.
15. Gómez-Lechón MJ, Guillén I, Ponsoda X, Fabra R, Trullenque R et al. Cell cycle progression proteins (cyclins), oncogene expression, and signal transduction during the proliferative response of human hepatocytes to hepatocyte growth factor. Hepatology 1996; 23: 1012-1019.
16. Donato MT, Gómez-Lechón MJ, Jover R, Nakamura T and Castell JV. Human hepatocyte growth factor down-regulates the expression of cytochrome P450 isozymes in human hepatocytes in primary culture. J Pharmacol Exp Ther 1998; 284: 760-767.
17. Ismail T, Howl J, Wheatley M, McMaster P, Neuberger JM and Strain AJ. Growth of normal human hepatocytes in primary cultures. Effect of hormones and growth factors on DNA synthesis. Hepatology 1991; 14: 1076-1082.
18. Nakamura T, Tomita Y, Hirai R, Yamaoka K, Kaji K and Ichiara A. Inhibitory effect of transforming growth factor α on DNA synthesis of adult rat hepatocytes in primary culture. Biochem Biophys Res Commun 1985; 133: 1042-1049.
19. Bucher NLR and McGowan JA. Regulatory mechanisms in hepatic regeneration. In: Wright R, Millward-Sadler GH, Alberti KGMM and Karran S, Eds. Liver and Biliary Disease. London: WB Saunders, 1985: 251-265.
20. Leffert HL, Koch KS, Moran T and Rubalcava B. Hormonal control of rat liver regeneration. Gastroenterology 1970; 76: 1470-1482.
21. Leffert HL, Koch KS, Lad PJ, Skelly H and de Hemtinne B. Hepatocyte regeneration, replication, and differentiation. In: Arias I, Popper H, Schachter D and Shafritz DA, Eds. The Liver: Biology and Pathobiology. New York: Raven Press, 1982: 601-614.
22. Burger CH, Wick M, Brusselbach S and Muller R. Differential induction of metabolic genes after mitogen stimulation and during normal cell cycle progression. J Cell Sci 1994; 107: 241-252.
23. Curran T and Franza BR. Fos and jun: the AP-1 connection. Cell 1988; 55: 395-397.
24. Kovary K And Bravo R. Existence of different fos/jun complex ensuring G0-to-G1 transition and during exponential growth in mouse fibroblasts: differential role of fos proteins. Mol Cell Biol 1992; 12: 5015-5023.
25. Wick M, Burger C, Brusselbach S, Lucibello FC and Muller R. Identification of serum-inducible genes: different patterns of gene regulation during G0-S and G1-S progression. J Cell Sci 1994; 107: 227-239.
26. Fausto N, Webber EM, Eaton EN, Arnold A and Weinberg A. Control of liver growth. Crit Rev Eukaryotic Gene Exp 1993; 3: 117-135.

27. Thompson NL, Mead JE, Braun L, Goyette M, Shank PR and Fausto N. Sequential protooncogene expression during rat liver regeneration. Cancer Res 1986; 46: 3111-3117.
28. Webber EM, Godowski PJ and Fausto N. *In vivo* response of hepatocytes to growth factors requires an initial priming stimulus. Hepatology 1994; 14: 489-497.
29. Fausto N and Mead JE. Regulation of liver growth: proto-oncogenes and transforming growth factors. Lab Invest 1991; 60: 4-13.
30. Loyer P, Glaise D, Cariou S, Baffet G, Meijer L and Guguen-Guillouzo C. Expression and activation of cdk (1 and 2) and cyclins in the cell cycle progression during liver regeneration. J Biol Chem 1994; 28: 2491-2500.
31. Sherr CHJ. Mammalian G1 cyclins. Cell 1993; 73: 1059-1065.
32. Van den Heuvel S and Harlow E. Distinct roles for cyclin-dependent kinases in cell cycle control. Science 1993; 262: 2050-2054.
33. Fang F and Newport JW. Evidence that G1-S and G2-M transition are controled by different cdc2 proteins in higher eukaryotes. Cell 1991; 66: 731-742.
34. Gelfant S. Cycling-noncycling cell transition in tissue aging, immunological surveillance, transformation and tumor growth. Int Rev Cytol 1981; 70: 1-12.
35. Gómez-Lechón MJ and Castell JV. Evidence for arrested G2 cell subpopulation in rat liver inducible to mitosis. Cell Tissue Kinet 1987; 20: 583-590.
36. Hinds PW, Dowdy S, Eaton EN, Arnold A and Weinberg A. Function of human cyclin gene as an oncogene. Proc Natl Acad Sci USA 1994; 91: 709-713.
37. Albretcht JH, Hu MY and Cerra FB. Distinct pattern of cyclin D1 regulation in models of liver regeneration and human liver. Biochem Biophys Res Commun 1995; 209: 648-655.
38. Lu XP, Koch KS, Lew DJ, Dulic V, Pines J, Reed SI, Hunter T et al. Induction of cyclin mRNA and cyclin-associated histone H1 kinase during liver regeneration. J Biol Chem 1992; 267: 2841-2844.
39. Etienne PL, Baffet Desvergne B, Boisnard-Rissel M, Glaise D and Guguen-Guillouzo C. Transient expression of c-fos and constant expression of c-myc in freshly isolated hepatocytes and cultured normal adult rat hepatocytes. Oncog Res 1988; 3: 255-262.
40. Tomomura A, Sawada N, Sattler G L, Kleinman H K And Pitot HC. The control of DNA synthesis in primary cultures of hepatocytes from adult and young rats: interactions of extracellular matrix components, epidermal growth factor, and cell cycle. J Cell Physiol 1987; 130: 221-227.
41. Colotta F, Polentarutti N, Staffico M, Fincato G and Montovani A. Heat-shock induces the transcriptional activation of c-fos protooncogenes. Biochem Biophys Res Commun 1990; 168: 1013-1019.
42. Jin P and Rigertz NR. Cadmium induces transcription of protooncogenes c-jun and c-myc in rat 16 myeloblasts. J Biol Chem 1990; 265: 14061-14068.
43. Mead JE, Braun L, Martin DA and Fausto N. Induction of replicative competence (priming) in normal liver. Cancer Res 1990; 50: 7023-7030.
44. Nurse P. Universal control mechanisms regulating onset of M-phase. Nature 1990; 244: 503-508.
45. Moghul A, Lin L, Kanbour-Shakir A, De Frances M, Liu Y and Zarnegar R. Modulation of c-MET proto-oncogene (HGF receptor) mRNA abundance by cytokines and hormones: evidence for a rapid decay of the 8 kb c-MET transcript. Oncogene 1994; 9: 2045-2052.

46. Baffi G, Michalopoulos GK and Williamson JR. Hepatocyte growth factor induces calcium inisitol phosphate production in rat hepatocytes. J Cell Physiol 1992; 143: 332-339.
47. Kaneko A, Hayashi TM, Tanaka Y, Ito T, Sasaki Y et al. Intracellular calcium as a second messenger for human growth factor in hepatocytes. Hepatology 1992; 15: 1173-1178.
48. Michalopoulos GK. Liver regeneration: molecular mechanism of growth control. FASEB J 1990; 4: 176-187.
49. Fabregat I, De Juan C, Nakamura T and Benito M. Growth stimulation of fetal rat hepatocytes in response to hepatocyte growth factor: modulation of c-myc and c-fos expression. Biochem Biophys Res Commun 1992; 2: 648-690.
50. Thoresen GH, Refsnes M and Christoffersen T. Inhibition of hepatocyte DNA synthesis by transforming growth factor $\alpha 1$ and cyclic AMP: Effect immediately before the G1/S border. Cancer Res 1992; 52: 3598-3603.
51. Castell JV, Gómez-Lechón MJ, David M, Fabra R, Trullenque R and Heinrich PC. Acute phase response of human hepatocytes: Regulation of acute phase protein synthesis by Interleukin 6. Hepatology 1990; 12: 1179-1186.
52. Heinrich PC, Castell JV and Andus T. Interleukin 6 and the acute phase response. Biochem J 1990; 2645: 621-636.
53. Greuet J, Pichard L, Ourlin JC, Bonfils C, Domargge J, Le Truet P and Maurel P. Effect of cell density and epidermal growth factor on the inducible expression of CYP3A4 genes in human hepatocytes in primary culture. Hepatology 1997; 25: 1166-1175.
54. Liddle C, Murray M and Farrell GC. Effect of liver regeneration on hepatic cytochrome P450 isozymes and serum sex steroids in the male rat. Gastroenterology 1989; 96: 864-872.

Chapter 17

Growth control and cell cycle progression in cultured hepatocytes

Gennady Ilyin, Claude Rescan, Mickaël Rialland, Pascal Loyer, Georges Baffet and Christiane Guguen-Guillouzo
INSERM Unit 522, Hopital de Pontchaillou, 35033 RENNES Cedex, France

Key words: apoptosis, cell cycle, cyclin-dependent kinases, cyclins, cytokines, differentiation, induction of immediate early genes, MAP kinases, ploidy, proliferation, protooncogenes, restriction point, survival

17.1 Introduction

In contrast to unicellular organisms such as yeast or bacteria for which there is strong selective pressure for each individual cell to grow and divide as rapidly as possible, the cells of multicellular animals are specialized members of a complex society. Delicately adjusted and highly specific controls exist to govern the proliferation of each type of cell according to need and rates of renewal that greatly vary from one tissue to another. Thus, while some tissues are composed of permanently growth-arrested differentiated cells with a long life-span, others are formed by continuously renewing cells, which die after reaching a terminal differentiated stage. The liver is an intriguing example of a tissue between these two categories, in which normal rates of renewal are maintained albeit at a very low level. This organ can be stimulated to produce new cells at a higher rate by surgically removing a large mass of tissue or by destroying the cells by toxins in the diet. Renewal in this tissue takes place by simple duplication of existing differentiated hepatocytes that divide to give pairs of daughter cells of the same type. This process, triggered by environmental signals such as growth factors, is associated with a transient change of phenotype involving numerous genes that are regulated by a complex combination of control mechanisms: those responsible for mutually exclusive molecular events leading to differentiation or proliferation, and to survival or apoptosis. The regulatory mechanisms in hepatocytes may differ in part from those involved in differentiated cells of other tissues. This chapter updates data, derived from studies in primary culture, that have contributed to a better understanding of the sequences of the processes regulating hepatocyte proliferation during liver regeneration.

M.N. Berry and A.M. Edwards (eds.), The Hepatocyte Review, 263–280.

17.2 Hepatocyte entry to G1 during isolation, and survival signals during cell attachment

17.2.1 Induction of immediate early genes

Washing the liver with a Ca^{2+}-free buffer followed by a collagenase treatment results in disruption of cell-cell communications and cell isolation. It is clearly established that the G0/G1 transition takes place during this isolation process [1,2]. Expression of *c-fos*, *c-jun* and *jun-B* is undetectable in normal liver, but rapid induction occurs after a few minutes of washing, increases during collagenase perfusion, reaches a maximum in freshly isolated cells and drastically decreases thereafter. A sequential and transient appearance of these immediate early protooncogenes is observed, indicating a progression of hepatocytes into early G1 as assessed by the induction 3-4 h later of other early genes such as *jun-D* and *c-myc*, the expression of which is maintained for a prolonged period after cell spreading. This G0/G1 transition mimics well the entry to G1 of hepatocytes *in vivo* after partial hepatectomy, corresponding to the "priming step" in the regeneration process [3,4].

17.2.2 Cytokines and protooncogene induction

One hypothesis is that this process of G1 transition is the consequence of metabolic changes associated with alterations in cell-cell interactions and loss of signals related to the disruption of the extracellular matrix. Indeed, Liu *et al.* have shown that the partial degradation of the extracellular matrix *in vivo* could trigger the G0/G1 transition [5]. In addition, this G0/G1 transition could also be controlled by a release of cytokines, mainly TNF-α and IL6 [6,7]. TNF-α is produced promptly by liver cells after partial hepatectomy *in vivo* and is released during liver perfusion as well. It has been reported that after partial hepatectomy, TNF-α activates NF-κB, a transcription factor inactive in quiescent hepatocytes *in vivo* [8]. Its activation was found to result in translocation to the nucleus where it can regulate transcription of downstream target genes, one of which is that for IL6. This cytokine may act through its receptor to activate the JAK pathways (associated with the activation of the transcription factor STAT3) and cause induction of early protooncogenes such as *c-jun* and *c-fos*. Interestingly, those genes are components of the AP-1 complex known to be involved in the induction of several cell cycle-related genes [9]. This cascade of events could take place during liver dissociation and lead to the early protooncogene induction and G0/G1 transition.

17.2.3 Changes in the hepatocyte differentiation status

Entry to G1 is immediately accompanied by a change in the balance between proliferation and differentiation. As a consequence, a strong decrease in the transcription level of liver-specific genes is observed together with re-expression of genes that are extinguished in quiescent differentiated hepatocytes but highly expressed in fetal liver [10]. Thus, fetal isoforms of enzymes involved in glucose metabolism like pyruvate kinase or aldolase, are re-expressed. In the same way, drug metabolism is modified. Cytochrome P450 (CYP) levels decrease approximately 50% during the first 48 h in rodent hepatocytes. Induction of specific CYP isozymes by phenobarbital is difficult to achieve while other isozymes, although decreased, remain capable of responding to inducers [11]. Concerning the Phase II enzymes, for instance UDP-glucuronyl-transferases and glutathione-S-transferase (GST) isoforms, differential changes are observed favouring expression of the fetal forms [12]. These changes are associated with a shift in the balance of positive transcription factors (decrease of C/EBPα to C/EBPβ ratio) [13,14], or induction of negative transcription regulators Id1 and Id2 belonging to the helix-loop-helix (HLH) protein family [15,16]. These factors all contribute to modifying the differentiation status in close correlation with promotion of responsiveness to growth signals.

17.2.4 Early antagonism between survival signals and apoptotic reaction

Paralleling these metabolic changes, one fraction of cells dies during their isolation followed by a second fraction during seeding, while the major part of the remaining population attaches, spreads and survives for at least one week. It is increasingly evident that most mammalian cell types require complex and ill-defined survival signals to override their natural tendency to undergo programmed cell death [17,18]. Among them, interactions between heterologous cell types appear to be critical. These signals are lost when the architecture of the tissue *in vivo* is destroyed during collagenase perfusion and when pure hepatocyte populations are selected, mimicking the extracellular matrix disruption occurring *in vivo* early after partial hepatectomy. Secretion of cytokines, including TNF-α, as described above, should also contribute to this death programme occurring during isolation and after seeding [19,20]. Indeed, TNF-α is known to induce the apoptotic pathway. However, the fact that most cells survive indicates that survival signals antagonize apoptotic signals in these cells, by mechanisms that still remain poorly understood but presumably include integrin-mediated signalling during cell attachment [21]. It is therefore reasonable to assume that the conditions chosen for collecting and seeding of cells strongly influence their survival in culture. Intriguingly, some genes that harbour a dual capacity for influencing growth or cell death can ensure the balance [22,23]. Thus, the Myc protein, belonging to the bHLH family of transcriptional regulators, has recently been described as necessary and sufficient for inducing cell survival and proliferation when its dominant apoptotic action is blocked by survival factors [17,23]. Early expression of this protooncogene in isolated hepatocytes and during spreading would contribute to determining whether cell death or survival occurs during monolayer establishment [23,24]. In addition, other proteins, including Akt, that are capable of blocking steps preceding caspase activation, could also play a part in this decision [25].

17.3 Location of a mitogen-dependent restriction point in mid-late G1

17.3.1 Blockage and shortened life-span of non-stimulated hepatocytes

In primary culture, rat hepatocytes progress through mid-G1 phase regardless of additional growth factor stimulation. As *in vivo* after partial hepatectomy, " priming" of hepatocytes corresponds to an initiating event that allows primed hepatocytes to progress into G1. Their progression is ascertained by the sequential activation of early and delayed protooncogenes like *c-myc*, *ras* and *p53*, the two latter being assumed to occur in mid-G1 [24]. It is important to note that after attachment, even in the absence of growth factor, surviving cells do not return to the quiescent G0 status characteristic of normal adult hepatocytes *in vivo*. In contrast, they progress until a restriction point located two-thirds through G1. This point is reached about 42 h after plating cells on plastic but may be earlier (*i.e.* G1 is shortened) when cells are seeded onto extracellular matrix components such as collagen type I. Non-stimulated cells remain arrested at this point, do not replicate and die early.

A large number of proteins involved in the cell cycle have been identified. Of the proteins characterized to date, the cyclin-dependent serine/threonine kinases (CDKs) and their cyclin partners play a crucial role in controlling the progression through a series of checkpoints during the cell cycle. Indeed, their catalytic sub-units act by phosphorylating various downstream target proteins that allow progression to the next step of the cell cycle. Unexpectedly, maintenance of non-stimulated cells is accompanied by characteristic changes in the expression of genes related to the cell cycle, mainly accumulation of cyclin D3 and CDK4 throughout the culture period, while neither cyclin D1 nor cyclin A could be detected. It may be postulated that accumulation of these proteins reflects perturbation of cell function associated with cell decision to go to death [24].

At early times in culture, (*i.e.* around 4 days after seeding) these non-stimulated hepatocytes begin to detach and die. Akaike's group reported DNA oligonucleosome fragmentation in hepatocyte primary cultures, suggesting that programmed cell death is occurring [26]. This short life-span could, at least in part, be explained by the "Dual Signal" hypothesis proposed by Evan's group [17,23]: they argue that proliferation and cell death are coupled and that the apoptotic programme, unless forestalled by survival signals, precipitates cell suicide. Under the conditions in which no growth factor is provided it can be assumed that supply of survival factors is missing for these cells which, as a consequence, die. Substantial evidence from other cell models suggests that different genes including *c-myc*, *ras* and *p53*, can participate in this dual control. Indeed, the Evan's group has revealed that *ras* is an effective promoter of apoptosis through the Raf pathway. The expression of Ras observed in mid G1 could therefore antagonize the apoptotic signal mediated by Myc [27]. Interestingly, these two genes are expressed in cultured hepatocytes maintained without mitogen and arrested at the Restriction Point (R-point).

17.3.2 Overriding the R-point and cyclin D1 induction in mitogen stimulated cells

In contrast, the presence of growth factor, *i.e.* hepatocyte growth factor (HGF), transforming growth factor α (TGF-α), or epidermal growth factor (EGF), induces hepatocytes to cross this R-point and progress to late G1 and S phase [24]. This R-point has been precisely localized in G1 phase by adding EGF at various times after plating and monitoring DNA synthesis. In our conditions of culture, *i.e.* in absence of any extracellular matrix components as cell attachment support, DNA synthesis starts at 56 to 60 h post-plating when stimulation is performed within 42 h after plating. Later additions of mitogen delay the initiation of DNA replication indicating that the R-point in mid-late G1 is reached approximately 42-45 h after seeding.

Expression of cyclin D1 is very low up to mid G1 but greatly increases in mitogen-stimulated cells in mid-late G1, showing that its induction is mitogen-dependent and suggesting that it is associated with the ability of the cells to override the R-point and to progress up to late G1 (Figure 1). This hypothesis is further supported by the fact that when the mitogen stimulation is delayed, the induction of cyclin D1 is also postponed [24]. These observations lead to the postulate that induction of cyclin D1 is an important regulatory pathway that triggers the R-point transition and further hepatocyte cell cycle progression.

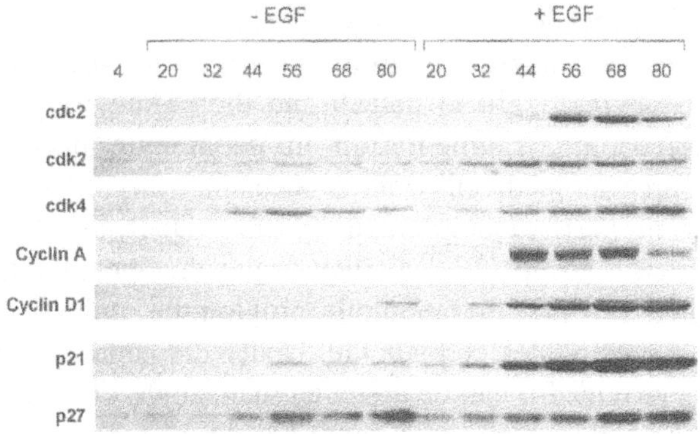

Figure 1. Sequential expression of cell cycle associated proteins in cultured hepatocytes whether stimulated or not with EGF.

17.3.3 Sequential induction of cyclins and cyclin/cdk complex formation during cell cycle progression of mitogen-stimulated hepatocytes

It is now well established that cell cycle control is provided by sequential assembly, activation and destruction of the specific complexes containing the cyclins and cyclin-dependent kinases and cyclin partners (Figure 2). We have determined the sequence of cyclin/CDK expression and complex formation in proliferating hepatocytes. The appearance of cyclin D1 at the R-point, a key event in cell cycle progression, is followed by a cascade of cyclin induction, cyclin E in late G1, then cyclin A in S phase and, finally cyclin B in G2/M. Some changes of CDK expression are characteristic for mitogen-stimulated hepatocytes. CDK1, also referred to as Cdc2, which is never observed in the absence of mitogen, is expressed during S and M phases with peaks of kinase activity in both phases, while CDK2 and CDK4 are expressed throughout the cell cycle with an increase from mid-late G1 phase (Figure 1). We have recently found that in mitogen-stimulated hepatocytes, cyclin D1 is associated with CDK2, CDK4 and CDK6 and formation of these complexes is correlated with cyclin D1 kinase activity also detected in mid-late G1.

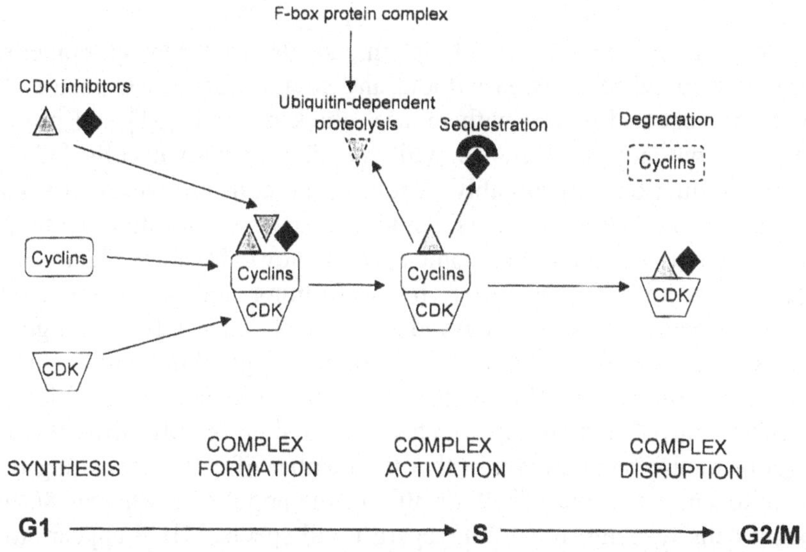

Figure 2. Schematic representation of the main steps of Cyclin/Cdk complex regulation.

Besides the regulatory step involving synthesis of positive cell cycle control molecules, there are further events related to assembly and activity of cyclin-cdk complexes that involve a number of small proteins termed cdk inhibitors. Interestingly, expression of a cdk inhibitor p21, which is barely detectable in adult liver, strongly increases in mitogen-stimulated hepatocytes and this induction is correlated with cyclin D1 accumulation. In addition, cyclin D1 complexes contain CDK inhibitors p21 and p27. It is likely that cyclin D1-CDK4/CDK6 complexes, accumulating in mitogen-stimulated hepatocytes from mid-late G1 phase, sequester cdk inhibitors and make them inaccessible to cyclin E-CDK2 and cyclin A-CDK2 complexes, the activity of which is essential for S phase progression [28].

Finally recent data allows identification of a group of proteins involved in the regulation of stability of some cell cycle control proteins (Figure 2). These proteins, characterized by the presence of an F-box domain, act as adaptors that link a core ubiquitin ligase with protein substrates and target them to ubiquitination and subsequent proteolysis [29]. SKP2, one of the

F-box proteins, induced and phosphorylated in late G1 phase of proliferating hepatocytes, is associated with the cyclin A-CDK2 complex [30]. At present we are looking for the role in cell cycle control of several other F-box proteins recently cloned in our laboratory.

17.3.4 Modulation of the mitogen response according to the species source of hepatocytes

Depending on the species of origin, hepatocytes can spontaneously undergo one round of division without any addition of growth factor. This property has been observed with mouse, chick or duck [31] and pig [32] hepatocytes but not for instance with rat, dog or human cells [33]. An autocrine induction, presumably resulting in cell production of some cytokines such as TGF-α or others, could be responsible for this spontaneous cell cycle activity observed with some species. However, all of them respond similarly to a mitogen thereafter, by performing additional cell cycles. Another difference may be observed in the cell sensitivity to exogenous mitogens. Thus, Fetal Calf Serum (FCS) alone is a good mitogen for mouse hepatocytes while it is inefficient for rat or human hepatocyte proliferation [24]. Behaviour of human hepatocytes is also significantly different from their rat or mouse counterparts in that the number of cells responding to an active mitogen, never exceeds 40 or 50% of the population whereas 80-90% of rodent cells respond. Finally in regard to all species, HGF appears to be one of the most efficient growth factors for hepatocyte proliferation *in vitro*. However, it is interesting that combination of HGF and EGF has the strongest proliferative effect. Indeed, several rounds of division can be achieved over a period of 15 days in the presence of this cocktail [34]. In addition several non-protein medium supplements were shown to improve dramatically the proliferation capacity of hepatocytes. In particular, addition of Cu^{2+}, Fe^{2+} and Zn^{2+} to chemically defined medium induced long-term replication of rat hepatocytes [35].

17.4 A main signalling pathway through MEK/ERK kinase activation related to cyclin D1 induction in mid-G1

17.4.1 MEK/ERK cascade activation associated with mitogen signal

The regulatory signals to the nucleus that specifically control one of the main steps of hepatocyte progression, in mid-late G1 phase at the R-point, are beginning to be elucidated. Among the signalling pathways responsible for this transmission of extracellular signals, the MAP kinases seem to play an important role. The general assumption that the activated MAPK cascades may influence the expression of some early genes, such as *c-fos* and *c-jun*, supports this assessment [36]. Three MAP kinase pathways have been described: the JNK and p38 cascades and the MEK/ERK cascade. They are all activated in response to external stimuli in numerous cell types, and cyclin D1 expression for instance appears to be regulated either positively by MEK/ERK kinase or negatively by the p38/HOG MAPK pathways [37]. The three distinct MAP kinase pathways seem important for the progression of hepatocytes in primary culture toward S phase [38, 39].

In a recent study we have clearly established that a sustained activation of ERK1/2, for at least 12 h, is associated with the replicative activity of hepatocytes stimulated by EGF [40]. The specificity of this signal is suggested by the observation that phosphorylation of these kinases is much less and only transient after stimulation with FCS which, as mentioned above, is a poor mitotic inducer in rat hepatocytes (Figure 3).

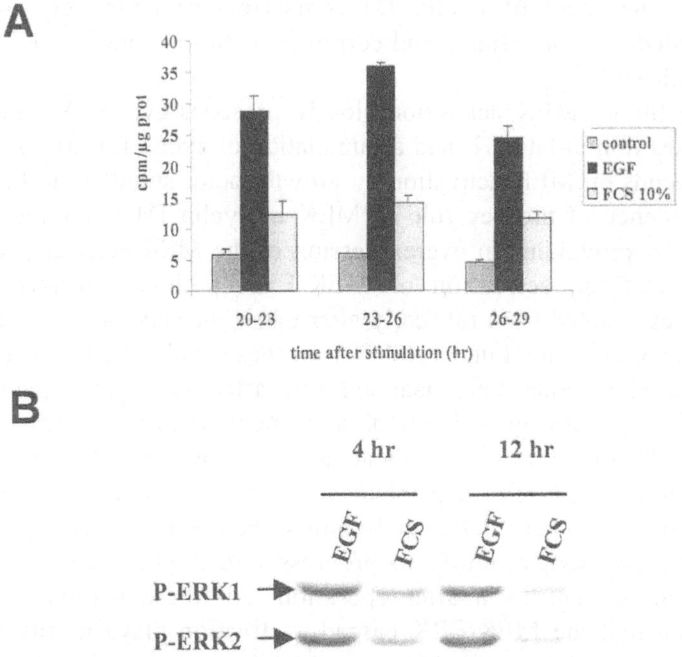

Figure 3. Effect of EGF and FCS stimulations on DNA synthesis and ERK1/2 phosphorylation. Hepatocytes were cultured for 48 h in basal medium and then stimulated or not by EGF or FCS and analyzed at the indicated times after stimulation. (A) Time-course of ^{3}H-methyl-thymidine incorporation into DNA. (B) Western blotting performed using anti-phospho-ERK1/2 antibody.

Evidence has also been provided of the role of MEK activation in hepatocyte progression through the R-point and S phase by using the MEK inhibitor PD 98059, known to specifically inhibit phosphorylation of the MEK protein [41]. When both PD 98059 and EGF are added to 48 h-old hepatocyte cultures, hepatocyte DNA replication is abolished. In parallel, phosphorylation of MEK/ERK kinases is greatly inhibited. By removing the inhibitor after 12 h of exposure, the cells are able to progress to S phase and undergo DNA synthesis with a corresponding 12 h delay indicating that cells treated by MEK inhibitor are unable to respond to EGF and behave as if they remain blocked at the R-point.

17.4.2 Cyclin D1 induction related to MEK/ERK Kinase activation

In different cell types, cyclin D1 induction has been found to depend directly on MEK/ERK activation after mitogenic stimulation. In 48 h-old unstimulated hepatocyte cultures, cyclin D1 mRNA rapidly accumulates after mitogenic addition. In contrast, in cells exposed to EGF and MEK inhibitor, the level of cyclin D1 transcripts remains very low as in unstimulated control cultures and correlates with the complete inhibition of DNA synthesis.

Interestingly, MEK activation closely precedes cyclin D1 induction in hepatocytes in mid-late G1, and accumulation of cyclin D1 mRNA at the R-point depends on MEK activation by growth factor stimulation. In addition, direct evidence of the key role of MEK in cyclin D1 induction has been provided by provoking an overexpression of the MEK activated form [40]. This mimics the activation of MEK/ERK pathway *in vivo*. Indeed, hepatocytes isolated from rats early after partial hepatectomy *i.e.* prior to 12 h post-operation, cannot undergo DNA synthesis without addition of EGF to the cultures, whereas cells isolated just after 15 h post-operation, are independent of any growth factor and spontaneously progress to DNA synthesis. These observations indicate an *in vivo* mitogen-dependent R-point between 9 and 12 h after hepatectomy. However, as expected, injection of MEK inhibitor to hepatectomized animals before they reach the R-point, blocks the hepatocyte ability to progress into S phase even if cells are isolated more than 15 h after operation. All these results lead to the conclusion that the MEK/ERK cascade activation plays a critical role in inducing cyclin D1 at the R-point which results in induction of DNA replication.

17.5 Reciprocal control regulations of proliferation, differentiation and apoptosis

17.5.1 Functional responsiveness of hepatocytes during progression in G1

It appears that progression through the first-third of Gl is required to make the cultured hepatocytes sensitive to the mitogen signal. Indeed, EGF stimulation during the first 24 h of culture induces a very low level of DNA synthesis while the maximal response is obtained when stimulation is performed between 24 and 48 h. The highest synchronization of the population entering the S phase is observed after 48 h. This leads us to postulate that hepatocytes cannot recognize or respond to the mitogen signal during the first 24 h and indicates that important molecular changes might occur at different steps of progression along G1. Various environmental signals including survival factors related to the monolayer establishment, would contribute to these initial changes, making the cells sensitive to growth factors after 24 h.

It is intriguing to observe that 24 h-old hepatocytes *in vitro* continue their progression in Gl up to the R-point in the absence of growth factors and cannot go back to the G0 state whereas hepatocytes from sham-operated animals, which do not receive mitogen signal, enter into Gl but return to G0 within 6 h after surgery [40].

17.5.2 Differentiation versus proliferation

There is an inverse relationship between cell density and the DNA replication rate, involving unidentified plasma membrane proteins [42]. Also, when heterotypic cell-cell interactions are re-established *i.e.* by coculture, hepatocytes are presumably blocked in Gl and continuously express Myc at a moderate and constant level, while they are unable to respond to growth factor signal and to enter into S phase, as if they were blocked before reaching the R-point [2]. In parallel, the differentiated status of cocultured hepatocytes is stabilized with maintenance of liver-specific transcription factor expression and transcription activity of liver-specific genes [43,44]. These observations lead to the conclusion that a high level of tissue-specific differentiation and proliferation activity are mutually exclusive in normal cells and that cell microenvironment plays a major role in this balance. Differentiation is accompanied by prolonged cell survival, indicating that these environmental factors also play a critical role in preventing programmed cell death. Extracellular matrix components, particularly matrigel rich in laminin and collagen IV, can play a similar role. Also sodium butyrate and dimethylsulphoxide (1.5-2%), a dipolar solvent, are powerful factors able to stabilise liver specific functions for a prolonged period when added to serum-free medium. Again, the cells do not divide and appear to be blocked in G1 [45]. In parallel, these factors appear to inhibit apoptosis by inducing survival signals that remain to be characterized [46].

17.5.3 Windows for optional decisions

After isolation and plating, hepatocytes have entered G1 phase and, as a consequence, must take a path either to differentiation or to further progression to late G1, depending on culture conditions. It is important to note that choice is limited in time: in fact, at least two windows for decisions in cell evolution might exist in G1. One, covering the two first days up to the R-point, allows most non-stimulated or stimulated cells progressing through G1, to stop this progression and to reverse to differentiation if appropriate conditions are provided; the other covering a maximum 2 day-period after the R-point, (*i.e.* up to 4 days after plating) allows the cells deprived of mitogen and consequently programmed to undergo apoptosis, to stop this progression and to switch to proliferation activity if this mitogen signal is provided (Figure 4). However, the number of hepatocytes able to reverse gradually diminishes with time within each window, indicating that hepatocyte population is only partially synchronous.

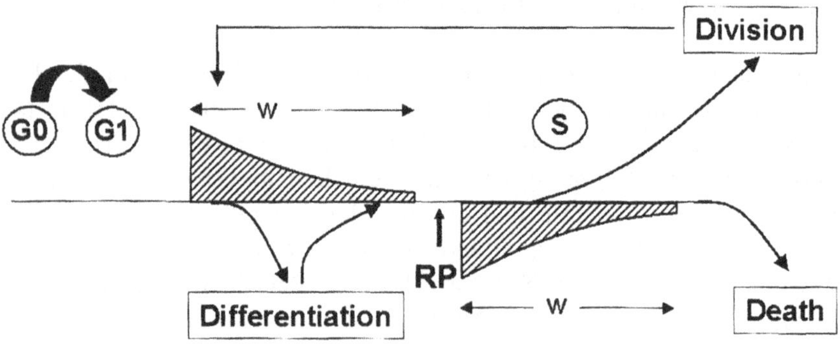

Figure 4. Location along G1 phase of the two windows of optional decisions in adult cultured hepatocytes.

17.6 Characteristics of control of S and G2/M phases in hepatocytes

17.6.1 Cell cycle control mechanisms in S and G2/M phases

Like other cell types after crossing the R-point, hepatocytes become independent of the mitogen signal, but progression in S phase remains controlled by different parameters. Cyclin A is a cell cycle regulatory protein that acts in S and mitotic phases. Inhibition of its expression blocks cell progression while over-expression advances entry to S phase in hepatocytes, suggesting that cyclin A can be a rate-limiting factor for the transition to S phase [47]. Unexpectedly, CDK1 induction strictly correlates with cyclin A induction at the G1/S transition in hepatocytes and is associated with kinase activation while CDK2 is present throughout the cell cycle. The role of CDK1 in the complex formation and activation of cyclins in hepatocytes is unclear. Whatever, its role, it represents a very specific marker of the entry to S phase in these cells *in vitro* (see Figure 1) and *in vivo* [48].

It is interesting to note that in hepatocytes, where iron metabolism is important, all iron chelators are able to induce cell cycle arrest of mitogen-stimulated hepatocytes but arrest may occur either in G1 [32] or S phase, indicating that different chelators may use different pathways of hepatocyte growth inhibition and, in some cases, may also induce apoptosis.

17.6.2 Nucleo- and cyto-kinesis regulations associated with ploidization

Hepatocytes have the ability to go through specialized cell cycles, which, during normal development of liver, result in the formation of binuclear and polyploid cells. In the adult rat liver, the majority of the hepatocytes are tetraploid (about 70%), 10-15% are octaploid and only 15-20% are diploid. One third of the hepatocytes in either rats or humans are binuclear with two diploid or two tetraploid nuclei. Among rat cultures, hepatocytes with the different ploidy status are present. However, following stimulation by a growth factor like EGF, one-half of the mitoses are of the binucleating type causing one-third of the post-mitotic cells to become binuclear. Furthermore, morphological observation of one week-old hepatocytes reveals a significant fraction of multinucleated cells. This contrasts with the *in vivo* situation where dividing cells after partial hepatectomy, are predominantly diploid. According to Seglen [49], the increased ploidy level in culture could account for the limited growth capacity of hepatocytes, but it may represent an adaptation for maximal metabolic efficacy and a protection against mutagenic change. Indeed, tetraploid or binucleated cells produce twice as much albumin as do diploid cells [50]. Presently, the mechanism that controls this ploidy status is poorly understood and is likely to involve a disregulation of mitosis control molecules such as polo-like kinases [51].

17.7 Reversible differentiation property of hepatocytes

Various cell types such as myocytes, adipocytes or neuronal and hematopoietic cells, undergo terminal differentiation after several rounds of division, *i.e.* they lose their ability to divide and die. Unlike these types of cells, mature hepatocytes, as described above, maintain a high proliferation potential after long-term quiescence, manifested by the ability of the liver *in vivo*, to regenerate at the adult stage. The mechanisms by which quiescent differentiated hepatocytes might reinitiate active proliferation is not completely understood. However, using a new human hepatocarcinoma cell line, HBG, a similar process of reversible differentiation has been observed and analysed [52]. These cells, growth-arrested at confluence in mid G1, can reinitiate active proliferation after several weeks of quiescence. Analysis of two members of the Cip/Kip family of cyclin-dependent kinase inhibitors, namely p27 and p21, proteins known to play a fundamental role in regulating cell cycle progression and terminal differentiation, suggests that these inhibitors are subject to specific regulation in liver cells. As expected, p27 accumulates with confluence whereas p21 levels, which generally remain elevated in terminally differentiated cells, are increased in HBG dividing cells but fall to very low levels during differentiation, reflecting to some extent, the *in vivo* situation where low levels of p21 are detected in adult liver [52].

17.8 Conclusions

Analysis of the different mechanisms that govern the hepatocyte evolution in culture has provided evidence that:

1. Several characteristics are shared with other highly differentiated cell types in culture in terms of survival and limited proliferating activity;
2. Some specific features are distinct from the other cell types, mainly those related to the control of cell cycle complex activity. They could account for the unique capacity of the liver to regenerate.

Nevertheless, increased understanding of control mechanisms influencing decisions between differentiation, proliferation and apoptosis is needed for further improving *in vitro* hepatocyte systems for application to treatment of human liver diseases. New strategies involving the modulation of expression in a time-dependent manner of the genes controlling the equilibrium between proliferation, differentiation and cell death are currently being set to establish hepatic cell lines, mainly from human origin.

Acknowledgments

Personal studies were supported in part by EEC Grant BIOT-CT 960052 and by the Ligue National Contre le Cancer, Comité d'Ille-et-Vilaine. C.R. and M.R. are recipients of fellowship from the Ministére de l'Education Nationale et de la Recherche and from the Association pour la Recherche contre le Cancer respectively.

References

1. Etienne PL, Baffet G, Desvergne B, Boisnard-Rissel M, Glaise D and Guguen-Guillouzo C. Transient expression of c-fos and constant expression of c-myc in freshly isolated and cultured normal adult rat hepatocytes. Oncogene Res 1988; 3: 255-262.
2. Corlu A, Ilyin G, Cariou S, Lamy I, Loyer P and Guguen-Guillouzo C. The coculture: A system for studying the regulation of liver differentiation/proliferation activity and its control. Cell Biol Toxicol 1997; 13: 235-242.
3. Dubois RD. Early changes in gene expression during liver regeneration: what do they mean? Hepatology 1990; 11: 1079-1082.
4. Fausto N. Liver regeneration: models and mechanisms. In: Bernuau D and Feldmann G, Eds. Trends and advances in liver diseases. Liver regeneration. London: John Libbey Eurotext, 1992: 1-6.
5. Liu ML, Mars WM, Zarnegar R and Michalopoulos GK. Collagenase pretreatment and the mitogenic effects of hepatocyte growth factor and transforming growth factor-α in adult rat liver. Hepatology 1994: 1521-1527.
6. Diehl AM, Yin M, Fleckenstein J, Yang SQ, Lin HZ et al. Tumor necrosis factor-α induces c-jun during the regenerative response to liver injury. Am J Physiol 1994; 267: G552-G561.
7. Diehl AM and Rai RM. Liver regeneration 3: Regulation of signal transduction during liver regeneration. FASEB J 1996; 10: 215-227.
8. Fitzgerald MJ, Webber EM, Donavan JR and Fausto N. Rapid DNA binding by nuclear factor κ B in hepatocytes at the start of liver regeneration. Cell Growth Diff 1995; 6: 417-427.
9. Ruff-Jamison S, Zhong Z, Wen Z, Chen K, Darnell JR and Cohen S. Epidermal growth factor and lipopolysaccharide activate Stat 3 transcription factor in mouse liver. J Biol Chem 1994; 269: 21933-21935.
10. Guguen-Guillouzo C, Szajner MF, Glaise D, Gregori C and Shapira F. Isozyme differentiation of aldolase and pyruvate kinase in fetal, regenerating, preneoplastic and malignant rat hepatocytes during culture. In Vitro 1980; 17: 369-377.

11. Padgham CR and Paine AJ. Altered expression of cytochrome P-450 mRNAs, and potentially of other transcripts encoding key hepatic functions, are triggered during the isolation of rat hepatocytes. Biochem J 1993; 621-624.
12. Vandenberghe Y, Morel F, Pemble S, Taylor JB, Rogiers V et al. Changes in expression of mRNA coding for glutathione S-transferase subunits 1-2 and 7 in cultured rat hepatocytes. Mol Pharmacol 1990; 37: 372-376.
13. Hanson RW. Biological role of the isoforms of C/EBP. J Biol Chem 1998; 273: 28543.
14. Greenbaum LE, Li W, Cressman DE, Peng Y, Ciliberto G, Poli V and Taub R. CCAAT enhancer-binding protein β is required for normal hepatocyte proliferation in mice after partial hepatectomy. J Clin Invest 1998; 102: 996-1007.
15. Sun XH, Copeland NG, Jenkins NA and Baltimore D. Id proteins Id1 and Id2 selectively inhibit DNA binding by one class of helix-loop-helix proteins. Mol Cell Biol 1991; 11: 5603-5611.
16. Le Jossic C, Ilyin G, Loyer P, Glaise D, Cariou S and Guguen-Guillouzo C. Expression of helix-loop-helix factor Id-1 is dependent on the hepatocyte proliferation and differentiation status in rat liver and in primary culture. Cancer Res 1994; 54: 6065-6068.
17. Evan G and Littlewood T. A matter of life and cell death. Science 1998; 281: 1317-1321.
18. Taub R. Blocking NF-κB in the liver: the good and bad news. Hepatology 1998; 27: 1445-1446.
19. Schulte-Hermann R, Bursch W, Grasl-Kraupp B, Mullauer L and Ruttkay-Nedecky B. Apoptosis and multistage carcinogenesis in rat liver. Mutat Res 1995; 333: 81-87.
20. Feldmann G. Liver apoptosis. J Hepatol 1997; 26 (Suppl 2): 1-11.
21. White E. Life, death and the pursuit of apoptosis. Genes Dev 1996; 10: 1-15.
22. Spector MS, Auer KL, Jarvis WD, Ishac EJ, Gao B, Kunos G and Dent P. Differential regulation of the mitogen-activated protein and stress-activated protein kinase cascades by adrenergic agonists in quiescent and regenerating adult rat hepatocytes. Mol Cell Biol 1997; 17: 3556-3565.
23. Hueber AO and Evan G. Traps to catch unwary oncogenes. Trends Genet 14: 1998; 364-367.
24. Loyer P, Cariou S, Glaise D, Bilodeau M, Baffet G and Guguen-Guillouzo C. Growth-factor dependence of entry and progression through G1 and S phases of adult rat hepatocytes *in vitro*. J Biol Chem 1996; 271: 11484-11492.
25. Eves EM, Xiong W, Bellacosa A, Kennedy SG, Tsichlis PN, Rosner MR and Hay N. Akt, a target of phosphatidylinositol 3-Kinase, inhibits apoptosis in a differentiating neuronal cell line. Mol Cell Biol 1998; 18: 2143-2152.
26. Kano A, Watanabe Y, Takeda N, Aizawa S and Akaike T. Analysis of IFN-γ-induced cell cycle arrest and cell death in hepatocytes. J Biochem 1997; 121: 677-683.
27. Kauffmann-Zeh A, Rodriguez-Viciana P, Ulrich E, Gilbert C, Coffer P et al. Suppression of c-Myc-induced apoptosis by Ras signalling through PI(3)K and PKB. Nature 1997; 385: 544-548.
28. Poon RYC, Toyoshima H and Hunter T. Redistribution of the CDK inhibitor p27 between different cyclin-CDK complexes in the mouse fibroblast cell cycle and in cells arrested with lovastatin or ultraviolet irradiation. Mol Biol Cell 1995; 6: 1197-1213.
29. Elledge SJ and Harper JW. The role of protein stability in the cell cycle and cancer. Biochim Biophys Acta 1998; 1377: M61-M70.

30. Bilodeau M, Talarmin H, Ilyin G, Rescan C, Glaise D et al. Skp2 induction and phosphorylation is associated with the late G1 phase of proliferating rat hepatocytes. FEBS Lett 1999; 452: 247-253.
31. Turin F, Borel C, Benchaid M, Kay A, Jamard C, Guguen-Guillouzo C and Trepo C. n-Butyrate, a cell cycle blocker, inhibits early amplification of duck hepatitis virus covalently closed circular DNA after *in vitro* infection of duck hepatocytes. J Virol 1996; 70: 2691-2696.
32. Chenoufi N, Baffet G, Drenou B, Cariou S, Desille M et al. Deferoxamine arrests *in vitro* the proliferation of porcine hepatocytes in G1 phase of the cell cycle. Liver 1998; 18: 60-66.
33. Rumin S, Gripon P, Le Seyec J, Corral-Debrinski M and Guguen-Guillouzo C. Long-term productive episomal hepatitis B virus replication in primary cultures of adult human hepatocytes infected *in vitro*. J Virol Hepatitis 1996; 3: 227-238.
34. Block GD, Locker J, Bowen WC, Petersen BE, Katyal S et al. Population expansion, clonal growth and specific differentiation patterns in primary cultures of hepatocytes induced by HGF/SF, EGF and TGF-α in a chemically defined (HGM) medium. J Cell Biol 1996; 132: 1133-1149.
35. Cable EE and Isom HC. Exposure of primary rat hepatocytes in long-term DMSO culture to selected transition metals induces hepatocyte proliferation and formation of duct-like structures. Hepatology 1997; 26: 1444-1457.
36. Karin M. The regulation of AP-1 activity by mitogen-activated protein kinases. J Biol Chem 1995; 270: 16483-16486.
37. Lavoie JN, L'Allemain G, Brunit A, Muller R and Pouyssegur J. Cyclin D1 expression is regulated positvely by the p42/p44MAPK and negatively by the p38/HOGMAPK pathway. J Biol Chem 1996; 271: 20608-20616.
38. Auer KL, Contessa J, Brenz-Verca S, Pirola L, Rusconi S et al. The Ras/Rac1/Cdc42/SEK/JNK/c-jun cascade is a key pathway by which agonists stimulate DNA synthesis in primary culture of hepatocytes. Mol Biol Cell 1998; 9: 561-573.
39. Gines P, Li X, Brown SES, Nakamura T, Guzelian PS, Heasley LE et al. Inhibitory actions of cyclic adenosine monophosphate and pertussis toxin define two distinct epidermal growth factor-regulated pathways leading to activation of mitogen-activated protein kinase in rat hepatocytes. Hepatology 1996; 23: 1167-1173.
40. Talarmin H, Rescan C, Cariou S, Glaise D, Bilodeau M et al. The MEK/ERK cascade activation is a key signalling pathway involved in the regulation of G1 phase progression in proliferating hepatocytes. Mol Cell Biol 1999; 19: 6003-6011.
41. Alessi DR, Cuenda A, Cohen P, Dudley DT and Saltiel AR. PD 98059 is a specific inhibitor of activation of mitogen-activated protein kinase kinase *in vitro* and *in vivo*. J Biol Chem 1995; 270: 27489-27494.
42. Ichihara A, Nakamura, Noda C and Tanaka K. Control of enzyme expression deduced from studies on primary cultures of hepatocytes. In: Guillouzo and Guguen-Guillouzo C, Eds. Isolated and cultured hepatocytes, London: John Libbey Eurotext Ltd/INSERM, 1986: 187-208
43. Fraslin JM, Kneip B, Voulont S, Glaise D, Munnich A and Guguen-Guillouzo C. Dependence of hepatocyte-specific gene expression on cell-cell interactions in primary culture. EMBO J 1985; 4: 2487-2491.

44. Corlu A, Kneip B, Lhadi C, Leray G, Glaise D et al. A plasma membrane protein is involved in cell contact-mediated regulation of tissue-specific genes in adult hepatocytes. J Cell Biol 1991; 115: 505-515.
45. Kost DP and Michalopoulos GK. Effect of 2% dimethyl sulfoxide on the mitogenic properties of epidermal growth factor and hepatocyte growth factor in primary hepatocyte culture. J Cell Physiol 1991; 147: 274-280.
46. Bour ES, Cornman GA and Isom HC. Tumor necrosis factor-α-induced apoptosis in hepatocytes in long-term culture. Am J Pathol 1996; 148: 485-495.
47. Rosenberg AR, Zindy F, Le Deist F, Mouly H, Metezeau P, Brechot C and Lamas E. Overexpression of human cyclin A advances entry into S phase. Oncogene 1995; 10: 1501-1509.
48. Loyer P, Glaise D, Cariou S, Baffet G, Meijer L and Guguen-Guillouzo C. Expression and activation of cdks (1 and 2) and cyclins in the cell cycle progression during liver regeneration. J Biol Chem 1994; 269: 2491-2500.
49. Seglen PO. DNA ploidy and autophagic protein degradation as determinants of hepatocellular growth and survival. Cell Biol Toxicol 1997; 13: 301-315.
50. Le Rumeur E, Guguen-Guillouzo C, Beaumont C, Saunier A and Guillouzo A. Albumin secretion and protein synthesis by cultured diploid and tetraploid rat hepatocytes separated by elutriation. Exp Cell Res 1983; 147: 247-254.
51. Nigg EA. Polo-like kinases: positive regulators of cell division from start to finish. Curr Opin Cell Biol 1998; 10: 776-783.
52. Glaise D, Ilyin GP, Loyer P, Cariou S, Bilodeau M et al. Cell cycle gene regulation in reversibly differentiated new human hepatoma cell lines. Cell Growth Differ 1998; 9: 165-176.

Chapter 18

Mechanisms of active cell death in isolated hepatocytes

Sumio Maeda
Department of Chemistry, National Industrial Research Institute of Nagoya, 1-1 Hirate-cho, Kita-ku, Nagoya, Aichi, 462-8510 Japan

Key words: active cell death, apoptosis, bile salts, calmodulin, caspase, cell density, Fas, heparin, internucleosomal DNA fragmentation, kinase, mitochondria, necrosis, protease, redox, TGF-β1, TNF-α

18.1 Introduction

Active (or programmed) cell death is a generic term for "suicide-like" cell death. Its critical feature is the dependency on vital cell functions and inherent genetic programming. It occurs in the context of a broad range of physiological functions, including homeostatic regulation of tissue mass, removal of unwanted or harmful cells in immune systems, removal of (pre)neoplastic tissues, and morphogenesis during early development [1-3]. "Apoptosis" is the principal form of active cell death and is generally defined by its characteristic morphological and biochemical changes: nuclear condensation and fragmentation, formation of apoptotic bodies and internucleosomal DNA fragmentation [4,5]. At present, there are growing numbers of reports dealing with the molecular mechanisms of apoptosis (for detailed reviews, see [6-10]). Briefly, stimuli that induce apoptosis vary widely, but their respective signalling pathways frequently converge on one mechanism, in which a few common protein families play important roles: among them are the caspase (previously called the ICE/Ced-3 family) [10,11] and the Bcl-2/Ced-9 [10,12] families. Recent data suggest that mitochondria and cytochrome c play key roles in the regulation of caspases [6,7,9]. In addition, other mechanisms including non-caspase proteases [*e.g.* 13-18], protein phosphorylation [5,19], redox regulation [20], and Ca^{2+} signaling [21], may also participate in the process of apoptosis. Other forms of caspase-independent apoptosis [22-25] as well as non-apoptotic forms of active cell death [26-30] have also been reported.

Despite the early investigation of hepatocyte apoptosis in ischemic liver by Kerr *et al.* [1] (the discoverers who coined the term "apoptosis"), studies into hepatocyte apoptosis, particularly those using primary hepatocytes, have

M.N. Berry and A.M. Edwards (eds.), The Hepatocyte Review, 281–300.

been limited [31-36]. In recent years, however, several *in vitro*, apoptotic hepatocyte systems have been identified and are being characterized. This review summarizes the recent progress of the studies on typical systems of active cell death in isolated, cultured hepatocytes.

18.2 Active cell death in isolated hepatocytes

18.2.1 Apoptosis in isolated and cultured hepatocytes caused by various stimuli

The known characteristics of *in vitro*, hepatocyte apoptosis systems are summarized in Tables 1 and 2. (There are some excellent reviews of *in vivo* hepatocyte apoptosis [37-39]). Studies are generally carried out using

Table 1. Apoptosis systems in isolated, cultured hepatocytes induced by physiological stimuli

Trigger	Sources of hepatocytes	Factors or reagents that regulate or influence apoptosis	Features, etc.
TGF-β1 [15,44-58]	rat, adult rat, fetal	caspases, calpains, O_2+glucagon, antioxidants, DMSO, peroxisome proliferators, EGF, TNF-α, Inchin-koto, NO, pRB, c-fos *(Smad family transcription factors, TAK, FKBP12)	lack of IDF induction; late induction
Fas ligand (or Anti-Fas antibody) [8,41, 59-62]	mouse, adult rat, young	caspases, serine proteases, PKC, PKA, protein tyrosine kinases, NO, peroxisome proliferators, HGF *(FADD, MORT1, caspase-8, ceramide, c-abl tyrosine kinase, FAP tyrosine phosphatase, HSP 24)	requirement for the co-presence of inhibitors of gene expression or PKC
TNF-α [8,63-69]	rat, adult mouse, adult	HSP70, protein glycosylation inhibitor, hepatitis B virus HBx protein, NO, INF-γ, DMSO *(FADD, MORT1, TRADD, RIP, RAIDD, caspase-8, caspase-2, TRAF-2, NF-κB, ceramide)	enhancement by transcriptional inhibitors but suppression by a translational inhibitor
bile salts, hydrophobic [70-78]	rat, adult	Mg^{2+}, cathepsin B, PKC, hydrophilic bile salts, oxidative stress, fructose, tagarose, sorbose	induction of apoptosis at low concentration (50 μM) but necrosis at high concentration (100 μM)
Interferon-γ [79-81]	mouse, adult	p53, p21, linoleic acid, HGF, TNF- α	synergistic effect with TNF-α
activin A [82]	rat, adult	norepinephrine	weaker effect than TGF-β
EGF+ urogastrone [31]	rat, neonatal	Ca^{2+}, glucagon	by repeated exposure; inducing hyperplasia initially, and then apoptosis

Note: Regulatory factors shown as *( ) are factors that have been proven or suggested to participate in respective signalling systems in other cells.

Table 2. Apoptosis systems in isolated, cultured hepatocytes induced by non-physiological stimuli

Trigger reagent or stimulus	Sources of hepatocytes	Expected action of the reagent	Features, etc.
vinblastine or colchicine [83]	rat, adult	microtubule antagonism	requirements for protein synthesis, extracellular Ca^{2+}, and PKC activity
staurosporine [48]	rat, adult	broad spectrum inhibition of protein phosphorylation including PKC	high concentrations of staurosporine (5-10 μM) is a general apoptosis-inducer for most cells; co-induction of necrosis
reoxygenation after hypoxia or ethanol [84, 85]	rat, adult	oxidative stress	dependency on caspases co-induction of necrosis
galactosamine [86]	mouse, adult	depletion of uridine nucleotides pool and transcriptional arrest	co-induction of necrosis
leupeptin or chloroquine [87]	rat, adult	inhibition of proteolysis	involvement of lysosomal proteolysis; co-induction of necrosis
cisplatin [88]	mouse, adult	broad spectrum effects: eg. lipid peroxidation, formation of cisplatin-DNA adducts, alterations of many enzyme activities	enhanced sensitivity in hepatocytes isolated from metallothionein-null mice
okadaic acid or microcystin [40]	rat, adult	inhibition of protein phosphatases	in suspension culture
pravastatin [89]	rat, adult	inhibition of HMG-CoA reductase resulting in inhibition of mevalonic acid metabolism	suppression by mevalonic acid
UV or 2-acetyl-aminofluorene [90]	rat, adult	DNA damage	suppression by TCDD or phenobarbital
LPS + anti-oxidants [91]	rat, adult	NO production	attenuation by taurine; co-induction of necrosis
diterpenoids from germander [92]	rat, adult	depletion of cellular thiols by their electrophilic metabilites	involvement of cytochrome P450, glutathione, Ca^{2+}, p53, and caspases

Abbreviations: HMG-CoA, 3-hydroxy-3-methylglutaryl coenzyme A; TCDD, 2,3,7,8-tetrachloro-dibenzo-p-dioxin; LPS, lipopolysaccharide.

conventional monolayer cultures of primary adult rat or mouse hepatocytes (sometimes fetal or neonatal cells are used), but suspension cultures have also been used in a few systems [40,41]. Some variability in the sensitivity to apoptotic signals exists among hepatocytes in monolayer and suspension cultures as well as in liver (see 18.2.2. and 18.2.3.), probably reflecting differences in the physiology of cells in the respective environments [42,43]. Stimuli that act as the trigger for apoptosis in hepatocytes can be broadly classified into two categories: physiological (Table 1) and non-physiological (Table 2). Physiological stimuli may play critical roles in liver with respect to the maintenance of liver homeostasis, liver development, and the inflammatory responses in liver disease or injury. Non-physiological stimuli may be useful for investigating certain pathological responses in hepatocytes as well as specific parts of apoptotic mechanisms.

Several reports have suggested that a linkage may exist between apoptosis and growth. Hepatomitogens such as HGF, EGF, and peroxisome proliferators (*e.g.* nafenopin) and phenobarbital are able to suppress certain apoptotic pathways (Table 1) [93-96,79]. This effect may be partly attributable to the action of anti-apoptotic proteins (Bcl-XL [96] and BAG-1 [97]) and to certain transcription factors (c-Fos [54], Smad [98], PPAR-α [99], *etc.*).

A physiological radical, NO (nitric oxide), at moderate concentrations, acts as a potent inhibitor of hepatocyte apoptosis by TGF-β1, Fas, and TNF-α [51,61,67,91]; but in contrast at high concentrations, NO in contrast induces both apoptosis and necrosis [55]. Recently, Kim *et al.* have reported that NO prevents apoptosis in hepatocytes by either directly or indirectly inhibiting caspase-3-like activation via a cGMP-dependent mechanism and by direct inhibition of caspase-3-like activity through protein S-nitrosylation [61]. They also suggested the involvement of Heat Shock Protein 70 (HSP70) induction in the protective effects of NO [67].

The following sections consist of overviews of four well-studied apoptotic systems involving physiological stimuli as well as a unique active cell death system that we discovered. The primary features, mechanistic aspects and putative physiological significance of each system are summarized.

18.2.2 TGF-β1-induced apoptosis

TGF-β is a multifunctional cytokine with effects on cell proliferation and differentiation as well as on the synthesis and degradation of extracellular matrix (ECM) [100,101]. With respect to hepatocytes, TGF-β is known to be a potent growth inhibitor both in culture and *in vivo* [102,103]. TGF-β1-induced apoptosis in hepatocytes was initially described by Oberhammer *et al.* [44,104] in primary cultures of adult rat hepatocytes. Apoptosis induced by another TGF-β family protein, activin A, was also reported by Schwall *et al.* [82], and recently, Sanchez *et al.* [45] described TGF-β1-induced apoptosis in primary fetal rat hepatocyte cultures. Apoptosis induced by TGF-β1, combined with its capacity to arrest growth, is currently thought to play important roles in the maintenance of liver parenchymal mass, termination of liver regeneration, and prevention of tumor development [37,39].

The molecular mechanism of TGF-β-signalling is being actively investigated in other cell systems, and some of the signalling molecules involved have been identified [56,57]. These include TGF-β receptors I and II (transmembrane serine-threonine kinases), TAK (TGF-β activated kinase), pRB (retinoblastoma gene product), FKBP12 (FK506 binding protein) and Smad family transcriptional factors. At present no direct linkage between any of these factors and TGF-β1-induced apoptosis in hepatocytes has been found, although pRB may function as a negative regulator [58].

One notable feature of TGF-β1-induced hepatocyte apoptosis is the absence of induction of internucleosomal DNA fragmentation [46]; the reason and ramifications of this phenomenon are unclear. Another characteristic feature is the time lag between TGF-β1 stimulation and cell death. In simple monolayer cultures, cell death occurs approximately 2 days after TGF-β1 stimulation [44,52]. Other apoptotic stimuli, by contrast, typically evoke cell death within several hours to 1 day [59,63]. In cultures treated with cyproterone acetate or in regressing liver, however, cell death occurs within 12 hours after TGF-β1 stimulation [44,47], indicating that the process of TGF-β1-induced apoptosis is susceptible to the differences in cell physiology, in this case possibly cell cycle.

The involvement of caspases in TGF-β1-induced apoptosis in hepatocytes has been demonstrated using caspase inhibitors and substrates [48,49]. Similarly, experiments with calpain inhibitors showed that calpain also participates in the induction of the apoptosis [15].

Ohno *et al.* [50] reported that TGF-β1 induces apoptosis more frequently under pericentral-equivalent conditions (4% O_2, 1 nM glucagon, and 10 nM insulin) than periportal conditions (13% O_2, 10 nM glucagon, and 10 nM insulin), suggesting that cellular redox status or glycolytic activity may influence the susceptibility of hepatocytes to TGF-β1-induced apoptosis. Importance of the cellular redox status in TGF-β1-induced apoptosis or TGF-β1-signalling in hepatocytes is also supported by the results of experiments employing antioxidants [51,52] and by other data [105,106].

Using a fetal hepatocyte system, Sanchez *et al.* [45,53] showed that protein synthesis was required to initiate TGF-β1-mediated apoptosis, and that ROS (reactive oxygen species) production and glutathione depletion also contributed. These authors suggested that such oxidative stress may lead to the induction of c-fos expression and increased binding to DNA of the transcription factor, AP-1 [45]. Consistent with this conclusion, EGF blocks c-fos induction by TGF-β1 and suppresses TGF-β1-induced apoptosis in fetal hepatocytes [54]. In contrast, HGF, which does not block c-fos, likewise does not suppress TGF-β1-induced apoptosis [54].

18.2.3 Fas-mediated apoptosis

Fas (also known as APO-1 or CD-95) is a member of the TNF receptor family, and its sole identified function is to mediate transduction of the death signal in cells [107]. Its physiological ligand is Fas ligand (FasL), which belongs to the TNF family, but some specific anti-Fas monoclonal antibodies can also initiate Fas-mediated apoptosis [107]. The function of Fas in the immune system is well-known: it prevents autoreactivity and maintains immunohomeostasis [108,109]. Fas is also abundantly expressed in liver [110], where it may be involved in pathological phenomena such as fulminant hepatitis as well as in the maintenance of liver homeostasis [111-113].

The molecular mechanism of Fas-mediated apoptosis has been extensively investigated in other cell types and the following scheme has emerged [8]: FasL-binding causes trimerisation of the Fas receptor, which in turn recruits caspase-8 via an adapter protein, FADD/MORT1, resulting in initiation of the caspase cascades. Evidence indicates that other regulators, including ceramide, c-abl tyrosine kinase, FAP tyrosine phosphatase, HSP 24, p59fyn kinase, and FAF, are also involved [8]; but how these proteins regulate the process is currently unknown.

When intraperitoneally injected into mice, anti-Fas antibody efficiently induces hepatocyte apoptosis in the absence other stimuli [113]. In culture, however, FasL or anti-Fas antibody act only in conjunction with additional factors [59,60], which is indicative of the alteration of cell physiology that may take place in culture. Ni *et al.* [59] first demonstrated Fas-mediated apoptosis *in vitro* in mouse hepatocytes. They showed that inhibitors of gene expression (actinomycin D or cycloheximide) or protein kinase C inhibitors (H-7 or HA1004) both sensitize hepatocytes to Fas-mediated apoptosis. This suggests that in culture, mouse hepatocytes express proteins that inhibit the Fas signalling, and that protein phosphorylation by PKC is also involved. Rouquet *et al.* [60] suggested that Fas signalling consists of 2 independent pathways, involving different classes of serine proteases. The cycloheximide-enhanced pathway can be inhibited by TPCK or TAME but not by TLCK (TPCK, TAME, and TLCK are serine protease inhibitors with slightly different specificities [60]). The H-7-enhanced pathway is inhibited by TLCK but not by TPCK and TAME. These two pathways also differ with respect to their sensitivities to protein tyrosine kinases inhibitors. The former is inhibited by genistein and herbymicine A, the latter is not [60].

Mechanisms dependent on cyclic nucleotide second messengers also appear to modulate Fas-mediated hepatocyte apoptosis. Kim *et al.* [61] reported that cGMP and cGMP-dependent protein kinase participate in the NO-associated protection from Fas-mediated apoptosis. Using suspension cultures of primary rat hepatocytes, Fladmark *et al.* [41] found that activators of cAMP-dependent protein kinase (PKA) protected against Fas-mediated apoptosis, whereas a PKA inhibitor enhanced the apoptosis, suggesting the involvement of PKA in suppression of apoptosis. In monolayer cultures of mouse hepatocytes however, a PKA inhibitor was unable to affect Fas-mediated apoptosis [59]. This discrepancy in the apparent action of PKA probably reflects differences in cell physiology between suspension and monolayer cultures. Indeed, these cultures also differ markedly with respect to the time course of Fas-mediated apoptosis: cell death occurs significantly sooner in suspension culture (~1.5-2 hr) [41] than in monolayer culture (~8-16 hr) [59,60].

There is also an interesting report that insoluble ECM (*e.g.* type IV collagen) increases Fas expression in primary rat hepatocytes [62], which makes the point that interactions between cells and ECM may also be important regulators of apoptosis.

18.2.4 TNF-α-induced apoptosis

Although TNF-α and FasL belong to the same protein family, in contrast to the Fas system, TNF-α elicits a wide array of biological responses, including inflammation, tumor necrosis, cell proliferation, differentiation, and apoptosis [114]. In hepatocytes, it induces inflammatory responses, mitogenic activity and apoptosis [63,64,102,115]. TNF-α-signalling has been extensively investigated in other cell types [8], and is apparently mediated via activation of type 1 TNF receptor (TNFR1). The apoptosis pathway is essentially the same as that for Fas, except that the TNF-α system requires an additional molecule, TRADD, which mediates the interaction between TNFR1 and FADD/MORT1. TRADD also recruits RIP, which may trigger a second apoptosis-inducing pathway via its interaction with RAIDD and caspase-2. On the other hand, there is a suppressive pathway in which the TRADD-RIP-TRAF2 complex activates NF-κB thereby protecting cells from apoptosis. Indeed, the cytotoxicity of TNF-α can often be potentiated by inhibitors of gene expression (*e.g.* actinomycin D, cycloheximide) [8,63], which are thought to inhibit NF-κB-induced gene expression. In addition, the TNF-signalling pathway diverges downstream from TRAF2 resulting in activation of Jun N-terminal kinase and AP-1, but this pathway is not currently thought to be linked to TNF-α-induced apoptosis, at least in certain cells [116].

TNF-α-induced apoptosis in primary hepatocytes was first reported by Shinagawa *et al.* [64]. The effect of TNF-α was enhanced in the presence of INF-γ, which has itself been found to be an apoptotic stimulus for hepatocytes [79,80]. TNF-α-induced apoptosis is augmented by various transcription inhibitors (actinomycin D, α-amanitin, or D-galactosamine) [63] and by hepatitis B virus HBx protein [65]. Conversely, the apoptosis is suppressed by a translation inhibitor (cycloheximide) [63], by a protein glycosylation inhibitor (tunicamycin) [66], and by NO [67]. The suppressive effect of translation inhibitors on TNF-α-induced hepatocyte apoptosis forms a marked contrast to their enhancing effects on Fas-mediated hepatocyte apoptosis [59,60] (see section 18.2.3.) as well as on TNF-α-induced apoptosis in other cell types [8,63].

As a means of assessing the action of TNF-α, Bour *et al.* [68] adapted hepatocytes to long-term culture using EGF and DMSO. They observed that TNF-α induced apoptosis only after DMSO removal and concluded that in the absence of DMSO, hepatocytes dedifferentiated making them more susceptible to TNF-α-induced apoptosis than the differentiated hepatocytes found in the presence of DMSO. It is interesting to note that DMSO, conversely, sensitizes hepatocytes to apoptosis induced by TGF-β1 in short-term culture [52], suggesting the complexity and flexibility of the apoptotic mechanisms according to the variations of trigger stimuli and culture environments.

It is of interest that some studies have shown TNF-α to be protective against cell death. For instance, Rolfe *et al.* [117] observed that TNF-α suppressed apoptosis induced by TGF-β1, and Imanishi *et al.* [69] found that pretreating hepatocytes with TNF-α suppressed H_2O_2-induced cell death; although, in the latter case, simultaneous treatment with TNF-α and H_2O_2

increased cell death. In these cases, suppression of cell death was probably due to the activation of NF-κB and illustrates that the timing of TNF-α administration is an important factor in its regulation of cell death.

18.2.5 Bile salt-induced apoptosis

Although bile salts are secreted by hepatocytes, their detergent-like nature is toxic to the hepatocytes themselves. Indeed, hepatocellular retention of bile salts contributes to cholestatic liver disease [118,119]. Gores *et al.* [70] found that bile salt-induced hepatocyte injury is due in part to apoptosis. Hydrophobic bile salts (glycodeoxycholate and glycochenodeoxycholate) primarily induce apoptosis at low concentrations (50 μM), but at high concentrations (>100 μM) they cause necrosis [70]. Conversely, the hydrophilic bile salts, ursodeoxycholate and taurodeoxycholate, reduced both apoptosis and necrosis caused by the hydrophobic bile salts [71,72]. The protective effect of hydrophilic bile salts may be due to prevention of plasma membrane damage as well as to modulation of mitochondrial membrane perturbation [73].

In contrast to most other apoptotic pathways [5,21], Mg^{2+} rather than Ca^{2+} appears to be important in mediating bile salt-induced apoptosis [70]. Moreover, the effect of bile salts are not inhibited by gene expression inhibitors [70], indicating that the mechanism is independent of *de novo* protein synthesis. While the involvement of caspases has not yet been shown with respect to bile salts, there is evidence that cathepsin B may be involved [74]. In response to the action of bile salts, cathepsin B, which is normally localized in lysosomes, is thought to translocate to the nucleus where it contributes to the induction of apoptosis [74]. Recently, Webster *et al.* [75] have demonstrated that agents that increase intracellular cAMP protect against bile salt-induced apoptosis. From studies using several kinase inhibitors, they have also suggested that activation of MAPK by bile salts may be involved in the apoptotic process, and that cAMP-mediated cytoprotection appears to involve PKA, MAPK, and phosphatidylinositol 3-kinase. Finally, there is evidence that oxidative stress and activation of PKC also play roles in bile salt-induced apoptosis [76,77]; whereas fructose and the poorly metabolized ketohexoses, tagarose and sorbose, appear to attenuate bile salt-induced apoptosis by as yet unknown mechanisms [78].

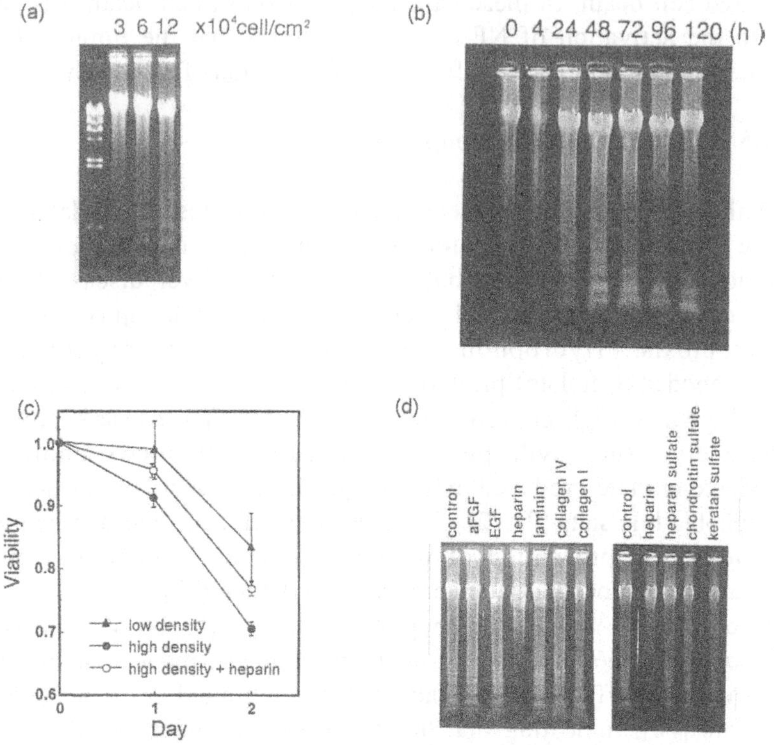

Figure 1. Cell density-dependent internucleosomal DNA fragmentation (IDF) in primary culture of rat hepatocytes. (a) Cell density-dependent IDF; (b) Time course of IDF in high density culture; (c) Time course of cell death at high and low densities and the effect of heparin on cell death; (d) Effect of heparin on IDF. Hepatocytes were prepared from Wistar rats (female, 7-9 weeks old) using an *in situ* collagenase perfusion method and four subsequent low-speed centrifugations [125]. Initial cell viability was assessed by Trypan blue exclusion, and was usually >90% (at least 86%). The isolated hepatocytes were resuspended in Williams' medium E supplemented with 10% FCS and 10^{-7} M dexamethasone, and then seeded into type I collagen-coated 6-well culture plates at densities between 3 x 10^4 cells/cm^2 and 12 x 10^4 cells/cm^2. The cells were allowed to attach for 4 h at 37°C under an atmosphere of 5% CO_2/95% air (v/v). The medium was then replaced with fresh Williams' medium E supplemented with 10^{-7} M insulin, 10^{-7} M dexamethasone, and 0.7 μg/ml aprotinin, which also served to remove non-attached cells. After an additional 24 h, 48 h or various other times of culture, the cells were used for DNA isolation and other analyses. Total DNA was isolated from cultured cells according to the conventional method with minor modifications [123, 124], and analysed by 0.8% agarose gel electrophoresis. The amount of DNA applied to each lane was normalized to the amount derived from 6 x 10^5 cells. Viability of cultured hepatocytes was assessed by MTT assay [126, 127]. (Reproduced with permission from S. Maeda, *et al.*: Biochem Biophys Res Commn 1993; 195: 270-275 [123] and S. Maeda, *et al.*: J Biochem (Tokyo) 1995; 118: 1161-1165 [124])

18.2.6 Cell death accompanied by internucleosomal DNA fragmentation in high-density cell cultures

I will now introduce the phenomenon of density-dependent cell death in hepatocyte cultures. Although it is not considered to be "typical" apoptosis, this form of cell death is clearly active in nature, and experimental evidence implies that it is mediated by a unique mechanism.

In primary cultures of hepatocytes, various cell functions are frequently modulated by the density of the cultures [120-122]. We recently observed that in ordinary monolayer cultures of adult rat hepatocytes, spontaneous internucleosomal DNA fragmentation and cell death occurred in a cell density-dependent manner (Figures 1A and 1C) [123,124]. In high-density cultures, DNA fragmentation occurred between 4 and 48 h after plating (Figure 1B). Essentially the same phenomenon was also observed by Shinzawa *et al.* [128,129] in primary cultures of mouse hepatocytes. Despite the occurrence of DNA fragmentation, no other features of apoptotic morphology were observed, suggesting that this cell death is at least not "typical" apoptosis [124].

Consistent with this idea, this form of cell death was not suppressed by caspase inhibitors (our unpublished data and [128]). It should be noted, however, that the process was dependent on the availability of cellular energy (mitochondrial respiratory function and ATP), gene expression (*de novo* protein synthesis), and calmodulin [124,129], indicating its active and controlled nature. Our experiments further showed that soluble additives (*e.g.* fetal bovine serum, dexamethasone, insulin, or growth factors) or conditioned media from cultures of varying density had little influence on the incidence of DNA fragmentation [123]. On the other hand, heparin (5 μg/ml), a component of ECM, specifically suppressed DNA fragmentation and cell death (Figures 1C and 1D) [123]. Thus, cell-cell or cell-ECM interactions, rather than soluble factors, appear to be involved in the triggering or control of this cell death. The importance of cell-cell contacts and plasma membrane physiology in priming cell death has also been pointed out by Shinzawa *et al.* [128], based on their experiments using in-gel culture or a cytoprotective amino acid, glycine.

Our current view is that cell death induced by high cell density, in the absence of a soluble trigger, involves a unique mechanism. While details of this death mechanism remain unknown for the most part, the calmodulin-dependent nuclear Ca^{2+} transport system [130-132] is most likely involved [124]. Because the cell death occurs specifically at early times of culture, high cell density is apparently not by itself sufficient to elicit the response. Instead, prior disruption of the inherent cell-cell and cell-ECM interactions associated with hepatocyte isolation may be required to trigger cell death. Given the suppressive effect of heparin (Figures 1C and 1D), it is noteworthy that heparan sulphate, which comprises part of the proteoglycan on the surfaces of hepatocytes *in situ*, contains a remarkable proportion of heparin-like structure [133]. Thus, the presence or absence of the heparin-like heparan sulphate on the cell surfaces may play a critical role in the suppression of this form of cell death.

Physiologically, density-dependent cell death mechanism would be expected to contribute to suppression of hepatocarcinogenesis. Indeed, Shinzawa *et al.* [129] demonstrated that density-dependent DNA fragmentation and cell death were absent in hepatoma cell lines.

18.3 Perspectives

In vitro, active cell death of hepatocytes appears to involve somewhat different mechanisms, depending upon which stimulus is used for the induction. Nevertheless, it appears that at least 3 classes of intracellular factors are typically involved: proteases, protein kinases, and factors altering cellular redox status. The lack of a unified scheme describing the mechanism of cell death is due in part to our incomplete understanding of the process; however, it probably also reflects the diversity and complexity of the mechanisms in hepatocytes. Further investigation, including precise dissection of each pathway leading to cell death, comparative analyses, and studies of the links between active cell death and other cell functions, will be necessary before a comprehensive understanding of the mechanisms of active cell death in hepatocytes can be achieved. This understanding may help in the development of novel techniques for controlling cell death both *in vivo* and *in vitro*, and should prove useful in a variety of medical and biotechnological applications. For example, simultaneous control of growth, differentiation, and cell death should enable future development of devices (*e.g.* bio-artificial liver and bio-reactors using isolated hepatocytes) that possess the kind of steady-state "cell-turnover" seen in liver.

Acknowledgements

I would like to thank Drs. Takao Saito, Kong Hua Lin, and Hidetoshi Inagaki (NIRIN) for helpful discussions; and Dr. Kouei Shinzawa (Tokyo Institute of Technology) for providing information in his thesis.

References

1. Kerr JF, Wyllie AH and Currie AR. Apoptosis: a basic biological phenomenon with wide-ranging implications in tissue kinetics. Br J Cancer 1972; 26: 239-257.
2. Tomei LD and Cope FO, Eds. Apoptosis II: the molecular basis of cell death in desease. In: Inglis J and Witkowski JA, Eds. Current Comminications in Cell & Molecular Biology. Cold Spring Harbor: Cold Spring Harbor Laboratory Press, 1994: vol 8.
3. Tomei LD and Cope FO, Eds. Apoptosis: the molecular basis of cell death. In: Inglis J and Witkowski JA, Eds. Current Communications in Cell & Molecular Biology. Cold Spring Harbor: Cold Spring Harbor Laboratory Press, 1991: vol 3.
4. Kerr JFR and Harmon BV. Definition and incidence of apoptosis: an historical perspective. In: Tomei LD and Cope FO, Eds. Apoptosis: the molecular basis of cell death. Cold Spring Harbor: Cold Spring Harbor Laboratory Press, 1991.
5. Schwartzman RA and Cidlowski JA. Apoptosis: the biochemistry and molecular biology of programmed cell death. Endocr Rev 1993; 14: 133-151.
6. Kroemer G, Dallaporta B and Resche-Rigon M. The mitochondrial death/life regulator in apoptosis and necrosis. Annu Rev Physiol 1998; 60: 619-642.
7. Mignotte B and Vayssiere JL. Mitochondria and apoptosis. Eur J Biochem 1998; 252: 1-15.
8. Nagata S. Apoptosis by death factor. Cell 1997; 88: 355-365.
9. Reed JC. Cytochrome c: can't live with it - can't live without it. Cell 1997; 91: 559-562.
10. Salvesen GS and Dixit VM. Caspases: intracellular signaling by proteolysis. Cell 1997; 91: 443-446.
11. Ellis RE, Yuan JY and Horvitz HR. Mechanisms and functions of cell death. Annu Rev Cell Biol 1991; 7: 663-698.
12. Newton K and Strasser A. The Bcl-2 family and cell death regulation. Curr Opin Genet Dev 1998; 8: 68-75.
13. Wright SC, Schellenberger U, Wang H, Kinder DH, Talhouk JW and Larrick JW. Activation of CPP32-like proteases is not sufficient to trigger apoptosis: inhibition of apoptosis by agents that suppress activation of AP24, but not CPP32-like activity. J Exp Med 1997; 186: 1107-1117.
14. Zhivotovsky B, Gahm A, Ankarcrona M, Nicotera P and Orrenius S. Multiple proteases are involved in thymocyte apoptosis. Exp Cell Res 1995; 221: 404-412.
15. Gressner AM, Lahme B and Roth S. Attenuation of TGF-β-induced apoptosis in primary cultures of hepatocytes by calpain inhibitors. Biochem Biophys Res Commun 1997; 231: 457-462.
16. Lu Q and Mellgren RL. Calpain inhibitors and serine protease inhibitors can produce apoptosis in HL-60 cells. Arch Biochem Biophys 1996; 334: 175-181.
17. Drexler HC. Activation of the cell death program by inhibition of proteasome function. Proc Natl Acad Sci USA 1997; 94: 855-860.
18. Grimm LM, Goldberg AL, Poirier GG, Schwartz LM and Osborne BA. Proteasomes play an essential role in thymocyte apoptosis. EMBO J 1996; 15: 3835-3844.
19. Lavin MF, Watters D and Song Q. Role of protein kinase activity in apoptosis. Experientia 1996; 52: 979-994.
20. Wolf JT, Ross D and Cohen GM. A role for metals and free radicals in the induction of apoptosis in thymocytes. FEBS Lett 1994; 352: 58-62.

21. McConkey DJ and Orrenius S. The role of calcium in the regulation of apoptosis. Biochem Biophys Res Commun 1997; 239: 357-366.
22. De Maria R, Lenti L, Malisan F, d'Agostino F, Tomassini B, Zeuner A, Rippo MR et al. Requirement for GD3 ganglioside in CD95- and ceramide-induced apoptosis. Science 1997; 277: 1652-1655.
23. McCarthy NJ, Whyte MK, Gilbert CS and Evan GI. Inhibition of Ced-3/ICE-related proteases does not prevent cell death induced by oncogenes, DNA damage, or the Bcl-2 homologue Bak. J Cell Biol 1997; 136: 215-227.
24. Miller TM, Moulder KL, Knudson CM, Creedon DJ, Deshmukh M, Korsmeyer SJ and Johnson EM Jr. Bax deletion further orders the cell death pathway in cerebellar granule cells and suggests a caspase-independent pathway to cell death. J Cell Biol 1997; 139: 205-217.
25. Monney L, Otter I, Olivier R, Ozer HL, Haas AL, Omura S and Borner C. Defects in the ubiquitin pathway induce caspase-independent apoptosis blocked by Bcl-2. J Biol Chem 1998; 273: 6121-6131.
26. Clarke PG. Developmental cell death: morphological diversity and multiple mechanisms. Anat Embryol (Berl) 1990; 181: 195-213.
27. Gavrieli Y, Sherman Y and Ben-Sasson SA. Identification of programmed cell death *in situ* via specific labeling of nuclear DNA fragmentation. J Cell Biol 1992; 119: 493-501.
28. Schwartz LM, Smith SW, Jones ME and Osborne BA. Do all programmed cell deaths occur via apoptosis? Proc Natl Acad Sci USA 1993; 90: 980-984.
29. Schweichel JU and Merker HJ. The morphology of various types of cell death in prenatal tissues. Teratology 1973; 7: 253-266.
30. Szende B, Keri G, Szegedi Z, Benedeczky I, Csikos A, Orfi L and Gazit A. Tyrphostin induces non-apoptotic programmed cell death in colon tumor cells. Cell Biol Int 1995; 19: 903-911.
31. Armato U, Romano F, Andreis PG, Paccagnella L and Marchesini C. Growth stimulation and apoptosis induced in cultures of neonatal rat liver cells by repeated exposures to epidermal growth factor/urogastrone with or without associated pancreatic hormones. Cell Tissue Res 1986; 245: 471-480.
32. Benedetti A, Jezequel AM and Orlandi F. Preferential distribution of apoptotic bodies in acinar zone 3 of normal human and rat liver. J Hepatol 1988; 7: 319-324.
33. Columbano A, Ledda-Columbano GM, Rao PM, Rajalakshmi S and Sarma DS. Occurrence of cell death (apoptosis) in preneoplastic and neoplastic liver cells. A sequential study. Am J Pathol 1984; 116: 441-446.
34. Fesus L, Thomazy V and Falus A. Induction and activation of tissue transglutaminase during programmed cell death. FEBS Lett 1987; 224: 104-108.
35. McConkey DJ, Hartzell P, Nicotera P, Wyllie AH and Orrenius S. Stimulation of endogenous endonuclease activity in hepatocytes exposed to oxidative stress. Toxicol Lett 1988; 42: 123-130.
36. Pritchard DJ and Butler WH. Apoptosis - the mechanism of cell death in dimethylnitrosamine-induced hepatotoxicity. J Pathol 1989; 158: 253-260.
37. Feldmann G. Liver apoptosis. J Hepatol 1997; 2: 1-11.
38. Galle PR. Apoptosis in liver disease. J Hepatol 1997; 27: 405-412.
39. Schulte-Hermann R, Bursch W, Low-Baselli A, Wagner A and Grasl-Kraupp B. Apoptosis in the liver and its role in hepatocarcinogenesis. Cell Biol Toxicol 1997; 13: 339-348.

40. Boe R, Gjertsen BT, Vintermyr OK, Houge G, Lanotte M and Doskeland SO. The protein phosphatase inhibitor okadaic acid induces morphological changes typical of apoptosis in mammalian cells. Exp Cell Res 1991; 195: 237-246.
41. Fladmark KE, Gjertsen BT, Doskeland SO and Vintermyr OK. Fas/APO-1(CD95)-induced apoptosis of primary hepatocytes is inhibited by cAMP. Biochem Biophys Res Commun 1997; 232: 20-25.
42. Berry MN, Edwards AM and Barritt GJ. Hepatocyte isolation for primary culture and methods for non-adherent culture. Isolated hepatocytes: preparation, properties and applications. In: Burdon RH and Knippenberg PH, Eds. Laboratory techniques in biochemistry and molecular biology. Amsterdam: Elsevier, 1991: 237-265.
43. Berry MN, Edwards AM and Barritt GJ. Monolayer culture of hepatocytes. Isolated hepatocytes: preparation, properties and applications. In: Burdon RH and Knippenberg PH, Eds. Laboratory techniques in biochemistry and molecular biology. Amsterdam: Elsevier, 1991: 265-354.
44. Oberhammer FA, Pavelka M, Sharma S, Tiefenbacher R, Purchio AF, Bursch W and Schulte-Hermann R. Induction of apoptosis in cultured hepatocytes and in regressing liver by transforming growth factor β1. Proc Natl Acad Sci USA 1992; 89: 5408-5412.
45. Sanchez A, Alvarez AM, Benito M and Fabregat I. Apoptosis induced by transforming growth factor-β in fetal hepatocyte primary cultures: involvement of reactive oxygen intermediates. J Biol Chem 1996; 271: 7416-7422.
46. Oberhammer F, Fritsch G, Schmied M, Pavelka M, Printz D, Purchio T, Lassmann H et al. Condensation of the chromatin at the membrane of an apoptotic nucleus is not associated with activation of an endonuclease. J Cell Sci 1993; 104: 317-326.
47. Oberhammer FA and Qin HM. Effect of three tumour promoters on the stability of hepatocyte cultures and apoptosis after transforming growth factor-β1. Carcinogenesis 1995; 16: 1363-1371.
48. Cain K, Inayat-Hussain SH, Couet C and Cohen GM. A cleavage-site-directed inhibitor of interleukin-1 β-converting enzyme-like proteases inhibits apoptosis in primary cultures of rat hepatocytes. Biochem J 1996; 314: 27-32.
49. Inayat-Hussain SH, Couet C, Cohen GM and Cain K. Processing/activation of CPP32-like proteases is involved in transforming growth factor β1-induced apoptosis in rat hepatocytes. Hepatology 1997; 25: 1516-1526.
50. Ohno K, Ammann P, Fasciati R and Maier P. Transforming growth factor β1 preferentially induces apoptotic cell death in rat hepatocytes cultured under pericentral-equivalent conditions. Toxicol Appl Pharmacol 1995; 132: 227-236.
51. Huang YL, Shen CK, Luh TY, Yang HC, Hwang KC and Chou CK. Blockage of apoptotic signaling of transforming growth factor-β in human hepatoma cells by carboxyfullerene. Eur J Biochem 1998; 254: 38-43.
52. Maeda S, Miyazawa A, Lin KH, Inagaki H and Saito T. Promotive effect of DMSO on TGF-β1-induced apoptosis in primary culture of rat hepatocytes. J Biochem Mol Biol Biophys 1997; 1: 117-124.
53. Sanchez A, Alvarez AM, Benito M and Fabregat I. Cycloheximide prevents apoptosis, reactive oxygen species production, and glutathione depletion induced by transforming growth factor β in fetal rat hepatocytes in primary culture. Hepatology 1997; 26: 935-943.

54. Fabregat I, Sanchez A, Alvarez AM, Nakamura T and Benito M. Epidermal growth factor, but not hepatocyte growth factor, suppresses the apoptosis induced by transforming growth factor-β in fetal hepatocytes in primary culture. FEBS Lett 1996; 384: 14-18.
55. Martin-Sanz P, Diaz-Guerra MJ, Casado M and Bosca L. Bacterial lipopolysaccharide antagonizes transforming growth factor β1-induced apoptosis in primary cultures of hepatocytes. Hepatology 1996; 23: 1200-1207.
56. Baker JC and Harland RM. From receptor to nucleus: the Smad pathway. Curr Opin Genet Dev 1997; 7: 467-473.
57. Heldin CH, Miyazono K and ten Dijke P. TGF-β signalling from cell membrane to nucleus through SMAD proteins. Nature 1997; 390: 465-471.
58. Fan G, Ma X, Kren BT and Steer CJ. The retinoblastoma gene product inhibits TGF-β1 induced apoptosis in primary rat hepatocytes and human HuH-7 hepatoma cells. Oncogene 1996; 12: 1909-1919.
59. Ni R, Tomita Y, Matsuda K, Ichihara A, Ishimura K, Ogasawara J and Nagata S. Fas-mediated apoptosis in primary cultured mouse hepatocytes. Exp Cell Res 1994; 215: 332-337.
60. Rouquet N, Carlier K, Briand P, Wiels J and Joulin V. Multiple pathways of Fas-induced apoptosis in primary culture of hepatocytes. Biochem Biophys Res Commun 1996; 229: 27-35.
61. Kim YM, Talanian RV and Billiar TR. Nitric oxide inhibits apoptosis by preventing increases in caspase-3-like activity via two distinct mechanisms. J Biol Chem 1997; 272: 31138-31148.
62. Fine A, Miranda K, Farmer SR and Anderson NL. Effect of insoluble extracellular matrix molecules on Fas expression in epithelial cells. J Cell Physiol 1998; 174: 285-292.
63. Leist M, Gantner F, Bohlinger I, Germann PG, Tiegs G and Wendel A. Murine hepatocyte apoptosis induced *in vitro* and *in vivo* by TNF-α requires transcriptional arrest. J Immunol 1994; 153: 1778-1788.
64. Shinagawa T, Yoshioka K, Kakumu S, Wakita T, Ishikawa T, Itoh Y and Takayanagi M. Apoptosis in cultured rat hepatocytes: the effects of tumour necrosis factor α and interferon γ. J Pathol 1991; 165: 247-253.
65. Su F and Schneider RJ. Hepatitis B virus HBx protein sensitizes cells to apoptotic killing by tumor necrosis factor α. Proc Natl Acad Sci USA 1997; 94: 8744-8749.
66. Leist M and Wendel A. Tunicamycin potently inhibits tumor necrosis factor-induced hepatocyte apoptosis. Eur J Pharmacol 1995; 292: 201-204.
67. Kim YM, de Vera ME, Watkins SC and Billiar TR. Nitric oxide protects cultured rat hepatocytes from tumor necrosis factor-α-induced apoptosis by inducing heat shock protein 70 expression. J Biol Chem 1997; 272: 1402-1411.
68. Bour ES, Ward LK, Cornman GA and Isom HC. Tumor necrosis factor-α-induced apoptosis in hepatocytes in long-term culture. Am J Pathol 1996; 148: 485-495.
69. Imanishi H, Scales WE and Campbell DA Jr. Tumor necrosis factor α alters the cytotoxic effect of hydrogen peroxide in cultured hepatocytes. Biochem Biophys Res Commun 1997; 230: 120-124.
70. Patel T, Bronk SF and Gores GJ. Increases of intracellular magnesium promote glycodeoxycholate-induced apoptosis in rat hepatocytes. J Clin Invest 1994; 94: 2183-2192.

71. Benz C, Angermuller S, Tox U, Kloters-Plachky P, Riedel HD, Sauer P, Stremmel W et al. Effect of tauroursodeoxycholic acid on bile-acid-induced apoptosis and cytolysis in rat hepatocytes. J Hepatol 1998; 28: 99-106.
72. Rodrigues CM, Fan G, Wong PY, Kren BT and Steer CJ. Ursodeoxycholic acid may inhibit deoxycholic acid-induced apoptosis by modulating mitochondrial transmembrane potential and reactive oxygen species production. Mol Med 1998; 4: 165-178.
73. Rodrigues CMP, Fan G, Ma X, Kren BT and Steer CJ. A novel role for ursodeoxycholic acid in inhibiting apoptosis by modulating mitochondrial membrane perturbation. J Clin Invest 1998; 101: 2790-2799.
74. Roberts LR, Kurosawa H, Bronk SF, Fesmier PJ, Agellon LB, Leung WY, Mao F et al. Cathepsin B contributes to bile salt-induced apoptosis of rat hepatocytes. Gastroenterology 1997; 113: 1714-1726.
75. Webster CR and Anwer MS. Cyclic adenosine monophosphate-mediated protection against bile acid-induced apoptosis in cultured rat hepatocytes. Hepatology 1998; 27: 1324-1331.
76. Jones BA, Rao YP, Stravitz RT and Gores GJ. Bile salt-induced apoptosis of hepatocytes involves activation of protein kinase C. Am J Physiol 1997; 272: G1109-G1115.
77. Patel T and Gores GJ. Inhibition of bile-salt-induced hepatocyte apoptosis by the antioxidant lazaroid U83836E. Toxicol Appl Pharmacol 1997; 142: 116-122.
78. Zeid IM, Bronk SF, Fesmier PJ and Gores GJ. Cytoprotection by fructose and other ketohexoses during bile salt-induced apoptosis of hepatocytes. Hepatology 1997; 25: 81-86.
79. Morita M, Watanabe Y and Akaike T. Protective effect of hepatocyte growth factor on interferon-γ-induced cytotoxicity in mouse hepatocytes. Hepatology 1995; 21: 1585-1593.
80. Kano A, Watanabe Y, Takeda N, Aizawa S and Akaike T. Analysis of IFN-γ-induced cell cycle arrest and cell death in hepatocytes. J Biochem (Tokyo) 1997; 121: 677-683.
81. Liang JF and Akaike T. Protective effect of linoleic acid on IFN γ-induced cellular injury in primary culture hepatocytes. J Biochem (Tokyo) 1998; 123: 213-218.
82. Schwall RH, Robbins K, Jardieu P, Chang L, Lai C and Terrell TG. Activin induces cell death in hepatocytes *in vivo* and *in vitro*. Hepatology 1993; 18: 347-356.
83. Tsukidate K, Yamamoto K, Snyder JW and Farber JL. Microtubule antagonists activate programmed cell death (apoptosis) in cultured rat hepatocytes. Am J Pathol 1993; 143: 918-925.
84. Shimizu S, Eguchi Y, Kamiike W, Akao Y, Kosaka H, Hasegawa J, Matsuda H et al. Involvement of ICE family proteases in apoptosis induced by reoxygenation of hypoxic hepatocytes. Am J Physiol 1996; 271: G949-G958.
85. Kurose I, Higuchi H, Miura S, Saito H, Watanabe N, Hokari R, Hirokawa M et al. Oxidative stress-mediated apoptosis of hepatocytes exposed to acute ethanol intoxication. Hepatology 1997; 25: 368-378.
86. Tsutsui S, Hirasawa K, Takeda M, Itagaki S, Kawamura S, Maeda K, Mikami T et al. Galactosamine-induced apoptosis in the primary mouse hepatocyte cultures. Exp Toxicol Pathol 1997; 49: 301-306.
87. Maeda S, Lin KH, Inagaki H and Saito T. Induction of apoptosis in primary culture of rat hepatocytes by protease inhibitors. Biochem Mol Biol Int 1996; 39: 447-453.

88. Liu J, Liu Y, Habeebu SS and Klaassen CD. Metallothionein (MT)-null mice are sensitive to cisplatin-induced hepatotoxicity. Toxicol Appl Pharmacol 1998; 149: 24-31.
89. Satoh T, Isobe H, Ayukawa K, Sakai H and Nawata H. The effects of pravastatin, an HMG-CoA reductase inhibitor, on cell viability and DNA production of rat hepatocytes. Life Sci 1996; 59: 1103-1108.
90. Worner W and Schrenk D. Influence of liver tumor promoters on apoptosis in rat hepatocytes induced by 2-acetylaminofluorene, ultraviolet light, or transforming growth factor β1. Cancer Res 1996; 56: 1272-1278.
91. Redmond HP, Wang JH and Bouchier-Hayes D. Taurine attenuates nitric oxide- and reactive oxygen intermediate-dependent hepatocyte injury. Arch Surg 1996; 131: 1280-1288.
92. Fau D, Lekehal M, Farrell G, Moreau A, Moulis C, Feldmann G, Haouzi D et al. Diterpenoids from germander, an herbal medicine, induce apoptosis in isolated rat hepatocytes. Gastroenterology 1997; 113: 1334-1346.
93. Amicone L, Spagnoli FM, Spath G, Giordano S, Tommasini C, Bernardini S, De Luca V et al. Transgenic expression in the liver of truncated Met blocks apoptosis and permits immortalization of hepatocytes. EMBO J 1997; 16: 495-503.
94. Bayly AC, Roberts RA and Dive C. Suppression of liver cell apoptosis *in vitro* by the non-genotoxic hepatocarcinogen and peroxisome proliferator nafenopin. J Cell Biol 1994; 125: 197-203.
95. Gill JH, James NH, Roberts RA and Dive C. The non-genotoxic hepatocarcinogen nafenopin suppresses rodent hepatocyte apoptosis induced by TGF-β1, DNA damage and Fas. Carcinogenesis 1998; 19: 299-304.
96. Kosai K, Matsumoto K, Nagata S, Tsujimoto Y and Nakamura T. Abrogation of Fas-induced fulminant hepatic failure in mice by hepatocyte growth factor. Biochem Biophys Res Commun 1998; 244: 683-690.
97. Bardelli A, Longati P, Albero D, Goruppi S, Schneider C, Ponzetto C and Comoglio PM. HGF receptor associates with the anti-apoptotic protein BAG-1 and prevents cell death. EMBO J 1996; 15: 6205-6212.
98. Kretzschmar M, Doody J and Massague J. Opposing BMP and EGF signalling pathways converge on the TGF-β family mediator Smad1. Nature 1997; 389: 618-622.
99. Roberts RA, James NH, Woodyatt NJ, Macdonald N and Tugwood JD. Evidence for the suppression of apoptosis by the peroxisome proliferator activated receptor α (PPAR α). Carcinogenesis 1998; 19: 43-48.
100. Alevizopoulos A and Mermod N. Transforming growth factor-β: the breaking open of a black box. Bioessays 1997; 19: 581-591.
101. Lawrence DA. Transforming growth factor-β: a general review. Eur Cytokine Netw 1996; 7: 363-374.
102. Fausto N, Laird AD and Webber EM. Liver regeneration. 2. Role of growth factors and cytokines in hepatic regeneration. FASEB J 1995; 9: 1527-1536.
103. Michalopoulos GK and DeFrances MC. Liver regeneration. Science 1997; 276: 60-66.
104. Oberhammer F, Bursch W, Parzefall W, Breit P, Erber E, Stadler M and Schulte-Hermann R. Effect of transforming growth factor β on cell death of cultured rat hepatocytes. Cancer Res 1991; 51: 2478-2485.

105. Benedetti A, Di Sario A, Svegliati Baroni G and Jezequel AM. Transforming growth factor β1 increases the number of apoptotic bodies and decreases intracellular pH in isolated periportal and perivenular rat hepatocytes. Hepatology 1995; 22: 1488-1498.
106. Kayanoki Y, Fujii J, Suzuki K, Kawata S, Matsuzawa Y and Taniguchi N. Suppression of antioxidative enzyme expression by transforming growth factor-β1 in rat hepatocytes. J Biol Chem 1994; 269: 15488-15492.
107. Nagata S and Golstein P. The Fas death factor. Science 1995; 267: 1449-1456.
108. Nagata S. Fas-induced apoptosis, and diseases caused by its abnormality. Genes Cells 1996; 1: 873-879.
109. Suda T and Nagata S. Why do defects in the Fas-Fas ligand system cause autoimmunity? J Allergy Clin Immunol 1997; 100: S97-S101.
110. Watanabe-Fukunaga R, Brannan CI, Itoh N, Yonehara S, Copeland NG, Jenkins NA and Nagata S. The cDNA structure, expression, and chromosomal assignment of the mouse Fas antigen. J Immunol 1992; 148: 1274-1279.
111. Adachi M, Suematsu S, Kondo T, Ogasawara J, Tanaka T, Yoshida N and Nagata S. Targeted mutation in the Fas gene causes hyperplasia in peripheral lymphoid organs and liver. Nat Genet 1995; 11: 294-300.
112. Kondo T, Suda T, Fukuyama H, Adachi M and Nagata S. Essential roles of the Fas ligand in the development of hepatitis. Nat Med 1997; 3: 409-413.
113. Ogasawara J, Watanabe-Fukunaga R, Adachi M, Matsuzawa A, Kasugai T, Kitamura Y, Itoh N et al. Lethal effect of the anti-Fas antibody in mice. Nature 1993; 364: 806-809.
114. Tracey KJ and Cerami A. Tumor necrosis factor, other cytokines and disease. Annu Rev Cell Biol 1993; 9: 317-343.
115. Beyer HS and Theologides A. Tumor necrosis factor-α is a direct hepatocyte mitogen in the rat. Biochem Mol Biol Int 1993; 29: 1-4.
116. Liu ZG, Hsu H, Goeddel DV and Karin M. Dissection of TNF receptor 1 effector functions: JNK activation is not linked to apoptosis while NF-κB activation prevents cell death. Cell 1996; 87: 565-576.
117. Rolfe M, James NH and Roberts RA. Tumour necrosis factor α (TNF-α) suppresses apoptosis and induces DNA synthesis in rodent hepatocytes: a mediator of the hepatocarcinogenicity of peroxisome proliferators? Carcinogenesis 1997; 18: 2277-2280.
118. Greim H, Trulzsch D, Czygan P, Rudick J, Hutterer F, Schaffner F and Popper H. Mechanism of cholestasis. 6. Bile acids in human livers with or without biliary obstruction. Gastroenterology 1972; 63: 846-850.
119. Greim H, Trulzsch D, Roboz J, Dressler K, Czygan P, Hutterer F, Schaffner F et al. Mechanism of cholestasis. 5. Bile acids in normal rat livers and in those after bile duct ligation. Gastroenterology 1972; 63: 837-845.
120. Ichihara A, Nakamura T, Noda C and Tanaka K. Control of enzyme expression deduced from studies on primary cultures of hepatocytes. In: Guillouzo A and Guguen-Guillouzo C, Eds. Research in isolated and cultured hepatocytes. Montrouge: John Libbey Eurotext Ltd/INSERM, 1986: 187-208.
121. Kumatori A, Nakamura T and Ichihara A. Cell-density dependent expression of the c-myc gene in primary cultured rat hepatocytes. Biochem Biophys Res Commun 1991; 178: 480-485.

122. Shimbara N, Takashina M, Sato C, Iizuka M, Kobayashi S, Tanaka K and Ichihara A. c-myc expression is down-regulated by cell-cell and cell-extracellular matrix contacts in normal hepatocytes, but not in hepatoma cells. Biochem Biophys Res Commun 1992; 184: 825-831.
123. Maeda S, Kimura H, Koga N, Lin KH and Saito T. Cell density-dependent DNA fragmentation and its suppression by heparin in primary culture of adult rat hepatocytes. Biochem Biophys Res Commun 1993; 195: 270-275.
124. Maeda S, Suzuki A, Lin KH, Inagaki H and Saito T. DNA fragmentation induced in high-cell-density culture of primary rat hepatocytes is an active process dependent on energy availability, gene expression, and calmodulin. J Biochem (Tokyo) 1995; 118: 1161-1165.
125. Seglen PO. Preparation of isolated rat liver cells. Methods Cell Biol 1976; 13: 29-83.
126. Mosmann T. Rapid colorimetric assay for cellular growth and survival: application to proliferation and cytotoxicity assays. J Immunol Methods 1983; 65: 55-63.
127. Oka M, Maeda S, Koga N, Kato K and Saito T. A modified colorimetric MTT assay adapted for primary cultured hepatocytes: application to proliferation and cytotoxicity assays. Biosci Biotechnol Biochem 1992; 56: 1472-1473.
128. Shinzawa K. Thesis: Tokyo Institute of Technology, 1998.
129. Shinzawa K, Watanabe Y and Akaike T. Primary cultured murine hepatocytes but not hepatoma cells regulate the cell number through density-dependent cell death. Cell Death and Differentiation 1995; 2: 133-140.
130. Jones DP, McConkey DJ, Nicotera P and Orrenius S. Calcium-activated DNA fragmentation in rat liver nuclei. J Biol Chem 1989; 264: 6398-6403.
131. Nicotera P, Bellomo G and Orrenius S. Calcium-mediated mechanisms in chemically induced cell death. Annu Rev Pharmacol Toxicol 1992; 32: 449-470.
132. Nicotera P, McConkey DJ, Jones DP and Orrenius S. ATP stimulates Ca^{2+} uptake and increases the free Ca^{2+} concentration in isolated rat liver nuclei. Proc Natl Acad Sci USA 1989; 86: 453-457.
133. Lyon M, Deakin JA and Gallagher JT. Liver heparan sulfate structure. A novel molecular design. J Biol Chem 1994; 269: 11208-11215.

Chapter 19

The protective role of nitric oxide in hepatocytes during responses to inflammatory mediators and induction of apoptosis

Christopher A. Bombeck, Jianrong Li, Young-Myeong Kim and Timothy R. Billiar
Department of Surgery, University of Pittsburgh, Pittsburgh, PA 15261

Key words: acute phase reaction, apoptosis, caspase, cGMP, cytokines, iNOS, NO, NF-κB, NO synthase induction, s-Nitrosylation

19.1 Introduction

The acute phase response (APR) has evolved in multicellular organisms as a result of their ever-changing surroundings. It refers to the alterations in cellular metabolism and genetic expression that have allowed organisms to maintain homeostasis and survive environmental stress. Frequently the APR is a systemic reaction to external stimuli, such as trauma, burn or infection; however it can also result from internal stimuli, such as neoplasia, pregnancy, or immunopathology [1]. The gross changes include: fever, somnolence, anorexia, thrombocytopenia, alterations in the hormonal milieu, and a net-negative nitrogen balance as the result of decreased protein synthesis [2]. Many of these changes are now understood to be the result of signalling cascades that act through inflammatory mediators, on individual cells, to produce organ specific responses.

These mediators, or cytokines, along with glucocorticoids and bacterial products, have been used to investigate the acute phase response in the liver. Several cytokines, including interleukin-1 (IL-1), IL-6, IL-11, interferon gamma (IFN-γ), tumor necrosis factor-alpha (TNF-α), leukemia inhibitory factor, transforming growth factor-beta (TGF-β) and oncostatin M, are known to influence protein synthesis by the liver in models of the acute phase [2]. These induced proteins include antiproteases, proteins that bind heavy metals, clotting factors and opsonins [3]. Individual cytokines are known to elicit differing effects, and while the response may be specific to the host or insult, it is governed by the cytokines that give rise to it. An

M.N. Berry and A.M. Edwards (eds.), The Hepatocyte Review, 301–315.

appropriate analogy likens individual cytokines to words, which may have meaning in and of themselves, but whose significance may change in the context of a sentence. A mixture of these signals, acting together through multiple post-receptor and gene regulatory mechanisms, modulates plasma protein synthesis and alters the intracellular environment of hepatocytes in the liver, thus inducing the acute phase response [4].

Our group has maintained an interest in the protective functional response of hepatocytes to these cytokines. One area of concentration has been the involvement of nitric oxide (NO) in the acute phase response to inflammatory stimuli both *in vivo* and in culture systems. Characterisation of this role has revealed an intimate association with the modulation of apoptosis, or programmed cell death. This investigation has yielded an extensive experience with primary hepatocyte cultures and their phenotypic and functional response to many inflammatory stimuli.

19.2 The production of nitric oxide in cultured hepatocytes

19.2.1 Cellular production of nitric oxide

As outlined in Figure 1, L-arginine is metabolized to citrulline and NO by the nitric oxide synthase (NOS) enzymes. Three separate enzymes have been isolated within this family; these include the inducible NOS (iNOS) and two constitutively expressed NOS (cNOS) enzymes: endothelial NOS and neuronal NOS. Their product, NO, freely diffuses across cellular membranes but it is rapidly oxidized within seconds to the stable end-products nitrite and nitrate (NO_2^-, NO_3^-). Low levels of NO can activate soluble guanylate cyclase to increase intracellular levels of cGMP, which may serve as another marker for the presence of nitric oxide [5,6]. The production of NO has been shown in many different cell types, including macrophages, smooth muscle cells, cardiac myocytes, endothelial cells, neurons, fibroblasts and epithelial cells [7].

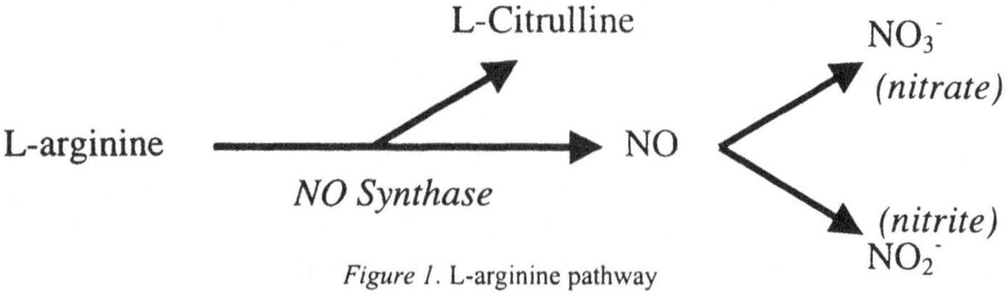

Figure 1. L-arginine pathway

19.2.2 Role of cytokines in nitric oxide production

Initial work in rat liver cell cultures demonstrated large amounts of nitrite and nitrate were produced when cocultures of hepatocytes and Kupffer cells were stimulated with lipopolysaccharide (LPS) [8]. Further investigation showed that transferable products from activated Kupffer cells induced the conversion of large quantities of L-arginine to NO_2^- and NO_3^- in primary cultures of hepatocytes isolated from rats [9]. This work suggested that inflammatory signalling led to the production of NO in hepatocytes. To identify these signalling factors, our group began experiments using recombinant cytokines to stimulate isolated hepatocytes. Using NO_2^- and NO_3^- supernatant concentrations as endpoints of the L-arginine pathway and markers of NO production, we determined that the highest output was achieved with a specific combination of cytokines, rather than any individual cytokine. The combination of endotoxin, TNF-α, IL-1, and IFN-γ, termed cytomix (CM) induced the maximal increase in hepatocyte NO production as well as significantly downregulating protein synthesis in the cultured hepatocytes [10]. Similar findings were obtained when these experiments were repeated in isolated human hepatocytes, suggesting an inducible pathway for NO production there as well [11].

19.2.3 Genetic induction of nitric oxide synthases

Our group then began experiments to characterize the events involved in the genetic induction of iNOS in primary cultures of hepatocytes. Using a mouse macrophage NOS cDNA clone and cross-species Northern blot hybridisation, we were able to show that CM induced the highest expression of iNOS mRNA in cultured rat hepatocytes [12]. This data correlated with levels of nitrite and nitrate in the culture supernatant, as well as cGMP release. Our group then isolated the first human iNOS cDNA from similar cultures of CM-stimulated primary human hepatocytes [13]. Using this clone, we were able to study the induction of iNOS in human hepatocytes. We found that when these cells were stimulated with CM, increased expression of iNOS mRNA preceded the formation of intracellular nonheme iron-nitrosyl complexes as well as maximal levels of supernatant NO_2^- and NO_3^- [14]. The iNOS mRNA first appeared 4 hours following CM stimulation, peaked at 8 hours then subsided by 48 hours; iNOS mRNA was not detectable in unstimulated hepatocytes. Subsequently, constitutive expression of iNOS has been shown in some epithelial cell types [15,16], however most cell types require exposure to inflammatory stimuli for iNOS expression [12,17,18].

To study the transcriptional regulation of the human iNOS gene, we characterized its promoter in the human liver epithelial cell line (AKN-1) [19]. Kinetics for the expression of iNOS mRNA in response to CM were similar to primary hepatocyte cultures, and increased levels of supernatant NO_2^- and NO_3^- correlated as well. Nuclear run-on analysis in the AKN-1 cells demonstrated a fivefold increase in the cytokine-induced transcriptional activity of the iNOS gene. Analysis of the first 4.7 kb of the 5′-flanking region of the AKN-1 promoter did not yield any cytokine-inducible increases in transcriptional activity [20]. However, promoter constructs extending to –5.8 kb and –7.2 kb induced a twofold to threefold, and a fourfold to fivefold increase, respectively. DNA sequence analysis from –3.8 to –7.2 kb demonstrated five NF-κB binding sites. While NF-κB has previously been implicated in the induction of iNOS in rodent macrophages and vascular smooth muscle cells [21,22], its role in the induction of human iNOS had remained uncertain. To investigate this we began experiments using promoter constructs with site-directed mutations at the NF-κB sites mentioned above. Within the first 7.2 kb of the promoter, we demonstrated that the NF-κB motif at –5.8 kb is required for both basal and inducible promoter activity. Three of the other sites were required for full promoter activity, however partial induction was still seen notwithstanding mutational alterations at these additional sites. The fifth site with a mutation at –6.8 kb, retained the full CM-inducibility seen with 7.2 kb AKN-1 promoter construct. In addition, we have found the cytokine-inducible expression of iNOS mRNA, in primary rodent and human hepatocyte cultures, to be regulated in an NF-κB-dependent manner (data not published). These data suggest that the cytokine-inducible expression of both human and rodent hepatocyte iNOS, and the resultant production of NO, is at least partially dependent on the transcription factor NF-κB.

19.3 The protective role of nitric oxide in cultured hepatocytes

19.3.1 Protective effects of nitric oxide

Since the time of its discovery, nitric oxide has been implicated in the cellular and subcellular pathways that determine survival for cells and microorganisms; although its ability to promote or limit cell death still remains controversial. Several reports suggest that NO and its metabolites induce tissue injury [23-27], as well as inhibiting enzymes such as aconitase and cytochrome c oxidase [28,29]. NO is also known to directly induce apoptosis in multiple cell lines, including macrophages, neurons, pancreatic β-cells and thymocytes [24,30-32]. However, other studies have shown that there is a protective role for NO as well [15,33-37]. Our group has maintained an interest in this investigation, specifically in characterising the role of nitric oxide in protecting hepatocytes from inflammatory mediators. In the acute phase, this is accomplished through the NO-mediated induction of protective factors and participation in signalling pathways as well as through its direct chemical reactions. A number of molecular targets have been identified for NO; these are typically sulfhydryl-containing molecules or metaloproteins that are sensitive to redox modification by NO. One of the first targets to be identified for NO was the heme protein, guanylate cyclase [38]. The quantities of NO produced by cNOS are adequate to dislocate the heme iron from soluble guanylate cyclase, thus inducing the activation of this enzyme [39]. Other signalling events, involving redox modulation by NO, include the inhibition of protein kinase C, activation of tyrosine kinase, and activation of G-proteins [40]. Hepatocytes also exhibit a profound decrease in total protein synthesis in response to the upregulation of NO production or its addition via exogenous donor [41]. This effect has been nonspecifically demonstrated across several cell lines, and is related to the phosphorylation of eIF-2 alpha and subsequent inhibition of 80S ribosomal complex formation, thus limiting protein translation [42]. This data is consistent with the fact that in hepatocytes, NO inhibits the production of the proteins, albumin and fibrinogen, without influencing the quantity of albumin mRNA [41]. While NO does limit the activity of enzymes involved in hepatocyte mitochondrial respiration in vitro, the net decrease in protein synthesis has been shown to be independent of this effect [43].

In vitro, hepatocytes are very resistant to NO toxicity, requiring millimolar concentrations of exogenous NO donors to kill them. Pre-exposure to an NO-donor, S-nitroso-N-acetyl-penicillamine (SNAP) or induced production via induction of iNOS with CM renders hepatocytes less susceptible to H_2O_2 or otherwise toxic concentrations of NO [44]. This is partially due to the induction of heme oxygenase, which prevents oxidative damage. Stimulation of hepatocytes with TNF-α, in the presence of a transcriptional inhibitor such as actinomycin-D (ActD) or D-galactosamine (GalN), has long been used as a model of hepatitis and fulminant hepatic failure [45]. Our group commonly uses TNF-α and ActD to stimulate primary hepatocytes in this culture model. Several groups in addition to ours have shown that the fulminant hepatic failure that follows *in-vivo* administration of these and similar agents is the result of increased hepatic apoptosis [46-48]. Preculture of hepatocytes with SNAP protects them from the onset of apoptosis, and this effect correlates with an increased expression of Heat Shock Protein-70 (HSP70) [49]. To investigate this effect our group found that heat exposure mimicked SNAP treatment in the protection of hepatocytes from TNF-induced toxicity, and this effect was blocked by the addition of an anti-sense oligonucleotide to HSP70. While NO induces the production of heme oxygenase, as mentioned above, the protection observed with the induction of HSP70 was not reversed with the heme oxygenase-inhibitor, protoporphyrin. Thus it is likely that both heme oxygenase and HSP70 are induced by NO exposure; however they offer protection from distinct insults namely, oxidative stress and receptor mediated toxicity respectively.

19.4 Nitric oxide and apoptosis

19.4.1 Introduction

Investigation into programmed cell death led to the recognition of apoptosis as a phenomenon that has been conserved across the species. The process of apoptotic cell death can be divided into three distinguishable phases: initiation, signalling and degradation [50]. There are many different stimuli that can induce the initiation phase of apoptosis. These include physiologic activators (*i.e.* TNF-α, Fas ligand, calcium, ceramide), damage-related inducers (*i.e.* Heat shock, p53, free radicals, cytolytic T cells), therapy-associated agents (*i.e.* chemotherapeutic drugs, gamma and UV radiation) and toxins (*i.e.* ethanol, β-amyloid peptide) [51-53]. It is in the signalling phase that common pathways toward cell death can be identified. Since the 1970's it has been known that mitochondrial Ca^{2+} overload leads to a non-specific permeabilisation of the inner mitochondrial membrane. Although a subject of controversy, the events associated with the mitochondrial membrane permeability transition (MMPT, MPT or PT) represent a dedicated step towards apoptosis. Most authors agree that this is the result of the opening of an inner membrane channel or pore, which allows the equilibration of solutes up to 1500 Da in mass, and is inhibited by cyclosporin A [54-58]. PT leads to the disruption of the mitochondrial transmembrane potential ($\Delta\psi$), cytochrome c redistribution into the cytosol, calcium flux and eventually activation of the Caspase-3-like family of cysteine proteases [59-63]. In other models, Caspase-3 activation occurs independently of cytochrome c release [64]. In either circumstance, these changes induce the events characteristic of the degradation phase of

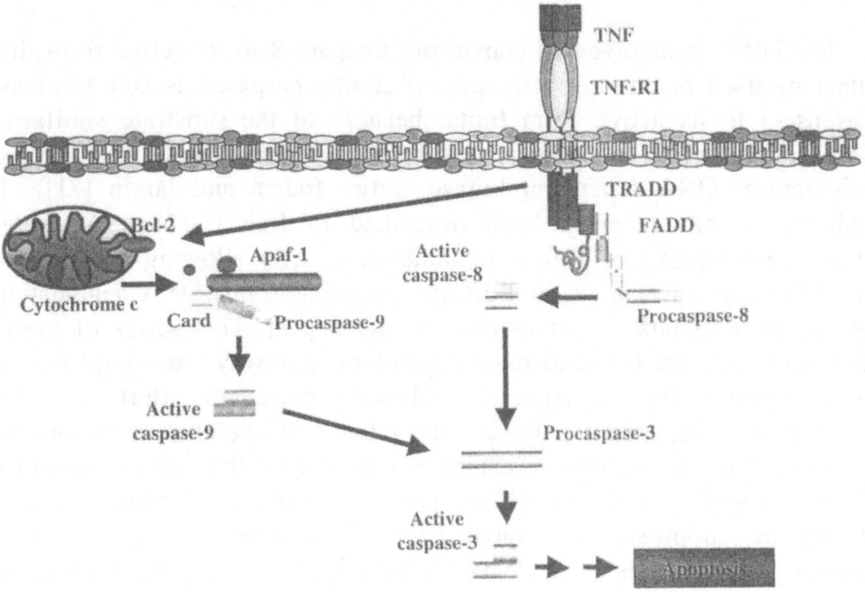

Figure 2. Suggested model for receptor mediated Caspase-3-activation

apoptosis: [1] changes in the cytoplasmic redox state, including decreased levels of reduced glutathione (GSH) and enhanced generation of free radicals; [2] membrane flipping of phosphatidylserine residues; [3] nuclear fragmentation and chromatin condensation [65-68].

19.4.2 Overview of apoptosis

While the signalling of apoptosis in hepatocytes is covered in Chapter 18, a brief overview will be included here to incorporate NO into the regulation of this pathway, (Figure 2). The oligomerisation of three TNF-α-receptor-ligand complexes establishes a docking site on the cytoplasmic side of the cellular membrane. Each of the TNF-R1 receptors contains a Death Domain within its cytoplasmic fraction and trimerisation of these receptors creates a functionally active unit. [51,53,64,69] The establishment of this site allows docking of the TNF Receptor Associated Death Domain (TRADD) molecule, followed by the Fas Associated Death Domain (FADD) molecule [70]. In this way the formation of the Death Inducing Signalling Complex (DISC) is completed. This allows receptor interaction with a group of cysteine proteases that preferentially cleave their proteinaceous substrates following the amino acid aspartate. They are cysteine aspases, hence the name caspases, of which ten have been cloned in humans to date.

The DISC is involved in converting Caspase-8 to its active form that, either by itself or through activation of similar caspases, is able to cleave Caspase-3 to its active form (note: because of the substrate similarity, Caspase-3 is known to cleave several substrates such as Poly(ADP-Ribose) polymerase, DNA-dependent kinase, actin, fodrin and lamin [71]). In addition, it has recently been described to free Caspase Activated Deoxyribonuclease (CAD) from its inhibitor (ICAD), allowing translocation of CAD to the nucleus where it effects intranucleosomal DNA degradation, one of the hallmarks of apoptotic cell death [72]. This series of events represents the mitochondrial-independent pathway to apoptosis in hepatocytes. By contrast evidence suggests that in the mitochondrial-dependent pathway, the release of cytochrome c from the mitochondrial intermembranous space is essential for the onset of apoptosis [73-76]. Together with dATP, cytochrome c induces a conformational change in Apoptosis Associated Factor-1 allowing the activation of Caspase-9, which goes on to cleave Caspase-3, followed by the downstream changes associated with apoptosis.

19.4.3 Effects of nitric oxide on signalling pathways associated with apoptosis

Using reconstituted systems as well as whole cell hepatocytes in culture, our group has been able to determine that NO inhibits these signalling pathways through two distinct mechanisms. We found that preculture of hepatocytes with exogenous donors of NO, or the induced production of NO through the upregulation of iNOS expression in response to CM, protected the cells from apoptosis and significantly reduced the spontaneous activation of Caspase-3-like proteases [77]. *In vitro* experiments with recombinant human Caspase-3 reveal that NO S-nitrosylates this protease at the cysteine residue within its active site, thus incapacitating its ability to cleave substrate. S-nitrosylation is determined through the use of dithiothreitol (DTT) which releases thiol-bound NO, allowing its measurement with the Greiss reaction [78,79]. Hepatocytes are known to undergo spontaneous apoptosis after several days in culture. This occurrence is inhibited by the exogenous addition of NO or through its endogenous production as a result of the induction of iNOS by CM. To investigate the S-nitrosylation of Caspase-3, we cultured hepatocytes with or without NO for four days. We then lysed these cells and assessed Caspase-3-like activity in the presence or absence of DTT (Figure 3). In each instance, Caspase-3-like activity was restored to at least half the activity of unprotected hepatocyte lysates when treated with DTT, suggesting that NO contributed to the inhibition of Caspase-3-like activity. To investigate this effect more specifically, recombinant human Caspase-3 was exposed to SNAP then added to a reconstituted cellular system consisting of nuclei isolated from untreated hepatocytes and suspended in a cytosolic fraction (S-100). In this experiment DNA fragmentation was induced by Caspase-3 alone, however it was not seen when SNAP was used to pretreat the Caspase-3. The addition of DTT reversed this effect and DNA fragmentation was once again seen. Taken together these data suggest that the S-nitrosylation of Caspase-3 is one possible mechanism for the protection of hepatocytes that are induced to undergo apoptosis. To further investigate this finding we conducted experiments using the recombinant human caspases 1,2,3,4,6,7, and 8. We found that in each instance, S-nitrosylation by NO was a reversible effect leading to the inactivation of these caspases [80]. However failure to completely restore caspase activity with DTT suggested that there were additional pathways for the inhibition of these proteases.

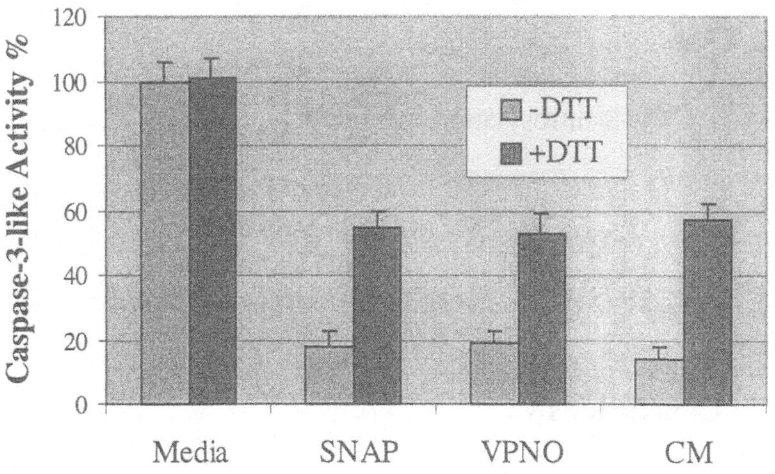

Figure 3. Caspase-3-like activity in cultured hepatocytes after treatment with the NO donors, SNAP and VPNO, or CM

As mentioned above, NO is known to induce cGMP through the activation of soluble guanylate cyclase, and we have recently shown the cGMP analog 8-bromo-cGMP protects cultured hepatocytes from TNF-α/ActD-induced toxicity [81]. When we conducted experiments similar to those mentioned above, we found that 8-bromo-cGMP also inhibited the activity of Caspase-3-like proteases. However the addition of DTT to 8-bromo-cGMP-treated hepatocyte lysates did not restore Caspase-3-like activity, as it had in the SNAP-treated cells. Furthermore, the addition of the soluble guanylate cyclase-inhibitor ODQ with SNAP to hepatocytes prevented approximately 50% of the SNAP-induced drop in Caspase-3-like activity [77]. The balance of the activity was restored when the lysates were incubated with DTT. Thus, NO-stimulated cGMP synthesis provides an additional pathway for the inhibition of Caspase-3 activation, or of some upstream event leading to that activation (Figure 4).

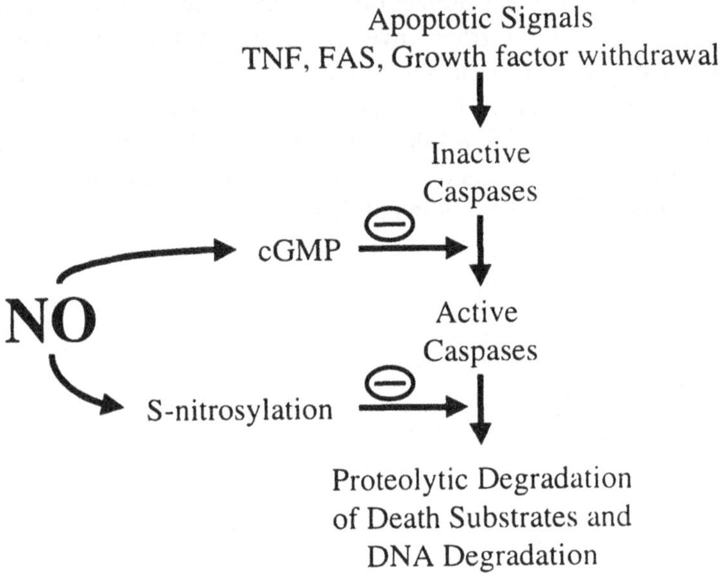

Figure 4. Mechanism of inhibition by NO of Caspase-3 activation

19.5 Conclusions

Over the past decade NO has emerged as an important molecule with a central role in modulating some of the key responses of cultured hepatocytes to inflammatory stimuli. Whether it is through redox regulation of metaloproteins, the induction of heat shock proteins or through s-Nitrosylation, NO has a dominant function in protecting hepatocytes from oxidative damage and the induction of apoptotic cell death. These observations raise the possibility that NO can be used as a therapeutic agent to suppress hepatocyte death [82]. Furthermore, iNOS-generated NO may limit hepatocyte death under conditions of cytokine stimulation such as sepsis, or during liver regeneration or allograft rejection.

References

1. Kushner I. The phenomenon of the acute phase response. Ann N Y Acad Sci 1982; 389: 39-48.
2. Mackiewicz A, Kushner I and Baumann H, Eds. Acute phase proteins: Molecular biology, biochemistry, and clinical applications. London: CRC Press, 1993.
3. Billiar TR and Curran RD, Eds. Hepatocyte and Kupffer cell interactions. Boca Raton: CRC Press, 1992.
4. Kushner I. Regulation of the acute phase response by cytokines. Perspect Biol Med 1993; 36: 611-622.
5. Azam M, Gupta G, Chen W, Wellington S, Warburton D and Danziger RS. Genetic mapping of soluble guanylyl cyclase genes: implications for linkage to blood pressure in the Dahl rat. Hypertension 1998; 32: 149-154.
6. Billiar TR, Curran RD, Harbrecht BG, Stadler J, Williams DL et al. Association between synthesis and release of cGMP and nitric oxide biosynthesis by hepatocytes. Am J Physiol 1992; 262: C1077-C1082.
7. Geller DA and Billiar TR. Molecular biology of nitric oxide synthases. Cancer Metastasis Rev 1998; 17: 7-23.
8. Billiar TR, Curran RD, Stuehr DJ, West MA, Bentz BG and Simmons RL. An L-arginine-dependent mechanism mediates Kupffer cell inhibition of hepatocyte protein synthesis *in vitro*. J Exp Med 1989; 169: 1467-1472.
9. Curran RD, Billiar TR, Stuehr DJ, Hofmann K and Simmons RL. Hepatocytes produce nitrogen oxides from L-arginine in response to inflammatory products of Kupffer cells. J Exp Med 1989; 170: 1769-1774.
10. Curran RD, Billiar TR, Stuehr DJ, Ochoa JB, Harbrecht BG, Flint SG and Simmons RL. Multiple cytokines are required to induce hepatocyte nitric oxide production and inhibit total protein synthesis. Ann Surg 1990; 212: 462-469.
11. Nussler AK, Di Silvio M, Billiar TR, Hoffman RA, Geller DA et al. Stimulation of the nitric oxide synthase pathway in human hepatocytes by cytokines and endotoxin. J Exp Med 1992; 176: 261-264.
12. Geller DA, Nussler AK, Di Silvio M, Lowenstein CJ, Shapiro RA et al. Cytokines, endotoxin, and glucocorticoids regulate the expression of inducible nitric oxide synthase in hepatocytes. Proc Natl Acad Sci 1993; 90: 522-526.
13. Geller DA, Lowenstein CJ, Shapiro RA, Nussler AK, Di Silvio M et al. Molecular cloning and expression of inducible nitric oxide synthase from human hepatocytes. Proc Natl Acad Sci 1993; 90: 3491-3495.
14. Nussler AK, Geller DA, Sweetland MA, Di Silvio M, Billiar TR et al. Induction of nitric oxide synthesis and its reactions in cultured human and rat hepatocytes stimulated with cytokines plus LPS. Biochem Biophys Res Commun 1993; 194: 826-835.
15. Mannick JB, Asano K, Izumi K, Kieff E and Stamler JS. Nitric oxide produced by human B lymphocytes inhibits apoptosis and Epstein-Barr virus reactivation. Cell 1994; 79: 1137-1146.
16. Asano K, Chee CB, Gaston B, Lilly CM, Gerard C, Drazen JM and Stamler JS. Constitutive and inducible nitric oxide synthase gene expression, regulation, and activity in human lung epithelial cells. Proc Natl Acad Sci 1994; 91: 10089-10093.

17. Geller DA, Di Silvio M, Nussler AK, Wang SC, Shapiro RA, Simmons RL and Billiar TR. Nitric oxide synthase expression is induced in hepatocytes *in vivo* during hepatic inflammation. J Surg Res 1993; 55: 427-432.
18. Xie QW, Whisnant R and Nathan C. Promoter of the mouse gene encoding calcium-independent nitric oxide synthase confers inducibility by interferon γ and bacterial lipopolysaccharide. J Exp Med 1993; 177: 1779-1784.
19. de Vera ME, Shapiro RA, Nussler AK, Mudgett JS, Simmons RL et al. Transcriptional regulation of human inducible nitric oxide synthase (NOS2) gene by cytokines: initial analysis of the human NOS2 promoter. Proc Natl Acad Sci 1996; 93: 1054-1059.
20. Taylor BS, de Vera ME, Ganster RW, Wang Q, Shapiro RA et al. Multiple NF-κB enhancer elements regulate cytokine induction of the human inducible nitric oxide synthase gene. J Biol Chem 1998; 273: 15148-15156.
21. Xie QW, Kashiwabara Y and Nathan C. Role of transcription factor NF-κ B/Rel in induction of nitric oxide synthase. J Biol Chem 1994; 269: 4705-4708.
22. Griscavage JM, Wilk S and Ignarro LJ. Inhibitors of the proteasome pathway interfere with induction of nitric oxide synthase in macrophages by blocking activation of transcription factor NF-κ B. Proc Natl Acad Sci 1996; 93: 3308-3312.
23. Billiar TR, Curran RD, West MA, Hofmann K and Simmons RL. Kupffer cell cytotoxicity to hepatocytes in coculture requires L-arginine. Arch Surg 1989; 124: 1416-1420.
24. Brune B and Sandau K. the role of NO in cell injury. Toxicol Lett 1995; 82/83: 233-237.
25. Hortelano S, Dallaporta B, Zamzami N, Hirsch T, Susin SA et al. Nitric oxide induces apoptosis via triggering mitochondrial permeability transition. FEBS Lett 1997; 410: 373-377.
26. Thomae KR, Joshi PC, Davies P, Pitt BR, Billiar TR, Simmons RL and Nakayama DK. Nitric oxide produced by cytokine-activated pulmonary artery smooth muscle cells is cytotoxic to cocultured endothelium. Surgery 1996; 119: 61-66.
27. Szabo C, Southan GJ and Thiemermann C. Beneficial effects and improved survival in rodent models of septic shock with S-methylisothiourea sulfate, a potent and selective inhibitor of inducible nitric oxide synthase. Proc Natl Acad Sci 1994; 91: 12472-12476.
28. Stadler J, Billiar TR, Curran RD, Stuehr DJ, Ochoa JB and Simmons RL. Effect of exogenous and endogenous nitric oxide on mitochondrial respiration of rat hepatocytes. Am J Physiol 1991; 260: C910-C916.
29. Cleeter MW, Cooper JM, Darley-Usmar VM, Moncada S and Schapira AH. Reversible inhibition of cytochrome c oxidase, the terminal enzyme of the mitochondrial respiratory chain, by nitric oxide. Implications for neurodegenerative diseases. FEBS Lett 1994; 345: 50-54.
30. Messmer UK, Ankarcrona M, Nicotera P and Brune B. p53 expression in nitric oxide-induced apoptosis. FEBS Letters 1994; 355: 23-26.
31. Fehsel K, Kroncke KD, Meyer KL, Huber H, Wahn V and Kolb-Bachofen V. Nitric oxide induces apoptosis in mouse thymocytes. J Immunol 1995; 155: 2858-2865.
32. Albina JE, Cui S, Mateo RB and Reichner JS. Nitric oxide-mediated apoptosis in murine peritoneal macrophages. J Immunol 1993; 150: 5080-5085.

33. Wink DA, Hanbauer I, Krishna MC, DeGraff W, Gamson J and Mitchell JB. Nitric oxide protects against cellular damage and cytotoxicity from reactive oxygen species. Proc Natl Acad Sci 1993; 90: 9813-9817.
34. Wink DA, Cook JA, Pacelli R, Liebmann J, Krishna MC and Mitchell JB. Nitric oxide (NO) protects against cellular damage by reactive oxygen species. Toxicol Lett 1995; 82-83: 221-226.
35. Tenneti L, D'Emilia DM and Lipton SA. Suppression of neuronal apoptosis by S-nitrosylation of caspases. Neurosci Lett 1997; 236: 139-142.
36. Nishida J, McCuskey RS, McDonnell D and Fox ES. Protective role of NO in hepatic microcirculatory dysfunction during endotoxemia. Am J Physiol 1994; 267: G1135-G1141.
37. Mueller AR, Platz KP, Langrehr JM, Hoffman RA, Nussler AK et al. The effects of administration of nitric oxide inhibitors during small bowel preservation and reperfusion. Transplantation 1994; 58: 1309-1316.
38. Moncada S and Higgs A. The L-arginine-nitric oxide pathway. N Engl J Med 1993; 329: 2002-2012.
39. Ignarro LJ. Heme-dependent activation of soluble guanylate cyclase by nitric oxide: regulation of enzyme activity by porphyrins and metalloporphyrins. Semin Hematol 1989; 26: 63-76.
40. Stamler JS. Redox signaling: nitrosylation and related target interactions of nitric oxide. Cell 1994; 78: 931-936.
41. Curran RD, Ferrari FK, Kispert PH, Stadler J, Stuehr DJ, Simmons RL and Billiar TR. Nitric oxide and nitric oxide-generating compounds inhibit hepatocyte protein synthesis. FASEB Journal 1991; 5: 2085-2092.
42. Kim YM, Son K, Hong SJ, Green A, Chen JJ, Tzeng E, Hierholzer C and Billiar TR. Inhibition of protein synthesis by nitric oxide correlates with cytostatic activity: nitric oxide induces phosphorylation of initiation factor eIF-2 α. Mol Med 1998; 4: 179-190.
43. Stadler J, Curran RD, Ochoa JB, Harbrecht BG, Hoffman RA et al. Effect of endogenous nitric oxide on mitochondrial respiration of rat hepatocytes *in vitro* and *in vivo*. Arch Surg 1991; 126: 186-191.
44. Kim YM, Bergonia H and Lancaster JRJ. Nitrogen oxide-induced autoprotection in isolated rat hepatocytes. FEBS Lett 1995; 374: 228-232.
45. Keppler D, Lesch R, Reutter W and Decker K. Experimental hepatitis induced by D-galactosamine. Exp Mol Pathol 1968; 9: 279-290.
46. Leist M, Gantner F, Bohlinger I, Germann PG, Tiegs G and Wendel A. Murine hepatocyte apoptosis induced *in vitro* and *in vivo* by TNF-α requires transcriptional arrest. J Immunol 1994; 153: 1778-1788.
47. Leist M, Gantner F, Bohlinger I, Tiegs G, Germann PG and Wendel A. Tumor necrosis factor-induced hepatocyte apoptosis precedes liver failure in experimental murine shock models. Am J Pathol 1995; 146: 1220-1234.
48. Leist M, Gantner F, Jilg S and Wendel A. Activation of the 55 kDa TNF receptor is necessary and sufficient for TNF-induced liver failure, hepatocyte apoptosis, and nitrite release. J Immunol 1995; 154: 1307-1316.
49. Kim YM, deVera ME, Watkins SC and Billiar TR. Nitric oxide protects cultured rat hepatocytes from tumor necrosis factor-α-induced apoptosis by inducing heat shock protein 70 expression. J Biol Chem 1997; 272: 1402-1411.
50. Kroemer G, Petit P, Zamzami N, Vayssiere JL and Mignotte B. The biochemistry of programmed cell death. FASEB J 1995; 9: 1277-1287.

51. Thompson C. Apoptosis in the pathogenesis and treatment of disease. Science 1995; 267: 1456-1462.
52. Leist M, Gantner F, Naumann H, Bluethmann H, Vogt K et al. Tumor necrosis factor-induced apoptosis during the poisoning of mice with hepatotoxins. Gastroenterology 1997; 112: 923-934.
53. Nagata S. Apoptosis by death factor. Cell 1997; 88: 355-365.
54. Hunter DR and Haworth RA. The Ca^{2+}-induced membrane transition in mitochondria. I. The protective mechanisms. Arch Biochem Biophys 1979; 195: 453-459.
55. Hunter DR and Haworth RA. The Ca^{2+}-induced membrane transition in mitochondria. III. Transitional Ca^{2+} release. Arch Biochem Biophys 1979; 195: 468-477.
56. Haworth RA and Hunter DR. The Ca^{2+}-induced membrane transition in mitochondria. II. Nature of the Ca^{2+} trigger site. Arch Biochem Biophys 1979; 195: 460-467.
57. Fournier N, Ducet G and Crevat A. Action of cyclosporine on mitochondrial calcium fluxes. J Bioenerg Biomembr 1987; 19: 297-303.
58. Halestrap AP and Davidson AM. Inhibition of $Ca^{2(+)}$-induced large-amplitude swelling of liver and heart mitochondria by cyclosporin is probably caused by the inhibitor binding to mitochondrial-matrix peptidyl-prolyl cis-trans isomerase and preventing it interacting with the adenine nucleotide translocase. Biochem J 1990; 268: 153-160.
59. Zamzami N, Marchetti P, Castedo M, Zanin C, Vayssiere JL, Petit PX and Kroemer G. Reduction in mitochondrial potential constitutes an early irreversible step of programmed lymphocyte death *in vivo*. J Exp Med 1995; 181: 1661-1672.
60. Vander Heiden MG, Chandel NS, Williamson EK, Schumacker PT and Thompson CB. Bcl-xL regulates the membrane potential and volume homeostasis of mitochondria. Cell 1997; 91: 627-637.
61. Marchetti P, Castedo M, Susin SA, Zamzami N, Hirsch T et al. Mitochondrial permeability transition is a central coordinating event of apoptosis. J Exp Med 1996; 184: 1155-1160.
62. Li P, Nijhawan D, Budihardjo I, Srinivasula SM, Ahmad M, Alnemri ES and Wang X. Cytochrome c and dATP-dependent formation of Apaf-1/caspase-9 complex initiates an apoptotic protease cascade. Cell 1997; 91: 479-489.
63. Costantini P, Chernyak BV, Petronilli V and Bernardi P. Modulation of the mitochondrial permeability transition pore by pyridine nucleotides and dithiol oxidation at two separate sites. J Biol Chem 1996; 271: 6746-6751.
64. Nagata S and Golstein P. The Fas death factor. Science 1995; 267: 1449-1456.
65. Kroemer G, Zamzami N and Susin SA. Mitochondrial control of apoptosis. Immunol Today 1997; 18: 44-51.
66. Vercesi AE, Kowaltowski AJ, Grijalba MT, Meinicke AR and Castilho RF. The role of reactive oxygen species in mitochondrial permeability transition. Biosci Rep 1997; 17: 43-52.
67. Castedo M, Hirsch T, Susin SA, Zamzami N, Marchetti P, Macho A and Kroemer G. Sequential acquisition of mitochondrial and plasma membrane alterations during early lymphocyte apoptosis. J Immunol 1996; 157: 512-521.
68. Kerr JF, Wyllie AH and Currie AR. Apoptosis: a basic biological phenomenon with wide-ranging implications in tissue kinetics. Br J Cancer 1972; 26: 239-257.
69. Liu ZG, Hsu H, Goeddel DV and Karin M. Dissection of TNF receptor 1 effector functions: JNK activation is not linked to apoptosis while NF-κB activation prevents cell death. Cell 1996; 87: 565-576.

70. Schulze-Osthoff K, Ferrari D, Los M, Wesselborg S and Peter ME. Apoptosis signaling by death receptors. Eur J Biochem 1998; 254: 439-459.
71. Casciola-Rosen L, Nicholson DW, Chong T, Rowan KR, Thornberry NA et al. Apopain/CPP32 cleaves proteins that are essential for cellular repair: a fundamental principle of apoptotic death. J Exp Med 1996; 183: 1957-1964.
72. Sakahira H, Enari M and Nagata S. Cleavage of CAD inhibitor in CAD activation and DNA degradation during apoptosis. Nature 1998; 391: 96-99.
73. Stridh H, Kimland M, Jones DP, Orrenius S and Hampton MB. Cytochrome c release and caspase activation in hydrogen peroxide- and tributyltin-induced apoptosis. FEBS Lett 1998; 429: 351-355.
74. Scarlett JL and Murphy MP. Release of apoptogenic proteins from the mitochondrial intermembrane space during the mitochondrial permeability transition. FEBS Lett 1997; 418: 282-286.
75. Petit PX, Goubern M, Diolez P, Susin SA, Zamzami N and Kroemer G. Disruption of the outer mitochondrial membrane as a result of large amplitude swelling: the impact of irreversible permeability transition. FEBS Lett 1998; 426: 111-116.
76. Kuwana T, Smith JJ, Muzio M, Dixit V, Newmeyer DD and Kornbluth S. Apoptosis induction by caspase-8 is amplified through the mitochondrial release of cytochrome c. J Biol Chem 1998; 273: 16589-16594.
77. Kim YM, Talanian RV and Billiar TR. Nitric oxide inhibits apoptosis by preventing increases in caspase-3-like activity via two distinct mechanisms. J Biol Chem 1997; 272: 31138-31148.
78. Mohr S, Stamler JS and Brune B. Posttranslational modification of glyceraldehyde-3-phosphate dehydrogenase by S-nitrosylation and subsequent NADH attachment. J Biol Chem 1996; 271: 4209-4214.
79. McDonald LJ and Moss J. Stimulation by nitric oxide of an NAD linkage to glyceraldehyde-3-phosphate dehydrogenase. Proc Natl Acad Sci 1993; 90: 6238-6241.
80. Li J, Billiar TR, Talanian RV and Kim YM. Nitric oxide reversibly inhibits seven members of the caspase family via S-nitrosylation. Biochem Biophys Res Commun 1997; 240: 419-424.
81. Saavedra JE, Billiar TR, Williams DL, Kim YM, Watkins SC and Keefer LK. Targeting nitric oxide (NO) delivery *in vivo*. Design of a liver-selective NO donor prodrug that blocks tumor necrosis factor-α-induced apoptosis and toxicity in the liver. J Med Chem 1997; 40: 1947-1954.
82. Ou J, Carlos TM, Watkins SC, Saavedra JE, Keefer LK et al. Differential effects of nonselective nitric oxide synthase (NOS) and selective inducible NOS inhibition on hepatic necrosis, apoptosis, ICAM-1 expression, and neutrophil accumulation during endotoxemia. Nitric Oxide 1997; 1: 404-416.

Chapter 20

Short-term regulation by insulin of glucose metabolism in isolated and cultured hepatocytes

Loranne Agius, Nasreen Alam and Susan Aiston
Human Diabetes and Metabolism Research Centre, School of Clinical Medical Sciences, University of Newcastle upon Tyne, The Medical School, Newcastle upon Tyne NE2 4HH, UK

Key words: adenylate cyclase, AMP activated protein kinase, Ca^{2+}, calmodulin-dependent protein kinase II, cAMP, cAMP phosphodiesterase, cAMP-dependent protein kinase, casein kinases I and II, cell swelling, diacylglycerol, ERK1/ERK2, ERK2, glucagon, glucokinase, glucose, glucose 6-phosphate, glycogen, glycogen synthase, glycogen synthase kinase-3, glycogen synthase kinases, glycogen synthesis, glycogen targetting subunit, glycogenolysis, glycosyl-phosphatidylinositol, inositol phosphates, insulin, IRS-1, IRS-2, MAP-kinase pathway, inositol phosphoglycan, p70rsk, p90rsk, phosphatidylinositol, phosphatidylinositol 3-kinase, phospholipase Cγ, phosphorylase, phosphorylase kinase, phosphorylase phosphatase, pp120, protein phosphatases, protein kinases A, B and C, rapamycin, rsk-2, synthase phosphatase

20.1 Background: direct and counter-regulatory effects of insulin on glucose metabolism

The detailed characterisation of the action of insulin on hepatic carbohydrate metabolism is one of the most important problems in understanding the regulation of blood glucose homeostasis. Early studies on the perfused rat liver and later work on freshly isolated hepatocyte suspensions established that insulin counteracts the stimulatory effects on glycogenolysis and gluconeogenesis of low concentrations of glucagon or α-adrenergic agonists. However the direct effects of insulin on glucose metabolism in the absence of catabolic hormones were initially the subject of great controversy, since direct effects of insulin on either the perfused liver or on freshly isolated hepatocyte suspensions were observed in only a minority of studies [1]. Later studies on hepatocyte monolayers showed unequivocally that insulin increases glycogen deposition by more than twofold at physiological glucose concentrations. However, arguments prevailed as to whether this role of insulin was a "minor" one in comparison

M.N. Berry and A.M. Edwards (eds.), The Hepatocyte Review, 317–341.

with the stimulatory effect of an increase in glucose concentration [2]. It is still a widespread but incorrect assumption that whereas insulin rather than glucose is the main stimulus for glycogen storage in muscle, the converse is true for liver [2,3].

In addition to stimulating glycogen storage, insulin exerts a variety of other direct effects on hepatocytes including stimulation of glycolysis, lipogenesis, protein and DNA synthesis, and regulation of long-chain fatty acid metabolism. Growth factors, such as epidermal growth factor (EGF) and transforming growth factor α, mimic some effects of insulin, such as stimulation of glycolysis, but potently counteract the glycogenic effects of insulin [4,5]. Since several protein kinases are activated by growth factors as well as by insulin, the characterisation of differences between insulin and growth factors in protein kinase activation has provided a useful approach for identifying signalling components that are involved in control of glycogen metabolism and glucose homeostasis as distinct from cell growth and proliferation [6]. This review focuses on the metabolic actions of insulin, both direct and counter-regulatory, in the short-term regulation of glucose metabolism in hepatocytes; and on the putative signalling mechanisms involved in mediating these effects.

20.2 Counter-regulatory actions of insulin on glucose and glycogen metabolism

20.2.1 Counteraction by insulin of the effects of glucagon and cAMP

Before the development of the technique for high yield preparation of isolated hepatocytes by collagenase perfusion [7], the metabolic actions of insulin were studied using the perfused rat liver [8]. This work led to the conclusion that insulin action on carbohydrate metabolism in liver is primarily due to inhibition of glycogenolysis and gluconeogenesis in the presence of glucagon or cAMP, and not to stimulation of glucose utilisation at elevated glucose concentrations [8]. Studies on freshly isolated hepatocyte suspensions by Hers and colleagues [9,10] corroborated the findings from the perfused liver. They showed that the activation of glycogenolysis by glucagon is associated with acute activation of phosphorylase, phosphorylase kinase and cAMP-dependent protein kinase [9] and that insulin does not affect these enzymes in the basal state but suppresses the effects of sub-maximal concentrations of glucagon. Likewise insulin counteracted the inactivation of pyruvate kinase by glucagon [10]. Accordingly, studies on insulin signalling in hepatocyte suspensions focused on the mechanisms by which insulin modulates glucagon action.

Rat hepatocyte suspensions express two types of glucagon receptors with dissociation constants of 0.7 nM and 13 nM [11], and two signalling pathways have been implicated, one coupled to cAMP production and the other to inositol phospholipid metabolism leading to an increase in inositol phosphates and a rise in Ca^{2+} [12]. The latter mechanism is in part independent of the rise in cAMP [13]. Several studies showed that insulin suppresses the cAMP response to glucagon in hepatocytes [14,15]. The inhibition of glucagon-stimulated phosphorylase can be explained by the lowering of cAMP and does not appear to involve changes in cell Ca^{2+} [14]. It has been attributed both to decreased cAMP production by inhibition of adenylate cyclase and to increased cAMP degradation by activation of cAMP phosphodiesterases. Hormonal control of adenylate cyclase is mediated by two distinct heterotrimeric G protein complexes designated G_s (stimulatory) and G_i (inhibitory). In hepatocyte suspensions, insulin in the absence of other hormones causes a rapid decrease in the phosphorylation of the G_i2, which may account for the inactivation of adenylate cyclase [16].

Multiple isoforms of cAMP phosphodiesterase have been identified in mammalian cells and are classified into five types based on their affinity for cGMP. Types I, II, III and IV have been shown to be present in hepatocyte suspensions on the basis of kinetic studies on subcellular fractions. Of these, type I is estimated to have a minor role in cAMP degradation [17]. Type IV phosphodiesterase (insensitive to cGMP), present in the plasma membrane fraction, is stimulated by insulin and a type III activity (inhibited by cGMP) associated with the endoplasmic reticulum (dense vesicle fraction), is stimulated by both insulin and glucagon [17]. The latter may have a role in the desensitisation mechanism. In adipocytes, insulin activation of type IIIB phosphodiesterase has been suggested to be mediated by protein kinase B [18]. The kinase involved in hepatocytes remains to be identified.

Insulin also antagonizes the glycogenolytic effects of exogenous cAMP [14] and of the non-metabolisable analogue 8-bromo-cAMP [19]. In the original reports, the former effect was attributed to increased cAMP degradation and the latter to an inhibitory effect of insulin on cAMP-dependent protein kinase (protein kinase A). A direct but transient inhibitory effect of insulin on the activity of protein kinase A, that was maximal after 5 min, was demonstrated in hepatocyte suspensions incubated in the absence of other hormones [20]. This effect was due to a decrease in the sensitivity to cAMP. Protein kinase A exists as an inactive tetramer of two catalytic (C) and two regulatory (R) subunits. Cyclic AMP binds to the R-units and causes dissociation of active catalytic units. Whether insulin alters the affinity of the R-subunit for cAMP is not known. Studies with a cAMP antagonist (the Rp diastereomer of adenosine cyclic 3',5'-phosphorothioate), which binds to the R-subunit but does not cause dissociation of the C-units, showed that the inhibitory effects of this antagonist on glucagon-induced activation of phosphorylase in intact hepatocytes are synergistic with the effects of insulin. From this it was inferred that insulin counteraction of the glucagon effects involves mechanisms distal to protein kinase A [21]. However a detailed study of effects of insulin on cAMP-analogue mediated glycogenolysis, using several analogues that cover a wide range of affinity for the low K_m phosphodiesterase, showed that the inhibitory effect of insulin correlated with phosphodiesterase hydrolysis. This suggests a major role for insulin-activation of low K_m phosphodiesterase in the counter regulatory effects of insulin [22].

20.2.2 Counteraction by insulin of the effects of α-adrenergic agonists

In contrast with the effect of glucagon [9], stimulation of glycogenolysis by the α-adrenergic agonist, phenylephrine, and also by vasopressin, is associated with an increase in the activation state of phosphorylase without a change in the activation state (phosphorylation) of phosphorylase kinase. The activation of phosphorylase was attributed to stimulation of phosphorylase kinase by a rise in cytosolic Ca^{2+}, based on the observations that the activation was markedly suppressed in the absence of extracellular Ca^{2+}. Insulin partially counteracts the effects of α-adrenergic agonists on activation of phosphorylase [9,23], inactivation of pyruvate kinase [10] and glycogen synthase [24] and suppression of fructose 2,6-bisphosphate levels [15]. The catabolic actions of vasopressin in contrast, are not affected by insulin [24,25]. The mechanism of action of α-adrenergic agonists involves increased phosphatidylinositol turnover [23] and Ca^{2+} efflux from hepatocyte suspensions [14].

Since insulin reversed the effect of phenylephrine on Ca^{2+} mobilisation [14] but had no effect on the increased rate of phosphatidylinositol turnover [23], the site(s) of action of insulin were suggested to be distal to phosphatidylinositol breakdown but prior to changes in Ca^{2+} fluxes. The lack of effect of insulin on vasopressin-induced glycogenolysis as compared with the counteraction of the α-adrenergic mechanism may be related to the different mechanisms of action of these hormones. Although vasopressin and epinephrine produce a similar activation of phosphorylase and increase in cell Ca^{2+}, vasopressin caused a fivefold greater increase in inositol trisphosphate (determined isotopically) [26] and also a threefold to fourfold increase in diacylglycerol accumulation in hepatocytes [27].

20.3 Effects of insulin on cell Ca^{2+}

Isolated hepatocytes have provided a very useful model for studying hormone-induced changes in Ca^{2+} concentrations and fluxes both in whole cell suspensions and at the single cell level. In the perfused liver the only measurement that is feasible for studying Ca^{2+} fluxes is the accumulation of Ca^{2+} in the perfusate [27]. Since Ca^{2+} is omitted from the medium to enable accurate determination of the incremental changes in Ca^{2+}, the responses observed may not be physiological. Studies on hepatocyte suspensions used the fluorescent indicator Quin-2, which has a K_d for Ca^{2+} of 115 nM and a fivefold higher fluorescence in the Ca-bound state. It is taken up by cells as the membrane-permeant tetraacetoxymethyl ester, and is then hydrolysed by esterases to the impermeant free acid. This technique enabled the monitoring of hormone-induced changes in cell Ca^{2+} in hepatocyte suspensions at varying extracellular Ca^{2+} concentrations [28]. Using Quin-2, Thomas and colleagues [29] showed that insulin lowers the increase in Ca^{2+} caused by phenylephrine but not that by vasopressin.

The monitoring of cell Ca^{2+} at the single cell level enabled more detailed studies of hormone action. Initial studies used the photoprotein aequorin, which was micro-injected into hepatocytes [30], and later studies employed the fluorescent indicator Fura-2 [31,32], which has a much greater fluorescence intensity on Ca^{2+} binding than Quin-2. These showed that the increase in Ca^{2+} by hormones that stimulate the hydrolysis of phosphatidylinositol is due to repetitive Ca^{2+} oscillations or transients. The concentration of the agonist determines the frequency but not the amplitude of the oscillations [32]. Insulin partially antagonizes the effects of low concentrations of phenylephrine by decreasing the frequency of the oscillations [33]. However, the effects of insulin were rather variable and it was suggested that the intercellular heterogeneity of some of the hormonal effects might depend on whether the hepatocytes were spherical or flattened [33].

A study using hepatocyte couplets, which are suggested to be a better model for electrophysiological studies than single isolated cells, showed that in the absence of other hormones insulin caused a transient and monophasic increase in Ca^{2+} [34]. This effect of insulin was blocked in the absence of Ca^{2+} and by non-selective cation blockers (Ni^{2+} and Gd^{3+}) but not by verapamil, an inhibitor of voltage-gated cation channels. The effect of insulin on cell Ca^{2+} differs from the EGF-induced rise in Ca^{2+} which is attributed to activation of phospholipase Cγ and not abolished in the absence of extracellular Ca^{2+} [35,36]. However, the effects of insulin and EGF have not been compared in the same study. The physiological significance of the rise in Ca^{2+} by insulin is not known. Haddad and colleagues suggest that it may be involved in the activation of ERK1/ERK2 ($p42/44^{MAPK}$), based on observations that activation of these kinases by insulin is abolished in conditions that block the rise in Ca^{2+}, and that a Ca-ionophore activates ERK1/ERK2 in hepatocytes [34]. Since the stimulation of glycogen synthesis by insulin is not abolished in the absence of extracellular Ca^{2+} [5] or by inhibition of the MAP-kinase pathway [6] it is unlikely that the rise in Ca^{2+} has a role in the metabolic effects of insulin.

20.4 The direct effects of insulin on glycogen storage

20.4.1 The relative roles of glucose and insulin in stimulating glycogen synthesis

The long-sustained notion that insulin does not have a role in stimulating glucose utilisation by the liver emerged from two sets of observations. First, the difficulty in demonstrating stimulation of glycogen deposition by insulin in the perfused liver or freshly isolated hepatocyte suspensions [1]. Second, in situations where activation of glycogen synthase by insulin was observed, the degree of stimulation was apparently low in comparison with the effects of glucose [2].

The stimulation of glycogen deposition by glucose is the composite result of both sequential and convergent mechanisms which include: (i) inhibition of phosphorylase by glucose as a result of increased dephosphorylation of phosphorylase *a* by phosphorylase phosphatase [37]; (ii) activation of glycogen synthase by synthase phosphatase as a result of the decreased activity of phosphorylase *a*, which is a potent inhibitor of synthase phosphatase [38]; (iii) glucose-induced increase in glucose 6-phosphate through translocation of glucokinase from an inactive bound state in the nucleus to an active state in the cytoplasm [39]; (iv) glucose 6-phosphate-induced activation of glycogen synthase by allosteric regulation and by favouring the dephosphorylation of synthase by synthase phosphatase and by translocation of synthase [37]. The overall effect of these mechanisms is that glycogen synthesis in hepatocytes is an almost linear function of glucose concentration [39].

Several studies found no effect of insulin on glycogen deposition in freshly isolated hepatocyte suspensions from fed or fasted rats [1]. Others however, reported stimulation of glycogen deposition in hepatocyte suspensions from fasted rats [40,41]. It was suggested that minor modifications in the hepatocyte isolation technique may be critical for studies on insulin action on glycogen deposition. However, a fundamental difficulty with most of the previous studies on hepatocyte suspensions from fed rats is that these were found to be in negative glycogen balance up to 20 mM glucose [42]. This negative glycogen balance could not be reversed by insulin [42]. One of the main advantages in the use of monolayer-cultured hepatocytes for studying glycogen metabolism is that by pre-culturing the cells at low glucose concentrations, in the absence of insulin, the glycogen stores decline to a steady state level that is dependent on the medium glucose concentration and the physiological state of the donor rat [43,44]. When the hepatocytes are subsequently incubated at a higher glucose concentration for up to 4 h, the rate of glycogen synthesis from glucose correlates closely with the rate of glycogen deposition determined analytically [43,45]. Under these conditions insulin stimulates glycogen synthesis or deposition by more than twofold at 10-15 mM glucose and by 30-50% at glucose concentrations from 15-30 mM [43,45-47]. This stimulatory effect of insulin on glycogen deposition is progressive with time [43,46] ranging from around 30% stimulation after 30 min (at 10-15 mM glucose) to more than a twofold stimulation after 3 h exposure to insulin. This effect of insulin in hepatocyte cultures is in good agreement with the twofold difference in hepatic glycogen storage *in vivo* between normal and insulin-deficient rats after a meal [48].

20.4.2 Effects of insulin on glycogen synthase and phosphorylase

The effects of insulin on glycogen synthase have been studied in both freshly isolated hepatocyte suspensions and in monolayer cultures. Witters and Avruch [49], using freshly isolated hepatocyte suspensions from fed rats, found that insulin potentiated the effects of glucose (10-30 mM) on synthase activation. The insulin effect was between 30 and 50% of total synthase activation and was maximal after 10-15 min and totally reversed by 30 min. There was no effect of insulin in the absence of glucose.

Ciudad and colleagues [50], using a modified glycogen synthase assay with low concentrations of glucose 6-phosphate (0.25 mM) to improve assay sensitivity, found a small and transient activation of glycogen synthase by insulin in the absence of glucose in hepatocyte suspensions from fasted rats. The conclusion was that the activation by insulin is "marginal" in comparison with the stimulation by glucose [2].

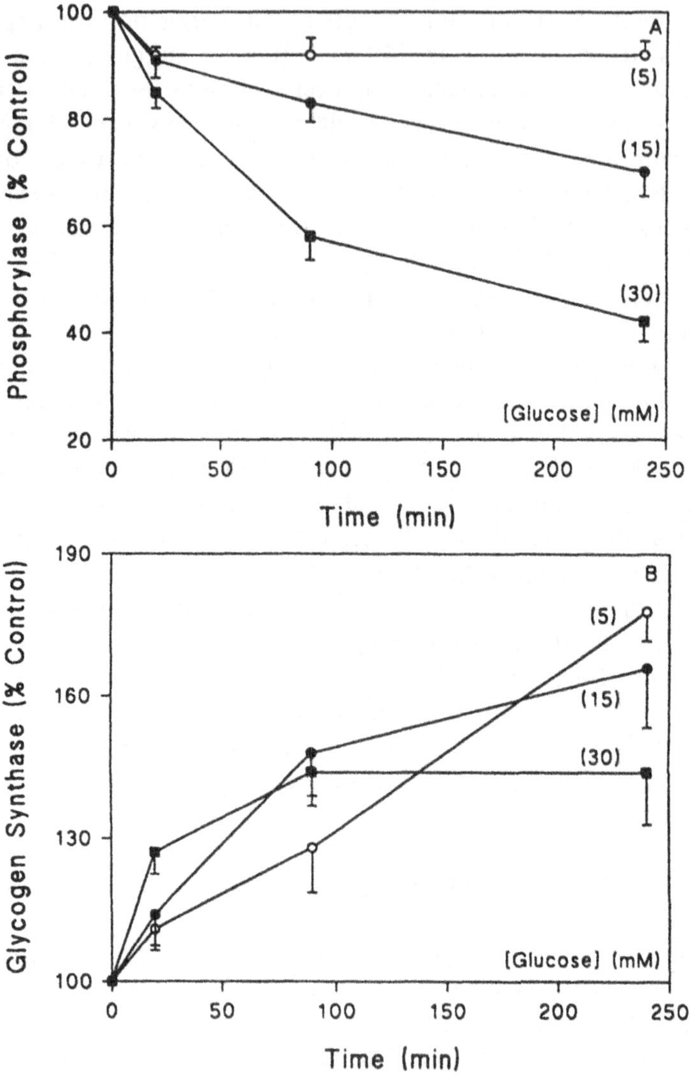

Figure 1. Time course of the effects of insulin on phophorylase a and glycogen synthase activity in hepatocyte monolayers. Hepatocytes isolated from fed rats were cultured as in [6]. After culture for 16 h in MEM containing 5 mM glucose and 10 nM dexamethasone, they were incubated for 4 h in MEM containing either 5, 15 or 30 mM glucose and with insulin for the time intervals indicated. They were then extracted for determination of glycogen synthase (% active: 0 / 10 mM glucose 6-phosphate × 100) and phosphorylase *a*. The effects of insulin on phosphorylase *a* (A) and glycogen synthase (B) are expressed as a percentage of control incubations without insulin. Values are means ± SEM for 4 cultures

In contrast with these transient effects of insulin on glycogen synthase, later studies on hepatocyte suspensions from either fasted [41] or fed rats [29] reported more sustained effects of insulin on synthase activation. The latter study is of particular interest because it demonstrated firstly, that the activation of glycogen synthase by insulin in the absence of other hormones has a slower time course than the antagonism by insulin of the effects of α-adrenergic agonists on glycogen metabolism, and secondly that the effect of insulin was glucose-dependent. The activation of glycogen synthase by insulin in hepatocyte cultures is not transient, and is sustained for at least 2-4 h [47,51]. The extent by which insulin activates glycogen synthase in cultures was found to be strongly dependent on the endogenous glycogen content with maximum activation (two to three times after 60 min) being observed in hepatocytes with the lowest glycogen content [46].

A key question when evaluating hormonal control of glycogen synthase in hepatocytes is whether the changes in glycogen synthase activity are coupled to inverse changes in phosphorylase *a*, which is a potent inhibitor of synthase phosphatase [38] through an allosteric site on the glycogen-targetting subunit PPP1R4. Few of the studies on glycogen synthase have included measurements of phosphorylase [29,47,49,52]. Based on reciprocal changes in synthase and phosphorylase activity in the presence of insulin and glucose, Witters and Avruch [49] concluded that the activation of synthase by insulin can be entirely accounted for by inactivation of phosphorylase. However, other studies on both suspensions [29] and cultures [52] found activation of synthase by insulin without changes in phosphorylase *a*.

In view of the role of glucose in the regulation of both phosphorylase *a* and glycogen synthase we investigated the time course of the actions of insulin on these enzymes at different glucose concentrations in hepatocyte cultures (Figure 1). Insulin inhibited phosphorylase more markedly at elevated glucose concentration with 58, 30 and 8% inhibition at 30 mM, 15 mM and 5 mM glucose respectively, and this inhibition was sustained over 4 h. The rate of activation of synthase by insulin was more rapid at 30 mM and 15 mM glucose than at 5 mM glucose. However at low glucose concentrations the activation of synthase by insulin was progressive and attained the same maximum stimulation as at elevated glucose concentrations. The inactivation of phosphorylase by insulin may therefore be a contributing factor to the more rapid activation of synthase at elevated glucose.

The notion that insulin has a minor role in comparison with glucose in activation of synthase in hepatocytes [2] stems from several studies that have shown a small activation of synthase by insulin in both hepatocyte suspensions and cultures at an early time point (10-15 min) and at low [glucose] [6,47,53-55]. However, the study by Fleig and colleagues [46] showed a large activation of synthase by insulin after 60 min in hepatocyte cultures with low glycogen stores. The data in Figure 1 support the conclusions of Fleig and colleagues in showing that the effect of insulin is time-dependent and also glucose-dependent. Since hepatic glycogen storage in the absorptive state extends over 4 h and occurs in response to elevated glucose levels in the portal vein, the activation of synthase by insulin at steady state or at elevated glucose concentrations rather than at 10-15 min and at low glucose concentrations appears more relevant to the physiological situation. From this it can be inferred that insulin has a major role in the activation of hepatocyte glycogen synthase.

20.4.3 Role of synthase and phosphorylase phosphatases in insulin action

Glycogen phosphorylase is activated by phosphorylation on a single residue (serine-14) catalysed by phosphorylase kinase. Glycogen synthase however, is regulated by multi-site phosphorylation involving nine serine residues and several kinases [56,57]. The hepatocyte phosphatases that act on glycogen synthase and phosphorylase are predominantly type-1, whereas both type-1 and type-2A phosphatases act on phosphorylase kinase [58]. Okadaic acid, an inhibitor of type-1 and type-2A phosphatases, blocks the stimulation of glycogen synthesis by insulin in hepatocyte cultures at low inhibitor concentrations (<2.5 nM) that do not affect the basal rate of synthesis, suggesting involvement of type-1 and/or 2A protein phosphatases in insulin action [59].

A rapid and sustained (5-40 min) stimulatory effect of insulin on phosphorylase phosphatase in hepatocyte cultures has been reported [60]. Effects of insulin on synthase phosphatase (using muscle synthase as substrate) have been demonstrated only after culture of hepatocytes with insulin for 2-3 days [61]. However, Toth and colleagues showed activation within 10 min of both synthase phosphatase (using liver synthase as substrate) and phosphorylase phosphatase by insulin or glucose in the perfused liver [58]. These effects were observed only in concentrated extracts and were lost on dilution, which may explain why acute activation of synthase phosphatase was not observed in hepatocyte cultures. Both effects of insulin are attributed to type-1 phosphatases and are apparently due to an increase in V_{max}. Based on fractionation studies and the loss of activation on dilution of extracts it was suggested that the mechanism may involve interaction of the phosphatase with other protein(s) which co-elute on Sephadex G-25. Vanadate was found to mimic the acute effects of insulin on these phosphatases providing a potential tool for further studies [62]. Three protein phosphatase type-1 binding proteins containing a glycogen-targeting domain, designated PPP1R4, 5 and 6, are expressed in liver [63-65]. Effects of insulin on the association of type-1 phosphatases with these binding proteins are as yet uncharacterized.

20.4.4 Glycogen synthase kinases

Several kinases have been shown to phosphorylate hepatic glycogen synthase including protein kinases A and C, phosphorylase kinase, calmodulin-dependent protein kinase II, glycogen synthase kinase-3, casein kinases I and II and AMP-activated protein kinase [56,66]. During incubation of hepatocyte suspensions with [^{32}P]phosphate and without added hormones, glycogen synthase incorporates label indicating that synthase kinases are active in the absence of hormones [56]. Since phosphorylation of glycogen synthase by these kinases leads in most instances to a decrease in activity, inhibition of kinases is a possible contributory mechanism to the activation of synthase by insulin.

Although insulin counteracts the activation of phosphorylase kinase by low concentrations of glucagon in hepatocyte suspensions [9], an inhibitory effect of insulin in the absence of glucagon has not been demonstrated *in vitro*. The activity of calmodulin-dependent protein kinases, may be modulated through changes in the phosphorylation state of calmodulin. In hepatocyte suspensions insulin increases the phosphorylation state of calmodulin (on serine/threonine and tyrosine residues) by approximately threefold within 10 min and both the insulin receptor kinase and casein kinase II may contribute to this phosphorylation [67]. Transient inhibition of AMP-activated protein kinase by insulin has been reported in Fao hepatoma cells [68] but no data are available for isolated hepatocytes. The function of this kinase is thought to be as a "low-fuel sensor" of the cell rather than a response to extracellular signals [69]. Since changes in substrate concentrations have a major role in the regulation of hepatic metabolism and gene expression, an effect of insulin on this kinase that is secondary to the metabolic effects but might have a role in amplification mechanisms, cannot be excluded.

Glycogen synthase kinase-3 (GSK-3) is one of the most intensely studied synthase kinases in terms of hormone and growth factor signalling. Based on two sets of observations, it is thought to have a major role in the activation of glycogen synthase by insulin. Firstly, GSK-3 is one of the more effective kinases involved in phosphorylation of glycogen synthase in terms of both stoichiometry of phosphorylation and enzyme inactivation. Secondly, GSK-3 is inactivated by a wide range of growth factors as well as by insulin suggesting its involvement in several protein kinase cascades.

In hepatocyte cultures GSK-3 activity (determined by immunoprecipitation of both α and β isoforms) is inactivated by both insulin and EGF by about 30% [6]. Since EGF does not mimic the effects of insulin on glycogen synthesis or glycogen synthase [70], three possibilities could be considered. Firstly, the inactivation of GSK-3 by 30% may make a negligible contribution to activation of glycogen synthase by insulin in hepatocytes. Secondly, additional components of the insulin signalling mechanism are required for activation of glycogen synthase, such as inactivation of other synthase kinases and phosphorylase or activation of synthase phosphatases [58,60] but inactivation of GSK-3 is nevertheless a prerequisite for enabling activation. Thirdly, inactivation of GSK-3 by 30% may be sufficient to cause activation of synthase, however, the lack of activation by EGF may be the result of other (inhibitory) components of EGF signalling, such as mechanisms that lie downstream of phospholipase C activation [70].

Recent studies showed that Li^+ inhibits GSK-3β with a K_i of 2-3 mM [71]. Li^+ had previously been shown to cause a marked activation of glycogen synthase [54,72] in hepatocyte suspensions, suggesting that inactivation of GSK-3 may in itself be sufficient to activate glycogen synthase (assuming that the effect of Li^+ is specific for GSK-3).

20.4.5 Protein kinase cascades leading to inactivation of GSK-3

Cohen & colleagues [74] have proposed a number of putative signalling kinase cascades that lead to inactivation of GSK-3 (Figure 2). GSK-3 can be phosphorylated and inactivated *in vitro* by p90rsk (also known as rsk-2, insulin-stimulated protein kinase or MAPKAP-kinase 1) by p70rsk [73] and by protein kinase B (Akt/RAC kinase) [74]. These three kinases are activated by growth factors as well as by insulin in various cell types. Inhibitors are available that prevent the activation of p70rsk (rapamycin) and the activation of the MAP-kinase pathway (PD98059, which inhibits activation of *ERK2*/*ERK*1 by MEK) but not for protein kinase B.

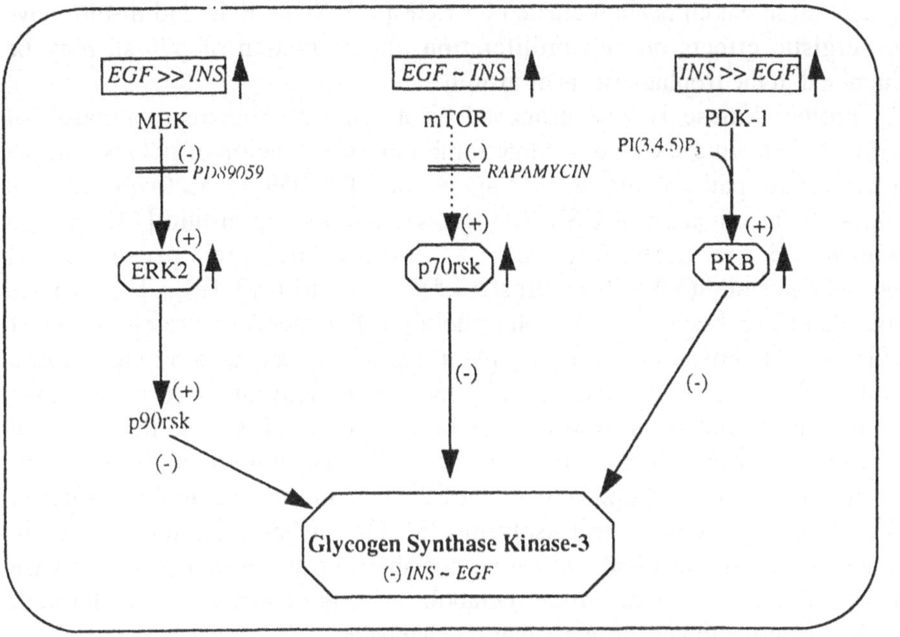

Figure 2. Protein kinases involved in the inactivation of GSK-3 in hepatocytes. Glycogen synthase is inactivated by phosphorylation by GSK-3 (glycogen synthase kinase-3). Phosphorylation (inactivation) of GSK-3 can be mediated by p90rsk, which is activated via the MAP-kinase pathway involving MEK (mitogen activated protein kinase kinase) and ERK-2 (extracellular regulated kinase), or by p70rsk, or by protein kinase B (which can be activated by PDK1, 3-phosphoinositide-dependent protein kinase 1). PD98059 inhibits activation of ERK-2 by MEK. Rapamycin inhibits activation of p70rsk by inhibiting the function of a kinase designated mammalian target of rapamycin (mTOR). Insulin and EGF cause a similar inactivation of GSK-3 and a similar activation of p70rsk. However, EGF causes a greater activation of ERK-2 than insulin whereas insulin causes a greater activation of protein kinase B than EGF (for details see [6]).

Studies using these inhibitors in hepatocytes have shown that neither p70rsk nor the MAP-kinase pathway are involved in the stimulation of glycogen synthesis by insulin [6,75]. Furthermore, the activation of ERK2 by insulin in hepatocyte cultures is much lower than the activation by EGF, and both effects are transient [6]. Insulin and EGF cause a similar activation of p70rsk (twofold to threefold) in hepatocytes [6]. Although rapamycin totally suppresses p70rsk activity in hepatocytes it does not block the stimulation by insulin of glycogen synthesis [6]. Rapamycin does however, inhibit the insulin-induced expression of glucose 6-phosphate dehydrogenase [76]. This effect of insulin is potentiated by EGF, unlike the expression of glucokinase which is counteracted by EGF [77]. Since EGF and insulin have synergistic effects on cell proliferation, the activation of p70rsk may be involved in the trophic effects of insulin.

Protein kinase B was proposed as a putative "upstream kinase" of GSK-3, following the findings that inhibition of activation of p70rsk and the MAP-kinase pathways (with rapamycin and PD98059, respectively) did not block the inactivation of GSK-3 by growth factors and insulin [74]. Protein kinase B is activated by a kinase in the presence of phosphoinositidyl(3,4,5)P_3 (PtdIns(3,4,5)P_3) or PtdIns(3,4)P_2. This kinase has therefore been named 3-phosphoinositide-dependent protein kinase-1 (PDK-1). In hepatocyte cultures protein kinase B is activated by both insulin and EGF. However the activation by insulin is about threefold greater than that by EGF and is sustained for at least 30 min [6]. Of the three known isoforms of protein kinase B, only the α and β isoforms are activated by insulin in hepatocytes, and activation of the α-isoform is about three times to four times higher than the β-isoform [78]. Only when inhibitors of protein kinase B become available will it be possible to test whether protein kinase B isoforms are involved in the metabolic effects of insulin, either through GSK-3 inactivation or through other mechanisms.

It is noteworthy that although GSK-3 is shown downstream of p90rsk in Figure 2, it also phosphorylates this kinase [79] as well as other "upstream" signalling components including IRS-1. The ability of GSK-3 to both phosphorylate and be phosphorylated by p90rsk may represent a feedback mechanism that accounts for the relatively modest effects of insulin or EGF on this kinase.

20.5 Upstream signalling elements: insulin receptor substrates and PI 3-kinase

Following insulin binding to the receptor, activation of the insulin receptor tyrosine kinase causes phosphorylation of a number of intracellular insulin receptor substrates (IRS) such as IRS-1 (pp185, 130 kDa), which is expressed ubiquitously, and IRS-2, an isoform that is upregulated in IRS-1 knockout mice. IRS-2 serves as an alternative substrate for insulin signalling [80] and may have an essential role in hepatocytes [81]. In addition to IRS-1 and IRS-2, other proteins are tyrosine-phosphorylated following insulin binding. In hepatocytes, a membrane glycoprotein, pp120, which is localized in the hepatocyte apical membrane [82] is tyrosine-phosphorylated following stimulation with insulin. Several putative functions have been ascribed to this protein, which is an ecto-ATPase, can transport bile acids and is related to cell adhesion molecules. It may have a role in receptor endocytosis and degradation [83] and additionally has endogenous phosphatase activity [84]. This protein is also tyrosine-phosphorylated by the EGF receptor [85], which suggests that a role specific for the metabolic effects of insulin is unlikely. However, it could have a role in hormone degradation since the liver is a major site of clearance of insulin and EGF.

IRS-1 and IRS-2 contain multiple tyrosine-phosphorylation motifs that function as binding sites for src-homology-2 (SH2) domains. A widely studied SH2 domain protein is the regulatory subunit (p85) of phosphatidylinositol (PI) 3-kinase (p110). The association of PI 3-kinase with tyrosine phosphorylated IRS-1 or tyrosine phosphorylated peptides with YMXM motifs increases its activity. Studies using inhibitors of PI-3-kinase (wortmannin and LY294002), which inhibit both basal and insulin-stimulated activity, have shown that virtually all metabolic actions of insulin, with the possible exception of activation of pyruvate dehydrogenase (PDH) phosphatase in adipocytes, are blocked by these inhibitors [86]. In hepatocytes, these inhibitors block the stimulation of glycogen synthesis [6,75], activation of glycogen synthase [6], translocation of glucokinase by insulin (L. Agius, unpublished data) and inhibition of apolipoprotein A secretion [87].

Three classes of PI 3-kinases have been identified in mammalian cells. Class I, comprise the p110 isoforms that bind to the p85 regulatory subunit. These enzymes phosphorylate PtdIns, PtdIns(4)P and PtdIns(4,5)P_2 on the 3-position. Class II, PI 3-kinases (p170-190) phosphorylate predominantly PtdIns and to a small extent PtdIns(4)P. They are much less sensitive to wortmannin than the class I isoenzymes and lack homology to the class I N-terminus, which is responsible for binding to the p85 subunit [88]. Class III PI 3-kinases phosphorylate only PtdIns and are thought to be involved in transfer of proteins from the Golgi compartment rather than in hormone signalling, since only PtdIns(3,4,5)P3 and PtdIns(3,4)P2 and not PtdIns(3)P increase in response to insulin or growth factor stimulation [89]. In hepatocyte monolayers insulin causes a threefold increase in PI 3-kinase determined in antiphosphotyrosine immunoprecipitates [87]. This is associated with an increase in binding of p85 to IRS-1 and the activity is inhibited by 1 μM wortmannin. Whether the activation by insulin involves exclusively the class I isoforms or also class II, which are thought to be involved in differentiated functions of hepatocytes [90], remains to be elucidated.

20.6 The role of cell swelling in insulin action: an upstream event of PI 3-kinase?

Regulation of potassium homeostasis by insulin shares several similarities with the regulation of glucose homeostasis [91]. Insulin deficiency results in an increase in plasma K^+ and insulin treatment acutely lowers plasma K^+ due to rapid uptake of K^+ by insulin-sensitive tissues. *In vivo* studies established that liver and muscle are the prime sites of insulin-stimulated K^+ uptake [92], and *in vitro* studies using hepatocyte monolayers, showed that insulin promotes net accumulation of K^+ in part by increasing the K^+/Na^+ ratio but predominantly by increasing the cell K^+ plus Na^+ content with a consequent increase in cell volume [53,93].

The effects of insulin on cell volume are of interest from the perspective of signalling mechanisms, because cell swelling *per se*, whether it is caused by hypotonic media or by solute accumulation, stimulates glycogen synthesis [94] (see also Chapter 11) and mimics several of the actions of insulin. Furthermore, ionic conditions that prevent cell swelling by insulin, such as absence of extracellular Na^+, block the metabolic effects of insulin, indicating that swelling is an essential component of the mechanism by which insulin stimulates glycogen synthesis [53]. Activation of protein phosphatases by an increase in intracellular glutamate (amino acid induced swelling) or a decrease in Cl^- (hypotonic swelling) was proposed as a contributory mechanism to activation of glycogen synthase by swelling [95].

An important development in the understanding of the mechanisms by which cell swelling alters hepatocyte metabolism was the finding by Krause and colleagues [96] that hypotonic swelling as well as amino acid-induced swelling in hepatocytes is associated with activation of PI 3-kinase, as determined by immunoprecipitation of the p85 subunit, and that inhibition of PI 3-kinase with wortmannin markedly counteracts the activation of glycogen synthase by cell swelling. These findings introduce a new dimension in the role of cell swelling in insulin signalling. Since the degree of cell swelling caused by insulin is similar to that by hypotonic swelling [53,94], the morphological changes and/or cytoskeletal rearrangements resulting from swelling may be an essential contributory mechanism to the recruitment of p85 to IRS-1 or other phosphorylated proteins with YMXM motifs. Thus cell swelling consequent to net K^+ accumulation may be an essential upstream event for activation of PI 3-kinase. The time course of activation of protein kinase B by insulin in hepatocyte monolayers [6,78] (which is dependent on 3-phosphoinositides generated by class I PI 3-kinases, Figure 2) is similar to that for K^+ uptake by the hormone [53], and therefore supports the hypothesis that cytoskeletal changes triggered by swelling may be necessary for activation of PI 3-kinase. Stimulation of K^+ uptake by insulin was recognized as an important component of insulin action as early as 1923 (see [91]).

20.7 The inositol phosphoglycan (IPG) mediator: an alternative signalling pathway or an upstream event?

Some of the metabolic actions of insulin have been proposed to be mediated by inositol phosphoglycans (IPGs) that are chemically related to the glycosyl-phosphatidylinositol (GPI) anchors of cell surface proteins [97]. The IPGs are thought to be generated extracellularly by phosphodiester hydrolysis (through the actions of phospholipase C or phospholipase D) of either GPI anchors or free GPI, and are then transported into the cell and exert their effects on various targets.

Studies characterising the metabolic actions of IPG (also described as insulin-induced phospho-oligosaccharide) have been performed predominantly on preparations obtained by extracting GPI (which accounts for <1% of total membrane phospholipid) from liver membranes followed by hydrolysis with bacterial PtdIns-Phospholipase C to IPG [98]. These IPG preparations, which contain phosphate, glucosamine, galactose and D-chiroinositol, [99] have various insulin-like effects in isolated hepatocyte suspensions. They mimic the counter-regulatory effects of insulin on glucagon-induced activation of phosphorylase and inactivation of pyruvate kinase [100], they inhibit cAMP-dependent protein kinase [101], antagonize the effects of α-adrenergic agonists on Ca^{2+} oscillations [102] and counteract the induction of phosphoenolpyruvate carboxykinase by cAMP [103]. They also mimic the stimulatory effects of insulin on glycogen synthesis [97,104,105] and stimulate amino acid transport [106]. The latter effect differed from the effect of insulin in that it was not additive with the stimulation by glucagon.

Insulin-like effects in hepatocytes have also been reported with IPG preparations from other sources. A preparation from the GPI anchor of trypanosomes was reported to inhibit gluconeogenesis (pyruvate incorporation into glucose) in hepatocytes as well as glucose 6-phosphatase and fructose-1,6-bisphosphatase in hepatocyte extracts [107]. An antibody raised against the surface glycoprotein of *Trypanosoma brucei* blocks the activation by IPG of PDH in myocytes [108] and glycogen synthesis in hepatocytes [104] and partially counteracts the stimulation of glycogen synthesis by insulin [104].

Various labelling techniques have been used to study the effects of insulin on GPI turnover. Using H35 hepatoma cells, Mato and colleagues [98] showed the incorporation of [^{3}H]glucosamine into a GPI fraction that also incorporated palmitic acid but not myo-inositol, and was hydrolysed by PtdIns-phospholipase C. They suggested that insulin affects the turnover of this GPI fraction. Subsequent studies on hepatocytes using [^{3}H]glucosamine labelling [101,109] and a similar extraction procedure [98] reported either no effect of insulin on the label in the GPI peak [109] or a transient decrease in labelling in hepatocytes from adult but not from fetal rats [101]. An alternative labelling approach has used [^{14}C]acetimidate (which labels phospholipids containing free amino groups) to label the GPI on termination of the hepatocyte incubation [110]. Both ethyl and isethionyl [^{14}C]acetimidate have been used as permeable and non-permeable probes, respectively, to determine the total GPI and that facing the extracellular surface. It was shown that most of the GPI faces the extracellular surface of hepatocytes [110] and that insulin decreases the GPI peak by 20-60% [104,105,110]. Hepatocytes from streptozotocin-treated diabetic rats and 24-month old rats, which show decreased stimulation of glycogen synthesis by insulin, were shown to have low levels of GPI (30-40% of controls) which were not decreased by insulin [104,105].

Whilst there is compelling evidence that IPGs prepared by enzymic treatment of GPI extracts mimic both the direct stimulation of glycogen synthesis by insulin and the counter-regulatory actions, their chemical structures are as yet uncharacterized. The possibility that there may be diverse IPG mediators with different functions has been suggested. The evidence in support of a role for IPG in insulin action in hepatocytes derives from: (i) the effect of insulin on GPI levels in hepatocytes [104,105]; (ii) the inhibition by an antibody raised against GPI of the stimulatory effects of both IPG and insulin on glycogen synthesis [104]. Studies on other physiological targets of insulin suggest that IPG-peptides derived from yeast promote the phosphorylation of IRS-1 and the recruitment of PI 3-kinase, and that wortmannin blocks the stimulation of glucose transport by IPG, suggesting that it acts upstream of PI 3-kinase [111]. This is a rather surprising finding in the context of the hypothesis that the actions of IPG's are mediated directly on pathway enzymes [107]. However, effects on protein kinases have also been reported [96]. It is of particular interest in the context of PDH phosphatase, which is reputedly not downstream of PI-3-kinase [86], but is regulated directly by IPG [96,108]. The activation of GPI hydrolysis by insulin appears to involve G_i-proteins [112]. However, the detailed mechanism remains to be elucidated.

20.8 Perspectives

Insulin signalling involves a wide repertoire of mechanisms that include tyrosine phosphorylation of the β-subunit of the receptor and of cellular substrates with protein-protein interaction domains, recruitment and activation of PI 3-kinase, changes in phosphorylation state of G-proteins, generation of intracellular and extracellular mediators, changes in ion flux and cell swelling, activation of protein kinase cascades and activation of protein phosphatases. Most of these mechanisms have been demonstrated in hepatocytes. This is consistent with the major role of insulin in the regulation of hepatic glucose utilisation. Activation of PI 3-kinase is now recognized as a component of the insulin-signalling mechanism that is essential for most of the metabolic actions of insulin. The recent findings that both cell swelling and IPG mediators activate PI 3-kinase and that inhibitors of this kinase block the metabolic actions of cell swelling and the IPG mediator, suggest that tyrosine phosphorylation of IRS proteins may be one of several mechanisms that exert a concerted action on the activation of PI 3-kinase. Understanding the relative contributions of diverse signalling mechanisms that either converge at a single site or alternatively activate different isoforms that may be involved in the regulation of different metabolic effects is important for identifying the metabolic defect(s) in insulin-resistant states. Isolated hepatocytes either in suspension or in short-term primary culture provide a unique experimental model for studying insulin signalling. In addition to expressing differentiated liver functions and responses to insulin that are similar to the physiological effects *in vivo*, they can also be used to study defects in insulin-signalling in animal models of insulin-resistance. This is because hepatocytes in suspension or in short-term monolayer culture retain metabolic differences that are related to the endocrine or nutritional state of the donor rat.

Acknowledgements

Our work has been funded by the British Diabetic Association, the Medical Research Council and Juvenile Diabetes Foundation, International. We are grateful to colleagues for stimulating discussions and particularly to Drs. Matthew Peak, Joan Guinovart and Susanna Baqu.

References

1. Katz J and McGarry JD. The glucose paradox. Is glucose a substrate for liver glycogen metabolism? J Clin Invest 1984; 74: 1901-1909.
2. Villar-Palasi C and Guinovart JJ. The role of glucose-6-phosphate in the control of glucogen synthase. FASEB J 1997; 11: 544-558.
3. Lawrence JC and Roach PJ. New insights into the role and mechanism of glycogen synthase activation by insulin. Diabetes 1997; 46: 541-547.
4. Chowdhury MH and Agius L. Epidermal growth factor counteracts the glycogenic effect of insulin in parenchymal hepatocyte cultures. Biochem J 1987; 247: 307-314.
5. Peak M and Agius L. Transforming growth factor α inhibits glycogen synthesis and counteracts the stimulation by insulin in hepatocytes. Biochem Biophys Res Commun 1992; 183: 560-565.
6. Peak M, Rochford JJ, Borthwick AC, Yeaman SJ and Agius L. Signalling pathways involved in the stimulation of glycogen synthesis by insulin in rat hepatocytes. Diabetologia 1998; 41: 16-25.
7. Berry MN and Friend DS. High yield preparation of isolated rat liver parenchymal cells: a biochemical and fine structural study. J Cell Biol 1969; 43: 5506-5520.
8. Exton JH and Park CR. Interaction of insulin and glucagon in the control of liver metabolism. In: Handb Physiol Section 7 Endocrinol 1 1972: 437-455.
9. Van de Werve G, Hue L and Hers HG. Hormonal and ionic control of the glycogenolytic cascade in rat liver. Biochem J 1977; 162: 135-143.
10. Feliu JE, Hue L and Hers HG. Hormonal control of pyruvate kinase activity and of gluconeogenesis in isolated hepatocytes. Proc Natl Acad Sci 1976; 73: 2762-2766.
11. Sonne O, Berg T and Christoffersen T. Binding of 125I-labelled glucagon and glucagon stimulated accumulation of adenosine 3':5'-monophosphate in isolated intact rat hepatocytes. J Biol Chem 1978; 253: 3203-3210.
12. Wakelam MJO, Murphy GJ, Hruby VJ and Houslay MD. Activation of two signal transduction systems in hepatocytes by glucagon. Nature 1986; 323: 68-71.
13. Mine T, Kojima I and Ogata E. Evidence of cAMP independent action of glucagon on calcium mobilisation in rat hepatocytes. Biochim Biophys Acta 1988; 970: 166-171.
14. Blackmore PF, Assimacopoulos-Jeannet F, Chan TM and Exton JH. Studies on α-adrenergic ativation of hepatic glucose output. J Biol Chem 1979; 254: 2828-2834.
15. Pilkis SJ, Claus TH, Johnson RA and Park CR. Hormonal control of cyclic 3':5'-AMP levels and gluconeogenesis in isolated hepatocytes from fed rats. J Biol Chem 1975; 250: 6328-6336.
16. Morris NJ, Young P and Houslay MD. Insulin inhibits the phosphorylation of α-Gi-2 in intact hepatocytes. Biochem J 1995; 308: 693-696.
17. Houslay MD, Griffiths SL, Horton YM and Livingstone C. Regulation of intracellular cAMP concentrations in hepatocytes involves integrated activation and desensitisation of adenyl cyclase coupled with the action and activation of specific isoforms of cAMP phosphodiesterase. Biochem Soc Trans 1992; 20: 140-146.
18. Wijkander J, Landstrom TR, Manganiello V, Belfrage P and Degerman E. Insulin-induced phosphorylation and activation of phosphodiesterase 3B in rat adipocytes: possible role of protein kinase B but not mitogen-activated protein kinase or p70 S6 kinase. Endocrinology 1998; 139: 219-227.
19. Gabbay RA and Lardy HA. Site of insulin inhibition of cAMP-stimulated glycogenolysis. J Biol Chem 1984; 259: 6052-6055.
20. Mor AM, Vila J, Ciudad CJ and Guinovart, JJ. Insulin inactivation of rat hepatocyte cyclic AMP dependent protein kinase. FEBS Lett 1981; 136: 131-134.

21. Marks JS and Botelho LHP. Synergistic inhibition of glucagon-induced effects on hepatic glucose metabolism in the presence of insulin and a cAMP antagonist. J Biol Chem 1986; 261: 15895-15899.
22. Beebe SJ, Redmon JB, Blackmore PF and Corbin JD. Discriminative insulin antagonism of stimulatory effects of various cAMP analogs on adipocyte lipolysis and hepatocyte glycogenolysis. J Biol Chem 1985; 260: 15781-15788.
23. Thomas AP and Williamson JR. Effects of insulin on phenylephrine-induced activation of phosphorylase and phosphatidylinositol turnover in isolated rat hepatocytes. J Biol Chem 1983; 258: 1411-1414.
24. Strickland WG, Blackmore PF and Exton JH. The role of calcium in α-adrenergic inactivation of glycogen synthase in rat hepatocytes and its inhibition by insulin. Diabetes 1980; 29: 617-622.
25. Dehaye JP, Hughes BP, Blackmore PF and Exton JH. Insulin inhibition of α-adrenergic actions in liver. Biochem J 1981; 194: 949-956.
26. Charest R, Prpic V, Exton JH and Blackmore PF. Stimulation of inositol trisphosphate formation in hepatocytes by vasopressin, adrenaline and angiotensin II and its relationship to changes in cytosolic free Ca^{2+}. Biochem J 1985; 227: 79-90.
27. Bocckino SB, Blackmore PF and Exton JH. Stimulation of 1,2-diacylglycerol accumulation in hepatocytes by vasopressin, epinephrine and angiotensin II. J Biol Chem 1985; 260: 14201-14207.
28. Charest R, Blackmore PF, Berthon B and Exton JH. Changes in free cytosolic Ca^{2+} in hepatocytes following α1-adrenergic stimulation. Studies on Quin-2 loaded hepatocytes. J Biol Chem 1983; 258: 8769-8773.
29. Thomas AP, Martin-Requero A and Williamson JR. Interactions between insulin and α1-adrenergic agents in the regulation of glycogen metabolism in isolated hepatocytes. J Biol Chem 1985; 260: 5963-5973.
30. Woods NM, Cuthbertson KSR and Cobbold PH. Repetitive transient rises in cytoplasmic free calcium in hormone-stimulated hepatocytes. Nature 1986; 319: 600-602.
31. Kawanashi T, Blank LM, Harootunian AT, Smith MT and Tsien RY. Ca^{2+} oscillations induced by hormonal stimulation of individual fura-2 loaded hepatocytes. J Biol Chem 1989; 264: 12859-12866.
32. Rooney TA, Sass EJ and Thomas AP. Characterisation of the cytosolic calcium oscillations induced by phenylephrine and vasopressin in single fura-2 loaded hepatocytes. J Biol Chem 1989; 264: 17131-17141.
33. Somogyi R, Zhao M and Stucki JW. Modulation of cytosolic-[Ca^{2+}] oscillations in hepatocytes results from cross talk among second messengers. Biochem J 1992; 286: 869-877.
34. Benzeroual K, Van De Werve G, Meloche S, Mathe L, Romanelli A and Haddad P. Insulin induces Ca^{2+} influx into isolated rat hepatocyte couplets. Am J Physiol 1997; 272: G1425-G1432.
35. Mine T, Kojima I, Ogata E and Nakamura T. Comparison of effects of HGF and EGF on cellular calcium in rat hepatocytes. Biochem Biophys Res Commun 1991; 181: 1173-1180.
36. Tanaka Y, Hayashi N, Kaneko A, Ito T, Miyoshi E, Sasaki Y, Fusamoto H and Kamada K. Epidermal growth factor induces dose-dependent calcium oscillations in single fura-2 loaded hepatocytes. Hepatology 1992; 16: 479-486.
37. Guinovart JJ, Gomez-Foix AM, Seoane J, Fernandez-Novell JM, Bellido D and Vilaro S. Bridging the gap between glucose phosphorylation and glycogen synthesis in the liver. Biochem Soc Trans 1997; 25: 157-160.
38. Stalmans W, DeWulf H and Hers HG. The control of liver glycogen synthase phosphatase by phosphorylase. Eur J Biochem 1971; 49: 415-427.
39. Agius L and Peak M. Intracellular binding of glucokinase in hepatocytes and translocation by glucose, fructose and insulin. Biochem J 1993; 296: 785-796.

40. Beynen AC and Geelen MJH. Control of glycogen metabolism by insulin in isolated hepatocytes. Horm Metab Res 1981; 13: 376-387.
41. Nyfeller F, Fasel P and Walter P. Short-term stimulation of glycogen production by insulin in rat hepatocytes. Biochim Biophys Acta 1981; 675: 17-23.
42. Boyd ME, Albright EB, Foster DW and McGarry JD. *In vitro* reversal of the fasting state of liver metabolism in the rat. Reevaluation of the roles of glucose and insulin. J Clin Invest 1981; 68: 142-152.
43. Agius L, Peak M and Alberti KGMM. Regulation of glycogen synthesis from glucose and gluconeogenic precursors by insulin in periportal and perivenous rat hepatocytes. Biochem J 1990; 266: 91-102.
44. Tosh D, Beresford G and Agius L. Glycogen synthesis from glucose by direct and indirect pathways in hepatocyte cultures from different nutritional states. Biochim Biophys Acta 1994; 1224: 205-212.
45. Agius L, Tosh D and Peak M. The contribution of pyruvate cycling to loss of [6-^{3}H]glucose during conversion of glucose to glycogen in hepatocytes: effects of insulin, glucose and acinar origin of hepatocytes. Biochem J 1993; 289: 255-262.
46. Fleig WE, Enderle D, Steudter S, Nother-Fleig G and Ditschuneit H. Regulation of basal and insulin-stimulated glycogen synthesis in cultured hepatocytes. J Biol Chem 1987; 262: 1155-1160.
47. Schudt C. Influence of insulin, glucocorticoids and glucose on glycogen synthase activity in hepatocyte cultures. Biochim Biophys Acta 1980; 629: 499-509.
48. Van de Werve G, Sestoft L, Folke M and Kristensen LO. The onset of liver glycogen synthesis in fasted-refed rats. Effects of streptozotocin diabetes and of peripheral insulin replacement. Diabetes 1984; 33: 944-949.
49. Witters LA and Avruch J. Insulin regulation of hepatic glycogen synthase and phosphorylase. Biochemistry 1978; 17: 406-410.
50. Ciudad CJ, Bosch F and Guinovart JJ. Insulin activation of basal hepatic glycogen synthase. FEBS Lett 1981; 129: 123-126.
51. Ciaraldi TP and Brady D. Comparisons of insulin and biosynthetic human proinsulin actions in cultured hepatocytes. Horm Metab Res 1990; 22: 283-288.
52. Baqu S, Guinovart JJ and Gomez-Foix AM. Amylin impairment of insulin effects on glycogen synthesis and phosphoenolpyruvate carboxykinase gene expression in rat primary cultured hepatocytes. Biochem J 1994; 304: 449-453.
53. Al-Habori M, Peak M, Thomas TH and Agius L. The role of cell swelling in the stimulation of glycogen synthesis by insulin. Biochem J 1992; 282: 789-796.
54. Bosch F, Gomez-Foix AM and Guinovart JJ. Effects of lithium ions on glycogen synthase and phosphorylase in rat hepatocytes. J Biol Chem 1986; 261: 16927-16931.
55. Gomez-Foix AM, Rodriguez-Gil JE and Guinovart JJ. Anti-insulin effects of amylin and calcitonin-gene related peptide on hepatic glycogen metabolism. Biochem J 1991; 276: 607-610.
56. Ciudad CJ, Camici M, Ahmad Z, Wang Y, DePaoli-Roach AA and Roach PJ. Control of glycogen synthase phosphorylation in isolated rat hepatocytes by epinephrine, vasopressin and glucagon. Eur J Biochem 1984; 142: 511-520.
57. Pugazhenthi S and Khandelwal RL. Regulation of glycogen synthase activation in isolated hepatocytes. Mol Cell Biochem 1995; 149/150: 95-101.
58. Toth B, Bollen M and Stalmans W. Acute regulation of hepatic protein phosphatases by glucagon, insulin and glucose. J Biol Chem 1988; 263: 14061-14066.
59. Agius L and Peak M. Interactions of okadaic acid with insulin action in hepatocytes: role of protein phosphatases in insulin action. Biochim Biophys Acta 1991; 1095: 243-248.
60. Pugazhenthi S and Khandelwal RL. Insulin action on protein phosphatase-1 activation is enhanced by the antidiabetic agent pioglitazone in cultured diabetic rat hepatocytes. Mol Cell Biochem 1998; 182: 185-191.

61. Miller TB, Garnache A and Cruz J. Insulin regulation of glycogen synthase phosphatase in primary cultures of hepatocytes. J Biol Chem 1984; 259: 12470-12474.
62. Stalmans W, Bollen M, Toth B and Gergely P. Short-term hormonal control of protein phosphatases involved in hepatic glucose metabolism. Adv Enz Regul 1990; 30: 305-327.
63. Doherty MJ, Moorhead G, Morrice N, Cohen P and Cohen PTW. Amino acid sequence and expression of the hepatic glycogen-binding (GL)-subunit of protein phosphatase-1. FEBS Lett 1995; 375: 294-298.
64. Doherty MJ, Young PR and Cohen PTW. Amino acid sequence of a novel protein phosphatase 1 binding protein (R5) which is related to the liver and muscle-specific glycogen binding subunits of protein phosphatase 1. FEBS Lett 1996; 399: 339-343.
65. Armstrong CG, Browne GJ, Cohen P and Cohen PTW. PPP1R6, a novel member of the family of glycogen targetting subunits of protein phosphatase 1. FEBS Lett 1997; 418: 210-214.
66. Imazu M, Strickland WG, Chrisman TD and Exton JH. Phosphorylation and inactivation of liver glycogen synthase by liver protein kinases. J Biol Chem 1984; 259: 1813-1821.
67. Joyal JL and Sachs DB. Insulin-dependent phosphorylation of calmodulin in rat hepatocytes. J Biol Chem 1994; 269: 30039-30048.
68. Witters LA and Kemp BE. Insulin activation of acetyl-CoA carboxylase accompanied by inhibition of 5'-AMP activated protein kinase. J Biol Chem 1992; 267: 2864-2867.
69. Carling D and Hardie DG. The substrate and sequence specificity of the AMP-activated protein kinase. Phosphorylation of glycogen synthase and phosphorylase kinase. Biohim Biophys Acta 1989; 1012: 81-86.
70. Peak M and Agius L. Inhibition of glycogen synthesis by epidermal growth factor in hepatocytes. Eur J Biochem 1994; 221: 529-536.
71. Klein PS and Melton DA. A molecular mechanism for the effect of lithium on development. Proc Natl Acad Sci 1996; 93: 8455-8459.
72. Nyfeller F and Walter P. Effect of lithium on glycogen synthesis in hepatocytes from rat liver. FEBS Lett 1979; 108: 197-199.
73. Sutherland C and Cohen P. The α-isoform of glycogen synthase kinase-3 from rabbit skeletal muscle is inactivated by p70 S6 kinase or MAP kinase-activated protein kinase-1 *in vitro*. FEBS Lett 1994; 338: 37-42.
74. Cross DAE, Alessi DR, Cohen P, Andjelkovich M and Hemmings BA. Inhibition of glycogen synthase kinase-3 by insulin mediated by protein kinase B? Nature 1995; 378: 785-789.
75. Carlsen J, Christiansen K and Vinten J. Insulin stimulated glycogen synthesis in isolated hepatocytes: effects of protein kinase inhibitors. Cellular Signalling 1997; 9: 447-450.
76. Wagle A, Jirvaj S, Garlock GL and Stapleton SR. Insulin regulation of glucose-6-phosphate dehydrogenase gene expression is rapamycin-sensitive and requires phosphatidylinositol 3-kinase. J Biol Chem 1998; 273: 14968-14974.
77. Beresford GW and Agius L. Epidermal growth factor counteracts insulin-induced expression of glucokinase in hepatocytes. Biochem Biophys Res Commun 1994; 201: 902-908.
78. Walker KS, Deak M, Paterson A, Hudson K, Cohen P and Alessi DR. Activation of protein kinase B β and γ isoforms by insulin *in vivo* and by 3-phosphoinositide-dependent protein kinase-1 *in vitro*: comparison with protein kinase B α. Biochem J 1998; 331: 299-308.
79. Wang QM, Vik TA, Ryder RW and Roach PJ. Phosphorylation and activation of p90rsk by glycogen synthase kinase 3. Biochem Biophys Res Commun 1995; 208: 796-801.

80. White MF. The insulin signalling system and the IRS proteins. Diabetologia 1997; 40: S2-S17.
81. Rother KI, Imai Y, Caruso M, Beguinot F, Formisano P and Accili D. Evidence that IRS-2 phosphorylation is required for insulin action in hepatocytes. J Biol Chem 1998; 273: 17491-17497.
82. Margolis RN, Taylor SI, Seminara D and Hubbard AL. Identification of pp120, an endogenous substrate for the hepatocyte insulin receptor tyrosine kinase, as an integral membrane glycoprotein of the bile canalicular domain. Proc Natl Acad Sci 1988; 85: 7256-7259.
83. Formisano P, Najjar SM, Gross CN, Philippe N, Oriente F, Kern-Buell CL, Accili D and Gorden P. Receptor-mediated internalisation of insulin. Potential role of pp120/HA4, a substrate of the insulin receptor kinase. J Biol Chem 1995; 270: 24073-24077.
84. Najjar SM. Pp120, a substrate for the insulin receptor tyrosine kinase, is associated with phosphatase activity. Biochem Biophys Res Commun 1998; 247: 457-461.
85. Phillips SA, Perrotti N and Taylor SI. Rat liver membranes contain a 120kDa glycoprotein which serves as a substrate for the tyrosine kinase of the receptors for insulin and epidermal growth factor. FEBS Lett 1987; 212: 141-144.
86. Moule SK and Denton RM. Multiple signalling pathways involved in the metabolic effects of insulin. Am J Cardiol 1997; 80: 41A-49A.
87. Phung TL, Roncone A, de Mesy Jensen KL, Sparks CE and Sparks JD. Phosphoinositide 3-kinase activity is necessary for insulin-dependent inhibition of apolipoprotein B secretion by rat hepatocytes and localises to the endoplasmic reticulum. J Biol Chem 1997; 272: 30693-30702.
88. Virbasius JV, Guilherme A and Czech MP. Mouse p170 is a novel phosphatidylinositol 3-kinase containing a C2 domain. J Biol Chem 1996; 271: 13304-13307.
89. Van der Kaay J, Batty IH, Cross DA, Watt PW and Downes CP. A novel rapid and highly sensitive mass assay for phosphatidylinositol 3,4,5-triphosphate (PtdIns(3,4,5)P_3) and its application to measure insulin stimulated (PtdIns(3,4,5)P_3) production in rat skeletal muscle *in vivo*. J Biol Chem 1997; 272: 5477-5481.
90. Ono F, Nagawa T, Saito S, Owada Y, Sakagami H, Goto K, Suzuki M, Matsuno S and Kondo H. A novel class II phosphoinositide 3-kinase predominantly expressed in the liver and its enhanced expression during liver regeneration. J Biol Chem 1998; 273: 7731-7736.
91. Moore RD. Effects of insulin upon ion transport. Biochim Biophys Acta 1983; 737: 1-49.
92. DeFronzo RA, Felig P, Ferrannini E and Wahren J. Effect of graded doses of potassium on splanchnic and peripheral potassium metabolism in man. Am J Physiol 1980; 238: E421-E427.
93. Agius L, Peak M, Beresford G, Al-Habori M and Thomas TH. The role of ion content and cell volume in insulin action. Biochem Soc Trans 1994; 22: 516-522.
94. Baquet A, Hue L, Meijer AJ, van Woerkom GM and Plomp PJAM. Swelling of rat hepatocytes stimulates glycogen synthesis. J Biol Chem 1990; 265: 955-959.
95. Meijer AJ, Baquet A, Gustafson L, van Werkom GM and Hue L. Mechanism of activation of liver glycogen synthase by swelling. J Biol Chem 1992; 267: 5823-5828.
96. Krause U, Rider M and Hue L. Protein kinase signalling pathways triggered by cell swelling and involved in the activation of glycogen synthase and acetyl-CoA carboxylase in isolated rat hepatocytes. J Biol Chem 1996; 271: 16668-16673.
97. Alvarez JF, Sanchez-Arias JA, Guadano A, Estevez F, Varela I, Feliu JE and Mato JM. Transport in isolated rat hepatocytes of the phosphooligosaccharide that mimics insulin action. Biochem J 1991; 274: 369-374.

98. Mato JM, Kelly KL, Abler A and Jarett L. Identification of a novel insulin-sensitive glycophospholipid from H35 Hepatoma cells. J Biol Chem 1987; 262: 2131-2137.
99. Mato JM, Kelly KL, Abler A, Jarett L, Corkey BE, Cashel JA and Zopf D. Partial structure of an insulin-sensitive glycophospholipid. Biochem. Biophys Res Commun 1987; 146: 764-770.
100. Alvarez JF, Cabello MA, Feliu JE and Mato JM. A phosphooligosaccharide mimics insulin action on glycogen phosphorylase and pyruvate kinase activities in isolated rat hepatocytes. Biochem Biophys Res Commun 1987; 147: 765-771.
101. Ruiz-Albusac JM, Zueco JA, Velazquez E and Blazquez E. Insulin does not induce the hydolysis of a glycosyl phosphatidylinositol in rat fetal hepatocytes. Diabetes 1993; 42: 1262-1272.
102. Sanchez-Bueno A, Greenwood MR, Valero-Nieto I, Marrero I, Gil B, Mato JM and Cobbold PH. Inositol phosphoglycan inhibits calcium oscillations in hepatocytes by reducing calcium entry. Cell Calcium 1997; 21: 125-133.
103. Alvarez L, Avila MA, Mato JM, Castano JG and Varela-Nieto I. Insulin-like effects of inositol phosphate glycan on messenger RNA expression in rat hepatocytes. Mol Endocrinol 1991; 5: 1062-1068.
104. Sanchez-Arias JA, Sanchez-Gutierrez JC, Guadano A, Alvarez JF, Samper B, Mato JM and Feliu JE. Impairment of glycosyl-phosphatidylinositol-dependent insulin signalling system in isolated rat hepatocytes by streptozotocin-induced diabetes. Endocrinology 1992; 131: 1727-1733.
105. Sanchez-Arias JA, Sanchez-Gutierrez JC, Guadano A, Alvarez JF, Samper B, Mato JM and Feliu JE. Changes in the insulin-sensitive glycosyl-phosphatidyl-inositol signalling system with aging in rat hepatocytes. Eur J Biochem 1993; 211: 431-436.
106. Varela I, Avila M, Mato JM and Hue L. Insulin-induced phospho-oligosaccharide stimulates amino acid transport in isolated rat hepatocytes. Biochem J 1990; 267: 541-544.
107. Misek DE and Saltiel AR. An inositol phosphate glycan derived from a Trypanosoma brucei glycosyl-phosphatidylinositol mimics some of the metabolic actions of insulin. J Biol Chem 1992; 267: 16266-16273
108. Romero G, Gamez G, Huang LC, Lilley K and Luttrell L. Anti-inositol glycan antibodies selectively block some of the action of insulin in intact BC3H1 cells. Proc Natl Acad Sci 1990; 87: 1476-1480.
109. Whatmore AJ, Bate AJ and Dickson AJ. Glycolipids isolated from cutlured rat hepatocytes: analysis of their role in insulin signal transduction. Cellular signalling 1993; 5: 21-32.
110. Varela I, Alvarez JF, Clemente R, Ruiz-Albusac JM and Mato JM. Asymmetric distribution of the phosphatidylinositol linked phosphooligosaccharide that mimics insulin action in the plasma membrane. Eur J Biochem 1990; 188: 213-218.
111. Kessler A, Muller G, Wied S, Crecelius A and Eckel J. Signalling pathways of insulin-mimetic phosphoinositolglycan-peptide in muscle and adipose tissue. Biochem J 1998; 330: 277-286.
112. Luttrell LM, Hewlett EL, Romero G and Rogol AD. Pertussis toxin treatment attenuates some of the effects of insulin in BC3H-1 murine myocytes. J Biol Chem 1988; 263: 6134-6141.

Chapter 21

Studies on mitochondrial-cytoplasmic interactions in isolated hepatocytes

John W. Phillips, Marlene B. Grivell, Anthony R. Grivell and Michael N. Berry

Department of Medical Biochemistry, Flinders University School of Medicine, GPO Box 2100, Adelaide 5001, Australia

Key words: cytoplasmic redox potential, electrochemical proton gradient, energy-dependency, enzyme-enzyme interaction, ethanol metabolism, glycerophosphate shuttle, malate-aspartate shuttle, metabolic rate, mitochondrial redox potential, redox potential difference, reducing equivalents, reducing-equivalent shuttles, thyroid hormone, transistor-like behaviour

21.1 Introduction

The original reason for attempting to prepare suspensions of intact isolated hepatocytes was to produce a preparation that would facilitate the study of hepatic metabolism. Forty years on, the great majority of metabolic pathways have been delineated and interest in metabolism has waned. Nevertheless, the manner in which flux is regulated still appears to be a fruitful area for study. In this chapter we review the results of metabolic experiments carried out in our laboratory during the thirty years that intact hepatocyte suspensions have been available for experimental work. Much of our effort during that time has been directed towards the study of mitochondrial-cytoplasmic interactions, in particular the intercompartmental transfer of reducing equivalents. We have endeavoured to elucidate the factors that determine the direction of flow, and have concluded that the relationships we observe cannot be explained solely on the basis of molecular diffusion and random collision. Instead, we put forward the hypothesis that the critical regulatory processes that determine the rate and direction of intercompartmental reducing-equivalent transfer in living cells are dependent on flow through organized systems, which is controlled by energy-dependent mechanisms.

M.N. Berry and A.M. Edwards (eds.), The Hepatocyte Review, 343–363.

21.2 Energy-dependent generation of reducing power

A major and challenging unresolved question relating to the field of metabolism is how do living systems maintain far-from-equilibrium steady states in apparent defiance of the second law of thermodynamics. A specific example of the problem is how do cells perform reductive synthetic processes such as gluconeogenesis and lipogenesis when, according to the current paradigm, all pathways for reducing-equivalent flow under aerobic conditions lead ultimately, albeit indirectly, to O_2 via the mitochondrial electron transfer chain. Under such circumstances it would be predicted that reductive syntheses would not be feasible since reduction of O_2 would inevitably be thermodynamically favoured over any other reductive process available to the hepatocyte. An obvious but insufficient explanation is that both animals and plants harness a continual supply of energy to counteract the tendency towards increasing entropy. Our recent studies now suggest a more complete explanation. Namely that intracellular enzyme organisation and the utilisation of energy to regulate proton and electron flow through the NADH dehydrogenase enables segregation of the pathways for reductive syntheses from those involved in energy transduction.

Our hypothesis is in keeping with the concept that has been gaining ground in recent years that many metabolites do not react by diffusion and random collision [1,2]. Rather they do so by means of enzyme-enzyme interactions so that the product of one reaction is passed directly from one enzyme active site to become the substrate of the next enzyme in the sequence. Much evidence in support of the concept that a high degree of enzyme organisation exists within the cell has been presented and a number of monographs on the subject have recently been published [3-5]. In arguing our hypothesis we will focus attention on the various reducing-equivalent shuttle mechanisms that must form an integral part of the conducting systems involved in both energy transduction and reductive syntheses.

21.3 Shuttle properties and functions

The major known intercompartmental reducing-equivalent transferring systems in the hepatocyte are the glycerophosphate (GP) and malate-aspartate (M-A) shuttles, which transfer reducing equivalents from cytoplasm to mitochondria [6-8]. A malate (aspartate-independent) shuttle (designated here as the malate shuttle) transfers reducing equivalents in the opposite direction and provides oxaloacetate for gluconeogenesis (Figure 1) [9].

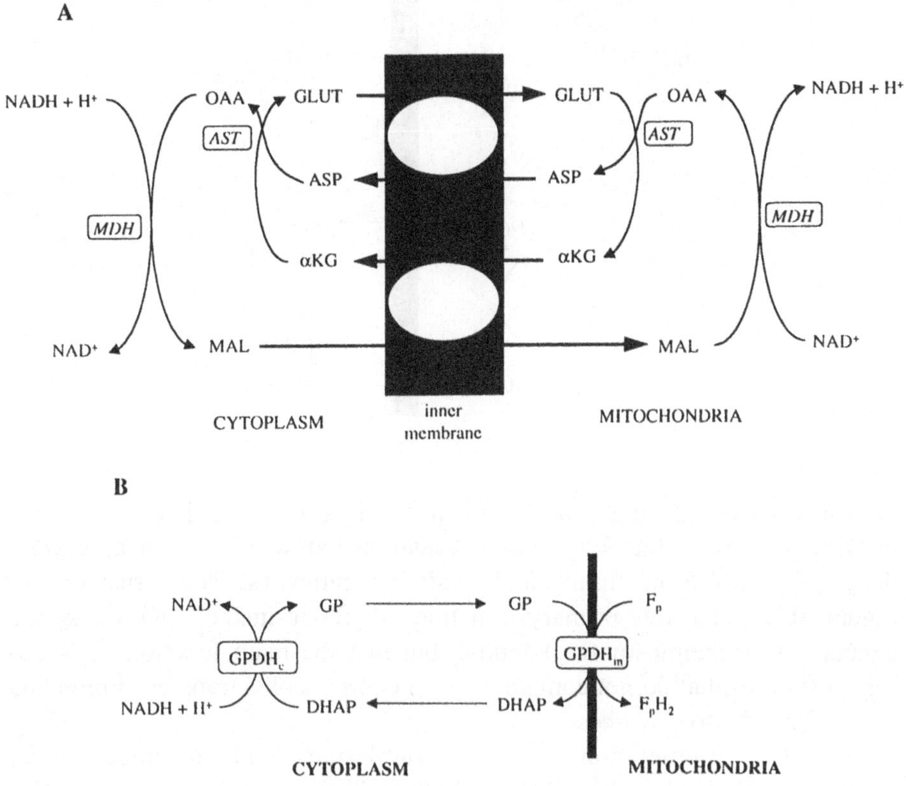

Figure 1. Mechanisms for the transfer of reducing equivalents into and out of the mitochondria. (a) Malate-aspartate shuttle, (b) glycerophosphate shuttle, (c) malate shuttle. $GPDH_c$, cytoplasmic glycerol 3-phosphate dehydrogenase; $GPDH_m$, mitochondrial glycerol 3-phosphate dehydrogenase; DHAP, dihydroxyacetone phosphate; GLUT, glutamate; ASP, aspartate; α-KG, α-ketoglutarate; OAA, oxaloacetate; MAL, malate; MDH, malate dehydrogenase; AST, aspartate aminotransferase; PEP, phosphoenolpyruvate; GLUC, glucose.

Within the mitochondria is a shuttle comprising an FAD-containing protein (ETF) that transfers electrons between at least 9 matrix-based primary flavoprotein dehydrogenases and the respiratory chain via the iron-sulfur flavoprotein, ETF-ubiquinone oxidoreductase (ETF-QO) [10]. The

C

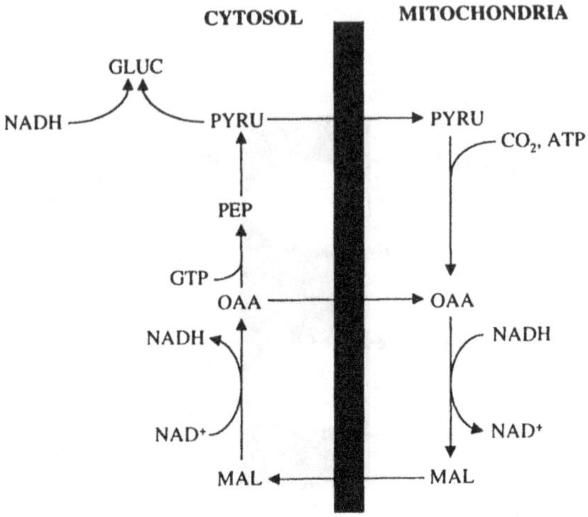

dehydrogenases include the short-medium and long-chain acyl-CoA dehydrogenases of the fatty acid β-oxidation pathway, as well as several dehydrogenases from amino acid oxidative pathways. The thrust of our argument is that the primary function of flavin-linked shuttles is the generation of chemiosmotic potential, but that the main function of NAD-linked shuttles that do not contain a flavin component is transfer of reducing power for reductive syntheses.

The GP shuttle comprises a cytoplasmic and a mitochondrial glycerophosphate dehydrogenase (GPDH). The former enzyme is NAD-linked, but the latter has a flavin coenzyme. The shuttle is highly active in brain and insect flight muscle, and also shows substantial activity in heart and skeletal muscle, kidney and liver. Its main function is to transfer reducing equivalents arising during aerobic glycolysis to the mitochondria. The capacity of the shuttle is limited by the activity of its mitochondrial limb, a flavoprotein located on the outside of the inner membrane. In mammalian tissues other than brain and testes the rate of transcription and activity of this enzyme is markedly stimulated by T_3 [11]. The M-A shuttle plays an important role in the transfer of reducing equivalents to the mitochondria during oxidation of lactate and its conversion to glucose. It is highly sensitive to aminooxyacetate, an inhibitor of aspartate aminotransferase (AST). Less information is available about the malate shuttle responsible for transferring reducing equivalents out of the mitochondria for reduction of pyruvate to lactate, or diphosphoglycerate (DPG) to glyceraldehyde 3-phosphate (GAP). The stimulation of gluconeogenesis by fatty acid oxidation and its depression by inhibitors of this process suggest that malate is generated within the mitochondrial matrix, by a coupled reaction between oxaloacetate that has been generated by pyruvate carboxylation and hydroxyacyl CoA.

21.3.1 Previous studies on shuttle properties and function

Over a period of 30 years members of our group have studied the function and regulatory role of the intercompartmental shuttles that transfer reducing equivalents between cytoplasmic and mitochondrial substrates. We have accrued much evidence that each of these shuttles can be limiting for metabolic flux from various substrates under certain physiological and pathological conditions [8,12-18]. Recently we have examined the role of shuttles in the competition between hepatic glycolysis and ethanol oxidation [18]. This work revealed that reducing equivalents, arising in ethanol oxidation, could be transported to the mitochondria by both the M-A and GP shuttles and that the degree of inhibition of glycolysis during ethanol oxidation was closely related to the rate of reducing-equivalent transfer. Ethanol oxidation induced an accumulation of GP but no rise in hexose phosphate and its effects were present even when aminooxyacetate was employed to block the M-A shuttle (Table 1). We inferred that these effects

Table 1. Effect of aminooxyacetate on the rates of ethanol oxidation and glycolysis and the cellular concentrations of glycerophosphate and glucose 6-phosphate

Addition	AOAC	*J*glucose	*J*ethanol	[GP]	[Glu6P]
		µmol/min/g wet wt		µmol/g wet wt	
None	-		-	0.29 ± 0.02	0.05 ± 0.01
Ethanol	-		0.58 ± 0.03	0.68 ± 0.04	0.05 ± 0.01
	+		0.41 ± 0.02	0.78 ± 0.06	0.05 ± 0.01
Glucose	-	1.98 ± 0.16		0.59 ± 0.06	0.07 ± 0.01
	+	1.95 ± 0.18		1.01 ± 0.08	0.17 ± 0.01
Ethanol +	-	0.42 ± 0.06	1.26 ± 0.06	1.96 ± 0.09	0.22 ± 0.02
	+	0.48 ± 0.04	0.85 ± 0.04	4.37 ± 0.11	0.20 ± 0.01

Hepatocytes from fasted rats were incubated for 30 min at 37°C [19] with the additions indicated at the following initial concentrations: glucose, 40 mM; ethanol 13 mM; aminooxyacetate (AOAC), 100 µM. The data are presented as mean ± SEM of 7 separate experiments. Glycerophosphate (GP), Glucose 6-Phosphate (Glu6P).

were a consequence of a successful competition for the GP shuttle by reducing equivalents arising during ethanol oxidation versus those generated in the cytoplasm during oxidation of GAP to DPG.

21.3.2 Energy-dependence of shuttles involving malate

In an early study of pyruvate reduction by isolated hepatocytes we demonstrated the energy-dependence of the shuttle responsible for the export of malate from the mitochondria [12], a result confirmed by others [9]. Later we observed that the rate of lactate oxidation, coupled to the M-A shuttle is also energy-dependent [17], showing a close relationship to the magnitude of the mitochondrial membrane potential ($\Delta\Psi$) [15,19]. Thus, there appear to be at least two intercompartmental shuttles through which flux is energy-dependent. This is in accord with the Mitchell chemiosmotic hypothesis [20], which postulates that the flow of electrons to O_2 via the mitochondrial carriers results in the generation of an electrochemical proton gradient ($\Delta\mu H^+$) across the inner mitochondrial membrane.

21.4 Roles of the electrochemical proton gradient

21.4.1 ATP synthesis

The $\Delta\mu H^+$ provides a vectorial protonmotive force sufficient to drive the phosphorylation of ADP to ATP, a reaction that does not take place spontaneously, and thus poises the cellular [ATP]/[ADP][Pi][H^+], from which the phosphorylation potential is derived, in a far-from-equilibrium steady state. As a consequence of this, free energy made available from the oxidation of metabolic fuels is utilized to produce ATP, which in turn serves as an energy-coupling device for driving a multiplicity of synthetic reactions.

21.4.2 Generation of reducing power

Since ATP is not an electron acceptor or donor, it remains to be explained how living cells can carry out syntheses that involve one or more thermodynamically unfavourable redox reactions. This ability can also be attributed to the generation and maintenance of $\Delta\mu H^+$, which in animals is represented almost entirely by an electric field ($\Delta\Psi$) across the inner membrane (matrix face negative). For reasons given by Kell [21] we prefer to envisage the existence of a number of localized fields. Thus $\Delta\mu H^+$ is recognized to be not only the primary high-energy intermediate of oxidative phosphorylation but also a force for driving reductive syntheses (as well as other functions such as ion and metabolite transport). It has been shown that the mitochondrial energy-linked transhydrogenase, when energized by $\Delta\mu H^+$, is capable of increasing the [NADPH]. [NAD^+]/[$NADP^+$].[NADH] ratio up to five hundredfold [22] and greatly stimulating the rate of NADPH formation. A $\Delta\mu H^+$-dependent mechanism also exists for increasing the mitochondrial [NADH].[H^+]/[[NAD^+] [23]. The ability of $\Delta\mu H^+$ to drive redox reactions in the thermodynamically unfavourable direction is termed energy-dependent "reversed electron transfer" and is especially associated with flavin-linked oxidations [24]. It follows that shuttles involved in reducing-equivalent transfer for reductive syntheses will be depressed by agents that decrease energy state.

21.5 Maintenance of potential difference between cytoplasm and mitochondria

We determined the redox potentials E_{hLDH} and E_{hHBDH} derived from the steady-state mass action ratio of the reactions catalysed by cytoplasmic lactate dehydrogenase (LDH) and mitochondrial 3-hydroxybutyrate dehydrogenase (HBDH) respectively [15,18], and showed that these potentials respond to agents that diminish energy-state. Thus, the normal potential difference between E_{hLDH} and E_{hHBDH} of 40-60 mV (E_{hdiff}) is depressed to less than 10 mV by treatment of hepatocytes with high concentrations of the uncoupling agent, FCCP - as E_{hHBDH} becomes more oxidized and E_{hLDH} more reduced (Table 2). This finding demonstrates for the first time that an energy-dependent process maintains the difference between these cytoplasmic and mitochondrial matrix redox potentials. A comparison of the effects of uncoupling agents on E_{hLDH} and E_{hGPDH} revealed that they had no effect on the latter potential. This was consistent with the observation that sorbitol oxidation to fructose, which feeds reducing equivalents into the flavin-linked GP shuttle is largely unaffected by FCCP [13]. At low concentrations of FCCP, O_2-uptake, Krebs cycle turnover and fatty acid oxidation were stimulated by uncouplers [25] and E_{hdiff} was maintained although E_{hHBDH} became more oxidized, indicating that the hepatocyte was able to compensate to some extent for a decline in $\Delta\mu H^+$ induced by uncouplers (Table 2). Furthermore, E_{hdiff} was maintained in the presence of high concentrations of rotenone, but not antimycin (Figure 2), implying the involvement of flavin-linked reactions in its maintenance.

Table 2. Effect of FCCP addition on the rate of oxygen utilisation, $-E_{hc}$, $-E_{hm}$, E_{hdiff}, ATP concentration and $\Delta\Psi$

FCCP (μM)	ΔΨ (mV)	$-E_{hc}$ (mV)	$-E_{hm}$ (mV)	E_{hdiff} (mV)	[ATP] (μmol/g wet wt)	JO_2 (μmol/min/g wet wt)
0	168±2	252±1	299±1	45±2	2.3±0.1	6.6±0.3
10	162±5	248±2	294±1	45±3	2.1±0.1	7.5±0.3
20	157±3	248±1	289±1	41±2	1.8±0.1	7.0±1.0
30	147±5	249±2	285±0	36±2	1.5±0.1	8.5±0.3
40	137±3	255±1	278±1	23±2	1.0±0.1	8.5±0.4
60	118±11	267±1	269±1	2±2	0.4±0.1	6.9±0.1

Hepatocytes from fasted rats were incubated at 37°C for 30 min [19] with 10 mM lactate, 2 mM palmitate, 2 mM asparagine and FCCP (0-60 μM). Results are the mean ± SEM of 5 expts.

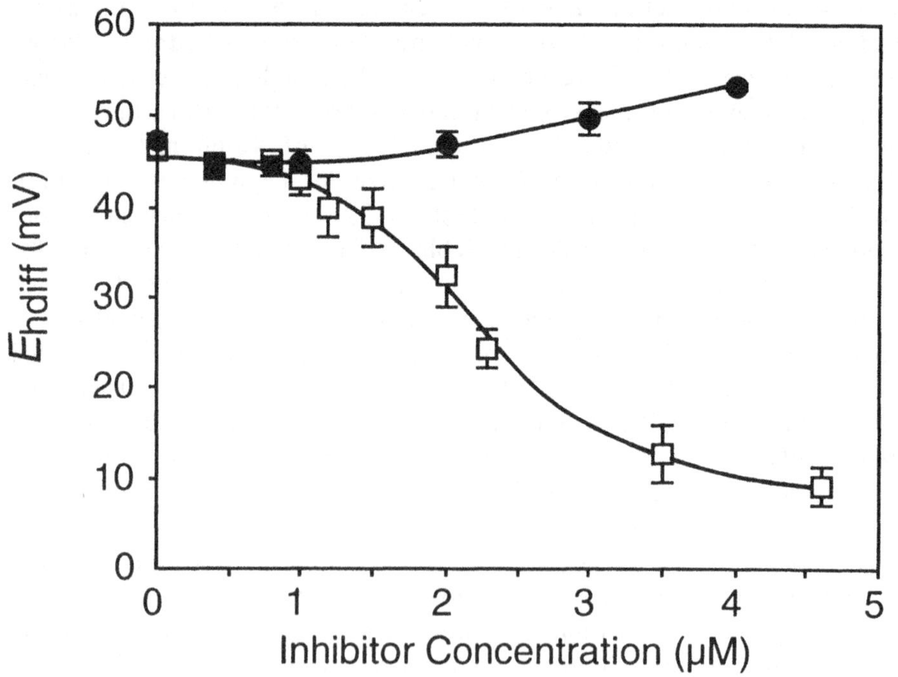

Figure 2. Effects of rotenone and antimycin on E_{hdiff}. Hepatocytes from 24 h-fasted rats were incubated with 10 mM lactate, 1 mM pyruvate and 1.75 mM palmitate for 30 min at 37°C as previously described [19] in the presence of either rotenone (●, 0-4 μM) or antimycin (❑, 0-4.6 μM). Data are presented as the mean ± SEM of 5 separate experiments.

21.5.1 Role of shuttles in maintenance of E_{hdiff}

The relationship between the two potentials is maintained by the operation of both the M-A shuttle (at [lactate]/[pyruvate] (L:P) ratios above about 8:1) and the malate shuttle at lower L:P ratios. This is shown by the effect of aminooxyacetate on E_{hdiff} when hepatocytes are incubated with lactate, as well as the ability of hepatocytes to correct E_{hdiff} when exposed to a pyruvate load (Figure 3). This correction is achieved even in the presence of high concentrations of aminooxyacetate, since this inhibitor comes into play only after a [lactate]/[pyruvate] ratio of 8:1 is attained.

21.6 Enzyme-enzyme interaction and shuttle function

Non-diffusional direct transfer of NADH between the active sites of two dehydrogenases can occur only when the enzymes have opposite chiral specificity for the transfer of hydrogen between the C-4 of the nicotinamide ring of NADH [26]. These two classes of enzymes have been designated as

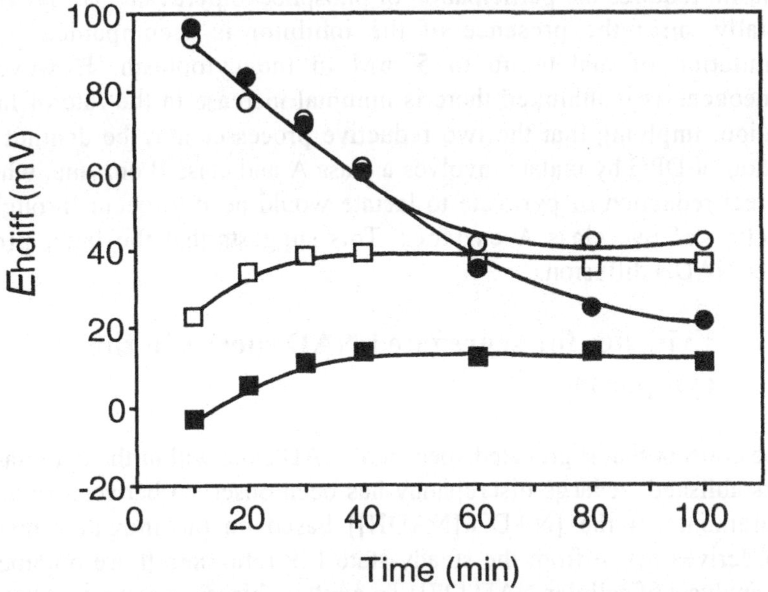

Figure 3. Effect of aminooxyacetate on E_{hdiff}. Hepatocytes from 24 h-fasted rats were incubated at 37°C under standard conditions [19] for up to 100 min with either 10 mM lactate (□,■) or 15 mM pyruvate (○,●) in the absence (□,○) and presence (■,●) of 100 μM aminooxyacetate. The data are from a representative experiment.

A and B. LDH, alcohol dehydrogenase and malate dehydrogenase (MDH) are all class A dehydrogenases: yet there is no doubt that a M-A shuttle transfers reducing equivalents from lactate (or ethanol) into the mitochondria. We conclude from this that the cytoplasmic transfer of reducing equivalents from lactate to malate via NADH is a diffusional process, *i.e.* it involves free NAD. Glyceraldehyde 3-phosphate dehydrogenase and GPDH are both class B enzymes from which it could be inferred that the entry into the GP shuttle of reducing equivalents generated in the oxidation of GAP must likewise be a diffusional process. However, we have circumstantial evidence that malate may participate as a carrier because it accumulates in concentrations over 1 mM when glycolysis is suppressed with palmitate (unpublished data).

Further studies are required to determine the nature of the interaction between malate that exits the mitochondria and participates in the reduction of DPG to GAP, and pyruvate to lactate. There is a stoichiometric, approximately 1:1 relationship, between the rates of reducing-equivalent transfer involved in these two processes and both reductive steps are inhibited by 3-mercaptopicolinate (3-MPA) and stimulated by palmitate. (Figure 4). The inhibition of pyruvate reduction by 3-MPA suggests the possibility that the malate shuttle, transferring reducing equivalents to the cytoplasm, requires the participation of phosphoenolpyruvate carboxykinase, especially since the presence of the inhibitor is accompanied by the accumulation of malate up to 5 mM in the cytoplasm. However, if gluconeogenesis is inhibited there is minimal increase in the rate of lactate formation, implying that the two reductive processes may be distinct. The reduction of DPG by malate involves a class A and class B enzyme, whereas the direct reduction of pyruvate to lactate would need to occur through the interaction of two class A enzymes. This suggests that the latter process requires NADH diffusion.

21.7 Evidence for segregated NAD species in the cytoplasm

The concept that segregated species of NAD exist within the cytoplasm is well established. A large discrepancy has been observed between results of measurements of the [NAD$^+$]/[NADH]] based on the indicator method, which derives E_{hLDH} from the steady-state L:P ratio, and those obtained by direct analysis of cellular NAD [27]. To explain this it is postulated that only a small quantity of the cytoplasmic NAD is free, most being bound to dehydrogenases (as NADH) and metabolically inert. The free (soluble) NAD pool is believed to equilibrate with all the major NAD-linked

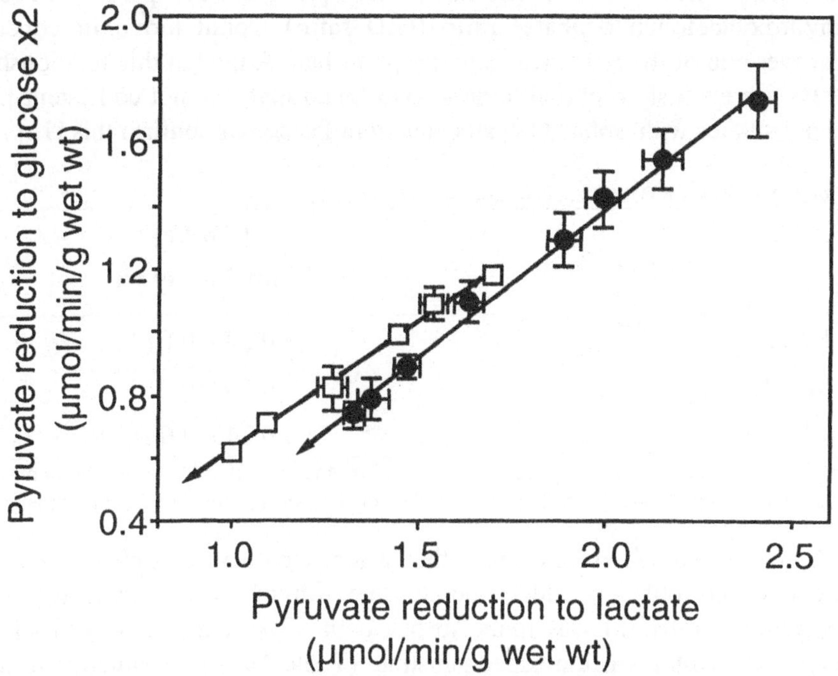

Figure 4. Effects of 3-mercaptopicolinate on the transfer of reducing equivalents from the mitochondria for pyruvate reduction. Hepatocytes from 24 h fasted rats were incubated as previously described [19] for 30 min at 37°C with either 15 mM pyruvate (❑) or 15 mM pyruvate + 1.75 mM palmitate (●) in the presence of a concentration range of 3-mercaptopicolinate (0-200 µM). The arrowhead indicates increasing concentration of 3-mercaptopicolinate. The rate of reducing equivalent transfer from the mitochondria is represented as the rate of pyruvate reduction to lactate + (rate of glucose formation × 2). The data are shown as the mean ± SEM from 4 separate experiments. For some points the error bars lie within the symbol.

dehydrogenases of the cytoplasm [27]. Experiments by ourselves [16] and others [3] cast doubt on the validity of this model. We observed that in cells from fasted rats the L:P ratio was twice as oxidized in the presence of glucose than when lactate was added [14,18]. These results mirrored data obtained with Ehrlich tumour cells [28], where a much higher L:P ratio was noted in the presence of added lactate than added glucose. In seeking an explanation for this, hepatocytes were exposed to either added lactate or glucose, and a time-course study conducted of the changes in the redox ratios of the components of the reactions catalysed by LDH and GPDH [16]. On the assumption that only a single cytoplasmic NAD pool, poised near-equilibrium, exists within the cytoplasmic compartment it would be expected, in view of the high activities of cytoplasmic LDH and GPDH, that the steady-state L:P ratio and the [α-glycerophosphate]/[dihydroxyacetonephosphate] ratio (G:D ratio) would move in concert whenever one of the ratios was initially perturbed. Although this relationship has been previously reported, others have found that it is not consistent [3], and our studies with isolated hepatocytes from fasted rats confirm this [16].

Table 3. Steady-state DHAP concentrations in hepatocytes

Addition	[DHAP] μmol/g wet wt
Lactate	0.145±0.013
Sorbitol	0.232±0.010
Glucose	0.240±0.008

Hepatocytes from fasted rats were incubated with 10 mM lactate, 10 mM sorbitol or 40 mM glucose for 30 min at 37°C [19]. The data are presented as mean ± SEM of 20 experiments.

A noteworthy observation was that when hepatocytes from fasted rats were exposed only to added lactate, the dihydroxyacetone phosphate concentration (DHAP) was found to be about 60% of that observed when glucose or sorbitol was the added substrate (Table 3). It is assumed that this DHAP is generated during glucose synthesis from lactate. Nevertheless, this gluconeogenic DHAP is apparently not available for reduction to GP by reducing equivalents arising as a result of lactate oxidation to pyruvate, and which therefore cannot enter the GP shuttle unless glycolysis is occurring. The inability of gluconeogenic DHAP to act as a substrate for cytoplasmic GPDH, which is obviously a desirable metabolic feature because it prevents a wasteful redox cycle, implies a segregation of the glycolytic and gluconeogenic pathways at the level of triose-P [29].

21.8 Segregation of NAD species within the mitochondria - the central role of NADH dehydrogenase

Opinion remains divided on whether matrix NAD is entirely free in solution or at least partly bound. Our own view favours the latter circumstance. HBDH is tightly membrane associated, but is thought to be able to interact with matrix NAD. We and others [30] have observed that the redox potential, E_{hGDH} for glutamate dehydrogenase (GDH), an enzyme located entirely in the matrix, is sometimes consistent with, but at other times different from, E_{hHBDH}. We suggest therefore that similar circumstances prevail in the mitochondrial matrix as in the cytoplasm;

namely, that there is a pool of free NAD as well as enzyme-bound NAD. There is much evidence that the matrix water contains at least 50% protein and that many of the enzymes of the matrix, including the M-A shuttle enzymes MDH and AST, exist in multi-enzyme complexes some of which are bound to the inner membrane in close association with NADH dehydrogenase [31,32]. This has been thought to provide a direct route for reducing equivalents, arising from the oxidation of mitochondrial NAD-linked substrates, into the electron transport chain. If so, how can these reducing equivalents be channelled into reductive syntheses?

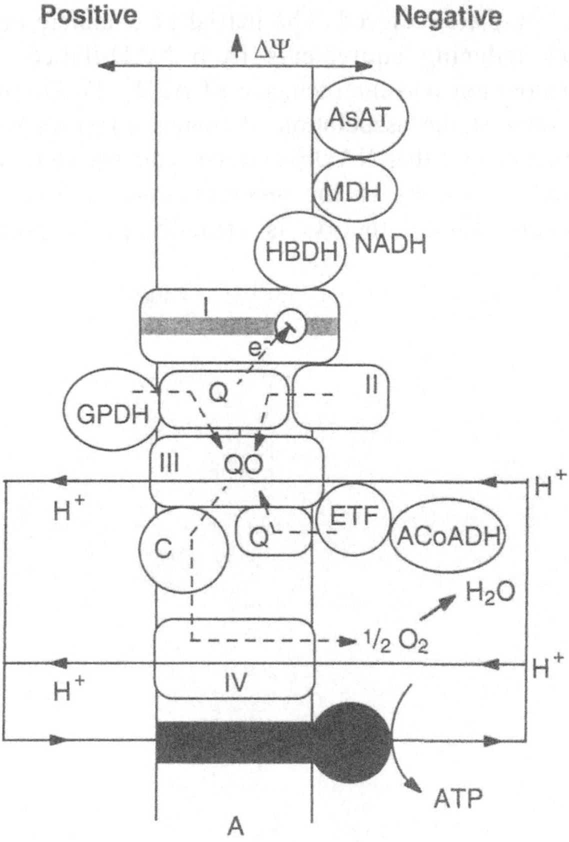

Figure 5. Figures 5A and 5B are schematic diagrams of the inner mitochondrial membrane, showing the four Complexes (I,II,III and IV) and associated enzymes. Electron flow is indicated by dashed lines, with the arrowheads showing the direction of flow. In Fig. A, $\Delta\mu H^+$ is at peak value. In Fig. B $\Delta\mu H^+$ is diminished and electron flow occurs from Complex I to ubiquinone, as protons are translocated.

We hypothesize that the proton pumping channel of NADH dehydrogenase is voltage-gated and remains closed at peak $\Delta\mu H^+$ (Figure 5A), a process possibly mediated by $\Delta\mu H^+$-driven reversed electron transfer. Because proton pumping is normally tightly coupled to electron flow, closure of the channel will inhibit reducing-equivalent flow through the NADH dehydrogenase, and reducing equivalents will flow instead into pathways for reductive syntheses. The channel will open (Figure 5B) when $\Delta\mu H^+$ declines in response to physiological energy demands or pathological inhibition of electron transport, and under the former circumstances is probably never completely closed. The partial or complete opening of the channel enables reducing equivalents from NAD-linked substrates to contribute to restoration and maintenance of $\Delta\mu H^+$. Furthermore it seems likely, on the basis of the association of matrix enzymes with the inner mitochondrial membrane, that NADH-derived reducing equivalents will be delivered in close proximity to the inner membrane. Since this region, according to chemiosmotic theory, is virtually proton-free, the redox

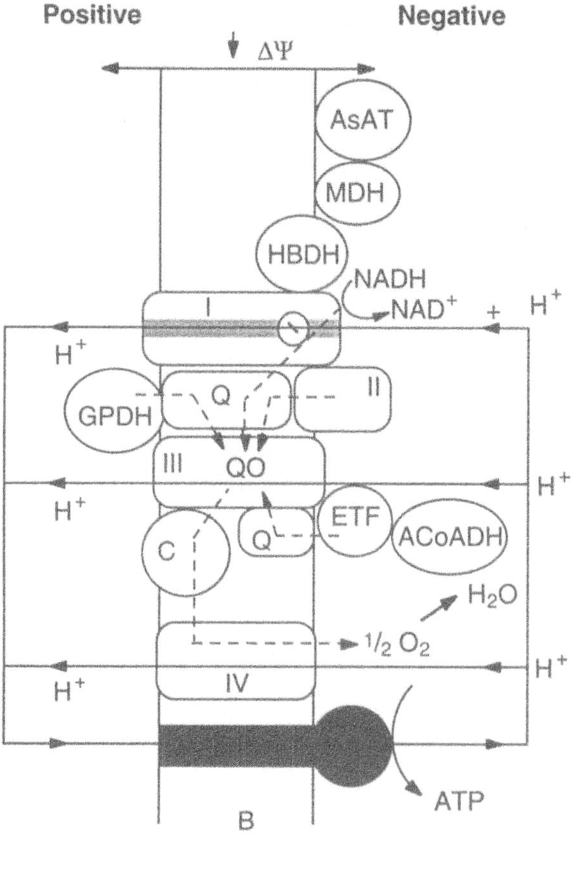

potential of these reducing equivalents may well be boosted about −200 mV, but this increased negativity would not be detected by measurement of redox potential by the indicator method, which determines free NAD^+. This mechanism would explain the 50-90% reduction of matrix NAD measured spectroscopically in isolated mitochondria [23], and would also account for the correlation between E_{hHBDH} and $\Delta\Psi$ observed in hepatocytes exposed to uncoupling agents [19].

21.8.1 The transistor-like behaviour of NADH dehydrogenase

Szent-Gyorgyi wrote extensively of the likelihood of electronic conduction through proteins [33], but his views have not been widely accepted. Rather it has been generally assumed that the high water content of living cells would preclude transmission of electrons. Hence the modulation of voltage-gated protein channels for ion conduction is thought to be brought about by the interaction of specific ions with receptor sites accessible to the aqueous phase [34]. In fact all these protein channels span hydrophobic membranes, so the possibility that the channels are regulated by electron density within the membrane, as well as by ions external to it, is by no means implausible. A likely site where control of this nature could be exerted is the inner mitochondrial membrane. Electron flow between components of the respiratory chain within this membrane is known to occur and the need for proton gating in the complexes of the chain such as cytochrome oxidase has been convincingly argued [35-38].

In keeping with this Cope, as early as 1974 postulated that cytochrome oxidase demonstrated transistor-like behaviour [39]. We propose here that NADH dehydrogenase likewise shows properties analogous to that of a field effect transistor. Thus, when $\Delta\Psi$ is near peak values, the flow of protons through the proton channel within the enzyme will be greatly decreased due to electronic constriction of the gate, and the associated electron flow to ubiquinone will also be impeded. The restriction of flow within this ATP-producing circuit, together with the driving force of $\Delta\Psi$, will inevitably enhance flow of reducing equivalents into pathways providing reducing power for energy-dependent reductive syntheses.

21.9 The calorigenic action of thyroid hormone

21.9.1 Discounted theories of thyroid action

It has been known for many years that thyroid hormone stimulates metabolism and heat production, and that liver and muscle are quantitatively the most important targets [11,40], but no consensus has emerged in regard to the mechanisms of thyroid hormone action. Theories that have been developed have concentrated on catabolic processes, a circumstance arising from early work on isolated mitochondria from hyperthyroid rats, which appeared to show that these organelles were uncoupled, *ie* their O_2-uptake was not tightly linked to ATP synthesis. These studies led to the assumption that the increased respiration and oxidation of cellular fuels, observed in the hyperthyroid state were required to maintain normal cellular ATP levels in the face of wasteful energy dissipation. However, it was later shown that physiological concentrations of the hormone did not uncouple and the hypothesis was abandoned [40], to be replaced some years later by a theory that proposed that thyroid hormone stimulates an increased leak of potassium ions from the cell, thereby promoting increased activity of the plasma membrane Na^+/K^+ ATPase and consequent enhanced ATP turnover [41]. This theory too has been discounted on the grounds that the energy requirement of the "sodium pump" is too low to account for the increased metabolic rate associated with hyperthyroidism [42]. More recently, it has been suggested that because of the high proton pressure across the inner mitochondrial membrane there is a constant backflow of protons to the matrix (essentially a form of uncoupling) and that this energy-wasting leakage may be greater in the hyperthyroid state [43], but no direct evidence for a "proton leak" has yet been forthcoming. *The key point is that the increase in O_2-consumption caused by thyroid hormone is associated with a stimulation not only of catabolic but also anabolic processes* [11], *and must thus be accompanied by enhanced rates of ATP synthesis and energy-dependent reducing-equivalent transfer.*

21.9.2 The anabolic actions of thyroid hormone in liver and their effects on energy metabolism

Liver, as the chief organ of metabolism, synthesizes and degrades protein, carbohydrate, and fat. The turnover of both carbohydrate and lipid [11] has been shown to be stimulated by thyroid hormone. In liver there is increased production of certain enzymes *e.g.* mitochondrial GPDH [44], which forms part of an important NAD- and flavin-linked intercompartmental reducing-equivalent shuttle and malic enzyme [45] which is believed to play a key role in the transfer of reducing-equivalents during lipogenesis, while other enzymes (such as alcohol dehydrogenase) decrease in activity. The outcome is that the major synthetic pathways of gluconeogenesis in the fasted state [46,47] and lipogenesis in the fed state [48] are increased by thyroid hormone administration and decreased in its absence. Moreover, it is now appreciated that the substantial anabolic effects of the hormone on carbohydrate and lipid metabolism may place considerable energy demands on the hepatocyte. As a result of these findings, together with the recognition of the importance of the hormone in growth and development [11], the concept that thyroid hormone is primarily catabolic in its activity has fallen into disfavour [40], to be replaced by the view that the hormone stimulates both anabolic and catabolic processes. In normal adult animals a steady state, reflecting a balance between these two processes, is maintained. In the hyperthyroid condition, where rates of metabolic turnover are increased, this balance and the associated steady-state is nevertheless maintained except in severe disease. A key question yet to be answered is what causes this increased anabolic and catabolic flux occur in the hyperthyroid state.

21.9.3. Our studies of the metabolic actions of thyroid hormone in hepatocytes

There is now a considerable body of information on the action of thyroid hormone, to which our laboratory has contributed. A major feature of thyroid hormone action is the stimulation of O_2-uptake, and this has focused attention on the mitochondria, though whether or not thyroid hormone acts directly on these organelles, or indirectly through its effects on nuclear DNA, remains a matter of some controversy [11]. A striking feature of hormone action is an increase in the capacity of intercompartmental shuttles for the bi-directional transport of reducing equivalents between mitochondria and cytoplasm [49]. Of special interest in relation to this is our finding that thyroid hormone stimulates gluconeogenesis from substrates such as lactate, pyruvate or sorbitol where reducing-equivalent transfer is required, but has no effect on glucose synthesis from fructose or dihydroxyacetone whose metabolism does not involve such transfer [9]. Other processes such as urea and glycogen synthesis, which do not involve reducing-equivalent transfer are also not stimulated. Intercompartmental shuttle activity is essential, and indeed can be rate-limiting, for reductive syntheses such as lipogenesis and gluconeogenesis. It is therefore not unexpected to find that these synthetic processes are stimulated in the hyperthyroid state. What is more surprising is the finding from our recent studies that hepatocytes from hyperthyroid, in contrast to control cells, retain the capacity for high rates of gluconeogenesis, fatty acid oxidation and lipogenesis regardless of nutritional state (unpublished observations).

21.9.4 An explanation for the calorigenic action of thyroid hormone

All these observations, together with the work of others, lead us to a straightforward and readily testable hypothesis regarding thyroid hormone action. We propose that a major action of thyroid hormone is to induce the synthesis of components of the respiratory chain involved in oxidative phosphorylation or in reductive syntheses, as well as the constituents of shuttle systems responsible for intercompartmental reducing-equivalent transfer between cytoplasm and mitochondria. The anticipated outcome is a stimulation of both catabolic, energy-yielding, processes and anabolic processes requiring ATP consumption, and particularly reductive syntheses which, because of their tight coupling, will inevitably lead to an increased rate of turnover of hepatic carbohydrate and lipid, and a corresponding stimulation of energy metabolism and O_2-uptake. The converse is anticipated to occur in the hypothyroid state.

21.10 Valediction

This book goes to press shortly after the announcement of the death of Paul Srere. Paul's work on the "metabolon" and his advocacy of the importance of enzyme-enzyme interactions has been an inspiration for all of us engaged in studying the relationship of structure to function. As expressed succinctly by Claude Aflalo, his spirit will always remain with us and continue to have a profound impact not only on modern biology, but on the philosophy of Science at large.

Another scientist and friend, whose premature death has occurred since this book was begun, is Dermot Williamson. Dermot gave me (MNB) enormous support when I first arrived in Oxford in 1961, and had a great influence on my subsequent choice of research interests, as well as introducing me to, and instructing me in many techniques, in particular enzymatic analyses. Much of the work in this chapter has been stimulated by his seminal studies on hepatic metabolism and cellular redox state. His passing is deeply mourned.

Acknowledgements

This work has been supported by the NH&MRC, the Drug and Alcohol Services Council and the Flinders Medical Centre Foundation.

References

1. Microenvironments and Metabolic Compartmentation. Srere PA and Estabrook RW, Eds. New York: Academic Press, 1978.
2. Organized Multienzyme Systems. Welch GR, Ed. New York: Academic Press, 1985.
3. The organisation of cell metabolism. Welch GR and Clegg JS, Eds. New York: Plenum Press, 1986.
4. Ovadi J. Cell Architecture and Metabolic Chanelling. New York: Springer-Verlag, 1995.
5. Channelling in Intermediary Metabolism. Agius L and Sherratt HAS, Eds. London: Portland Press, 1997.
6. Sacktor B and Dick A. Pathways of H^+ transport in the oxidation of extramitochondrial reduced diphosphopyridine nucleotide in flight muscle. J Biol Chem 1962; 237: 3259-3263.
7. Borst P. In: Karlson P, Ed. Funktionelle und Morphologische Organisation der Zelle. Berlin: Springer-Verlag, 1963: 137-158.
8. Berry MN and Kun E. Rate-limiting steps of gluconeogenesis in liver cells as determined with the aid of fluoro-dicarboxylic acids. Eur J Biochem 1972; 27: 395-400.
9. Meijer AJ and Williamson JR. Transfer of reducing equivalents across the mitochondrial membrane. I. Hydrogen transfer mechanisms involved in the reduction of pyruvate to lactate in isolated liver cells. Biochim Biophys Acta 1974; 333: 1-11.
10. Frerman FE. Acyl-CoA dehydrogenases, electron transfer flavoprotein and electron transfer flavoprotein dehydrogenase. Biochem Soc Trans 1988; 16: 416-418.
11. Freake HC and Oppenheimer JH. Thermogenesis and thyroid function. Annu Rev Nutr 1995; 15: 263-291.
12. Berry MN. Energy-dependent reduction of pyruvate to lactate by intact isolated parenchymal cells from rat liver. Biochem Biophys Res Commun 1971; 44: 1449-1456.
13. Berry MN, Kun E and Werner HV. Regulatory role of reducing-equivalent transfer from substrate to oxygen in the hepatic metabolism of glycerol and sorbitol. Eur J Biochem 1973; 33: 407-417.
14. Berry MN. The function of energy-dependent redox reactions in cell metabolism. FEBS Lett 1980; 117 (Suppl): K106-K120.
15. Berry MN, Gregory RB, Grivell AR et al. Linear relationships between mitochondrial forces and cytoplasmic flows argue for the organized energy-coupled nature of cellular metabolism. FEBS Lett 1987; 224: 201-207.
16. Berry MN, Phillips JW and Grivell AR. Interactions between mitochondria and cytoplasm in isolated hepatocytes. Curr Top Cell Regul 1992; 33: 309-328.
17. Berry MN, Phillips JW, Gregory RB, Grivell AR and Wallace PG. Operation and energy dependence of the reducing-equivalent shuttles during lactate metabolism by isolated hepatocytes. Biochim Biophys Acta 1992; 1136: 223-230.
18. Berry MN, Gregory RB, Grivell AR, Phillips JW and Schön A. The capacity of reducing-equivalent shuttles limits aerobic glycolysis during ethanol oxidation. Eur J Biochem 1994; 225: 557-564.

19. Berry MN, Gregory RB, Grivell AR et al. Intracellular mitochondrial membrane potential as an indicator of hepatocyte energy metabolism: further evidence for thermodynamic control of metabolism. Biochim Biophys Acta 1988; 936: 294-306.
20. Mitchell P. Vectorial chemistry and the molecular mechanics of chemiosmotic coupling: power transmission by proticity. Biochem Soc Trans 1976; 4: 399-430.
21. Kell DB. The protonmotive force as an intermediate in electron transport-linked phosphorylation: problems and prospects. Curr Top Cell Regul 1992; 33: 279-289.
22. Rydstrom J. Energy-linked nicotinamide nucleotide transhydrogenases. Biochim Biophys Acta 1977; 463: 155-184.
23. Klingenberg M. The Respiratory Chain. In: Singer TP, Ed. Biological Oxidations. New York: Interscience Publishers, 1968: 3-54.
24. Chance B, Hollunger G. The interaction of energy and electron transfer reactions in mitochondria. VI. The efficiency of the reaction. J Biol Chem 1961; 236: 1577-1584.
25. Berry MN, Clark DG, Grivell AR and Wallace PG. The calorigenic nature of hepatic ketogenesis: An explanation for the stimulation of respiration induced by fatty acid substrates. Eur J Biochem 1983; 131: 205-214.
26. Srivastava DK, Bernhard SA, Langridge R and McClarin JA. Molecular basis for the transfer of nicotinamide adenine dinucleotide among dehydrogenases. Biochemistry 1985; 24: 629-635.
27. Sies H. Nicotinamide nucleotide compartmentation. In: Sies H, Ed. Metabolic compartmentation. London: Academic Press, 1982: 205-231.
28. Grivell AR, Berry MN. The effects of phosphate- and substrate-free incubation conditions on glycolysis in Ehrlich ascites tumour cells. Biochim Biophys Acta 1996; 1291: 83-88.
29. Threlfall CJ and Heath DF. Compartmentation between glycolysis and gluconeogenesis in rat liver. Biochem J 1968; 110: 303-312.
30. Söling HD and Kleineke J. Species dependent regulation of hepatic gluconeogenesis in higher animals. In: Hanson RW and Mehlman MA, Eds. Gluconeogenesis. Its regulation in mammalian species. New York: John Wiley and Sons, 1976: 369-462.
31. Teller JK, Fahien LA and Valdivia E. Interactions among mitochondrial aspartate aminotransferase, malate dehydrogenase, and the inner mitochondrial membrane from heart, hepatoma, and liver. J Biol Chem 1990; 265: 19486-19494.
32. Ushiroyama T, Fukushima T, Styre JD and Spivey HO. Substrate channeling of NADH in mitochondrial redox processes. Curr Top Cell Regul 1992; 33: 291-307.
33. Szent-Györgyi A. The study of energy levels in biochemistry. Nature 1941; 148: 157-159.
34. Karpefors M, Adelroth P, Aagaard A, Sigurdson H, Svensson Ek M and Brzezinski P. Electron-proton interactions in terminal oxidases. Biochim Biophys Acta 1998; 1365: 159-169.
35. Krab K and Wikstrom M. Principles of coupling between electron transfer and proton translocation with special reference to proton-translocation mechanisms in cytochrome oxidase. Biochim Biophys Acta 1987; 895: 25-39.
36. Wilson MT and Bickar D. Cytochrome oxidase as a proton pump. J Bioenerg Biomembr 1991; 23: 755-771.
37. Nicholls P and Butko P. Protons, pumps, and potentials: control of cytochrome oxidase. J Bioenerg Biomembr 1993; 25: 137-143.
38. Calhoun MW, Thomas JW and Gennis RB. The cytochrome oxidase superfamily of redox-driven proton pumps. Trends Biochem Sci 1994; 19: 325-330.

39. Cope FW. Solid state physical mechanisms of biological energy transduction. Ann N Y Acad Sci 1974; 227: 636-640.
40. Soboll S. Thyroid hormone action on mitochondrial energy transfer. Biochim Biophys Acta 1993; 1144: 1-16.
41. Ismail-Beigi F and Edelman IS. The mechanism of the calorigenic action of thyroid hormone. Stimulation of Na^+ + K^+-activated adenosinetriphosphatase activity. J Gen Physiol 1971; 57: 710-722.
42. Sestoft L. Metabolic aspects of the calorigenic effect of thyroid hormone in mammals. Clin Endocrinol 1980; 13: 489-506.
43. Brand MD, Steverding D, Kadenbach B, Stevenson PM and Hafner RP. The mechanism of the increase in mitochondrial proton permeability induced by thyroid hormones. Eur J Biochem 1992; 206: 775-781.
44. Lee Y-P and Lardy HA. Influence of thyroid hormones on L-α-glycerophosphate dehydrogenases in various organs of the rat. J Biol Chem 1965; 240: 1427-1436.
45. Oppenheimer JH, Schwartz HL, Lane JT and Thompson MP. Functional relationship of thyroid hormone-induced lipogenesis, lipolysis, and thermogenesis in the rat. J Clin Invest 1991; 87: 125-132.
46. Berry MN, Werner HV and Kun E. Regulatory function of intercompartmental metabolite and reducing-equivalent translocation in hepatic gluconeogenesis: role of thyroid hormone. In: Lundquist F and Tygstrup N, Eds. Regulation of hepatic metabolism (A. Benzon symposium 6). Copenhagen: Munksgaard, 1974: 501-519.
47. Bartels PD and Sestoft L. Thyroid hormone-induced changes in gluconeogenesis and ketogenesis in perfused rat liver. Biochim Biophys Acta 1980; 633: 56-67.
48. Freake HC, Schwartz HL and Oppenheimer JH. The regulation of lipogenesis by thyroid hormone and its contribution to thermogenesis. Endocrinology 1989; 125: 2868-2874.
49. Gregory RB, Phillips JW and Berry MN. Reducing-equivalent transfer to the mitochondria during gluconeogenesis and ureogenesis in hepatocytes from rats of different thyroid status. Biochim Biophys Acta 1992; 1137: 34-38.

Chapter 22

The importance of hepatocytes in drug metabolism studies: an industrial perspective

Martin K. Bayliss and David M. Cross
Division of Bioanalysis and Drug Metabolism, GlaxoWellcome, Park Road, Ware, Hertfordshire, SG12 0DP, England

Key words: coculture, cryopreservation, cytochrome P450, drug metabolism, enzyme kinetics, hepatic clearance, hepatic extraction ratio, hepatocyte, inter-species scaling, intrinsic clearance, in vivo/in vitro correlations, loxtidine, physiological models, primary culture, ranitidine,suspension culture

22.1 Drug metabolism and the role of the isolated hepatocyte

Historically, the study of drug metabolism [1] in the pharmaceutical industry has of necessity concentrated on the practical issues concerning the toxicological response of selected animal species to candidate drug compounds, and a balancing of the presence of metabolites in one species against those found in another. The responses in terms of metabolite production have been compared to those expected or demonstrated in man. Although understanding the mechanism underlying metabolite formation was erstwhile the province of the academic world, advances in technology, especially molecular biology of the cytochrome P450 (CYP) enzymes, have significantly influenced the pragmatic industrial approach. It is now widely recognized that *in vitro* screening for metabolism and/or specific CYP interactions can foster a rational approach to the selection of drug molecules.

22.2 Models for studying drug metabolism

Current regulatory requirements specify that toxicity studies must be carried out in one rodent and one non-rodent species irrespective of the merits of the two species as "good" models for man. Thus, regulation and a rational approach to testing appear at odds with each other. To help resolve this dilemma, early drug metabolism and kinetic studies in various animal species are undertaken in order to provide an indication of which species

M.N. Berry and A.M. Edwards (eds.), The Hepatocyte Review, 365–389.

should be selected for optimum evaluation of a particular new chemical entity. The recognition by pharmaceutical scientists of the inherent limitations of animal experimentation in drug research, coupled with heightened awareness of animal welfare issues, has increased the pressure for early information on the behaviour of candidate drugs in the target therapeutic species, man.

The relatively recent prospect of whole animal studies actually being replaced by *in vitro* techniques that provide equivalent data, as opposed to merely supplementing traditional animal studies, has provided the pharmaceutical industry with an incentive to validate such models with a view to eventual replacement of certain animal studies. Currently, animals cannot entirely replace human beings in studies trying to predict all responses in man, but it is clear, for ethical reasons [2], that the risk to human volunteers participating in early drug development studies should be minimized. Thus, a compromize between increasing familiarity with the behaviour of a new compound, as it progresses through drug development, and premature exposure to man must be reached. It is in this context that hepatocyte models of drug metabolism have been set.

The type of model used will no doubt be a compromize between convenience and relevance. In the past, drug metabolism has been investigated in increasingly purified liver fractions culminating in single enzyme systems. Conversely, perfusion of the whole liver *in situ* has been

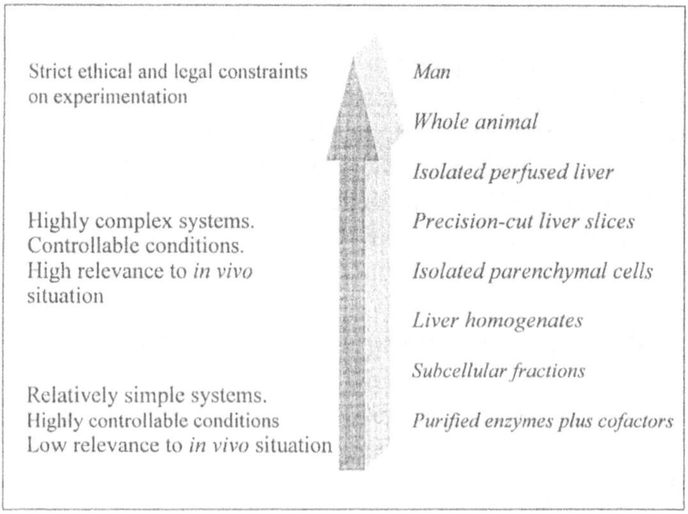

Figure 1. A hierarchy of models used to investigate drug metabolism *in vivo* and *in vitro*.

employed to investigate biotransformation. The former suffers from lack of relevance to the *in vivo* situation and the latter, while as close as possible to the liver *in vivo*, suffers from complex methodology and limited multiple usage. As described in Figure 1, isolated hepatocytes provide a convenient link between subcellular fractions and the complex architecture of the intact organ whilst encouraging multiple experiments on the same batch of cells or individual liver. Thus, the isolation and maintenance of viable hepatocytes has been desirable so that liver function can be studied in a controlled environment.

22.2.1 Isolated cells in suspension and culture: an overview

As a good model of liver metabolism, isolated hepatocytes must ideally be able to undertake the full range of known *in vivo* biotransformations and predict known endpoints. Problems encountered during extended culture of hepatocytes are circumvented by use of freshly isolated cell suspensions. Hepatocytes used in short-term suspension culture have been shown to retain the intact biochemical functions of the *in vivo* liver cell. Most drug metabolic processes occur in the liver parenchymal cells which are preferentially isolated by traditional collagenase perfusion techniques [3,4]. Thus suspensions of isolated parenchymal cells offer a powerful tool for the investigation of many aspects of drug metabolism. In fact, good *in vivo/in vitro* correlations of metabolic patterns have been reported using both isolated animal and human hepatocytes [5-10], whilst studies reporting the failure of hepatocytes to predict metabolic pathways have been few [11]. However, there have been only a small number of reports on the comparative metabolism of single substrates in hepatocytes from the three species central to drug development, namely rat, dog and man [12-16].

Perhaps one of the main advantages of suspension culture is the rapid dispersion of drug throughout the incubation: this potentially results in rapid availability of test compound to cellular enzymes and facilitates sampling, which simultaneously incorporates drug substrate and products from both the media and the cells. Preparations of hepatocytes immobilized in "liver beads" also fulfil these requirements. However, non-parenchymal cells can be important in specific metabolic pathways and appear to help in maintaining enzyme activities. In a typical hepatocyte preparation, non-parenchymal cells account for <1% of the isolated cell population [17] in contrast to 20-40% of the cell number *in vivo* [4]. Thus, whilst freshly isolated suspensions have become widely adopted as the standard for drug metabolism studies carried out over short durations, it is recognized that they represent an incomplete model. Although significant steps have been taken towards overcoming these deficits, the need to carry out investigations over extended periods *in vitro*, without differential decline in drug metabolising-enzyme activity, still remains to be fully satisfied.

One motivation behind attempts to prolong the culture of hepatocytes has been the desire to model CYP induction *in vitro*. Data is relatively scarce for human hepatocytes, but studies with omeprazole and rifampicin have demonstrated that induction observed in man *in vivo* can also be reproduced in human hepatocyte cultures *in vitro* [18-20]. Similarly, induction of cyclosporin A metabolism by a variety of compounds has been reported in human hepatocyte cultures [21]. It has, however, proved notoriously difficult to maintain parenchymal cells in culture for extended periods without loss of specialized functions [22] (see also Chapter 7). In long term (>24 h) cultured hepatocytes, rapid changes in the fidelity of xenobiotic metabolism and alteration in phenotype are widely acknowledged [19,23,24] although some evidence that human hepatocytes are more stable than rat has been presented [3]. Various attempts to ameliorate the problem of differential decline in drug metabolising capabilities have been made, including modification of the culture medium [20,25], co-culture with rat non-parenchymal cells [23,26-28] and investigation of the influence of the attachment substrata [22,29-31]. These problems have been reviewed in detail by a number of workers [22,32-34] (see also Chapters 7 and 8). Investigation of hepatocyte co-cultures has focused almost exclusively on using rat epithelial, epithelial-like and endothelial cells and cell lines to support rat hepatocytes in extended culture. Rare examples involving human hepatocytes have been cited in the literature, but few relate to the study of drug metabolism [31,35]. Human hepatocyte cocultures have been shown to support the *O*-deethylation of ethoxycoumarin for up to 29 days [35] although the inducibility of this pathway by dibenz(*a,h*)anthracene was slightly lower than that seen in collagen-sandwich cultures (see below). However, in typical extended hepatocyte cultures, metabolic profiles have been shown to vary with time in culture and according to the type of cocultured cell, additives and substrata used in the culture. Comparison of results from cultured hepatocytes with those from freshly isolated cells may be necessary to define the correct baseline.

Immobilisation of liver parenchymal cells in alginate [36,37] has facilitated commercialisation of isolated hepatocyte technology but has not succeeded in significantly progressing the key issue of dedifferentiation. Culture of hepatocytes as spheroids showed promising results with the peroxisome proliferator nafenopin [38], lidocaine metabolism, CYP1A1 induction by 3-methylcholanthrene (3MC) [39], ethoxycoumarin- and ethoxyresorufin-*O*-deethylase and glutathione S-transferase activity [17]. However, use of spheroids still awaits realisation of their full potential as models of drug metabolism.

Traditional culture methods are characterized by an unstirred medium layer overlying cells which are attached to a gas impermeable substratum. For liver cells, which are highly perfused *in vivo*, such conditions are clearly susceptible to hypoxia or anoxia with consequent reduced cell functionality and viability. In addition it has been assumed that various other factors influence dedifferentiation of hepatocytes *in vitro*, for example, both homotypic and heterotypic cell-cell interactions, the type and geometry of extracellular matrix, exposure to soluble factors such as cytokines [17,40] and supply of nutrients. Some recent attempts to address these issues by construction of increasingly complex *in vitro* culture models are related below.

22.2.2 Complex in vitro culture models

Various collagen sandwich cultures ("double gel culture") have been evaluated that support significant improvements in the fidelity of drug metabolism processes over conventional culture (see Chapter 8). Several probes and drug substrates have been used to assess drug metabolism competency during culture: phenobarbital, paracetamol [24], urapidil [41], 3,3',4,4'-tetrachlorobiphenyl, ethoxyresorufin [35,42], pimobendan [8], dexamethasone [25], rifampicin, ethoxycoumarin, testosterone, sulphobromophthalein, p-nitrophenol [19] and tacrolimus [43]. In general, both Phase I and II processes, measured in terms of enzyme concentration and activity, including induction typical of *in vivo* responses in the target species, were reported over anything from 3 days to 4 weeks. Morphology was demonstrated to be more akin to the *in vivo* situation, which may be important in maintaining drug metabolising enzyme activity profiles [34], and convincing *in vivo/in vitro* correlations in metabolic profiles have been established in rat and man. However, metabolic competency still declined to various extents indicating that continued development of the techniques is required.

Aggregates of parenchymal and non-parenchymal liver cells have been successfully established in an attempt to promote *in vitro* the kind of cellular interactions which are likely to be found *in vivo* [17]. Amongst the cell types incorporated into the culture aggregates are non-parenchymal liver cells, including Kupffer cells that allow modulation of metabolic profiles by non-parenchymal cells to be investigated. Aggregates were sensitive to induction by phenobarbital and 3MC as measured using ethoxyresoroufin, ethoxycoumarin, aldrin, p-nitrophenol and 1-chloro-2,4-dinitrobenzene (CDNB), although different Phase I and Phase II enzymes responded differently over the culture period. Issues relating to supply of nutrients, oxygen and drug molecules to the center of cell aggregates are partly being addressed by reducing the size of aggregates and preventing coalescence by encapsulation of aggregates and spheroids in polysaccharide [44] or alginate [45].

A combination of techniques has been adopted by Bader and co-workers [46] (see also Chapter 8) wherein parenchymal cells are encased in a collagen sandwich and the non-parenchymal cells overlaid on the outer surface of the collagen gel. Investigation of drug metabolising enzyme activity was limited to the standard probe ethoxyresorufin and CYP1A1 protein levels were measured using Western blotting. It may be that cultures that were supported on a gas permeable membrane to circumvent problems of limited gas exchange in aggregates suffered from oxygen toxicity during extended culture. Nevertheless, such cultures appear to have improved characteristics in terms of morphology and functional stability, including drug metabolism capabilities, when compared to traditional culture methods.

Attempts to overcome these limitations have resulted in the development of dynamic culture models such as the "bioreactor". In this system, hepatocytes may be cultured within a cartridge containing numerous hollow fibers around which the cells grow. The hollow fibers ensure a continuous flow of medium, and therefore oxygen, nutrients and test substances, to the cells [47]. However, oxygen toxicity was thought to be responsible for a decline in metabolism of diazepam beyond 10 days in culture, although the inclusion of various supplements in the perfusion medium, including selenium and dexamethasone, were thought to help maintain some CYP activities. Phase II reactions were also compromised after extended operation of the bioreactor as indicated by following phenolsulphonphthalein glucuronidation. Such a technique may however have much to offer as a "factory" for the production of large amounts of metabolites or as a bioartificial liver.

Systems of hepatocyte culture that appear to offer longevity also add a layer of complexity to interpretation of metabolic studies where outcomes can depend on co-factor levels, cell-cell interactions, nature of the matrix, cellular transport and dedifferentiation of higher functions, such as competency in drug metabolism. The complex nature of the culture technology needed to foster cell-cell and cell-matrix interactions reduces its appeal to the industrial scientist investigating drug metabolism. Nevertheless, metabolic competency appears to be retained to a greater or lesser extent over a period of weeks and, if the methods are sufficiently robust to withstand transfer between laboratories, they offer an exciting addition to the presently available battery of techniques. This is especially so if they are applicable to man and the animal species more commonly employed during drug discovery and development.

22.3 Cryopreservation

Cryopreservation has been repeatedly proposed as an effective method of preserving isolated hepatocytes [48-50]. CYP activities measured in cell homogenates using aminopyrine, ethoxy- and pentoxy-resorufin were well maintained after cryopreservation but some Phase II reactions, such as glutathione S-transferase activity measured with CDNB, had declined after thawing [50]. Various cryoprotectants have been advocated in pursuit of high viability upon thawing [50]. Furthermore, some of the advantages accrued by culturing freshly isolated hepatocytes using collagen sandwich methods can be imparted to cryopreserved cells [51], which may encourage wider acceptance of this preservation technique. Although the technique is still not fully developed, cryopreservation is particularly useful for species where the source material is less readily available, especially human, and remains a process eagerly awaited by the industrial drug metabolism scientist.

22.4 Case histories illustrating in vivo/in vitro correlations

Some investigations from our own laboratories, which serve to illustrate the practical issues associated with examining cross-species comparisons of metabolite profiles in isolated hepatocytes and problems associated with the further characterisation of enzymology underlying metabolic processes, are discussed in the two case histories below.

22.4.1 Ranitidine

Ranitidine (Figure 2) is a histamine H_2-receptor antagonist used in the treatment of diseases associated with gastric acid hypersecretion. The metabolism of ranitidine in animals and man appears to be limited to the Phase I processes of N-oxidation, S-oxidation, N-demethylation and oxidative deamination to the furoic acid metabolite [15,52-54].

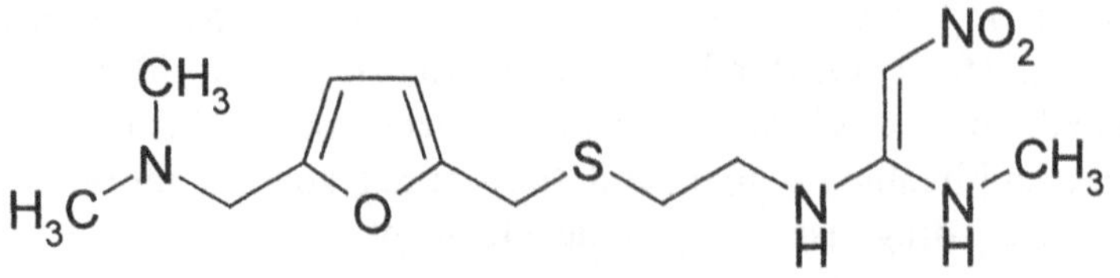

Figure 2. The chemical structure of ranitidine: N-[2-({[5(dimethylaminomethyl)-furan-2-yl]methyl}thio)ethyl]N[1]methylnitroethene 1,1 diamine (as HCl salt)

Using isolated hepatocytes in short-term suspension incubations, we demonstrated the various major pathways found *in vivo* for the formation of non-acidic ranitidine metabolites; although the furoic acid analogue itself could not be detected in suspensions of isolated hepatocytes from rat, dog and man [15,55]. In this case the absence of a metabolite expected from *in vivo* studies may reflect the absence of extrahepatic contributions to metabolism *in vitro*. In this work [15,55], two previously unidentified metabolites were detected in isolated rat and dog hepatocytes. It should, however, be considered that the formation of apparently novel metabolites *in vitro* may be a function of incubation at relatively high non-physiological concentrations of a compound in a closed system where metabolite concentrations can increase to abnormally high levels. The identity of these two *in vitro* metabolites could not be established in the original work [15,55]. However, two further *in vivo* metabolites were subsequently characterized from rat urine and found to be the single N-desmethyl analogue from the N^1-methyl-nitroethene functional group and its corresponding *bis*-N-desmethyl derivative [54]. There was no clear evidence that the *in vitro* and *in vivo* metabolites corresponded but the utility of *in vitro* systems in directing further investigations has been illustrated. Reduction of ranitidine N-oxide to the parent compound was also observed in hepatocytes isolated from the Wistar rat, but not the Random Hooded rat, suggesting a strain-specific mechanism. The activity of reductases in hepatocytes incubated under high oxygen tension may seem surprising at first but perhaps serves to draw attention to the fact that the actual oxygen tension at the site of metabolism may be dramatically lower than that in the gas phase and incubation medium. This observation is especially pertinent when considering precision-cut liver slices, aggregates and spheroids, where it is not possible for all hepatocytes to be directly exposed to media, and oxygen gradients are inevitable.

Table 1. Estimates of Michaelis-Menten parameters and the evaluation of intrinsic and hepatic clearance of ranitidine in rat, dog and human hepatocytes

Species and Pathway	**K_m (mM)**	**V_{max} (pmol/min/ 10^6 cells)**	**CL_{int} [1,2] (ml/min/g liver)**	**CL_{int} [3] (ml/min/kg)**	**CL_{int} [3] (ml/min/kg) total**	**Predicted *in vitro* CL_H [4] (ml/min/kg)**	**Determined *in vivo* CL_H [5] (ml/min/kg)**
Rat							
N-oxidation	0.29 ± 0.02	150 ± 13	0.062	2.44			
S-oxidation	0.29 ± 0.03	85 ± 9	0.035	1.39	6.4	5.6	23.9 ± 2.7[6]
N-demethylation	0.30 ± 0.02	160 ± 9	0.064	2.52			
Rat							
Disappearance	0.47 ± 0.15	1200 ± 208	0.306		12.1	9.9	
Dog							
N-oxidation	0.85 ± 0.03	470 ± 21	0.133	3.2			
S-oxidation	2.80 ± 0.38	110 ± 8	0.009	0.2	3.5	3.2	7.7[7]
N-demethylation	2.10 ± 0.32	55 ± 5.1	0.006	0.1			
Human							
N-oxidation	7.0	195	0.003	0.08			
S-oxidation	15.5	175	0.001	0.03	0.16	0.15	2.8[8]
N-demethylation	3.9	70.5	0.002	0.05			

[1] Number of hepatocytes per gram of liver determined from the literature or by assessment of nmol CYP content of hepatocyte preparations and microsomal protein concentrations; rat and human 120×10^6, dog 240×10^6. [2] CL_{int} calculated from $Vmax_{(app)}/ Km_{(app)}$. [3] CL_{int} determined after scaling for liver weight: rat 10 g; dog 265 g; human 1694 g, calculated from [76], and body weight: rat 0.245 kg; dog 11.0 kg; human 70.0 kg. [4] CL_H *in vitro* calculated using the 'well stirred' model [74]. [5] CL_H *in vivo* calculated from the relationship $CL_p = CL_R + CL_H + CL_B$. [6] Calculated from [75]. [7] Calculated from [54]. [8] Calculated from [53].

It is also possible to further investigate mechanistic issues using *in vitro* systems [15,55]. The enzyme kinetics for the formation of ranitidine metabolites were studied in isolated hepatocytes from various species and the *in vivo* species differences in metabolite profiles were found to be consistent with the fractional intrinsic clearance to the various metabolites (Table 1). Substantial ranitidine metabolism typically occurred only at high substrate concentration. Rat hepatocytes also exhibited a higher affinity enzyme system with $Km_{(app)}$ approximately tenfold below the low affinity systems detectable in dog, human or rat hepatocytes. These observations are in line with the relatively minor importance of metabolic clearance *in vivo* in these species. Indeed, disappearance of ranitidine *in vitro* was generally too slow to enable precise determination of global kinetic parameters which would be useful in calculating total intrinsic clearance for the compound, so the kinetics of formation for each metabolite had to be measured and summed (Table 1). Note that only primary metabolite kinetics can be summed in this way because only these contribute to the metabolic clearance of the parent compound. The influence of multiphasic kinetics due to the presence of more than one competing enzyme within a pathway, can make both of these approaches difficult: in some cases only the "high affinity" enzyme kinetic values are appropriate for calculation of intrinsic clearance. (Section 22.5)

22.4.2 Loxtidine

Loxtidine hemisuccinate (Figure 3) is a potent histamine H_2-receptor antagonist with a relatively long duration of action [56]. The drug is a member of the aminomethyl phenoxypropyl family of the H_2-antagonists in which the dimethyl nitrogen present in ranitidine has become part of the piperidine heterocycle.

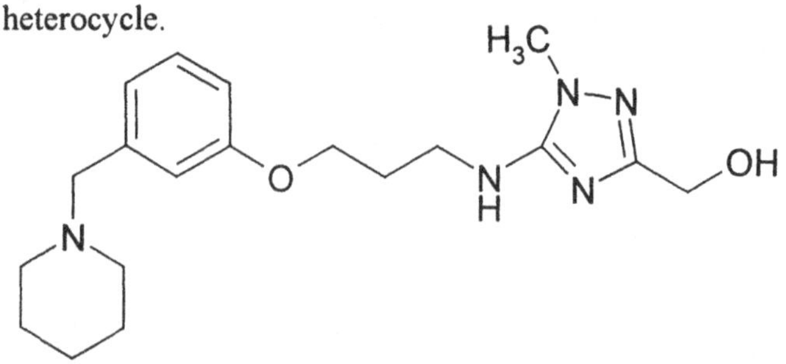

Figure 3. The chemical structure of loxtidine: 1-methyl-S-[3-[3-[(1-piperidinyl)methyl]phenoxy]propyl]amino-1H-1,2,4-triazole-3-methanol

Table 2. Estimates of Michaelis-Menten parameters and the evaluation of intrinsic and hepatic clearance of loxtidine in rat, dog and human hepatocytes

Species	**K_m (μM)**	**V_{max} (nmol/min/ 10^6 cells)**	**CL_{int} [1,2] (ml/min/g liver)**	**CL_{int} [3] (ml/min/kg)**	**Predicted *in vitro* CL_H [4] (ml/min/kg)**	**Determined *in vivo* CL_H [5] (ml/min/kg)**
Rat	9.9 ± 4.0	0.114 ± 0.032	1.45 ± 0.28	51.4 ± 12.4	24.5	26.6
Dog	441 + 257	0.512 ± 0.341	0.276 ± 0.028	8.0 ± 0.8	3.1	6.6
Human	56.3 ± 33.6	0.015 ± 0.006	0.041 ± 0.024	1.0 ± 0.58	0.2	0.4

[1] Number of hepatocytes per gram of liver determined from the literature or by assessment of nmoles CYP content of hepatocyte preparations and microsomal protein concentrations; rat and human 120×10^6, dog 240×10^6. [2] CL_{int} calculated from $Vmax_{(app)}$ / $Km_{(app)}$. [3] CL_{int} determined after scaling for liver weight: rat 9.72 ± 1.75 g; dog 348.8 ± 87.4 g; human 1694 g, calculated from [76], and body weight: rat 0.275 ± 0.04 kg; dog 11.9 ± 3.0 kg; human 70.0 kg. [4] CL_H *in vitro* calculated using the 'well stirred' model [74]. [5] CL_H *in vivo* calculated from the relationship $CL_p = CL_R + CL_H + CL_B$. (Reprinted from [16] with permission of Taylor & Francis)

In laboratory animals and human volunteers loxtidine undergoes metabolism by several routes and shows significant species differences [57]. In rat and dog loxtidine is cleared by metabolism. The major route of loxtidine metabolism in the rat was N-dealkylation to form the proprionic acid and hydroxymethyl triazole metabolites, and oxidation to the carboxylic acid derivative. In the dog, the major pathway was glucuronidation with only minor amounts of the above Phase I metabolites being detected. In contrast to animal species, loxtidine in man is eliminated predominantly unchanged by renal clearance and metabolism was limited to formation of the carboxylic acid derivative and trace amounts of the N-dealkylated products. No evidence for the presence of a glucuronide metabolite was observed.

Using isolated hepatocytes from rat, dog and man, we have been able to demonstrate that quantitative metabolic profiles in these species closely matched those found *in vivo* [16]. Furthermore, the non-linear dose dependency of loxtidine metabolism observed in the rat *in vivo* was found to be reproduced in isolated hepatocytes from the same species. Likewise, the observed linear dose-dependency in the dog *in vivo* was confirmed *in vitro*, thus adding to the growing database of *in vivo/in vitro* correlations in this species. Detailed calculation of intrinsic clearance in these three key species *in vitro* demonstrated an excellent correlation with *in vivo* (Table 2). In this case it was possible to derive enzyme kinetic parameters for loxtidine metabolism in terms of disappearance of parent compound. This approach has the advantage that all pathways contributing to metabolic clearance, even those as yet unknown, are incorporated into calculation of intrinsic clearance, and ultimately pharmacokinetic parameters.

22.5 An introduction to intrinsic clearance in vitro

The potential of the hepatocyte system for the study of drug metabolism pathways under defined conditions has been established for a variety of chemical entities as discussed previously. However, in recent years there has been a growing interest in the use of hepatocytes from animals, mainly the rat [58-62] and from man [13,14,63-65] to predict quantitative pharmacokinetic parameters of drug biotransformation. The prediction of human pharmacokinetic parameters from *in vitro* studies could be of great significance in the drug discovery and development process. If an accurate assessment of these parameters could be made before the compound is administered to man for the first time, then it may be possible to make an informed decision on the initial dose, which could result in a reduction in failure during phase I clinical trials [66]. Currently, the hepatic clearance of a new chemical entity in man cannot be predicted prior to undertaking expensive clinical trials. Thus prior knowledge that a new chemical entity may be rapidly eliminated in man due to metabolic clearance would be important in the decision making process of compound selection. There has been an increase in the number of published studies where the *in vivo* intrinsic clearance in rat, dog and man have been modelled successfully from *in vitro* data [12-14,63,67]. The success in modelling *in vivo* pharmacokinetic parameters from *in vitro* data represents a significant advance, as traditionally *in vitro* drug metabolism data were confined to qualitative analysis only.

The following discussion attempts to summarize the current situation for the extrapolation of *in vitro* kinetic data to the *in vivo* situation using hepatocyte models. The discussion focuses predominantly on work with hepatocytes in short-term suspension and monolayer culture. The use of short-term suspension culture currently appears to be most extensively studied and as such has been the system at the center of much debate [68]. Debate has focused on the accuracy of enzyme kinetic parameters and scaling factors. The reader is referred to the work of Houston and co-workers [62,68,69] who have completed an exhaustive investigation of the literature with respect to data produced in rats *in vitro* and *in vivo*. These workers collated data on the intrinsic clearance (CL_{int}) of some 21 compounds in rat hepatocytes, a number of which have comparable data in microsomes. Compounds were substrates for the rat CYPs and covered a range of CL_{int} values. For two compounds, diazepam and phenytoin, microsomal preparations appeared to under-predict both the hepatocyte and *in vivo* CL_{int} values. The authors [59,61] suggested that the under prediction in microsomes was in fact due to product inhibition which could affect determinations of Michaelis-Menton parameters and thus influence the calculation of CL_{int}. Both compounds form Phase I metabolites that are subsequently conjugated. Clearly, the hepatocyte system has advantages as functioning Phase II enzymes are in operation to conjugate the hydroxylated metabolites and thus prevent the build up of Phase I metabolites that may be the source of inhibition. Alternatively, some drugs will be concentrated in hepatocytes thereby altering the effective substrate concentration in the vicinity of the drug metabolising enzymes and influencing the apparent kinetic parameters, a phenomenon less likely to occur with microsomes but more reflective of the situation *in vivo*.

A number of drugs have very slow rates of metabolism and short-term suspension cultures may not be suitable to determine accurately a measurable metabolic rate. Primary hepatocyte cultures and perhaps liver slices provide systems that may be suitable. The utility of primary hepatocyte cultures [13,14,63,65,67] to predict pharmacokinetic parameters *in vivo* will be discussed briefly below. To our knowledge, hepatocytes in the collagen sandwich systems, beads, spheroids, aggregates or "bioreactors" have not been used as yet to model and predict pharmacokinetic parameters of drug entities.

22.5.1 Factors to consider when determining intrinsic clearance

The key parameter that enables the incorporation of *in vitro* drug metabolism data into pharmacokinetic models is CL_{int}. CL_{int} is a measure of an enzyme's capacity to metabolize drugs in the absence of physiological constraints such as liver blood flow and protein binding which may mask the true *in vivo* capacity.

Under first order conditions CL_{int} can be calculated from the apparent Michaelis-Menten parameters V_{max} and K_m by the following relationship: V_{max}/K_m [69]. Using this approach, conditions that support linear rates of metabolism with respect to time and cell density must be established. The potential for non-specific binding to cellular material can influence the apparent kinetic parameters by altering the effective substrate concentration available to the drug metabolising enzymes. Therefore it is important to establish both the non-specific binding of drug to cellular proteins and to the incubation vessel.

Enzyme parameters can be determined by measuring the rate of disappearance of drug [16,60] or the appearance of metabolites [59,61,62] in hepatocyte systems. Estimates of apparent Michaelis-Menten constants can then be obtained graphically. If the rate of metabolism is particularly slow or limited data points are available, an approach using an integrated form of the Michaelis-Menten equation can be used [12]. When measuring disappearance of drug it is important to establish that disappearance is in fact due to metabolism and not either chemical instability or some non-specific binding process. It is worth noting that if metabolite production is used to determine apparent enzyme parameters then one must consider the sum of the individual primary pathways when modelling CL_{int}. Alternatively, rather than measuring initial rates of metabolism, the time course of the disappearance can be determined using a single concentration and the area under the time profile curve assessed in a similar fashion to that of an *in vivo* pharmacokinetic profile. This approach has been used by a number of investigators in rat hepatocytes [70,71] and more recently when using primary culture systems to compare a number of species, some of which have very low rates of metabolism [13,63,64,67].

Thus, using hepatocytes in either suspension, primary culture or as slices it is possible to directly measure CL_{int}. However, no direct estimate of hepatic clearance (CL_H) can be made from these *in vitro* systems without adjustments using appropriate scaling factors which are discussed below.

22.5.2 Scaling factors and liver models

To utilize CL_{int} values determined *in vitro* to model data *in vivo*, the ratio $V_{max(app)} / Km_{(app)}$ has to be scaled, usually to give units of ml/min/kg: thus scaling factors form the link between enzyme kinetics and pharmacokinetics. Scaling requires consideration of factors such as numbers of hepatocytes per gram of liver, animal liver weight and body weight. The definition and utility of scaling factors has been discussed and reviewed extensively by Houston and co-workers [68,69]. To model the CL_H, CL_{int} is used with a liver model that encompasses physiological factors such as liver blood flow (Q) and unbound fraction of drug (f_u) in plasma or blood. It is the uncertainty of these scaling factors, particularly the number of cells per g liver that can make precarious the prediction of CL_{int} and CL_H from *in vitro* data.

Several workers have discussed the appropriate methods to determine the number of hepatocytes per g liver. Approaches taken have included relating cell number to DNA and CYP content of the liver [62]. The value used by different workers has been in the order of $110\text{-}150 \times 10^6$ hepatocytes per g rat liver [62,70,72]. The methods employed to determine and rationalize cell number per gram of liver have been discussed in detail for the rat by Houston and Carlile [68] but will inevitably remain a point of discussion for the future, particularly when comparing data between species.

The number of hepatocytes per g of dog and human liver have not been individually defined. However, figures of 240×10^6 and 120×10^6 hepatocytes per g for dog and human [73] liver have been indirectly calculated. These data were confirmed in our own laboratories by relating the microsomal protein content of a gram of liver, the nmoles CYP per mg of microsomal protein and the nmoles CYP per 10^6 hepatocytes for each species.

In order to employ CL_{int} to calculate the hepatic clearance of a drug the use of a physiologically based liver model is required to determine the *in vivo* clearance process. A number of physiological models have been extensively studied; these are the venous equilibrium (well stirred), the undistributed sinusoidal (parallel tube) and the dispersion model [74]. A recent publication [68] has compared hepatic clearance data from rat hepatocytes using the three liver models in an attempt to define the most appropriate model for predicting *in vivo* CL_H. The models exhibit no substantial differences in their predictions with compounds that have low clearance values and therefore low hepatic extraction ratios. Differences become apparent as clearance values increase, however, the observed differences were found to be inconsistent across the models. On the basis of the evidence obtained [68], the venous equilibrium model would appear to be most widely applicable for the description of the hepatic elimination of drugs. Investigation of data in our own laboratories when using limited data across a number of species revealed no obvious differences between the venous equilibrium and the undistributed sinusoidal models.

22.6 Case histories: the intrinsic clearance of ranitidine and loxtidine

Data from our own laboratories investigating the prediction of hepatic clearance of ranitidine and loxtidine across a range of species were encouraging and are presented in Tables 1 and 2. These examples serve to highlight certain issues and practical difficulties.

22.6.1 Ranitidine

The mean $Km_{(app)}$ and $V_{max(app)}$ values for the metabolism of ranitidine by rat, dog and human hepatocytes are detailed in Table 1 and have been determined for disappearance of ranitidine (rat) and appearance of metabolites (rat, dog and man). Using the appropriate scaling, final mean values for the CL_{int} of ranitidine by rat, dog and human hepatocytes calculated from *in vitro* data were 12.1, 3.5 and 0.16 ml/min/kg respectively (Table 1).

Using the "well-stirred" liver model these values of CL_{int} provide mean CL_H values of 9.9, 3.2 and 0.15 ml/min/kg (Table 1). These values rank in the same order as CL_H derived *in vivo*, 23.9, 7.7 and 2.8 ml/min/kg in rat [75], dog [54] and human [53] (Table 1) demonstrating the utility of this approach in semi-quantitative ranking.

The modelled apparent hepatic extraction ratio (ER) (ER = CL_H/Q) of ranitidine derived from *in vitro* data was 0.14, 0.07 and 0.01 in rat, dog and man using blood flows of 69, 44, 25 ml/min/kg respectively (Table 3). These data compare well with the ER's of 0.35, 0.18 and 0.11 determined from *in vivo* studies in rat, dog and man. The ER was low in rat, dog and human hepatocytes, vindicating the use of the "well-stirred" model. However, the predicted ER in all species examined was low compared to the data produced *in vivo*.

Table 3. Comparison of isolated hepatocyte-derived and *in vivo* determined extraction ratios for the hepatic elimination of ranitidine and loxtidine

Species	Ranitidine		Loxtidine	
	Predicted *in vitro* ER[1]	Measured *in vivo* ER[1]	Predicted *in vitro* ER[1]	Measured *in vivo* ER[1]
Rat	0.14[2]	0.35	0.36	0.38
Dog	0.07	0.18	0.07	0.15
Man	0.01	0.11	0.01	0.02

[1] Extraction ratios (ER) were determined using liver blood flows of 69, 44 and 25 ml/min/kg for rat, dog and man respectively [76]. [2] Determined from the disappearance of ranitidine from rat hepatocyte incubations.

hepatocytes were different from those calculated for formation of metabolites. Global kinetic parameters for disappearance of ranitidine were difficult to measure due to the slow rate of metabolism but were preferable to metabolite formation kinetics which were rendered unreliable in calculation of CL_{int} because they required summation of the fractional CL_{int} for each pathway. The measurement of CL_{int}, using metabolite appearance, was shown to be vulnerable as two new metabolites were observed in the rat illustrating that undetected routes of metabolism can dramatically influence calculation of CL_{int}. Additionally, enzyme kinetics were found to be biphasic in rat hepatocytes. Thus, only the high affinity-low capacity data were used to calculate CL_{int} on the basis that these data relate more closely to the expected plasma concentrations of ranitidine in the rat after a modest dose, that is, one likely to be more in keeping with the human therapeutic dose.

The metabolism of ranitidine by dog hepatocytes could be modelled successfully using metabolite appearance because the kinetics appeared monophasic and the majority of the clearance was accounted for by one major metabolite (the N-oxide). In human hepatocytes, ranitidine metabolite formation was very slow, making determination of enzyme kinetics inaccurate. This, coupled with high inter-individual variability between preparations, has made the estimation of CL_{int} vulnerable.

22.6.2 Loxtidine

The mean $Km_{(app)}$ and $V_{max(app)}$ values for the metabolism of loxtidine by rat, dog and human hepatocytes are detailed in Table 2. Using the appropriate scaling factors, final mean values for the CL_{int} of loxtidine by rat, dog and human hepatocytes calculated from *in vitro* data were 51.4 ± 12.4, 8.0 ± 0.8 and 1.0 ± 0.6 ml/min/kg respectively [16] (Table 2).

Using the "well-stirred" liver model these values of CL_{int} provide CL_H values of 24.5, 3.1 and 0.2 ml/min/kg (Table 2). These values compare favourably with the CL_H derived *in vivo*, 26.6, 6.6 and 0.4 ml/min/kg in rat, dog and human [16] (Table 2) demonstrating the ability of the model system to predict quantitatively *in vivo* data.

The predicted hepatic ER of loxtidine derived from *in vitro* data was 0.36, 0.07 and 0.01 in rat, dog and man using blood flows of 69, 44, 25 ml/min/kg respectively (Table 3). These data compare well with the ER's of 0.38, 0.15 and 0.02 determined from *in vivo* studies in rat, dog and man. The ER was low in dog and human hepatocytes, vindicating the use of the "well-stirred" model. However, the predicted apparent hepatic ER in rat (0.36) was close to a value that is considered to be intermediate. Data were also evaluated using the "parallel tube" liver model. The use of the "parallel tube" model resulted in little change in the CL_H values determined for dog and human as expected. A small difference was observed for the CL_H value predicted for the rat when using the "parallel tube" model (29.2 ml/min/kg) compared to the "well-stirred" model (24.5 ml/min/kg), suggesting that, in the case of moderate extraction compounds, the choice of model to determine CL_H may be important.

In the absence of pharmacokinetic data in animals and man *in vivo*, the present studies with isolated hepatocytes would have accurately projected the species differences in the hepatic clearance of ranitidine and loxtidine. These data would have indicated that both ranitidine and loxtidine have increased metabolic stability in man compared with animal species due to a reduced hepatic clearance, which in turn provides information crucial for advancing the development of candidate compounds to pre-clinical and clinical evaluation. Furthermore, these data suggest that relatively little ranitidine or loxtidine would be eliminated by hepatic metabolism in man implying that the half-life of both compounds may be increased when compared with animal species, thus influencing choice of dose and dosing interval in man.

22.7 Experiences with hepatocyte culture

In an attempt to enhance the prospective prediction of the systemic clearance *in vivo* in man and thus optimize the pharamcokinetic protocol for phase I clinical studies, *in vitro* data have been used in combination with *in vivo* animal data. A number of studies by Lave and co-workers [13,14,63,67] have used hepatocyte data produced from culture systems to improve the modelling of *in vivo* human CL_{int}. This has been achieved by correcting human *in vitro* CL_{int} values by a factor describing the inaccuracy associated with forecasting a measured *in vivo* animal CL_{int} value using *in vitro* hepatocyte data from the same animal species. Intrinsic clearance data was produced in hepatocytes from a number of animal species including human by following the disappearance of the parent compound and assessing the area-under-the-time-profile curve. The ratio of the *in vivo* and *in vitro* CL_{int} values determined for the animal species was then used to obtain a corrected human *in vivo* CL_{int} value:

$$\text{human } CL_{int} \textit{ in vivo} = \frac{\textit{in vivo} \text{ animal } CL_{int} \times \textit{in vitro} \text{ human } CL_{int}}{\textit{in vitro} \text{ animal } CL_{int}}$$

The compounds investigated have different clearance values in different animal species, and this was reflected in hepatocytes isolated from the corresponding species. In all cases the animal hepatocyte modelled the CL_H of the animal *in vivo* and the combination of *in vitro* and *in vivo* animal data successfully enhanced the overall prediction of human CL_H for compounds cleared by metabolism in man.

22.8 Summary

This type of kinetic information produced from *in vitro* systems early in the drug development process could help to eliminate the possible pharmacokinetic shortfalls that are encountered when some new chemical entities are first administered to man.

The intrinsic clearance of ranitidine and loxtidine has been shown to be species dependent in both the hepatocyte model and *in vivo*, being greatest in the rat and somewhat less in both dog and man. The model-dependent hepatic clearance values and the subsequent hepatic extraction ratios derived from *in vitro* data successfully approximate the *in vivo* parameters derived from pharmacokinetic data. However, the use of *in vitro* data to calculate intrinsic and subsequently hepatic clearance in various species is vulnerable, largely because of the use of ill-defined scaling factors. Owing to the lack of experimental data from a range of species, a comprehensive approach has yet to be established and the correct scaling factors await clarification. However, it should be remembered that correlating rank orders of clearance values rather than determining specific values should be the primary aim of these types of studies.

22.9 Concluding remarks

Whilst isolated hepatocytes cannot replace the whole animal in metabolism studies, they provide a useful tool for the investigation of drug metabolism at a cellular level and additionally provide a vehicle for qualitative, semi-quantitative ranking and quantitative *in vivo/in vitro* predictions and extrapolations between species. The validation of the hepatocyte model for compounds of widely different structure and behaviour by independent laboratories around the world helps impart confidence in the utility of the model as a predictor of animal metabolism *in vivo*. At the same time, accurate *in vitro* predictions of *in vivo* metabolism in a variety of non-human animal species lends weight to the less readily available model of human isolated hepatocytes as a predictor of human metabolism *in vivo*. Early studies with *in vitro* systems have attempted with varying success to reproduce endpoints from pre-existing *in vivo* work, so as to judge their success against an established baseline. Nevertheless, it remains unclear whether first generation *in vitro* investigations alone are sufficiently well characterized and accurately predictive to exclude the need for a subsequent *in vivo* study to confirm the *in vitro* findings. This point is critical, for until *in vitro* models can achieve this exacting standard, they cannot achieve the many savings in animal usage, time and expense that they purport to offer.

In all the discussions above, it should be noted that qualitative similarities in metabolite profiles studied *in vivo* and *in vitro* depend mostly on correlating the detection of major metabolites. However, the differences between *in vivo* and *in vitro* behaviour depend often on analytical sensitivity for minor metabolites, method differences between laboratories and the availability of materials, such as radiolabelled analogues and non-commercially available discovery compounds and reference standards. This can result in the apparent lack of absolute correlation when the *in vitro* and *in vivo* phases are carried out in different laboratories, or validation of recent *in vitro* work is attempted by reference to older published *in vivo* studies that utilized less sensitive methods of analysis. Clearly the evolution and divergence of isolated hepatocyte techniques complicates the task of reviewing and interpreting published data. In fact, the wide diversification of hepatocyte-based methods, each of which has remained largely an art with each new worker in the field perfecting a new variant technique, has expanded the range of application beyond the bounds of reasonable review. Nevertheless, taken together this vast body of information, which has been painstakingly uncovered using isolated hepatocytes over the past 30 years, must serve to provide the scientific community with a clear and unmistakable indication of the immense value of these kind of studies. Thus, whilst the shortcomings of the techniques cannot be ignored, present activity suggest the subject is in good health!

References

1. Parkinson A. Biotransformation of Xenobiotics. In: Klassen CD, Ed. Casarette and Doulls Toxicology, 5th ed. New York: McGraw-Hill, 1996: 113-186.
2. Denk H. Human surgical and postmortem material. Ethically justifiable? In: Lembeck F, Ed. Scientific alternatives to animal experiments. Chichester: Ellis Horwood, 1989: 75-79.
3. Bayliss MK and Skett P. Isolation and culture of human hepatocytes. In: Jones GE, Ed. Human cell culture protocols. Totowa, New Jersey: Humana Press, 1996: 369-389.
4. Berry MN, Edwards AM and Barritt GJ. Isolated hepatocyte preparation; properties and applications. In Burdon RH and van Knippenberg PH, Eds. Laboratory Techniques in Biochemistry and Molecular Biology. Amsterdam: Elsevier, 1991: 1-460.
5. Lavrijsen K, van Houdt J, van Dyck D, Hendrickx J, Bockx M, Hurkmans W, Meuldermans W, Le Jeune L, Lauwers W and Heykants J. Comparative metabolism of flunarizine in rats, dogs and man: an *in vitro* study with subcellular liver fractions and isolated hepatocytes. Xenobiotica 1992; 22: 815-836.
6. Sandker GW, Vos RME, Delbressine PC, Slooff MJH, Meijer DKF and Groothuis GMM. Metabolism of three pharmacologically active drugs in isolated human and rat hepatocytes: analysis of interspecies variability and comparison with metabolism *in vivo*. Xenobiotica 1994; 24: 143-155.
7. Rahmani R, de Sousa G, Marre F, Nicolas F and Placidi M. Potential of freshly isolated and cryopreserved human hepatocytes in drug research and development. In Rogiers V, Sonck W, Shephard E and Vercruysse A, Eds. Human cells in *in vitro* pharmaco-toxicology. Brussels: Vub Press, 1993: 117-138.
8. Pahernik SA, Schmid J, Sauter T, Schildberg FW and Koebe H-G. Metabolism of pimobendan in long-term human hepatocyte culture: *in vivo-in vitro* comparison. Xenobiotica 1995; 25: 811-823.
9. de Souza G, Florence N, Valles B, Coassolo P and Rahmani R. Relationship between *in vitro* and *in vivo* biotransformation of drugs in humans and animals: pharmaco-toxicological consequences. Cell Biol Toxicol 1995; 11: 147-153.
10. Maurel P. The use of adult human hepatocytes in primary culture and other *in vitro* systems to investigate drug metabolism in man. Advanced Drug Delivery Reviews 1996; 22: 105-132.
11. Humpel M, Sostarek D, Gieschen H and Labitzky C. Studies on the biotransformation of ionazolac, bromerguride, lisuride and terguride in laboratory animals and their hepatocytes. Xenobiotica 1989; 19: 361-377.
12. Chenery RJ, Ayrton A, Oldham HG, Standring P, Norman SJ, Seddon T and Kirby R. Diazepam metabolism in cultured hepatocytes from rat, rabbit, dog, guinea pig and man. Drug Metab Dispos 1987; 15: 312-317.

13. Lave T, Dupin S, Schmitt M, Kapps M, Meyer J, Morgenroth B, Chou RC, Jaeck D and Coassolo P. Interspecies scaling of tolcapone, a new inhibitor of catechol-*O*-methyltransferase. Use of *in vitro* data from hepatocytes to predict metabolic clearance in animals and humans. Xenobiotica 1996; 26: 839-851.
14. Lave T, Coassolo P, Ubeaud G, Brandt BBR, Schmitt C, Dupin S, Jaeck D and Chou RC. Interspecies scaling of bosentan, a new endothelin receptor antagonist and integration of *in vitro* data into allometric scaling. Pharmaceutical Res 1996; 13: 97-101.
15. Cross DM. A consideration of the enzymology underlying ranitidine metabolism and drug interations. J of Biopharmaceut Sci 1991; 2: 339-377.
16. Bayliss MK, Bell JA, Wilson K and Park GR. The utility of hepatocytes to model species differences in metabolism and predict pharmacokinetic parameters in rat, dog and human hepatocyte suspensions. Xenobiotica 1999; 29:253-268.
17. Ammann P and Maier P. Preservation and inducibility of xenobiotic metabolism in long-term cultures of adult rat liver cell aggregates. Toxicology *in Vitro* 1997; 11: 43-56.
18. Diaz D, Fabre I, Daujat M, Saint-Aubert B, Borries P, Michel H and Maurel P. Omeprazole is an aryl hydrocarbon like inducer of human hepatic cytochrome P450. Gastroenterology 1990; 99: 737-747.
19. Kern A, Bader A, Pichmayr R and Sewing KF. Drug metabolism in hepatocyte sandwich cultures of rats and humans. Biochem Pharmacol 1997; 54: 761-772.
20. Ferrini J-B, Pichard L, Domergue J and Maurel P. Long-term primary cultures of adult human hepatocytes. Chemico-Biological Interactions 1997; 107: 31-45.
21. Pichard L, Fabre I, Fabre G, Domergue J, Aubert BS, Mourad G and Maurel P. Cyclosporin A drug interactions. Screening for inducers and inhibitors of cytochrome P450 (Cyclosporin A Oxidase) in primary cultures of human hepatocytes and in liver microsomes. Drug Metab Dispos 1990; 18: 595-606.
22. Skett P. Problems in using isolated and cultured hepatocytes for xenobiotic metabolism/metabolism-based toxicity testing – solutions? Toxicology *In Vitro* 1994; 8: 491-504.
23. Guillouzo A, Morel F, Ratanasavanh D, Chesne C and Guguen-Guillouzo C. Long-term cultures of functional hepatocytes. Toxicology *In Vitro* 1990; 4: 415-427.
24. Koebe H-G, Pahernik S, Eyer P and Schildberg F-W. Collagen gel immobilisation: a useful cell culture technique for long-term metabolic studies on human hepatocytes. Xenobiotica 1994; 24: 95-107.
25. Liu L, LeCluyse EL, Liu J and Klaassen CD. Sulfotransferase gene expression in primary cultures of rat hepatocytes. Biochem Pharmacol 1996; 52: 1621-1630.
26. Guillouzo A, Begue J-M, Campion J-P, Gascoin M-N and Guguen-Guillouzo C. Human hepatocyte cultures: a model of pharmaco-toxicological studies. Xenobiotica 1985; 15: 635-641.
27. Guillouzo A, Begue J-M, Maurer G and Koch P. Identification of metabolic pathways of pindolol and fluperlapine in adult human hepatocyte cultures. Xenobiotica 1988; 18: 131-139.
28. Donato MT, Castell, JV and Gomez-Lechon MJ. Co-cultures of hepatocytes with epithelial-like cell lines: expression of drug-biotransformation activities by hepatocytes. Cell Biol Toxicol 1991; 7: 1-14.
29. Sidhu JS, Farin FM and Omiecinski CJ. Influence of extracellular matrix overlay on phenobarbital-mediated induction of CYP2B1, 2B2 and 3A1 genes in primary adult rat hepatocyte culture. Arch Biochem Biophys 1993; 301: 103-113.

30. LeCluyse EL, Audus KL and Hochman JH. Formation of extensive canalicular networks by rat hepatocytes cultured in collagen-sandwich configuration. Amer J Physiol 1994; 266: C1764-C1774.
31. Oesch F and Diener B. Cell systems for use in studies on the relationship between foreign compound metabolism and toxicity. Pharmacol Toxicol 1995; 76: 325-327.
32. Rogiers V. Cultures of human hepatocytes in *in vitro* pharmaco-toxicology. In: Rogiers V, Sonck W, Shephard E and Vercruysse A, Eds. Human cells in *in vitro* pharmaco-toxicology. Brussels: Vub Press, 1993: 77-115.
33. Guillouzo A, Morel F, Fardel O and Meunier B. Use of human hepatocyte cultures for drug metabolism studies. Toxicology 1993; 82: 209-219.
34. Rogiers V and Vercruyees A. Rat hepatocyte cultures and co-cultures in biotransformation studies of xenobiotics. Toxicology 1993; 82: 193-208.
35. Kono Y, Yang S and Roberts E. Extended primary culture of human hepatocytes in a collagen gel sandwich system. *In Vitro* Cell Devel Biol-Animal 1997; 33: 467-472.
36. Guyomard C, Rialland L, Fremond B, Chesne C and Guillouzo A. Influence of alginate gel entrapment and cryopreservation on survival and xenobiotic metabolism capacity of rat hepatocytes. Toxicol Appl Pharmacol 1996; 141: 349-356.
37. Selden C, Roberts E, Stamp G, Parker K, Winlove P, Ryder T, Platt H and Hodgson H. Comparison of three solid phase supports for promoting three-dimensional growth and function of human liver cell lines. Artificial Organs 1998; 22: 308-319.
38. Roberts RA and Soames AR. Hepatocyte spheroids: Prolonged hepatocyte viability for *in vitro* modeling of nongenotoxic carcinogenesis. Fundamental and Applied Toxicology 1993; 21: 149-158.
39. Tong JZ, DeLagausie P, Furlan V, Cresteil T, Bernard O and Alvarez F. Long-term culture of adult rat hepatocyte spheroids. Exp Cell Res 1992; 200: 326-332.
40. Abdel-Razzak Z, Loyer P, Fautrel A, Gautier JC, Corcos L, Turlin B, Beaune P and Guillouzo A. Cytokines down-regulate expression of major cytochrome P450 enzymes in adult human hepatocytes in primary culture. Mol Pharmacol 1993; 44: 770-775.
41. Bader A, Zechs K, Crome O, Christians U, Ringe B, Pichlmayr R and Sewing K-FR. Use of organotypical cultures of primary hepatocytes to analyse drug biotransformation in man and animals. Xenobiotica 1994; 24: 623-633.
42. Rotem A, Matthew HWT, Hsiao PH, Toner M, Tompkins RG and Yarmush ML. The activity of cytochrome P450IA1 in stable cultured rat hepatocytes. Toxicology *in Vitro* 1995; 9: 139-149.
43. Bader A, Knop E, Boker KHW, Crome O, Fruhauf N, Gonschior AK, Christians U, Esselmann H, Pichlmayr R and Sewing K-F. Tacrolimus (FK 506) biotransformation in primary rat hepatocytes depends on extracellular matrix geometry. Naunyn-Schmiedberg's Arch Pharmacol 1996; 353: 461-473.
44. Surapaneni S, Pryor T, Klein D and Matthew HWT. Rapid hepatocyte spheroid formation: optimization and long term function in perfused microcapsules. ASAIO Journal 1997; 43: 848-853.
45. Takabatake H, Koide N and Tsuji T. Encapsulated multicellular spheroids of rat hepatocytes produce albumin and urea in a spouted bed circulating culture system. Artificial Organs 1991; 15: 474-480.
46. Bader A, Knop E, Kern A, Böker K, Frühauf N, Crome O, Esselmann H, Pape C, Kempka G and Sewing K-FR. 3-D coculture of hepatic sinusoidal cells with primary hepatocytes – design of an organotypical model. Exp Cell Res 1996; 226: 223-233.

47. Jauregui HO, Naik S, Santangini H, Pan J, Trenkler D and Mullon C. Primary cultures of rat hepatocytes in hollow fiber chambers. *In Vitro* Cell Dev Biol 1994; 30A: 23-29.
48. Powis G, Santone KS, Melder DC, Thomas L, Moore DJ and Wilke TJ. Cryopreservation of rat and dog hepatocytes for studies of xenobiotic metabolism and activation. Drug Metab Dispos 1987; 15: 826-832.
49. Loretz LJ, Li AP, Flye MW and Wilson AGE. Optimisation of cryopreservation procedures for rat and human hepatocytes. Xenobiotica 1989; 19: 489-498.
50. Coundouris JA, Grant MH, Engeset J, Petrie JC and Hawksworth GM. Cryopreservation of human adult hepatocytes for use in drug metabolism and toxicity studies. Xenobiotica 1993; 23: 1399-1409.
51. Pang JM, Zaleski J and Kauffman FC. Toxicity of allyl alcohol in primary cultures of freshly isolated and cryopreserved hepatocytes maintained on hydrated collagen gels. Toxicol Appl Pharmacol 1997; 142: 87-94.
52. Bell JA, Dallas FAA, Jenner WN and Martin LE. The metabolism of ranitidine in animals and man. Biochemical Society Transactions 1980; 8: 93.
53. Mitchard M, McIsaac RL and Bell JA. H_2-receptor antagonists. In: Damani LA, Ed. Sulphur-containing drugs and related organic compounds. Chichester: Ellis Horwood Limited, 1989: (3 part A): 53-86.
54. Eddershaw PJ, Chadwick AP, Higton DM, Fenwick SH, Linacre P, Jenner WN, Bell JA and Manchee GR. Absorption and disposition of ranitidine hydrochloride in rat and dog. Xenobiotica 1996; 26: 947-956.
55. Cross DM, Bell JA and Wilson K. Kinetics of ranitidine metabolism in dog and rat isolated hepatocytes. Xenobiotica 1995; 25: 367-375.
56. Brittain RT and Jack D. Histamine H_2-antagonists – past, present and future. J Clin Gastroent 1983; 5: 71-79.
57. Bell JA, Bradbury A, Jenner WN, Manchee GR and Martin LE. The metabolism of loxtidine in rat and dog . Biochem Soc Trans 1983; 11: 715-716.
58. Singh K, Tripp SL, Dunton AW, Douglas FL and Rakhit A. Determination of *in vivo* hepatic extraction ratio from *in vitro* metabolism by rat hepatocytes. Drug Metab Dispos 1991; 19: 990-995.
59. Ashforth EL, Carlile DJ, Chenery RJ and Houston JB. Prediction of *in vivo* disposition from *in vitro* systems: clearance of phenytoin and tolbutamide using rat hepatic microsomal and hepatocyte data. J Pharmacol Exp Therap 1995; 274: 761-766.
60. Hayes KA, Brennan B, Chenery RJ and Houston JB. *In vivo* disposition of caffeine predicted from hepatic microsomal and hepatocyte data. Drug Metab Dispos 1995; 23: 349-353.
61. Zomorodi K, Carlile DJ and Houston JB. Kinetics of diazepam metabolism in rat hepatic microsomes and hepatocytes and their use in predicting *in vivo* hepatic clearance. Xenobiotica 1995; 25: 907-916.
62. Carlile DJ, Zomorodi K and Houston JB. Scaling factors to relate drug metabolic clearance in hepatic microsomes, isolated hepatocytes and intact liver. Drug Metab Dispos 1997; 25: 903-911.
63. Lave T, Schmitt-Hoffman AH, Coassolo P, Valles B, Ubeaud G, Brandt BBR and Chou RC. A new extrapolation method from animals to man: application to a metabolised compound, mofarotene. Life Sciences 1995; 56: 473-478.
64. Lave T, Dupin S, Schmitt C, Valles B, Ubeaud G, Chou R, Jaeck D and Coassolo P. The use of hepatocytes to select compounds based on their expected extraction ratios in humans. Pharmaceutical Res 1997; 14: 152-155.

65. Iwatsubo T, Hirota N, Ooie T, Suzuki H, Shimada N, Chiba K, Ishizaki T, Green CE, Tyson CA and Sugiyama Y. Prediction of *in vivo* metabolism in the human liver from *in vitro* metabolism data. Pharmacol Therap 1997; 73: 147-171.
66. Boxenbaum H and di Lea C. First-time in human dose selection: allometric thoughts and perspectives. J Clin Pharmacol 1995; 35: 957-966.
67. Ubeaud G, Schmitt C, Jaeck D, Lave T and Coassolo Ph. Bosentan, a new endothelin receptor antagonist: prediction of the systemic plasma clearance in man from combined *in vivo* and *in vitro* data. Xenobiotica 1995; 25: 1381-1390.
68. Houston JB and Carlile DJ. Incorporation of *in vitro* drug metabolism data into physiologically based pharmacokinetic models. Toxicology *in Vitro* 1997; 11: 473-478.
69. Houston JB. Utility of *in vitro* drug metabolism data in predicting *in vivo* metabolic clearance. Biochem Pharmacol 1994; 47: 1469-1479.
70. Pang KS, Kong P, Terrell JA and Billings RE. Metabolism of acetaminophen and phenacetin by isolated rat hepatocytes. Drug Metab Dispos 1985; 13: 42-50.
71. Bodd E, Drevon CA, Kveseth N, Olsen H and Morland J. Ethanol inhibition of codeine and morphine metabolism in isolated rat hepatocytes. J Exp Therap 1986; 237: 260-264.
72. Mizuma T, Machida M, Hayashi M and Awazu S. Correlation of drug conjugative metabolism rates between *in vivo* and *in vitro*: glucuronidation and sulphation of *p*-nitrophenol as a model compound. J Pharmacobiodynamics 1982; 5: 811-817.
73. Arias IM. Introduction. In: Arias IM, Jakoby WB, Popper H, Scahacter D and Shafritz DA. The liver: biology and pathobiology. New York: Raven Press, 1988: 3-10.
74. Morgan DJ and Smallwood RA. Clinical significance of pharmacokinetics: models of hepatic elimination. Clin Pharmacokinetics 1990; 18: 61-76.
75. Varughese A and Lee C-SC. Disposition kinetics of ranitidine in rats. Drugs Experimental Clinical Research 1983; IX: 837-842.
76. Boxenbaum H. Interspecies variation in liver weight, hepatic blood flow and antipyrene intrinsic clearance: extrapolation of data to benzodiazepines and phenytoin. J Pharmacokinet Biopharm 1980; 8: 165-176.

Chapter 23

Human hepatocytes as an experimental system for the evaluation of xenobiotics

Albert P. Li
In Vitro Technologies, Inc. 1450 S. Rolling Rd, Baltimore, MD 21045

Keywords: clinical trials, cryopreservation, cytochrome P450, drug-drug interactions, drug screening, drug metabolism, enzyme induction, evaluation of human hepatotoxicity, hepatocyte culture, hepatocyte isolation, human hepatocytes, sandwich culture, xenobiotics

23.1 Xenobiotics and the human liver

The liver is the key organ for biotransformation of xenobiotics that enter the human body, either intentionally (*e.g.* pharmaceuticals), or unintentionally (*e.g.* environmental pollutants). All blood-borne xenobiotics are first metabolized by the liver, a process known as first-pass metabolism. The major consequence of the metabolic transformation is the formation of water-soluble metabolites that are removed from the body. In general, the xenobiotic is firstly oxidized (phase I metabolism) and then conjugated to highly polar molecules such as glucose, sulphate, cysteine or glutathione (phase II metabolism). The highly polar metabolites are then transported directly from the hepatocytes into the biliary canaliculi (phase III metabolism) to be excreted in the bile, or are released back into the blood stream to be excreted into the urine via the kidneys. In the intestine, the metabolites may be deconjugated by gut bacterial flora and reabsorbed, leading to a repeat of the metabolic processes (enterohepatic recirculation). The sequential oxidation-conjugation of xenobiotics is commonly known as metabolic detoxification, a process responsible for ridding the body of xenobiotic toxicants. However, it is now also known that many innocuous chemicals can be metabolically "activated" into highly reactive metabolites with toxicological consequences. As the organ where this metabolic "activation" occurs, the liver is therefore often the target organ of chemical toxicants (see Chapter 27).

M.N. Berry and A.M. Edwards (eds.), The Hepatocyte Review, 391–410.

23.2 Isolation and culturing of human hepatocytes

23.2.1 Introduction

The major enzymes for the biotransformation pathways, phase I oxidation and phase II conjugation, are present in the liver in significant amounts. It is now known that human drug metabolising enzymes, especially those related to cytochrome P450 (CYP), can be substantially different from those for nonhuman animals. This species difference in xenobiotic metabolism is one of the major reasons why results from laboratory animals may not be directly applicable to man. Human hepatocytes, which contain the drug-metabolising enzymes in the human liver, represent an experimental system with which human-specific drug metabolism can be studied. The key CYP isoforms in human liver include CYP1A2, CYP2A6, CYP2C9, CYP2C19, CYP2D6, CYP2E1 and CYP3A4. Of these isoforms, CYP3A4 is the isoform present in the highest quantity and is considered to catalyze the transformation of a large number of xenobiotics [1]. It is generally believed that over 50% of current human pharmaceuticals are substrates of CYP3A4.

23.2.2 Hepatocyte isolation

Procedures for the isolation of human hepatocytes are well established and are not significantly different from the procedures used to isolate hepatocytes from laboratory animals. The major difficulty that most researchers encounter is the availability of freshly procured human livers. In our laboratory, we employ a two-step collagenase digestion procedure that has been previously reported in detail [2]. The entire lobe of a three-side encapsulated liver segment is first perfused with an EGTA-containing isotonic solution at physiological pH (pH 7.2-7.5) to clear the blood. The EGTA serves to bind divalent ions, thereby preventing clotting. Also, EGTA loosens cell junctions to facilitate the subsequent enzyme digestion step. After the first perfusion, the relatively blood-free liver segment is perfused with a collagenase solution for cell dissociation. After dissociation, the digested liver is dissected and agitated or combed to release the dissociated cells. The fibrous connective tissue is separated from the dissociated cells by filtering through either a nylon screen or gauze. The hepatocytes can be harvested by low speed centrifugation ($50 \times g$). Most of the non-parenchymal cells remain in the supernatant. If necessary, the intact non-staining hepatocytes can be further purified from damaged hepatocytes and the non-parenchymal cells by density gradient sedimentation over Percoll™.

The critical factors for successful isolation of human hepatocytes are the condition of the human liver, the collagenase employed and the length of digestion. In general, the longer the length of cold storage and the higher the fat content, the lower the yield and viability. I prefer to use livers that have been stored for less than 24 h and have a fat content of less than 50%. It is important to select a batch of collagenase that is not cytotoxic to human hepatocytes. However, this is generally the case for commercially available collagenase. The timing of the digestion is critical: both under- and over-digestion will compromise the yield and viability of hepatocytes. In general, the viability of the isolated hepatocytes is over 80%. Hence the yield of hepatocytes is approximately 10 million cells per gram of human liver. Hence a large segment of human liver weighing 500 g would yield approximately 5 billion viable hepatocytes. In our laboratory, we are able to routinely procure relatively large liver segments for research. In general, our yield ranges from 5 to 20 billion hepatocytes per isolation. This high yield of human hepatocytes is one of the major motivating forces for the optimisation of the cryopreservation procedures (see below). The yield and viability of human hepatocytes isolated from a number of different donors is shown in Table 1.

Table 1. Viability and yield of hepatocytes from human livers. All the donors are Caucasian

Age (years)	Sex	Yield ($\times 10^9$)	Viability % (Dye exclusion)	Postclamp duration (h)
47	M	24	98	14
74	F	8	90	9
43	M	26	94	12
72	F	26	93	9
57	M	7	95	10
68	F	10	94	14
51	F	11	90	19
44	M	11	92	20
84	F	30	93	17
47	F	25	93	12

Postclamp time represents time from when the liver was removed from the donor.

23.2.3 Hepatocyte culture

After isolation, human hepatocytes can be used as a suspension culture for short-duration studies (hours) or as attached cultures for longer-term studies (days or weeks). In suspension, hepatocytes tend to rapidly lose viability; usually 50% after approximately 12 h and over 90% after 24 h at 37°C. As attached cultures, hepatocytes can remain viable for weeks. Most investigators culture hepatocytes on collagen-coated surfaces. While basement membrane extract (*e.g.* Matrigel) can be used to prolong the maintenance of differentiated properties for rat hepatocytes, it appears to have fewer effects on human hepatocytes than on rodent hepatocytes. When cultured on collagen-coated substratum, hepatocytes exhibit an epithelial morphology with prominent nuclei. The cells are often binucleated and, when cultured as confluent cultures or couplets, form bile canaliculi. On Matrigel the cells remain rounded. Overlaying hepatocytes, attached on collagen, with either collagen (collagen-gel sandwich) or Matrigel is believed to further maintain the functions of hepatocytes. While previously we performed our studies using hepatocytes cultured on collagen [9-11], we now routinely use the collagen-gel overlaying procedure in our laboratory.

23.3 Cryopreservation of human hepatocytes

23.3.1 Advantages of cryopreserved cells

For obvious reasons, cryopreservation can significantly enhance the convenience and ease of experimentation with human hepatocytes. During the past decade, it has generally been concluded that cryopreservation of hepatocytes is possible, but not yet highly reproducible. Formerly, it was generally believed that rodent hepatocytes were quite easily cryopreserved while human hepatocytes were more recalcitrant. Now we believe that it is just a matter of the quality of the starting material. Human hepatocytes isolated from fresh livers (*i.e.* less than 24 h after removal from the donor) can be cryopreserved equally as well as hepatocytes from other animals. DMSO is the most commonly used cryopreservant [3,4]. Frequently after cryopreservation, viability based on Trypan blue exclusion remains high and the cells are metabolically active as suspension cultures [5-7]. However, in general only a small percentage (less than 50%) of the cells attach. We observed that in some cases the cells would initially attach but would detach after overnight culturing. In the cases that the majority of the cryopreserved hepatocytes remain viable and able to attach after thawing, normal hepatocyte functions apparently are retained, as measured by metabolic activation of promutagens [3] and response to CYP inducers [6].

In our laboratory, we are continuing to examine approaches for improving the attachment efficiency of cryopreserved human hepatocytes.

For cryopreserved human hepatocytes to be used routinely, the cells should have adequate viability and metabolic activities. Data from our laboratory (Table 2) show that cryopreserved human hepatocytes after thawing consistently exhibit acceptable viability, albeit slightly lower than the values of freshly isolated hepatocytes [7]. We also observed that neither viability nor yield of viable cells decreases with cryopreservation duration for at least 12 months [7]. Storage of cryopreserved hepatocytes is only possible in the liquid or vapor phase of liquid nitrogen, at a temperature of <–150°C. We have observed that storage of cryopreserved hepatocytes at higher temperatures (*e.g.* at –70°C) leads to a rapid decrease in viability.

23.3.2 Concerns about use of cryopreserved cells

The major concerns about the use of cryopreserved hepatocytes in drug metabolism studies are that specific drug-metabolising enzymes may be compromised by cryopreservation, and that cryopreservation may lead to the selection of a specific subpopulation of hepatocytes which may have properties significantly different from those of the freshly isolated hepatocytes. Our results show that there is apparently no basis for these concerns [7]. We have compared results with hepatocytes isolated and cryopreserved from multiple donors and found that in all cases, the cells were shown to be competent in drug metabolism. Activities of the major CYP isoforms (CYP1A2, CYP2A6, CYP2C9, CYP2C19, CYP2D6, and CYP3A4) as well as the phase II enzymes UDP-glucuronosyl transferase and glutathione sulphotransferase are present in cryopreserved human hepatocytes (Table 2). No major differences in activities were found between freshly isolated and cryopreserved hepatocytes (Table 2). Our results indicate that the major drug-metabolizing pathways were not significantly altered as a result of cryopreservation, thereby supporting the validity of cryopreserved human hepatocytes as an experimental system for drug metabolism studies.

Table 2. Viablity and drug-metabolizing enzyme activities of freshly isolated (Fresh) and cryopreserved (Cryo.) human hepatocytes isolated from 10 donors.

Lot No.	Status	Viability (%)	CYP1A2	CYP2A6	CYP2C9	CYP2C19	CYP2D6	CYP2E1	CYP3A4	7HC-G	7HC-S	ECOD
72	Fresh	91	0.07	202	13	30.6	68	18	40	522	17	66
	Cryo.	65	0.06	187	13	39.5	72	39	112	495	10	53
73	Fresh	97	0.31	119	30	30.5	54	17	59	474	14	43
	Cryo.	73	0.43	95	19	21.5	42	35	93	277	16	32
74	Fresh	93	0.03	127	16	0.8	53	25	90	307	9	30
	Cryo.	66	0.05	140	16	0.5	57	29	70	175	8	21
75	Fresh	96	0.01	78	21	6.3	29	114	47	316	56	56
	Cryo.	85	0.02	55	25	6.0	33	150	44	235	28	40
76	Fresh	86	1.31	25	23	16.0	42	152	32	513	33	70
	Cryo.	71	1.49	24	17	20.0	46	300	45	380	27	67
77	Fresh	96	0.20	62	23	2.4	42	8	101	517	53	33
	Cryo.	56	0.22	74	20	2.2	36	11	90	388	25	24
78	Fresh	97	0.31	119	30	31.0	54	17	59	474	14	43
	Cryo.	71	0.50	99	19	24.0	48	31	100	314	8	31
79	Fresh	96	0.03	185	47	9.1	49	122	-	315	109	85
	Cryo.	76	0.04	175	69	9.4	51	156	-	265	140	79
80	Fresh	87	0.03	28	15	-	12	233	25	171	44	64
	Cryo.	71	0.01	56	8	-	16	502	24	137	41	70
Mean	Fresh	93	0.26	105	24	15.8	45	78	57	401	39	54
Mean	Cryo.	70	0.31	101	23	15.4	45	139	72	296	34	46
(Cryo/Fr)	%	76	123	96	94	97	100	178	128	74	87	85

Results of 10 sequential isolations are shown. Viability was determined based on trypan blue exclusion. Enzyme activities are: CYP1A2, ethoxyresorufin O-deethylation; 2A6, coumarin 7-hydroxylation; 2C9, tolbutamide 4-hydroxylation; 2C19, mephenytoin 4-hydroxylation; 2D6, dextromethorphan O-demethylation; 3A4, testosterone 6β-hydroxylation; UDPGT, umbelliferone glucuronidation; ST, umbelliferone sulfation; ECOD, 7-ethoxycoumarin O-deethylation. Activity units are pmol/min/10^6 viable hepatocytes.

23.4 Applications of primary human hepatocytes in drug metabolism

23.4.1 Advantages and limitations of primary hepatocytes in drug metabolism

Hepatocytes are of value for use in the study of drug metabolism mainly because of their intact cell properties (Table 3).

Table 3. A comparison of the limitations of human *in vitro* drug metabolism systems

System	Limitations
Hepatic microsomes	1. Disrupted system, lack ICP 2. Nonphysiological enzyme concentration and cofactors 3. Mainly phase I oxidation activity, limited phase II and no phase III
Hepatocytes	1. Experimentation dependent on availability of fresh human livers 2. Cannot use reagents (e.g. inhibiting antibodies) that cannot cross plasma membrane 3. Cannot study substrate at cytotoxic concentrations
Liver slices	1. Limited drug availability to the inner cell layers 2. Limited viability (< 24 h) 3. Generally considered not suitable for enzyme induction studies

Physiological enzyme, cofactor, and substrate concentrations: Freshly isolated hepatocytes contain complete, undisrupted enzymes and cofactors at the same level as the liver *in vivo*, therefore allowing function similar to that observed during hepatic metabolism *in vivo*.

Relevant drug concentrations: Hepatocytes in culture retain active uptake as well as biliary excretion similar to hepatocytes in the liver *in vivo* [8]. Cultured hepatocytes can maintain similar gradients to those observed *in vivo* for drugs that are either bioaccumulated and therefore have a higher intrahepatocyte concentration than that of plasma, or actively excreted and thus possess a lower intrahepatocyte concentration than that of plasma. For such drugs, measurements critical for making *in vitro*: *in vivo* extrapolation of drug-drug interaction, *e.g.* Ki values, are more relevant if derived from hepatocytes than those obtained with microsomes.

Inducible CYP: CYP inhibition and induction are the two major mechanisms of drug-drug interactions. While CYP inhibition can be studied with microsomes and hepatocytes, CYP induction can only be studied in a living cell, *i.e.* hepatocytes. Protocols for CYP induction, especially for CYP1A and 3A, are well established [9-11]. Known human *in vivo* inducers such as rifampin, phenobarbital, dexamethasone and phenytoin are all potent CYP inducers in primary human hepatocytes as reviewed in a recent consensus document [12]. The limitations of the use of human hepatocytes are in general related to the availability of freshly isolated hepatocytes and can be overcome by the use of cryopreserved cells (see above).

Limited availability: The major limitation of primary human hepatocytes is that viable hepatocytes can be obtained only from relatively fresh human livers, usually within 24 h after liver isolation. To establish a realistic human hepatocyte laboratory, one needs to have an efficient liver procurement system. For some applications (*e.g.* short-term drug metabolism and toxicity studies), this limitation can now be overcome by the used of cryopreserved hepatocytes. Human hepatocytes can be prepared in a laboratory which has access to human livers, cryopreserved, and then transported to other laboratories for experimentation. However, freshly isolated hepatocytes are still required for enzyme induction studies.

Limited storage options: Freshly isolated hepatocytes contain metabolic activities similar to those of intact liver. After culturing, the activities of the inducible isozymes, such as CYP1A and CYP3A, will decrease with time in culture to a "basal" level due to the lack of endogenous CYP inducers (see below). Because of this decrease, freshly isolated hepatocytes or cryopreserved hepatocytes are more appropriate for the evaluation of metabolite profiles and metabolic rate whereas cultured hepatocytes are useful for induction studies.

Changes in drug-metabolising enzymes in cultured cells: The levels of the multiple CYP isozymes change with the duration of culture [13]. Usually, a decline of activities is observed. This is in general considered as "dedifferentiation". However, CYP isozymes have evolved to be inducible, so that an animal is able to respond to the changes of the levels of endogenous substrates and environmental toxin levels. This response is prevented following removal of cells from the milieu *in vivo*. It is likely therefore that in the absence of endogenous "inducers" and substrates in the cell culture medium, CYP levels merely revert to the basal, "uninduced", level.

Because of the intactness of isolated hepatocytes, their use should, at least in theory, provide data more representative of the *in vivo* data than cell-free systems such as microsomes. This aspect was recently reviewed by Houston [14]. After reviewing literature values of *in vivo* metabolic clearance data and intrinsic clearance obtained using different *in vitro* hepatic systems, he concluded that the best correlation with *in vivo* data was made using hepatocytes, followed by liver slices, and then by liver microsomes. The conclusion is therefore consistent with the theoretical consideration as stated above. That hepatocytes performed better than liver slices is probably due to the artefactual presence of barriers arising from the multiple cell layers in the liver slices. The importance of the use of intact cells is illustrated by the fact that both hepatocytes and liver slices perform better than liver microsomes in the prediction of intrinsic clearance. A comparison of the three commonly used *in vitro* liver systems, namely microsomes, hepatocytes and liver slices, is presented in Table 3.

23.5 Species and sex differences in drug metabolism in hepatocytes

The validity of isolated hepatocytes as a metabolic model is further substantiated by studies on specific chemicals with known *in vivo* species and sex differences in their metabolism. In general, results with hepatocytes are consistent with *in vivo* results. In our laboratory, we observed that the known species differences in amphetamine metabolism *in vivo* are reproduced in primary hepatocyte cultures. The liver metabolizes amphetamine into two major metabolites: para-hydroxyamphetamine (pHA) by means of aromatic hydroxylation, and benzoic acid (BA) through oxidative deamination. Species differences in amphetamine metabolism have been established *in vivo*: (a) the rate of metabolism differs among different animal species, with rabbit > rat > monkey > human; and (b) the metabolite ratios differ among species: in the rat, aromatic hydroxylation leading to pHA is the major pathway while in all other species, side chain oxidation leading to BA is the predominant pathway. Our results on amphetamine metabolism in primary hepatocytes from rat, rabbit, rhesus monkey and human demonstrate species differences in both rate of metabolism and pathway preference, similar to those observed *in vivo* [15].

Both sex and species differences in phase II metabolism that are found *in vivo* are also reproduced by primary hepatocytes. In rodents, sex differences in hepatocyte metabolism of acetaminophen are known. The male rat has higher acetaminophen glutathione sulphotransferase activity than the female rat. The difference is believed to be a result of the stimulatory influence of testosterone and suppressive effect of estrogen on one of the two acetaminophen glutathione sulphotransferases. Upon acetaminophen administration, the male rat excretes more acetaminophen sulphate conjugate and less acetaminophen glucuronide conjugate than the female rat. For the human, however, both male and female excrete more glucuronide than sulphate. In our laboratory, we found primary hepatocyte cultures from male and female human and rat to reproduce accurately the known sex- and species-differences [16,17]: female rat hepatocytes produced glucuronide and sulphate from acetaminophen in similar quantities while male rat hepatocytes produced more of the sulphate than the glucuronide. Increasing the acetaminophen concentration led to increases in glucuronide formation while sulphate formation reached a plateau, suggesting the early saturation of the sulphation pathway that is a well-established phenomenon *in vivo* [16]. In humans, this sex difference is not observed either *in vivo* or in cultured hepatocytes. In our study, male and female patients were administered acetaminophen before abdominal surgery. Urine acetaminophen metabolites were compared to metabolites generated from hepatocytes from the same patients. Similar results were observed both *in vivo* and *in vitro*: acetaminophen glucuronidation predominated over acetaminophen sulphation with no sex differences [17].

23.6 Examples of applications

Human hepatocytes are now used in both applied and basic research as a model of the liver for drug metabolism studies. The most common applications include the following:

23.6.1 Screening for metabolic stability

Metabolic stability is one of the several criteria for the selection of lead drug candidates. Because of the known species differences in drug toxicity, it would be ideal to screen for human-specific drug toxicity. Using cryopreserved human hepatocytes, a high-throughput screening assay has been developed [7]. The assay involves the evaluation of cytotoxicity using hepatocytes in suspension in 96-well plate format. The rate of disappearance of the parent compound after incubation with human hepatocytes is determined by comparing the amount of parent chemical incubated with or without hepatocytes. LC/MS is in general used as the analytical procedure.

Determination of human metabolites and metabolic pathways: Incubation of human hepatocytes with a xenobiotic followed by analysis of metabolite profile and identification of the metabolites will allow one to ascertain the possible metabolites formed in human *in vivo*. Based on the structure of the metabolites, one can deduce the key pathways and sequence of metabolite formation. The key metabolic pathways (*e.g.* involvement of CYP isoforms) can be further evaluated using metabolic inhibitors (see below).

Evaluation of species differences: There are two major reasons for evaluating potential species differences in metabolic rate/profile between laboratory animals and human: (i) to evaluate if the observation made in laboratory animals can be extrapolated to humans; and (ii) to evaluate which animal species is most similar to humans in order to guide the decision when choosing the appropriate animal species to act as the human surrogate for pharmacokinetic, pharmacodynamic and toxicological studies.

The typical study involves incubation of the test chemical with hepatocytes isolated from man and multiple animal species. A comparison of metabolite profiles will allow one to deduce which animal species is similar to man in the metabolism of the test chemical. An example of the application of primary hepatocytes to evaluate species difference in drug metabolism is our study comparing, in rat and human hepatocytes, the metabolic fate of two structurally similar drug candidates, SC-42867 and SC-51089 [18]. We found that rat and human hepatocytes produced similar metabolites from SC-42867, therefore suggesting that the rat is a good model for human hepatic metabolism. For SC-51089, in spite of the structural similarity to SC-42867, the human hepatocytes and rat hepatocytes produced very different metabolites, therefore suggesting that the rat would not be a good experimental model for humans.

23.7 Application of human hepatocytes in the evaluation of drug-drug interaction potential

23.7.1 Mechanism-based approaches to the evaluation of clinical drug-drug interactions

Multiple drug therapy is widely practised either to treat a medical disorder or to treat several concurrently existing ailments in the same patient. It is now known that drugs may interact, resulting in serious pharmacological and/or toxicological consequences. Drugs interact mainly by a phenomenon known as pharmacokinetic drug-drug interactions - alteration of the metabolic clearance of a drug by a co-administered drug. While pharmacokinetic drug-drug interactions can occur during absorption, metabolism, disposition and elimination phases after initial drug administration, interference with drug metabolism appears to be the predominant mechanism. In the past, drug-drug interaction potential for specific drugs was discovered only because of the occurrence of adverse reactions in the patients. It is more desirable to evaluate drug-drug interaction potential of new drugs before the drugs reach the population.

Because it is impossible to evaluate the potential interaction of a new drug with all existing drugs, a more practical approach is to define the drug-drug interaction potential of the new drug based on mechanistic evaluations [19], followed by clinical trials specifically designed to further refine those observations. The key mechanism of pharmacokinetic drug-drug interactions is the ability of one drug to affect the metabolism of another drug. This can be due to inhibition or induction of the drug metabolising enzymes. At present, emphasis is placed on the induction and inhibition of CYP isoforms. Both mechanisms of pharmacokinetic drug-drug interactions can have serious clinical consequences.

23.8 Practical approaches

Based on our understanding of the mechanism of drug-drug interactions, three approaches have been adopted recently for the evaluation of new drugs [19].

23.8.1 CYP isozyme identification

Understanding the relative contribution of each CYP isozyme can provide valuable information on the drug-drug interaction potential of the drug in question. If one isozyme is found to be mainly responsible for its metabolism, it is likely that pharmacokinetic interactions would occur between this drug and drugs that are known to be inhibitors or inducers of the isozyme. For instance, if the new drug were metabolized mainly by CYP3A4, known CYP3A4 inhibitors like itraconazole, ketoconazole, and troleandomycin would inhibit the drug's metabolism. On the other hand, CYP3A4 inducers such as phenytoin, rifampin, and phenobarbital would enhance the drug's metabolism. Further, the new drug may have drug interaction with other drugs metabolized by the same isozyme because of competition for the same binding site of the CYP isozyme. A CYP3A4 substrate, for instance, may cause competitive inhibition of metabolism with other CYP3A4 substrates such as erythromycin and cyclosporin. Microsomes can be used for inhibitor studies, but only intact cells can be employed for examining induction.

23.8.2 CYP inhibitory potential

A drug that can inhibit a specific CYP isozyme would have the potential to inhibit the metabolism of drugs that are substrates of the specific isozyme. For instance, a drug that is a potent inhibitor of CYP3A4 would have the potential to inhibit the metabolism of a co-administered CYP3A4 substrate such as terfenadine, erythromycin, and cyclosporin. Intact cells, or microsomes derived from cells engineered, using cDNA, to express only one specific CYP isozyme can be used.

Human hepatic microsomes are most frequently used as an experimental model to evaluate the inhibitory mechanism of drug-drug interactions. The most effective application is to use isozyme-specific inhibitors to determine which CYP isozymes are responsible for the drug's metabolism. The principle is that drugs metabolized by the same isozymes should have inhibitory effects on the metabolism of each other. Primary hepatocytes represent an experimental system that can extend findings obtained with microsomes. For a drug to inhibit the metabolism of another drug, it needs to be present in the hepatocytes *in vivo*. The intracellular concentration is dependent on both membrane permeability as well as the activity of the multiple drug resistance protein that actively transports xenobiotics out of the hepatocytes. It is therefore necessary to confirm microsome findings with an intact cell system such as hepatocytes.

Inhibition of drug metabolism results in an increase in plasma/tissue drug concentration, which, especially for drugs with a narrow therapeutic index, can lead to toxicity. Pharmacokinetic drug-drug interactions via the inhibitory mechanism therefore are a safety concern. A well-established example of drug-drug interaction via an inhibitory mechanism is the occurrence of "*torsades de pointes*" ventricular arrhythmia in patients receiving concomitant therapy with the non-sedating antihistamine, terfenadine, and azole antifungals such as ketoconazole, itraconazole, clotrimazole, or macrolide antibiotics such as erythromycin [20-22]. Terfenadine is believed to be metabolized by CYP3A4 [23]. It has been suggested that the adverse drug interactions observed with terfenadine are related to inhibition of its metabolism by the co-administered drugs, leading to the accumulation of parent terfenadine to a cardiotoxic level [24].

In collaboration with Dr. Malle Jurima-Romet of Health Canada, we studied the effects of various CYP3A substrates and inhibitors on terfenadine metabolism using primary human and rat hepatocytes and compared the results with those from an immortalized cell line, HepG2. While all three-cell systems extensively metabolized terfenadine, human hepatocytes were found to produce both C-oxidation and N-dealkylation products. Rat hepatocytes and HepG2 cells produced only C-oxidation products. Furthermore, with CYP3A, the isozyme believed to be mainly responsible for terfenadine metabolism, the known inhibitors ketoconazole erythromycin and troleandomycin, inhibited terfenadine metabolism in human but not in rat hepatocytes. We thereby concluded that primary human hepatocytes represent an appropriate experimental model for the evaluation of drug-drug interactions by means of inhibitory mechanisms [9,25].

23.8.3 CYP induction

Human hepatocytes represent the only proven preclinical experimental system with which induction of human CYP can be studied. Multiple laboratories have demonstrated that known *in vivo* CYP inducers are also potent inducers in primary human hepatocytes [9]. Thus these hepatocytes represent the most relevant *in vitro* human system in which pharmacokinetic drug-drug interactions via the induction mechanism can be studied.

Induction studies are not possible in the microsomal subcellular fraction; and the viability of the liver slices cannot be maintained for the duration (usually over 48 h) that is required for induction. We have established a routine CYP induction assay using primary human hepatocyte cultures. This assay has been applied towards the evaluation of the rifamycin analogs rifampin, rifapentine, and rifabutin [11]. We found that rifampin and rifapentine were more potent inducers of CYP3A than rifabutin, a finding that is similar to what has been observed in humans *in vivo* [11].

Induction of drug metabolism can result in a decrease in drug concentration to a level that is no longer efficacious. A significant drug-drug interaction via an induction mechanism is the interaction between rifampin and oral contraceptives [26-28]. Rifampin is an antimicrobial agent which is a known inducer of CYP3A4 in the human [29] and in human hepatocytes [1,9-11]. The key active ingredients of oral contraceptives, estrogen and progesterone analogs, are substrates of CYP3A4 [30]. Rifampin administration was found to lead to uterine bleeding and pregnancies in women taking oral contraceptives, which is believed to have been due to the enhanced clearance of the active ingredients [31]. In addition to oral contraceptives, rifampin was reported to decrease the therapeutic effects and/or plasma concentration of other drugs, including verapamil [32], cyclosporin [33], doxycycline [34,35], itraconazole [36], prednisolone [18] and zidovudine [37]. While drug-drug interaction due to enzyme inhibition is a safety concern, therapeutic failure is the major concern for drug-drug interaction due to enzyme induction.

Using primary human hepatocytes, we evaluated the effects of rifampin on lidocaine metabolism. As lidocaine is known to be a CYP3A substrate, and rifampin is a known CYP3A inducer, we predicted that lidocaine metabolism in primary human hepatocytes would be induced by rifampin. This is indeed what we found. Treatment of primary human hepatocytes with rifampin led to a significant, dose-dependent induction of lidocaine metabolism. There was a significant increase in V_{max} with little effect on K_m, an observation consistent with increased expression of the CYP3A isozyme protein rather than a change in affinity for the substrate. Our results suggest that the dosage of lidocaine administration to patients on rifampin should be greater than that normally recommended.

We also evaluated CYP1A induction by polybrominated biphenyl in primary human and rat hepatocytes and found significant species-differences in sensitivity, with the rat being the more sensitive species [38]. A tenfold to a thousandfold higher concentration of the inducer level active in rat hepatocytes was required to induce CYP1A in human hepatocytes. Our finding suggests that humans may be similarly less susceptible to the induction effects of polybrominated biphenyl.

23.9 Applications of primary hepatocyte cultures in toxicology

23.9.1 Hepatocytes as a critical cell population for hepatotoxicity studies

A large number of naturally occurring and man-made chemicals are hepatotoxins. In many cases, the toxicity is a function of the metabolic conversion (bioactivation) of the parent compound into highly reactive metabolites. Acetaminophen, carbon tetrachloride, dimethylnitrosamine, halothane are examples of xenobiotics that are "bioactivated" by CYP mixed function monooxygenases in the liver (see Chapter 27). Species differences in xenobiotic metabolism are therefore important factors contributing to the known species differences in chemical toxicity.

As discussed above, most of the xenobiotic-metabolising enzymes of the liver reside in the parenchymal cells or hepatocytes. The hepatocytes therefore are usually the first cell type to be damaged upon hepatotoxic insult. Primary hepatocytes thus represent a useful experimental system to evaluate acute hepatic injury. The non-parenchymal cells are important in the progression of pathological events. For instance, inflammation of the liver is primarily a function of the cytokine production in the endothelial cells and Kupffer cells. Hepatic fibrosis and cirrhosis are believed to be due to the "activation" of lipocytes to collagen-synthesizing fibroblasts.

23.9.2 Advantages and disadvantages

In vitro toxicology, the evaluation of toxicological properties of physical and chemical agents using isolated cells and tissues, is an area that has evolved rapidly in the past decade. The initial purpose is to achieve the so-called "3Rs": replacement (of whole animals); reduction (of animal use); and refinement (of toxicity assays). Pragmatically, *in vitro* toxicology is mostly useful in the rapid screening of chemicals and in the mechanistic evaluation of toxicological phenomena.

The hepatocyte is an ideal *in vitro* system to evaluate hepatotoxicity. The advantages are, as stated earlier, the retention of species-specific metabolism, the defined experimental conditions common to most *in vitro* systems, and the requirement of relatively low amount of test materials. Molecular approaches now are often coupled with hepatocyte culture systems for mechanistic evaluation. Disadvantages are the lack of host factors and the absence of non-parenchymal cells. Reviews on the application of hepatocytes in toxicology include [9,39,40] (see also Chapter 27).

23.10 Evaluation of human hepatotoxicity

23.10.1 Toxicity during a clinical trial

During drug development, it is a rather frequent occurrence that a drug with no preclinical toxicity proves to have marked hepatotoxicity in humans during clinical trials. For compounds that are only toxic to humans but no other species, human hepatocytes are an invaluable model to aid in understanding the mechanism, and thereby the significance of the hepatotoxicity, and to be used as a screen for nonhepatotoxic structural analogs. Compound X was a proprietary compound which had no significant toxicological consequences observable during preclinical safety evaluations in laboratory animals including rats, dogs, and monkeys. During a phase I clinical trial, significant elevation of serum alanine aminotransferase and aspartate aminotransferase was observed, indicating hepatotoxicity. Using human hepatocytes, we found that compound X was a potent inhibitor of RNA synthesis. Over 80% inhibition of RNA synthesis was observed at 50 μM and higher concentrations. Under similar experimental conditions, less than 50% inhibition of RNA synthesis was observed in rat hepatocytes at concentrations as high as 200 μM. [15].

The implications of the study are extremely important to drug development. For a chemical with human-specific hepatotoxicity, it would be impossible to screen for nonhepatotoxic analogs using laboratory animals. Human hepatocytes, coupled with a relevant, mechanism-based assay for hepatotoxicity (like the [^{3}H]uridine incorporation assay) for compounds related to compound X, would allow rational selection of nonhepatotoxic analogs. This would be more practical and cost-efficient than the initiation of preclinical safety tests and clinical trials in the absence of human toxicity information for the analogs.

23.10.2 Evaluation of peroxisomal proliferation

Peroxisome proliferators, agents that can increase the number of peroxisomes, are also rodent hepatocarcinogens. Examples of these agents are the hypolipidemic agents clofibrate, cyprofibrate and the plasticiser di(2-ethylhexyl) phthalate. Primary hepatocyte cultures are found to respond to these agents. Both in our laboratory and others [41], species differences in response to these agents are observed: the most significant finding is that the agents cause peroxisomal proliferation in rodent, but not human hepatocytes [15]. While the generally accepted concept is that peroxisome proliferators are nongenotoxic carcinogens, Hwang [42] found that the potent peroxisome proliferator and rodent hepatocarcinogen [4-chloro-6-(2,3-xylidino)-2-pyridinylthio] acetic acid (Wy-14,643) could induce genotoxicity measured as sister-chromatid-exchange and chromosomal aberrations in primary cultures of both rodent and human hepatocytes.

23.10.3 Correlation of metabolism and toxicity

The use of a metabolically competent intact cell system like primary hepatocytes has another advantage: toxicity (cytotoxicity) can be studied simultaneously with metabolism, so that one can study the relationship between metabolism and toxicity.

23.11 Conclusion

Human hepatocytes represent an experimental system with which human-specific data can be obtained. The major applications are the evaluation of metabolic fate, drug-drug interaction potential, and toxicity of xenobiotics. The important advances in the past years in studies on human hepatocytes include the development of techniques for routine isolation and culturing of the cells [2], the development of data demonstrating that species-differences in xenobiotic metabolism and hepatotoxicity can be reproduced in hepatocytes [9], the development of mechanistic approaches for the evaluation of both inhibitory and inductive drug-drug interactions [19], and cryopreservation [7]. Advances in cryopreservation technologies definitely expand the applications of this important experimental system. The use of cryopreserved human hepatocytes now allows researchers to perform short-term studies without delays in availability of fresh human liver for hepatocyte isolation. Approaches to enhance the attachment efficiency of hepatocytes after cryopreservation are being established and would further minimise the need to use freshly isolated human hepatocytes for research.

References

1. Li AP, Rasmussen A and Kaminski DL. Substrates of human hepatic cytochrome P450 3A4. Toxicology 1995; 104: 1-8.
2. Li AP, Roque MA, Beck DJ and Kaminski DL. Isolation and culturing of hepatocytes from human liver. J Tiss Cult Meth 1992; 14: 139-146.
3. Loretz LJ, Wilson AGE and Li AP. Promutagen activation by freshly isolated and cryopreserved rat hepatocytes. Environ Mol Mutagen 1988; 12: 335-341.
4. Loretz LJ, Li AP, Flye MW and Wilson AGE. Optimization of cryopreservation procedures for rat and human hepatocytes. Xenobiotica 1989; 19: 489-498.
5. Moerlein SM, Weisman RA, Beck DL, Li A and Welch MJ. Metabolism *in vitro* of radioiodinated N-isopropyl-para-iodoamphetamine by cultured hepatocytes. Nucl Med Biol 1993; 20: 49-56.
6. Ruegg CE, Silbher PM, Mughal RA, Ismail J, Lu C and Li AP. P450 induction in primary human hepatocytes after cryopreservation. In Vitro Toxicol 1997; 10: 217-222.
7. Li AP, Lu C, Brent JA, Pham C, Fackett A, Ruegg CE and Silber P. Cryopreserved Human hepatocytes: Characterization of drug metabolizing activities and applications in higher throughput screening assays for hepatoxicity, metabolic stability, and drug-drug interaction potential. Chem-Biol Interact 1999; 121: 17-35.
8. Takenaka O, Horie T, Suzuki H and Sugiyama Y. Carrier-mediated active transport of the glucuronide and sulfate of 6- hydroxy-5,7-dimethyl-2-methylamino-4-(3-pyridylmethyl) benzothiazole (E3040) into rat liver: quantitative comparison of permeability in isolated hepatocytes, perfused liver and liver *in vivo*. J Pharmacol Exp Ther 1997; 280: 948-958.
9. Li AP and Jurima-Romet M. Applications of primary human hepatocytes in the evaluation of pharmacokinetic drug-drug interactions: Evaluation of model drugs terfenadine and rifampin. Cell Biol and Toxicol 1997; 13: 365-374.
10. Li AP. Evaluation of drug-drug interactions in primary human hepatocytes. In: Li AP, Ed. Drug-drug interactions: Scientific and regulatory perspectives. Advances in Pharmacology 1997; 43: 103-130.

11. Li AP, Reith MK, Rasmussen Kaminski DL and Cheng LK. *In vitro* evaluation of drug-drug interaction potential: A collection of rifampin, rifapentine, and rifabutin in cytochrome P450 3A induction potential: A comparison of rifampin, rifapentine, and rifabutin in cytochrome P450 induction potential in primary human hepatocytes. Chem-Biol Interact 1997; 107: 5-16.
12. Li AP, Maurel JP, Lechon-Gomez M, Cheng LC and Jurima-Romet Preclinical M. evaluation of drug-drug interactions: present status of the application of primary human hepatocytes in the evaluation of cytochrome P450 induction. Chem-Biol Interact 1997; 107: 5-16.
13. Donato MT, Gomez-Lechon MJ, Jover R, Nakamura T and Castell JV. Human hepatocyte growth factor down-regulates the expression of cytochrome P450 isozymes in human hepatocytes in primary culture. J Pharmacol Exp Ther 1998; 284: 760-767.
14. Houston JB. Utility of *in vitro* drug metabolism data in predicting *in vivo* metabolic clearance. Biochem Pharmacol 1994; 47: 1469-1479.
15. Li AP. Primary hepatocyte cultures as an *in vitro* experimental model for xenobiotic metabolism and toxicology. Comments on Toxicol 1998; 6: 199-219.
16. Kane RE, Tector J, Brems JJ, Li AP and Kaminski D. Sulfation and glucuronidation of acetaminophen by cultured hepatocytes reproducing *in vivo* sex differences in conjugation on Matrigel and Type 1 collagen. In Vitro Cell Dev Biol 1991; 27A: 953-960.
17. Kane RE, Li AP and Kaminski DR. Sulfation and glucuronidation of acetaminophen by human hepatocytes cultured on matrigel and type 1 collagen reproduces conjugation *in vivo*. Drug Metab Dispos 1995; 23: 303-307.
18. Lee K, Vandenberghe Y, Herin M, Herin R, Cavalier D et al. Comparitive metabolism of SC-42867 and SC-51089, two PGE2 antagonists, in rat and human hepatocyte cultures. Xenobiotica 1994; 24: 25-36.
19. Li AP. The scientific basis of drug-drug interactions: mechanism and preclinical evaluation. Drug Information Journal 1998; 32: 657-664.
20. Monahan BP, Ferguson CL, Killeavy ES, Lloyd BK, Troy J and Cantilena LR Jr. Torsades de pointes occurring in association with terfenadine use. JAMA 1990; 264: 2788-2790.
21. FDA. Warnings issued on nonsedating antihistamines terfenadine and astemizole. JAMA 268: 705.
22. Pohjala-Sintonen S, Viitsalo M, Toivonen L and Neuvonen P. Torsades de pointes after terfenadine-itraconazole interaction. Br Med J 1993; 306: 186.
23. Yun CH, Okerholm A and Guengerich FP. Oxidation of the antihistamine drug terfenadine in human liver microsomes. Role of cytochrome P4503A(4) in N-dealkylation and C-hydroxylation. Drug Metab Dispos 1993; 21: 403-409.
24. Woosley RL, Chen Y, Freiman JP and Gillis RA. Mechanism of the cardiotoxic actions of terfenadine. JAMA 1993; 269: 1532-1536.
25. Jurima-Romet M, Huang HS, Beck DJ and Li AP. Metabolism of terfenadine and its inhibition in primary human and rat hepatocytes and HepG2 cells. Toxicology In Vitro 1996; 10: 655-663.
26. D'Arcy PF. Drug interactions with oral contraceptives. Drug Intell Clin Pharm 1986; 20: 353-362.
27. Grange JM, Winstanley PA and Davies PD. Clinically significant drug interactions with antituberculosis agents. Drug Safety 1994; 11: 242-251.

28. Strolin Benedetti M and Dostert P. Induction and autoinduction properties of rifamycin derivatives: a review of animal and human studies. Environ Health Perspect 1994; 102: 101-105.
29. Ged C, Roullion JM, Pichard L, Combalbert J, Bressot N, Bories P, Michel H, Beaune P and Maurel P. The increase in urinary excretion of 6ß-hydroxycortisol as a marker of human hepatic cytochrome P450 IIIA induction. Br J Clin Pharmacol 1989; 38: 373-387.
30. Guengerich FP. Metabolism of 17α-ethynylestradiol in humans. Life Sci 1990; 47: 1981-1988.
31. Reimers D, Nocke-Finck L and Breuer H. Rifampicin and the "pill" do not go well together. JAMA 1974; 227: 608.
32. Fromm MF, Busse D, Kroemer HK and Eichelbaum M. Differential induction of prehepatic and hepatic metabolism of verapamil by rifampin. Hepatology 1996; 24: 796-801.
33. Herbert MF, Roberts JP, Prueksaritanont T and Benet LZ. Bioavailability of cyclosporin with concomitant rifampin administration is markedly less than predicted by hepatic enzyme induction. Clin Pharmacol Ther 1992; 52: 453-457.
34. Lang R, Shasha B and Rubinstein E. Therapy of experimental murine burcellosis with streptomycin alone and in combination with ciprofloxacin, doxycycline, and rifampin. Antimicrob Agents Chemother 1993; 37: 2333-2336.
35. Colmenero JD, Fernandez-Gallardo LC, Agundez JA, Sedeno J, Benitez J and Valverde E. Possible implications of doxycycline-rifampin interaction in the treatment of brucellosis. Antimicrob Agents Chemother 1994; 38: 2798-2802.
36. Drayton J, Dickinson G and Rinaldi MG. Coadministration of rifampin and itraconazole leads to undetectable levels of serum intraconazole. Clin Infect Dis 1994; 18: 266.
37. Burger DM, Meenhorst PL, Koks CH and Beijnen JH. Pharmacokinetic interaction between rifampin and zidovudine. Antimicrob Agents Chemother 1993; 37: 1426-1431.
38. Merrill JC, Beck DJ, Kaminski DA and Li AP. Polybrominated biphenyl induction of cytochrome P450 mixed function oxidase activity in primary rat and human hepatocytes. Toxicology 1995; 23: 147-152.
39. Li AP. Primary hepatocyte culture as an *in vitro* toxicological system. In: Gad S, Ed. In Vitro Toxicol. New York: Raven Press, 1994: 195-220.
40. Gomez-Lechon MJ, Monloya A, Lopez P, Donato T, Larrauri A and Castell JV. The potential use of cultured hepatocytes in predicting the hepatotoxicity of xenobiotics. Senobiotica 1988; 18: 725-735
41. Elcombe CR and Mitchell AM. Peroxisome proliferation due to di(2-ethylhexyl) phthalate (DEHP): Species differences and possible mechanisms. Environ Health Perspect 1986; 70: 211-219.
42. Hwang JJ, Hsia MT and Jirtle RL. Induction of sister-chromatid-exchange and micronuclei in primary cultures of rat and human hepatocytes by the peroxisome proliferator, Wy-14,643. Mutat Res 1993; 286: 123-133.

Chapter 24

Constitutive expression of cytochromes P450 in rat hepatocyte culture

Alan J. Paine
St Bartholomew's and Royal London School of Medicine & Dentistry, Charterhouse Square, London, EC1M 6BQ, U.K.

Key words: cytochrome P450 loss, hepatocyte culture, metapyrone, mRNA stability, NF-κB binding, pyridines, transcription, xenobiotic metabolism

24.1 Constitutive expression of cytochromes P450 in rat hepatocyte culture

24.1.1 Background

The various isoforms of cytochromes P450 (CYP) [1] are the major determinants of the pharmacological and toxicological activity of numerous drugs and many other foreign chemicals present in the human environment [2,3]. A major limitation to the use of hepatocyte cultures in pharmaco-toxicological studies is their rapid loss of CYP content [4]. This loss of CYP and associated xenobiotic metabolising activity occurs in hepatocyte cultures prepared from the common species of experimental [5] and farm animals [6] as well as in human hepatocyte culture [7] suggesting similar underlying mechanism(s). Thus while the focus in this chapter is on rat hepatocytes, the commentary on control of CYP gene expression in hepatocytes is likely to be of relevance to liver cell cultures prepared from other species.

24.1.2 The problem outlined

Bissell *et al* [8] were the first to demonstrate that rat hepatocytes, isolated by the well-established collagenase perfusion procedure, lost 80% of their CYP content during the first 24 hours of culture. CYP content is lost to the same extent in hepatocytes isolated non-enzymically with EDTA [9], suggesting that the use of collagenase *per se* may not be the culprit. In this respect, precision cut liver slices, that seemingly retain their intracellular contacts [9,10], or the livers of partially-hepatectomized rats [11], lose CYP

M.N. Berry and A.M. Edwards (eds.), The Hepatocyte Review, 411–420.

content to the same extent as cultured hepatocytes over a 24 h period. Collectively, these observations suggest that there may be more than one mechanism mediating loss of CYP and by the same token more than one way to prevent its loss. Nevertheless the decline in CYP occurs whatever commercially available culture medium is employed for hepatocytes [12], suggesting that once the loss of CYP has been triggered it is irreversible [13]. It should be noted here that this laboratory's definition of preventing the loss of CYP content is to maintain, throughout 24 hours of culture, its concentration (whether expressed per mg of cellular protein, per μg DNA or per cell) at the same concentration as present in the donor organ. Hence supplementation of the culture medium with the products of intermediary metabolism [14], 2% dimethylsulphoxide [15] or modification of the substratum [16], reported by others to prevent the loss of CYP content do not, in our experience, represent a true maintenance of hepatic CYP content. Quite simply such treatments might retard the loss of total CYP content but they never maintain its concentration at the same level as found in the donor organ [17] (McGill & Paine, unpublished results). Accordingly, the finding that the sole addition of any one of a number of pyridines [18] to the culture medium maintained total CYP content of cultured rat hepatocytes at the same level as found in the donor organ provided the means to investigate the mechanisms underlying its loss.

24.2 Effects of metyrapone

24.2.1 Effects of metapyrone on cytochrome P450s

The ability of pyridines to prevent the loss of CYP content is related to their ability to bind to the cytochrome and metyrapone is the most potent of these compounds [19]. Thus the addition of 0.5mM metyrapone to Williams Medium E maintains the CYP concentration at the same level found in the donor organ over a 120 hour period of culture [20] accompanied, as indicated below, by change in the isoenzyme profile.

Ligands are known to protect hepatic CYP from degradation *in vivo* [21] and so the effect of hepatocyte culture conditions on CYP degradation was determined. Indeed it was found [22] that metyrapone prevented the enhanced degradation of CYP that pertains in hepatocyte culture, and also stimulated the synthesis of CYP which in the absence of metyrapone does not occur. Thus, the failure of hepatocytes to synthesize CYP, as well as its accelerated degradation, underlie its loss in rat hepatocyte culture. It is now known that the failure of CYP synthesis is related to the extinction of transcription of CYP genes between 4 and 24 hours of culture [23] while its enhanced degradation is due to the presence of cyst(e)ine in the culture medium.[12,24]. However the term CYP has been used rather loosely up to now because rather than representing a single entity it is in fact a collective term for a gene superfamily of hemoproteins [1]. To date about 20 individual forms, which based on sequence homology are contained within 5 gene families, are known to comprise hepatic CYP content (Table 1). Some of these forms are so closely related that they can only be distinguished with anti-peptide antibodies [25]. Nevertheless the results presented in Table 2 demonstrate that the immunochemical quantification of hepatic CYP subfamily members with polyclonal antibodies agrees very well with the total amount of CYP determined spectrophotometrically. The results presented in Table 2 also demonstrate that metyrapone maintains the total CYP content of rat hepatocytes in culture by inducing CYP1A and CYP3A subfamily members, results which have been confirmed by molecular biology techniques [26,27]. Metyrapone also induces CYP1A and CYP3A subfamily members in human hepatocyte culture [28].

24.2.2 Drug metabolism in metyrapone treated hepatocytes

Although metyrapone is generally considered to be an inhibitor of CYP-mediated metabolism [29], hepatocytes cultured for 24 hours with 0.5 mM metyrapone display markedly increased activities towards CYP1A and CYP3A substrates [30,31], commensurate with the induction of these subfamilies. Similarly, since CYP3A is a major subfamily in human hepatocytes, the culture of metyrapone-treated rat hepatocytes with diazepam results in the production of oxazepam, a metabolite normally only observed in incubations with human liver [17]. In general, because of the absence of excretory pathways the majority of drug metabolites can be found in hepatocyte cultures, which reduces the multiplicity of systems normally required for their identification [32]. In contrast the lack of extrahepatic pathways of excretion can cause problems in the prediction of toxicity [12, 33-35] but this does not explain a marked difference in sensitivity between cultured hepatocytes and precision cut liver slices [36].

Table 1. Some rat hepatic CYPs

Gene family	Subfamily member	Model substrate/activity	Expression	Regulation
1	A1	3, 4 Benzpyrene	Not constitutive but inducible by pah	AhR [43] GH & TH suppress [44]
	A2	Acetylaminofluorene	Constitutive and inducible by pah	Induction via AhR [43]
2	A1	Testosterone 7α hydroxylase [45] Testosterone 15α hydroxylase [45]	Constitutive but suppressed in male at puberty [46]	Male/Female GH pattern and also TH [44]
	A2		Activated at puberty [46]	
2	B1/2	7 Ethoxycoumarin	Suppressed by GH and TH induced by phenobarbitone	Unknown [47]
2	C11	testosterone 2α, 16α and 17β hydroxylases	Constitutive in male rats absent in females [46,48]	Male/female GH pattern & testosterone
2	C12	testosterone 15 β hydroxylase	Constitutive in female rats absent in males [46,48]	Male/female GH pattern & estradiol
2	E1	Involved in gluconeogenesis, affected by diabetes, fasting & ethanol intake	Constitutive but suppressed by GH and TH [44,49]	mRNA & protein stability
3	A1/2	Erythromycin demethylation	Constitutive and inducible by glucocorticoids	Unknown but seemingly involves GR [25]
4	A1	Lauric acid 11,12 hydroxylation	Constitutive and induced by hypolipaedemic agents	PPAR [50]
7	-	Cholesterol 7α hydroxylase	Constitutive, induced by cholesterol & TH, suppressed by bile acids	Exhibits circadian rhythm [51]

Abbreviations: pah , polycyclic aromatic hydrocarbons; AhR, aromatic hydrocarbon receptor; GH, growth hormone; TH, thyroid hormone; GR, glucocorticoid receptor; PPAR, peroxisome proliferator activated receptor.

24.3 The problem refined

The "holy grail" of *in vitro* toxicology has been to maintain not only the CYP content of cultured hepatocytes but also the CYP isozyme profile at the same level as the donor organ. Although the former criterion can be achieved with metyrapone, the variety of physiological processes that regulate individual hepatic isozymes (Table 1) indicates the latter may be very difficult to achieve. The loss of particular isozymes involved in detoxification will inevitably result in false positives if these cultures are employed in toxicity tests. Conversely if a xenobiotic becomes active pharmacologically or as a toxin only after metabolism, then loss of relevant CYP isozymes(s) will lead to false negatives. Accordingly one option available at present is to prevent the loss of the major isozyme known as CYP2C11, since in adult male rats, conventionally employed in toxicity studies [37], this single form represents more than half of the constitutively expressed hepatic CYP content (Table 2). Furthermore, a consequence of understanding how CYP2C11 is lost from hepatocyte cultures has revealed that we still have much to learn about the control of hepatic gene expression *in vitro* (Figure 1).

Table 2. CYP subfamily abundance (pmoles/mg microsomal protein) in intact liver and hepatocytes cultured with (+) and without (-) 0.5mM metyrapone (M)

CYP subfamily	Intact liver	24 h -M	72 h -M	24 h +M	72 h +M
1A	n.d.	n.d.	n.d.	18	88
2A	62	11	8	11	8
2B	n.d.	22	n.d.	31	n.d.
2C	136	43	40	76	40
3A	18	18	8	84	84
4A	n.d.	n.d.	n.d.	n.d.	n.d.
Immunochemical total	216	94	56	220	220
Spectrophotometric total	222	107	51	209	222

The results are presented as the average of 2 separate experiments. For technical details of quantification see [52]. n.d. = below limit of detection = 5 pmoles CYP subfamily/mg microsomal protein.

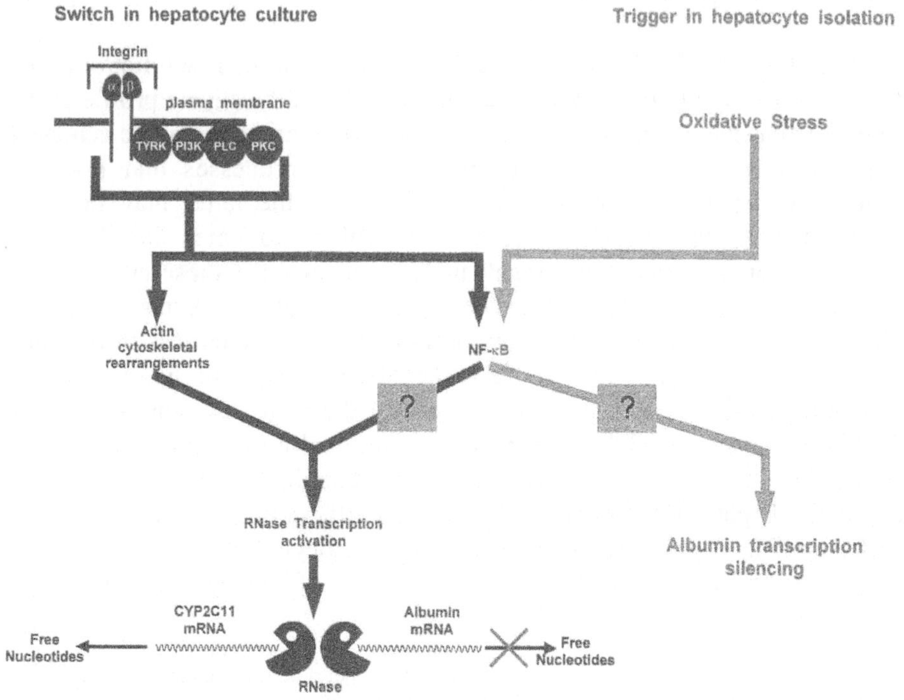

Figure 1. Pathways underlying the extinction of hepatocyte phenotype *in vitro.* The involvement of tyrosine kinases (TYRK) phosphatidyl inositol 3 kinase (PI3K) phospholipase C (PLC) and protein kinase C (PKC) have been deduced from inhibitor studies and therefore are not placed into any signalling hierarchy. Nuclear factor kappa B [NF-κB].

In brief, the well-established hepatocyte isolation procedure with collagenase itself triggers alterations in liver cell gene expression. This is evidenced by the silencing of albumin transcription [23,38,39] in freshly isolated hepatocytes. An earlier change, just 20 minutes into the conventional perfusion with collagenase is the induction of NF-κB binding to its consensus sequence as demonstrated by electrophoretic mobility shift assays (Rodriguez-Ariza & Paine unpublished results). At present it is not known how increased NF-κB binding relates to the silencing of albumin transcription but NF-κB is a marker of oxidative stress [40]. In contrast, CYP2C11 transcription is not affected by hepatocyte isolation with collagenase [23]. Indeed CYP2C11 mRNA abundance remains constant for the first 4 hours of culture in serum-free medium. Then as the hepatocytes spread in culture a second switch occurs so that by 6 hours of culture

CYP2C11 transcripts have declined to 20% of the donor organ value. This rapid loss of CYP2C11 mRNA can be accelerated by culturing rat hepatocytes with 1 μg/ml fibronectin, an attachment factor present in serum. The delay in CYP2C11 loss in serum-free culture medium is commensurate with fibronectin synthesis which takes 6 hours to produce a medium concentration of 1 μg/ml (Wright & Paine unpublished results). This effect of fibronectin, whether added exogenously or synthesized by the hepatocytes themselves, involves an $\alpha_5\beta_1$ integrin-mediated signal transduction pathway and cell shape changes that activate RNase gene expression and affect CYP2C11 mRNA, but not albumin mRNA, turnover [23]. The reason why albumin mRNA is resistant to degradation is unknown but its stability is in accord with the ability of rat hepatocytes to synthesize albumin at the same rate as the intact organ during the first 24 hours of culture [41]. Thus the selectivity between liver specific transcripts may reside in the specificity of the RNase induced.

24.4 Conclusions

The potential for rigorous studies of gene expression to explain the classic phenomenon of extinction of hepatocyte phenotype *in vitro* is much enhanced by the current observations that this process results from both transcriptional silencing and the rapid degradation of specific mRNAs. Here I believe that liver-specific CYP gene expression serves as a better marker than albumin because most liver derived cell lines and undifferentiated tumours synthesize albumin at low rates [41,42], but consensus indicates that none contain the CYP isozyme profile of the donor organ. Furthermore studies at a molecular level must be improved by appreciation of two distinct time-frames underlying these early events. It needs to be appreciated that the first mechanism extinguishing phenotype actually occurs during the procedure for hepatocyte isolation, while a second, distinct, switch operates during the first 4-6 hours of serum-free culture. This allows us to approach a long standing problem from a new perspective, rather than relying on empirical manipulations of medium composition/culture conditions or immortalisation of hepatocytes once their phenotype has been lost. Recognition of the roles played by transcription, mRNA stability and cell shape therefore provides the surest foundation for refining hepatocyte culture technology as well as for understanding the molecular basis for the role of the liver cell's unique polarized morphology in the maintenance of its differentiated phenotype.

References

1. Nelson DR, Koymans L, Kamataki T, Stegeman JJ et al. P450 superfamily: update on new sequences, gene mapping, accession numbers and nomenclature. Pharmacogenetics 1996; 6: 1-42.
2. Gonzalez FJ. The molecular biology of cytochrome P450s. Pharmacological Reviews 1989; 40: 243-288.
3. Paine AJ. Heterogeneity of cytochrome P450 and its toxicological significance. Human & Experimental Toxicology 1995; 14: 1-7.
4. Paine AJ and Legg RF. Apparent lack of correlation between the loss of cytochrome P-450 in hepatic parenchymal cell culture and the stimulation of haem oxygenase activity. Biochem Biophys Res Commun 1978; 81: 672-679.
5. Maslansky CJ and Williams GM. Primary cultures and the levels of cytochrome P450 in hepatocytes from mouse, rat, hamster and rabbit liver. In Vitro 1982; 18: 683-693.
6. Montesissa C, Anfossi P, van't Klooster G and Mengelers M. The use of cultured hepatocytes from goats and cattle to investigate xenobiotic oxidative metabolism. Vet Res Commun 1996; 20: 449-460.
7. Blaauboer BJ, van Holsteijn I, van Graft M and Paine AJ. The concentration of cytochrome P-450 in human hepatocyte culture. Biochem Pharmacol 1985; 34: 2405-2408.
8. Bissell DM. Study of hepatocyte function in cell culture. Progr Liver Dis 1976; 5: 69-82.
9. Wright MC and Paine AJ. Evidence that the loss of rat liver cytochrome P450 *in vitro* is not solely associated with the use of collagenase; the loss of cell contacts; and/or the absence of an extracellular matrix. Biochem Pharmacol 1992; 43: 237-243.
10. Lake BG, Beamand JA, Japenga AC, Renwick A, Davis S and Price RJ. Induction of cytochrome P-450 dependent enzyme activities in cultured rat liver slices. Food Chem Toxicol 1993; 31: 377-386.
11. Padgham CRW, Boyle CC, Wang XJ, Raleigh SM, Wright MC and Paine AJ. Alteration of transcription factor mRNAs during the isolation and culture of rat hepatocytes suggests the activation of a proliferative mode underlies their dedifferentiation. Biochem Biophys Res Commun 1993; 197: 599-605.
12. Paine AJ. The maintenance of cytochrome P-450 in rat hepatocyte culture: some applications of liver cell cultures to the study of drug metabolism,toxicity and the induction of the P-450 system. Chem Biol Interactions 1990; 74: 1-31.
13. Padgham CRW and Paine AJ. Altered expression of cytochrome P450 mRNAs and potentially other transcripts encoding key hepatic functions are triggered during the isolation of rat hepatocytes. Biochem J 1993; 289: 621-624.
14. Guzelian P and Bissell DM. Metabolic factors in the regulation of cytochrome P-450: studies in rat hepatocyte monolayer culture. Fed Proc 1974; 33: 545.
15. Muakkassah-Kelly SF, Beiri F, Waechter F, Bentley P and Staubli W. Long term maintenance of hepatocytes in primary culture in the presence of DMSO. Exp Cell Res 1987; 71: 37-51.
16. Scheutz EG, Li D, Omiecinski CJ, Muller-Eberhard U, Kleinman HK, Elswick B and Guzelian PS. Regulation of gene expression in adult rat hepatocytes cultured on a basement membrane matrix. J Cell Physiol 1988; 134: 309-323.
17. Blankson EAB, Chenery RJ and Paine AJ. Cytochrome P450 maintenance and diazepam metabolism in cultured rat hepatocytes. Biochem Pharmacol 1991; 42: 1241-1245.

18. Villa P, Hockin LJ and Paine AJ. The relationship between the ability of pyridine and substituted pyridines to maintain cytochrome P-450 and inhibit protein synthesis in rat hepatocyte cultures. Biochem Pharmacol 1980; 29: 1773-1777.
19. Paine AJ, Villa P and Hockin LJ. Evidence that ligand formation is a mechanism underlying the maintenance of cytochrome P450 in rat liver cell culture: Potent maintenance by metyrapone. Biochem J 1980; 188: 937-939.
20. Paine AJ, Hockin LJ and Allen CM. Long term maintenance and induction of cytochrome P-450 in rat liver cell culture. Biochem Pharmacol 1982; 31: 1175-1178.
21. Hockin LJ and Paine AJ. The role of 5-aminolaevulinate synthase, haem oxygenase and ligand formation in the mechanism of maintenance of cytochrome P-450 in hepatocyte culture. Biochem J 1983; 210: 855-857.
22. Paine AJ and Villa P. Ligands maintain cytochrome P-450 in liver cell culture by affecting P-450 synthesis and degradation. Biochem Biophys Res Commun 1980; 97: 744-750.
23. Wang X-J, Hodgkinson CP, Wright MC and Paine AJ. Temperature sensitive mRNA degradation is an early event in hepatocyte dedifferentiation. Biochem J 1997; 328: 937-944.
24. Paine AJ and Hockin LJ. Nutrient imbalance causes the loss of cytochrome P-450 in liver cell culture: formulation of culture media which maintain cytochrome P450 at *in vivo* concentrations. Biochem Pharmacol 1980; 29: 3215-3218.
25. Wright MC, Edwards R, Pimenta M, Ribeiro V, Ratra GS, Lechner MC and Paine AJ. Developmental changes in the constitutive and inducible expression of cytochrome P450 3A2. Biochemical Pharmacology 1997; 54: 841-846.
26. Wright MC, Wang X-J, Pimenta M, Ribeiro V, Paine AJ and Lechner MC. Glucocorticoid receptor independent transcriptional induction of cytochrome P450 3A1 and its potentiation by a glucocorticoid. Molecular Pharmacology 1996; 50: 856-863.
27. Harvey JL, Paine AJ and Wright MC. Evidence for the disruption of endogenous regulator homeostasis in the mechanism of CYP1A1 induction by metyrapone. Biochem J 1998; 331: 273-281.
28. Wright MC, Maurel P and Paine AJ. Induction of cytochrome P450 3A by metyrapone in human hepatocyte culture. Human & Experimental Toxicology 1996; 15: 203-204.
29. Wright MC, Paine AJ, Skett P and Auld R. Induction of rat hepatic glucocorticoid inducible cytochrome P450 3A subfamily by metyrapone. J Steroid Biochem Molec Biol 1994; 48: 271-276.
30. Lake BG and Paine AJ. The effect of hepatocyte culture conditions on cytochrome P-450 linked drug metabolising enzymes. Biochem Pharmacol 1982; 31: 2141-2144.
31. Lake BG and Paine AJ. Induction of hepatic cytochrome P-450 and drug metabolism by metyrapone in the rat: Relevance to its effects in rat liver cell culture. Xenobiotica 1983; 13: 725-730.
32. Paine AJ, Maurer B and von Wartburg BR. The application of hepatocyte culture to the identification of pathways of drug metabolism: Studies with Pindolol and Fluperlapine. Biochem Pharmacol 1984; 33: 3111-3114.
33. Paine AJ and Hockin LJ. The maintenance of cytochrome P-450 in liver cell culture: Recent studies on P-450 mediated mechanisms of toxicity. Toxicology 1982; 25: 41-45.

34. Paine AJ. *In vitro* toxicology: an alternative to animal testing in safety evaluation? In: *In Vitro* Toxicology: an alternative to animal testing? Jolles G and Cordier A, Eds. Academic Press: London. 1992: 33-38.
35. Paine AJ. Validity and reliability of *in vitro* systems in safety evaluation. Environmental Toxicology and Pharmacology 1996; 2: 207-212.
36. Wright MC and Paine AJ. Resistance of precision cut liver slices to the toxic effects of Menadione (vitamin K). Toxicology *In Vitro* 1992; 6: 475-481.
37. Paine AJ. The design of toxicology studies. In: General and Applied Toxicology. Turner P, Marrs T and Ballantyne, Eds. Macmillan: London. 1993: 231-244.
38. Clayton DF and Darnell JE. Changes in liver specific compared to common gene transcription during primary culture of mouse hepatocytes. Mol Cell Biol 1983; 3: 1552-1561.
39. Clayton DF, Harrelson AL and Darnell JE. Dependence of liver-specific transcription on tissue organisation. Mol Cell Biol 1985; 5: 2623-2632.
40. Pahl HL and Baeuerle PA. The ER-overload response: activation of NF-κ B. Trends Biochem Sci 1997; 22: 63-67.
41. Rothschild MA, Oratz M and Schreiber SS. Serum albumin. Hepatology 1988; 8: 385-401.
42. Clayton DF, Weiss M and Darnell JE. Liver-specific RNA metabolism in hepatoma cells: variations in transcription rates and mRNA levels. Mol Cell Biol 1985; 5: 2633-2641.
43. Nebert DW. The Ah locus: genetic differences in toxicity, cancer, mutation and birth defects. Critical Reviews in Toxicology 1989; 20: 153-174.
44. Yamazoe Y, Murayama N and Kato R. Suppressive regulation of rat hepatic cytochrome P-450s by growth hormone and thyroid hormone. In: Drug Metabolising Enzymes: Genetics, Regulation and Toxicology. Ingleman-Sundberg M, Gustafsson JA and Orrenius S, Eds. Karolinska Institutet: Stockholm, Sweden. 1990: 30.
45. Gonzalez FJ, Matsuna T and Nagata K. Structure and regulation of P-450s in the rat P450IIA gene subfamily. Drug Metabolism Reviews 1989; 20: 827-837.
46. Gonzalez FJ. Molecular genetics of the P450 superfamily. J Pharm Ther 1990; 45: 1-38.
47. Waxman DJ and Azaroff L. Phenobarbital induction of cytochrome P450 gene expression. Biochem J 1992; 281: 577-592.
48. Zaphiropoulos PG, Mode A, Norstedt G and Gustafsson J-A. Regulation of sexual differentiation in drug and steroid metabolism. Trends in Pharmacological Sciences 1989; 10: 149-153.
49. Yamazoe Y, Murayama N, Shimada M, Imaoka S, Funae Y and Kato R. Suppression of hepatic levels of an ethanol inducible P-450DM/j by growth hormone: relationship between increased level of P-450DM/j and depletion of growth hormone in diabetes. Mol Pharmac 1989; 36: 716-722.
50. Issemann I and Green S. Activation of a member of the steroid hormone receptor superfamily by peroxisome proliferators. Nature 1990; 374: 645-650.
51. Li YC, Wang DP and Chiang JY. Regulation of cholesterol 7 α hydroxylase in the liver: Cloning, sequencing and regulation of cholesterol 7 α hydroxylase mRNA. J Biol Chem 1990; 265: 12012-12019.
52. Shean K and Paine AJ. Immunochemical quantitation of cytochrome P450IA and IIB subfamilies in the livers of metyrapone treated rats: Relevance to its effects in liver cell culture. Biochem J 1990; 267: 715-719.

Chapter 25

Optimising conditions for induction of cytochrome P450 in primary hepatocyte cultures

Gabrielle M. Hawksworth[1] and Ieva Stupans[2]
[1]Departments of Medicine and Therapeutics and Biomedical Sciences, Polwarth Building, Foresterhill, Aberdeen AB25 2ZD, U.K. and [2]School of Pharmacy and Medical Sciences, North Terrace, Adelaide, 5000, SA, Australia

Key words coculture, cytochrome P450, dexamethasone, dimethyl sulphoxide, ethanol induction, isoforms, Matrigel, media, oxysterols, peroxisome proliferators, phenobarbitone, polycyclic hydrocarbons, primary culture, sodium butyrate, spheroids, substrata

25.1 Introduction

Isolated hepatocytes undergo rapid dedifferentiation in primary culture, one of the indices of this dedifferentiation being differential loss of the isoforms of cytochrome P450 (CYP) [1]. The loss of CYPs occurs to varying extents and at different rates across species. The most extensively investigated species and the one in which there is the most marked loss of CYPs is the rat [2]. Chapter 24 describes some strategies and mechanisms involved in the maintenance of CYP in rat hepatocyte cultures. Some of the same strategies have been used to enhance induction of CYP in culture. However, there is no correlation between maintenance of constitutive CYPs and optimising induction of CYPs. Since the inducible CYP isoforms are differently regulated [3], there is unlikely to be one set of conditions where all CYPs can be maximally induced. Similarly species differences in CYP induction necessitate different culture conditions across species [4]. The focus of this chapter will be optimisation of conditions for induction of CYP in rat hepatocyte cultures *i.e.* such that the extent of induction reflects that seen *in vivo*. It will also refer to human hepatocyte cultures and to dog and rabbit hepatocyte cultures where data are available. Although there is pressure to decrease the use of dogs in research, the dog is still used in toxicology studies for regulatory purposes and validation of the use of cultured dog hepatocytes could reduce the use of this species in research.

M.N. Berry and A.M. Edwards (eds.), The Hepatocyte Review, 421–432.

25.2 CYP1A1/1A2

CYP1A1 is induced forty to fifty times in male Sprague Dawley rats, when quantified by either ethoxyresorufin deethylase or by Western blotting [5]. This isoform is the most readily inducible and shows at least tenfold induction under almost any culture conditions. However induction can be enhanced when cells are cultured with 1-2% dimethylsulphoxide (DMSO) in the medium, or with Matrigel as the substratum or if the hepatocytes are cocultured with a liver epithelial cell line. This can markedly improve the sensitivity of detection of induction, which may be useful if multiwell plate techniques are to be used for high throughput screening. Hepatocytes cultured under all three conditions maintain a more rounded cell shape, but this in itself does not result in enhanced induction. Both DMSO and sodium butyrate induce differentiation in rat hepatocyte cultures. This can be seen from the cell shape in culture and prevention of the increase in γ-glutamyltranspeptidase with time in culture. However addition of sodium butyrate (5 mM) to the medium does not result in enhanced CYP1A1 induction. Hepatocytes cultured on Matrigel exhibit enhanced expression of many liver-specific functions and also decreased responsiveness to mitogenic stimuli.

Culturing hepatocytes in the presence of 1-2% DMSO elicits a pleiotropic response. The cells are more rounded and there are alterations in the cellular cytoskeleton [6]. However 2% DMSO is toxic to the cells, causing a significant decrease in hepatocyte attachment over a seven day period and the appearance of abnormal mitochondria. 1% DMSO is equally effective and is less toxic to the cells. Heme availability is increased due to both an increase in δ-aminolevulinic acid synthetase and a decrease in heme oxygenase [7]. However this does not lead to enhanced induction for all isoforms of rat CYP. Coculture of hepatocytes with a liver epithelial cell line also results in reciprocal changes, with enhanced epithelial cell growth occurring in the coculture system. Another difficulty with coculture is that the epithelial cell lines generated differ in their characteristics and in their ability to enhance induction. Cocultures in a perifusion system or culture between two layers of collagen gel demonstrate a further enhancement of induction, but these are significantly more complex systems [8,9]. The presence of 5% fetal calf serum in the medium does not have any significant effect on CYP1A1 induction, in contrast to CYP3A1/3A23 induction.

The polycyclic hydrocarbons do not induce CYP1A2 in human hepatocyte cultures to the same extent as CYP1A1 induction in rodents. Benzimidazoles such as omeprazole induce CYP1A2 in human hepatocytes, but are not inducers in rodents either *in vivo* or in cell culture [10]. Dexamethasone (10^{-7} M) is required for 3-methylcholanthrene-inducible CYP1A2 expression in primary rat hepatocyte cultures [11,12], but is not required for CYP1A2 expression in rabbit or human hepatocytes. Use of Waymouth MB-752 medium, rather than Dulbecco's modified Eagle's and Ham's F-12 (DME/F12) medium (1:1), results in suboptimal CYP1A2 mRNA expression in rat hepatocytes. In these studies DME/F12 was supplemented with 5% fetal bovine serum and 5 µg/ml insulin, in addition to dexamethasone.

25.3 CYP2B1/2

It is generally accepted that CYP2B1/2 are induced in the liver by phenobarbitone by means of a direct effect that enhances transcription of the 2B1/2 genes. It is important to note, however, that, to date no phenobarbitone receptor has been identified in the liver [13], although phenobarbitone-responsive elements have been identified within the CYP2B2 promoter [14,15]. It has been difficult to demonstrate expression of 2B1/2 genes either in the presence or absence of inducer in continuously replicating cell lines, or in primary, non-dividing cultures of adult rat hepatocytes. Table 1 summarizes data taken from key references which have examined phenobarbitone induction in primary rat hepatocyte cultures. There are obvious difficulties in comparing CYP2B-dependent activity in terms of the degree of increase over basal levels where basal levels are at, or below, the limit of detection.

An important advance for phenobarbitone induction was the finding that hepatocytes cultured on Matrigel in a serum-free medium formed rounded clusters which permitted phenobarbitone-mediated induction of CYP2B1/2 mRNAs and immunoreactive protein [16]. Studies investigating biophysical elements of the culture substratum suggest that extracellular factors such as laminin exert effects through transmembrane signal transduction [17].

Table 1. Comparison of induction protocols, detection methods and induction levels achieved for CYP2B1/2

Substratum/ culture method	Media	Fold induction and detection method	Reference
'Matrigel'	Serum-free medium (modification of Waymouth's)	Immunoreactive protein +++ mRNA +++	[16]
'Matrigel'	Serum-free medium (modification of Waymouth's)	mRNA +++	[26]
'Matrigel' or Vitrogen	Williams E	PROD +++ Immunoblotting +++	[18]
Collagen coated	Chee's medium	PROD +++ Immunoblotting +++	[19]
Not stated/co culture with epithelial cells	75% MEM + 25%199 medium	ECOD activity and CYP450 content higher than in pure culture	[22]
Nunc plastic flasks/co-culture with epithelial cells	75% MEM + 25% 199 medium	PROD activity reported, no activity found in reference pure culture	[23]
Not stated/Co-culture with rat liver epithelial cells	75% MEM + 25% 199 medium	mRNA ++	[24]
'Matrigel' added to medium	William's E	PROD ++ mRNA ++	[20]
Not stated/Co-culture with rat liver epithelial cells	75% MEM + 25% 199 medium	PROD ++ mRNA ++	[25]

Induction is shown as +, 2- to 5-fold; ++, 5- to 50-fold; and +++, greater than 50-fold

The original observations using Matrigel prompted many other studies which examined the role of both substratum and media composition on the induction of CYP2B1/2. Experiments comparing Matrigel with a collagen matrix showed that both substrata resulted in CYP2B1/2 induction when used with Williams' E medium [18], whereas the use of another highly enriched medium (Chee's) with "Vitrogen" coating also gave good induction over a comparatively long time period at induction levels approaching those seen *in vivo* [19]. A detailed examination of substratum composition and media composition subsequently suggested that the use of a collagen substratum together with either a single or daily overlay of Matrigel in Williams' E medium resulted in induction *in vitro* which closely resembled that obtained *in vivo* [20].

Inclusion of dexamethasone in media (10^{-6} M) has been demonstrated to enhance induction of CYP2B1/2 by phenobarbitone [19], although in these studies dexamethasone alone was not an effective inducing agent. More recent studies [21] demonstrated a biphasic regulation of CYP2B1/2 by dexamethasone. Using a Matrigel substratum, low concentrations of dexamethasone were found to enhance induction of CYP2B1/2 by phenobarbitone (2.5-fold at 10^{-7} M), whilst high doses of dexamethasone alone were found to induce CYP2B1/2 (4.1-fold at 10^{-5} M). Other studies [20] have suggested that the dexamethasone may help to maintain cell morphology and stabilise the expression of differentiated functions such as albumin mRNA.

Sustained CYP expression has also been achieved in primary hepatocytes by coculture of hepatocytes together with rat liver epithelial cells. Early studies [16] reported a lack of induction in coculture. However, comparison of hepatocyte cultures in the presence and absence of cocultured epithelial cells demonstrated superior maintenance and induction of enzyme activities by phenobarbitone in the presence of epithelial cells, although these do not approach the levels of induction seen *in vivo* [22,23]. More recent investigations, using Northern blotting or RNase protection assays reported induction by phenobarbitone of approximately 15 to 20 times [24,25] in hepatocytes cocultured with epithelial cells.

Most studies using phenobarbitone in isolated hepatocyte culture have used high concentrations of phenobarbitone (0.5-2 mM). However, the dose dependence of phenobarbitone induction has been determined, with ED_{50} values of 1.5×10^{-5} and 5.7×10^{6} M respectively for CYP2B1 and CYP2B2 [26]. Other experiments utilising phenobarbitone in hepatocyte culture have demonstrated ED_{50} values for induction of activity which represent a good qualitative and quantitative approximation of the hepatic ED_{50} value for phenobarbitone-type induction *in vivo* [27]. Recent studies [28] have shown pleiotropic effects of high phenobarbitone concentrations (3 mM), including suppression of TGF-α and down-regulation of transcription factors associated with proliferation, as well as the expression of CYP2B1/2.

Aside from the concentration of phenobarbitone used in isolated hepatocyte culture, the degree of induction has been shown to vary with alteration in oxygen tension, with higher induction at a higher oxygen tension of 13%, equivalent to the periportal oxygen tension [29]. The induction was affected by the insulin concentration and the insulin:glucagon ratio. In untreated hepatocyte cultures, maintained in the absence of insulin, increased CYP2B1/2 expression has been demonstrated [30]. The time of insulin addition, however, in these studies was shown to affect the degree of induction by phenobarbitone.

A wide range of inducers of CYP2B1/2 other than phenobarbitone has been reported. There is interest in both endogenous and exogenous regulation. For example, lovastatin, a widely used HMG-CoA reductase inhibitor, has been found to increase CYP2B1/2 mRNA levels at concentrations required for its inhibition of HMG-CoA reductase (3×10^{-5} M) [31]. Lovastatin was also found to potentiate the induction of CYP2B1/2 by phenobarbitone (10^{-4} M). These studies implicated oxysterols as possible intracellular modulators of CYP2B1/2. Evidence of regulation of CYP2B1/2 by intracellular levels of fatty acids and ketone bodies has also been found [32]. Ciprofibrate has also been found to induce CYP2B1/2 [33], albeit at much higher concentrations than those required to induce CYP2E1 (discussed below).

Organochlorine pesticides such as mirex, heptachlor and dieldrin have been found to be potent CYP2B1/2 inducers, although kepone showed selective induction of CYP2B2 [34]. The pesticide DDT (1,1,1-trichloro-2,2-bis(p-chlorophenyl)ethane) and its metabolites have been found to be inducers of CYP2B1/2 in the rat and have also been found to induce CYP2B and CYP3A in hepatocyte cultures [35]. However there are no examples where induction in rat hepatocyte cultures measured as an increase in CYP2B-dependent activity, as opposed to an increase in mRNA, reaches the extent of induction seen *in vivo* (40- to 50-fold) [5].

25.4 CYP2E1

Overall studies of CYP2E1 protein expression *in vivo* have not yielded consistent data on the mechanism(s) of CYP2E1 induction. Also the extent of transcriptional, post-transcriptional and post-translational regulation of CYP2E1 varies markedly across species. Treatment of rats with solvents such as ethanol increases CYP2E1 levels in the absence of increased

CYP2E1 mRNA levels, suggesting post-translational mechanisms of regulation or increased translational efficiency. The inability to maintain or enhance CYP2E1 expression in primary cultured hepatocytes has limited studies of CYP2E1 induction.

The induction of CYP2E1 by ethanol has not been readily demonstrable *in vitro* [33]. Without induction, constitutive CYP2E1 was recognized as having a very short half-life in primary culture [36]. Early studies reported that even with Matrigel substrata there was no induction of CYP2E1 with ethanol. However using Williams' E medium and Matrigel substrata, high concentrations of ethanol (200 mM) were found to increase para-nitrophenol hydroxylation threefold and also increase levels of immunoreactive protein [37].

A detailed comparison of CYP2E1 induction with both collagen and Matrigel substrata using Chee's medium found that CYP2E1 levels but not CYP2E1 mRNA levels were elevated by induction with ethanol and pyridine [33]. In this study ciprofibrate was found to enhance induction of CYP2E1 both at the protein and mRNA levels. In another study pyridine was not found to increase mRNA levels in cells maintained on Matrigel in Williams' E medium, in spite of increases in para-nitrophenol hydroxylation and CYP2E1 content [38].

Ethanol, acetone and pyrazine have recently been compared as inducers of CYP2E1. Using modified Chee's medium in flasks covalently coated with Vitrogen, and a 72 h hepatocyte maintenance period before exposure to the xenobiotics, concentrations of 1 mM of each xenobiotic were reported to increase CYP2E1 protein expression without any induction of mRNA [38]. The magnitude of induction of CYP2E1 protein (twofold to eightfold) was comparable to levels of induction seen *in vivo*.

Other components of media such as serum and insulin concentration have been examined. The content of CYP2E1, CYP2E1 mRNA levels and CYP2E1 catalytic activity was greater in fetal cultures grown in the presence of 2.5% fetal calf serum than in 15% fetal calf serum, suggesting that factors present in the serum limited expression or stability [40]. Another study that examined the effects of insulin in culture showed that decreasing insulin concentrations in culture media increased CYP2E1 mRNA levels by as much as elevenfold [30].

25.5 CYP3A

In the rat, marked sex differences are observed in the metabolism of xenobiotics by CYP3A. The cDNAs of five members of the rat CYP3A subfamily have been characterized - CYP3A1, CYP3A2, CYP3A9,

CYP3A18 and CYP3A23. Rat hepatic CYP3A genes are highly inducible by compounds such as dexamethasone, pregnenolone 16α-carbonitrile and phenobarbitone. Another characteristic feature of the physiological regulation of hepatic CYP3A genes is age- and sex-dependent expression. CYP3A2 and CYP3A18 are constitutively expressed in adult male rats, whereas the hepatic expression of CYP3A9 is higher in adult females than in males. Hence female rats are frequently used to investigate CYP3A1/3A23 induction by determining testosterone hydroxylation at the 6β-, 2β- or 15β-positions [41]. This sexual dimorphism is primarily mediated by the gender-specific profile of pituitary growth hormone secretion. It is not therefore surprising that CYP3A induction is influenced by the presence or absence of fetal calf serum (FCS). Dexamethasone-mediated CYP3A induction is 6.5 times higher when rat hepatocytes are cultured in the absence of FCS. This FCS-mediated effect is not observed in human hepatocytes, cultured in the presence or absence of FCS [42]. In the presence of FCS, DMSO increases CYP3A induction by synthetic steroids approximately twofold. In contrast to CYP1A and CYP2B, CYP3A1/3A23/3A2 induction is similar whether cells are cultured on collagen or Matrigel [43].

There are at least two functionally distinct CYP3A isoforms in the dog. Dexamethasone does not induce the formation of 6β-OH testosterone in the male or female Beagle dog [44], although phenobarbitone can induce CYP3A12 in dog hepatocytes [45].

In man, the major isoforms are 3A4 and 3A5. Is it well established that rifampicin is a potent inducer in man, rabbit and dog, but only induces in the rat at very high doses [4]. Induction by rifampicin in the dog is higher in the absence of FCS, but the differences are not significant. Pregnenolone-16α-carbonitrile does not significantly induce CYP3A in man or rabbit, in contrast to the induction seen in rat hepatocytes. Induction protocols for CYP3A induction *in vivo* and in primary cultures of human hepatocytes have been described by Daujat *et al* [46].

There has been interest recently in inhibitors of the mevalonic acid pathway, as possible inducers of CYP3A1. Lovastatin, an HMGCo-A reductase inhibitor, increases CYP3A1 transcription in cultured rat hepatocytes [31], and it has been suggested that this may be due to the increased level of oxysterols. However, whilst lovastatin increased phenobarbitone-mediated CYP3A1 induction, it had no effect on dexamethasone-mediated CYP3A1 induction [31]. ACAT inhibitors also result in CYP3A1 induction [47]. Induction by this ACAT inhibitor is significantly higher in human hepatocyte cultures than in rat hepatocyte cultures.

25.6 CYP4A

The most striking feature of CYP4A induction by peroxisome proliferators, such as the hypolipidemic drugs *e.g.* clofibrate, ciprofibrate, methyl clofenapate or di-(2-ethylhexyl)phthalate, is the marked species differences in response to these agents, both *in vivo* and *in vitro* [48]. This is suggested to be due to the lower expression of PPARα in nonresponsive species [49]. Bifonazole, a non-carboxylate inducer of CYP4A, is 160 and 40 times more potent than clofibrate for induction of the 11- and 12-hydroxylation of lauric acid respectively [50]. As CYP4A is so highly inducible in rat hepatocyte cultures, relatively little effort has been expended in optimising culture conditions for induction. Both RPMI 1640, containing 5% FCS, 1 μM insulin, 0.1 mM hydrocortisone, and Williams' E medium with standard additions, are used to investigate CYP4A induction [51].

25.7 Spheroids

Landry *et al* [52] observed that hepatocyte culture under conditions where attachment to the substratum was inhibited resulted in the formation of small sphere-shaped aggregates - "spheroids". Rat hepatocyte spheroids are reported to maintain differentiated function for prolonged periods in culture, possibly due to maintenance of cell shape and cell-cell contacts [53]. A number of publications from 1992 onwards have suggested that there is improved maintenance of CYP-dependent metabolism in spheroids compared with monolayer cultures. However activities are not maintained at levels found on initial cell isolation and most of these papers do not involve a direct comparison of monolayer cultures and spheroids. Inevitably the culture media are different in the two systems. Our laboratory and others have preliminary data suggesting that CYP3A induction by dexamethasone may be enhanced in spheroids compared with hepatocyte cultures, but this needs further rigorous evaluation.

References

1. Thurman RG and Kaufman FC. Factors regulating drug metabolism in intact hepatocytes. Pharmacol Rev 1980; 31: 229-251.
2. Guzelian PS, Bissell DM and Meyer UA. Drug metabolism in adult rat hepatocytes in primary monolayer culture. Gastroenterology 1977; 72: 1232-1329.
3. Gonzalez FJ. The molecular biology of cytochrome P450s. Pharmacol Rev 1989; 40: 243-288.
4. Kocarek TA, Schuetz EG, Strom SC, Fisher RA and Guzelian PS. Comparative analysis of cytochrome P450 induction in primary cultures of rat, rabbit and human hepatocytes. Drug Metab Disp 1995; 23: 415-421.

5. Burke MD, Thompson S, Elcombe CR, Halpert J, Haaparanta T and Mayer RT. Ethoxy-, pentoxy- and benzyloxyphenoxazones and homologues: a series of substrates to distinguish between different induced cytochromes P450. Biochem Pharmacol 1985; 34: 3337-3345.
6. Muakkassah-Kelly SF, Bieri F, Waechter F, Bentley P and Staubli W. Long-term maintenance of hepatocytes in primary culture in the presence of DMSO: further characterisation and the effect of nafenopin, a peroxisome proliferator. Exp Cell Res 1987; 171: 37-51.
7. Lindsay CK, Chenery RJ and Hawksworth GM. Primary culture of rat hepatocytes in the presence of dimethylsulphoxide: a system to investigate the regulation of cytochrome P4501A. Biochem Pharmacol 1991; 42: 517-525.
8. Gebhardt R, Wegner H and Alber J. Perifusion of co-cultured hepatocytes: optimization of studies on drug metabolism and cytotoxicity *in vitro*. Cell Biol Toxicol 1996; 12: 57-68.
9. Rotem A, Matthew HWT, Hsiao PH, Toner M, Tompkins RG and Yarmush ML. The activity of cytochrome P4501A1 in stable cultured hepatocytes. Toxic *In Vitro* 1995; 9: 139-149.
10. Li AP, Maurel P, Gomez-Lechon MJ, Cheng LC and Jurima-Romet M. Preclinical evaluation of drug-drug interaction potential: present status of the application of primary human hepatocytes in the evaluation of cytochrome P450 induction. Chem Biol Interactions 1997; 107: 5-16.
11. Silver G, Reid LM and Krauter KS. Dexamethasone-mediated regulation of 3-methylcholanthrene-induced cytochrome P450d mRNA accumulation in primary rat hepatocyte cultures. J Biol Chem 1990; 265: 3134-3138.
12. Shih H, Pickwell GV, Guenette DK, Bilir B and Quattrochi LC. Species differences in hepatocyte induction of CYP1A1 and CYP1A2 by omeprazole. Human Exp Toxicol 1999; 18: 95-105.
13. Waxman DJ and Azaroff L. Phenobarbital induction of cytochrome P-450 gene expression. Biochem J 1992; 281: 577-592.
14. Trottier E, Belzil A, Stoltz C and Anderson A. Localisation of a phenobarbital-responsive element (PBRE) in the 5'-flanking region of the rat CYP2B2 gene. Gene 1995; 158: 263-268.
15. Stoltz C, Vachon MH, Trottier E, Dubois S, Paquet Y and Anderson A. The CYP2B2 phenobarbital response unit contains an accessory factor element and a putative glucocorticoid response element essential for conferring maximal phenobarbatial responsiveness. J Biol Chem 1998; 273: 8528-8536.
16. Schuetz EG, Li D, Omiecinski CJ, Muller-Eberhard U, Kleinman HK, Elswick B and Guzelian PS. Regulation of gene expression in adult rat hepatocytes cultured on a basement membrane matrix. J Cell Physiol 1998; 134: 309-323.
17. Brown SE, Guzelian CP, Schuetz E, Quattrochi LC, Kleinman HK and Guzelian PS. Critical role of extracellular matrix on induction by phenobarbital of cytochrome P450 2B1/2 in primary cultures of adult rat hepatocytes. Lab Invest 1995; 73: 818-827.
18. Sinclair PR, Bement WJ, Haugen SA, Sinclair JF and Guzelian PS. Induction of cytochrome P-450 and 5-aminolevulinate synthase activities in cultured rat hepatocytes. Cancer Res 1990; 50: 5219-5224.
19. Waxman DJ, Morrisey JJ, Naik S and Jauregui HO. Phenobarbital induction of cytochromes P-450. High level long term responsiveness of primary rat hepatocyte cultures to drug induction, and glucocrticoid dependence of the phenobarbital response. Biochem J 1990; 271: 113-119.
20. Sidhu JS, Farin FM and Omiecinski CJ. Influence of extracellular matrix overlay on phenobarbital-mediated induction of CYP2B1, 2B2 and 3A1 genes in primary adult rat hepatocyte culture. Arch Biochem Biophys 1993; 301: 103-113.

21. Kocarek TA, Schuetz EG and Guzelian PS. Biphasic regulation of cytochrome P450 2B1/2 mRNA expression by dexamethasone in primary cultures of adult rat hepatocytes maintained on matrigel. Biochem Pharmacol 1994; 48: 1815-1822.
22. Rogiers V, Vandenberghe Y, Callaerts A, Verleye G, Cornet M et al. Phase I and Phase II xenobiotic biotransformation in cultures and co-cultures of rat hepatocytes. Biochem Pharmacol 1990; 40: 1701-1706.
23. Niemann C, Gauthier JC, Richert L, Ivanov MA, Melcion and Cordier A. Rat adult hepatocytes in primary pure and mixed monolayer culture. Comparison of the maintenance of mixed function oxidase and conjugation pathways of drug metabolism. Biochem Pharmacol 1991; 42: 373-379.
24. Akrawi M, Rogiers V, Vandenberghe Y, Palmer CN, Vercruysse A et al. Maintenance and induction in co-cultured rat hepatocytes of components of the cytochrome P450-mediated mono-oxygenase. Biochem Pharmacol 1993; 45: 1583-1591.
25. Lerche C, Fautrel A, Shaw PM, Glaise D, Ballet F, Guillouzo A and Corcos L. Regulation of the major detoxication functions by phenobarbital and 3-methylcholanthrene in co-cultures of rat hepatocytes and liver epithelial cells. Eur J Biochem 1997; 244: 98-106.
26. Kocarek TA, Schuetz EG and Guzelian PS. Differentiated induction of cytochrome P450b/e and P450p mRNA by dose of phenobarbital in primary cultures of adult rat hepatocytes. Mol Pharmacol 1990; 38: 440-444.
27. Nims RW, Sinclair PR, Sinclair JF, Thomas PE, Jones CR et al. Pharmacodynamics of cytochrome P450 2B induction by phenobarbital, 5-ethyl-5-phenylhydantoin, and 5-ethyl-phenyloxazolidinedione in the male rat liver or in cultured rat hepatocytes. Chem Res Toxicol 1993; 6: 188-196
28. Miyazaki M, Mars WM, Runge D, Kim TH, Bowen WC and Michalopoulos GK. Phenobarbital suppresses growth and accelerates restoration of differentiation markers of primary culture rat hepatocytes in the chemically defined hepatocyte growth medium containing hepatocyte growth factor and epidermal growth factor. Exp Cell Res 1998; 241: 445-457.
29. Saad B, Thomas H, Schawalder H, Waechter F and Maier P. Oxygen tension, insulin, and glucagon affect the preservation and induction of cytochrome P450 isoforms in cultured rat hepatocytes. Toxicol Appl Pharmacol 1994; 126: 372-379.
30. Woodcroft KJ and Novak RJ. Insulin effects on CYP2E1, 2B, 3A and 4A expression in primary cultured rat hepatocytes. Chem Biol Interactions 1997; 107: 75-91.
31. Kocarek TA, Scheutz EG and Guzelian PS. Regulation of phenobarbital-inducible cytochrome P450 2B1/2 mRNA by lovastatin and oxysterols in primary cultures of adult rat hepatocytes. Toxicol Appl Pharmacol 1993; 120: 298-307.
32. Zangar RC and Noval RF. Effects of fatty acids and ketone bodies on cytochromes P450 2B, 4A and 2E1 expression in primary cultured rat hepatocytes. Arch Biochem Biophys 1997; 337: 217-224.
33. Zangar RC, Woodcroft KJ, Kocarek TA and Novak RF. Xenobiotic-enhanced expression of cytochrome P450 2E1 and 2B1/2B2 in primary cultured rat hepatocytes. Drug Metab Dispos 1995; 23: 681-687.
34. Kocarek TA, Schuetz EG and Guzelian PS. Selective induction of cytochrome P450e by kepone (chlordecone) in primary cultures of adult rat hepatocytes. Mol Pharmacol 1991; 40: 203-210.
35. Nims RW, Lubet RA, Fox SD, Jones CR, Thomas PE, Reddy AB and Kocarek TA. Comparative pharmacodynamics of CYP2B induction by DDT, DDE and DDD in male rat liver and cultured rat hepatocytes. J Toxicol Environ Health 1998; 53: 455-477.
36. Eliasson E, Mkrtchian S and Ingleman-Sundberg M. Hormone- and substrate-regulated intracellular degradation of cytochrome P450 (2E1) involving MgATP-activated rapid proteolysis in the endoplasmic reticulum membranes. J Biol Chem 1992; 267: 15765-15769.

37. Sinclair JF, McCaffrey J, Sinclair PR, Bement WK, Lambrecht LK et al. Ethanol increases cytochromes P450IIE, IIb1/2, and IIIA in cultured rat hepatocytes. Arch Biochem Biophys 1991; 284: 360-365.
38. Wu D, Ramin SA and Cederbaum AI. Effect of pyridine on the expression of cytochrome P450 isozymes in primary rat hepatocyte culture. Mol Cell Biochem 1997; 173: 103-111.
39. Woodcroft KJ and Novak RF. Xenobiotic-enhanced expression of cytochromes P450 2E1 and 2B in primary cultured rat hepatocytes. Drug Metab Dispos 1998; 26: 372-378.
40. Wu D and Cederbaum AI. Expression of cytochrome P4502E1 in rat fetal hepatocyte culture. Mol Pharmacol 1996; 49: 802-807.
41. Williams JA, Chenery RJ and Hawksworth GM. Induction of CYP3A enzymes in human and rat hepatocyte cultures. Biochem Soc Trans 1994; 22: 131S.
42. Jackson FC, Chenery RJ and Hawksworth GM. Fetal calf serum decreases CYP3A induction in rat but not in dog or human hepatocyte cultures. Biochem Soc Trans 1997; 25: S607.
43. Silva JM, Morin PE, Day SH et al. Refinement of an *in vitro* cell model for cytochrome P450 induction. Drug Metab Disp 1998; 26: 490-496.
44. Jayyosi Z, Muc M, Erick J, Thomas PE and Kelley M. Catalytic and immunochemical characterisation of cytochrome P450 isozyme induction in dog liver. Fund Appl Toxicol 1996; 31: 95-102.
45. Ciaccio PJ and Halpert JR. Characterisation of a phenobarbital-inducible dog liver cytochrome P450 structurally related to rat and human enzymes of the CYPIIIA (steroid-inducible) gene subfamily. Arch Bichem Biophys 1989; 271: 284-299.
46. Daujat M, Pichard L, Pineau T, Fabre G, Bonfils C and Maurel P. Induction protocols for cytochromes P450IIIA *in vivo* and in primary cultures of animal and human hepatocytes. Methods Enzymol 1991; 206: 345-353.
47. Williams JA, Chenery RJ, Berkhout TA and Hawksworth GM. Induction of cytochrome P4503A by the antiglucocorticoid mifepristone and a novel hypocholesteraemic drug. Drug Metab Disp 1997; 25: 757-761.
48. Elcombe CR, Bell DR, Elias E, Hasmall SC and Plant MJ. Peroxisome proliferators: species differences in response of primary hepatocyte cultures. Ann NY Acad Sci 1996; 804: 628-635.
49. Palmer CN, Hsu MH, Griffin KJ, Raucy JL and Johnson EF. Peroxisome proliferator activated receptor-α expression in human liver. Mol Pharmacol 1998; 53: 14-22.
50. Sabsevari O, Hatcher M, O'Sullivan M, Kentish P and Gibson G. Comparative induction of cytochrome P4504A in rat hepatocyte culture by the peroxisome proliferators, bifonazole and clofibrate. Xenobiotica 1995; 25: 395-403.
51. Gibson GG and Lake BG. Induction protocols for the cytochrome P450IV sub family in animals and primary hepatocyte cultures. Meth Enzymol 1991; 206: 353-364.
52. Landry J, Bernier D, Ouellet C, Goyette R and Marceau N. Spheroidal aggregate culture of rat liver cells: histotypic reorganization, biomatrix disposition and maintenance of functional activities. J Cell Biol 1985; 101: 914-923.
53. Juillerat M, Marceau N, Coeytaux S, Sierra F, Kolodziejczyk E and Guigoz Y. Expression of organ-specific structures and functions in long-term cultures of aggregates from adult rat liver cells. Toxicology *In Vitro* 1997; 11: 57-69.

Chapter 26

The use of isolated hepatocytes to study the mechanisms of action of environmental contaminants

Stéphanie Reader and Francine Denizeau
Département de Chimie, Université du Québec à Montréal, Canada

Key words: albumin secretion, apoptosis, cadmium, copper metallothionein, O-6-methylguanine, trout hepatocytes, vitellogenin secretion, xenobiotic metabolism

26.1 Introduction

Understanding the impact of environmental toxic substances encompasses many challenges. Application of the basic concepts of toxicology, as developed in the medical sciences, has to be adapted and, in some instances, reinvented. In contrast to what is seen in laboratory or clinical settings when drugs are used, exposure to xenobiotics in the environment is most often poorly characterized. In addition, under environmental conditions, the poison is usually low in dose and repeatedly absorbed over prolonged periods of time. Because toxic substances almost invariably occur in the environment in the form of complex mixtures, simultaneous exposure to multiple agents is frequent and the biological responses seen are difficult to interpret.

A fundamental understanding of the mechanisms of toxicity of xenobiotics provides a sound basis for producing answers to the challenges of environmental toxicology. Knowing the cellular and molecular changes that are associated with the development of a pathological process, from toxic insult to disease, allows the identification of specific biomarkers that can be used as probes for toxic exposure or effects. Reliable biomarkers are very much needed for more efficient environmental surveillance.

Cellular models are particularly well suited for studying the mechanisms of action of environmental xenobiotics. Cells from various tissues and organisms can be used, and invaluable information that could not be obtained in any other system can be acquired. Hepatocytes stand out as cells of choice for the purpose of studying environmental toxicants because they are a major site of xenobiotic metabolism in living organisms. Hepatocytes represent a universal target for environmental contaminants.

This review shows how isolated hepatocytes can provide useful

M.N. Berry and A.M. Edwards (eds.), The Hepatocyte Review, 433–453.

mechanistic data on the mode of action of toxic agents ubiquitously found in the environment. The use of *in vitro* systems as a complementary approach to *in vivo* evaluation is a concept that has been particularly stimulating in the field of environmental hepato-cytotoxicology.

26.2 Biotransformation of environmental xenobiotics

26.2.1 Role of the liver

The liver is the major organ responsible for xenobiotic biotransformation. In general, this process consists of reactions of two types, Phase I and Phase II reactions. The former include oxidation, reduction, and hydrolysis, while the latter are characterized by conjugation of the exogenous compound with an endogenous hydrophilic moiety such as glutathione (GSH), glucuronic acid, sulphate, or amino acids [1]. In general, the metabolism of hydrophobic toxicants into hydrophilic molecules increases their elimination, thereby reducing their toxicity. Paradoxically however, biotransformation is often a major step in chemically-induced hepatotoxicity because metabolism can produce reactive species that bind to cellular constituents, in particular proteins, RNA and DNA, and affect their function. Biotransformation is then regarded as a bioactivation process. A family of enzymes that plays a crucial role in Phase I reactions and in the bioactivation of toxic agents is cytochrome P450 (CYP). In the field of environmental toxicology, special interest has been devoted to the isoforms CYP1A1 and CYP1A2 of CYP because these isoforms are predominantly involved in the metabolism and activation of environmental xenobiotics.

26.2.2 Use of hepatocytes for study of environmental xenobiotic transformation

Primary cultures of hepatocytes represent a unique *in vitro* model, as they retain both Phase I and Phase II enzyme activities. The biotransformation of toxic compounds, as well as the inducibility of the biotransformation enzymes by toxic compounds themselves, can be studied in this model. In addition, the use of primary cultures of hepatocytes allows investigation of the metabolism of toxic substances under strictly controlled conditions, avoiding systemic effects such as those of the nervous, hormonal or immune systems, making it possible to pinpoint direct effects on liver parenchymal cells. However, it is well known that hepatocytes dedifferentiate after some time in culture and lose their biotransformation capacity [1]. It is therefore important to select culture conditions that minimize this disadvantage. The use of a specifically designed culture medium and cell substratum can significantly improve the metabolic performance of hepatocytes *in vitro* [1].

Cultured hepatocytes have been used for following the metabolism of environmental agents and to show that chemical structure plays an important role in the biotransformation process. A good example is that of polychlorinated biphenyls (PCBs). PCBs are mixtures of aromatic chemicals, manufactured by the chlorination of biphenyl, that are known to have marked effects on the liver. The generic chemical formula of PCB molecules is $C_{12}H_{10-n}Cl_n$, where n is the number of chlorine atoms (varying from 1 to 10). Thus, it is chemically possible to form 209 congeners but only about 130 congeners are likely to occur in commercial products [2]. The rate of metabolism of PCB congeners is influenced by both the position and degree of chlorine substitution on the biphenyl nucleus. The ability of liver enzymes to biotransform the various PCBs to excretable products is a rate-limiting feature, decreasing to very low levels as the degree of chlorine substitution becomes elevated [3]. Vickers *et al.* [3] have used the three PCBs: 4,4'-dichlorobiphenyl (4-DCB), 2,2',3,3',6,6'-hexacholorobiphenyl (2,3,6-HCB) and 2,2',4,4',5,5'-hexacholorobiphenyl (2,4,5-HCB) as model compounds to examine the relationship between chlorine content and interactions with CYP. They have shown that PCBs are taken up rapidly by hepatocytes but that they are incompletely metabolized by the liver cells. After 60 min of exposure, only 32% of 4-DCB and of 2,3,6-HCB have been metabolized, while metabolism of 2,4,5-HCB is not detected. They proposed that the metabolism of PCB congeners is influenced by the affinity of the molecule for CYP since analysis of absorbance differences reveals that 2,3,6-HCB displays the greatest binding affinity for CYP followed by 4-DCB, while 2,4,5-HCB shows no binding. The affinity of PCB congeners for CYP can determine the excretion rate of metabolites from the cell. This conclusion points to the fact that elimination of PCBs *in vivo* could largely depend on the affinity of the congeners for the biotransformation enzyme.

In addition to PCBs, several other environmental compounds have been studied from the point of view of their biotransformation in the hepatocyte culture model. These include 2,3,7,8-tetrachloro-dibenzo-p-dioxin (TCDD) [4], the organochlorine pesticide aldrin [5] and the antifouling agent tributyltin (TBT) [6].

Table 1. Environmental pollutants modulating CYP activity and expression in cultured hepatocytes

Species	Chemicals	Isoforms/ activity	Effects	Reference
Human	TCDD	EROD	+	[10]
Human	*3-Mc, BaP	EROD	+	[11]
Human and rat	*HBB	ECOD	+	[12]
Rat	TCDD	EROD	+	[13]
Rat	BaP, 3-MC	ECOD	+	Menard et al.
Rat	Aroclor 1254	ECOD	+	[14]
		EROD	+	[15]
		CYP2B	no effect	
		CYP3A	no effect	
Rat	Antracene Benz[a]anthracene	CYP1A	+	[16]
	TCDD, 3-MC	CYP2C11	down-regulation	
Rat	*TCB	EROD/ECOD	+	[17]
Mouse	Aroclor 1254	EROD	+	[18]
		PROD	+	
	*HCB, *HBB, *TCB	EROD	+	
	2,2',4,4'-TCB	EROD	no effect	
Mouse	TCDD	EROD	+	[19]
Embryos of chickens, pheasants, turkeys, ducks and herring gull	TCDD, TCDF, *PCB 126, *PCB 81, *PCB 77[a], *PCB 169, *PCB 105[b], *PCB 118[c]	EROD	+	[20,21]
Embryo of chick	BaP, Aroclor 1254	EROD	+	[22]
Fetal quail	Aroclor 1254, benz[a]anthracene	EROD/ECOD	+	[23]
Quail	*TCB	EROD/ECOD	+	[17]
Rainbow trout	TCDD	EROD	+	[24,25]
Zebrafish	TCDD	EROD	+	[26]

Where + = increase, - = inhibition. a, No induction in h. gull; b, No induction in pheasant, duck and h. gull; c, No induction in h. gull. *: PCB 81, 3,4,4',5-tetrachlorobiphenyl; PCB 126, 3,3',4,4',5-pentachlorobiphenyl; PCB 105, 2,3,3',4,4'-pentachlorobiphenyl; PCB 118, 2,3',4,4',5-pentachlorobiphenyl; PCB 77 or TCB:, 3,3',4,4'-tetrachlorobiphenyl; PCB 169 or HCB, 3,3',4,4',5,5'-hexachlorobiphenyl; HBB, 3,3',4,4',5,5'-hexabromobiphenyl; Aroclor 1254, Mixture of PCB congeners containing approximately 54% chlorine; 3-MC, 3-methylcholanthrene; BaP, benzo[a]pyrene; EROD, Ethoxyresorufin-O-deethylase activity; ECOD, Ethoxycoumarin-O-deethylase activity; PROD, Pentoxyresorufin-O-dealkylase activity.

26.2.3 Regulation of xenobiotic-metabolizing enzymes

CYP1A1 and CYP1A2 are weakly expressed at constitutive levels in tissues but they can be induced by compounds belonging to major classes of environmental pollutants, namely polycyclic aromatic hydrocarbons (PAHs), dioxins, furans, and PCBs. The expression of CYP1A1 and CYP1A2 enzymes is regulated at the transcriptional level via the cytosolic aryl hydrocarbon (Ah) receptor [7-9]. A number of studies on cultured hepatocytes have focused on the induction of CYP by environmental xenobiotics. Table 1 lists a series of publications that address this topic.

The induction of CYP1A1 and CYP1A2 by xenobiotics can be modulated by hormones, a phenomenon that has been investigated in isolated hepatocytes. In our laboratory, we have used primary cultures of rat hepatocytes to understand the influence of growth factors and hormones on the regulation of CYP. The effects of the two representative PAHs, benzo[a]pyrene and 3-methylcholanthrene (3-MC), on CYP-catalyzed 3-cyano-7-ethoxycoumarin O-deethylase activity (ECOD), and their modulation by hydrocortisone (HC) and epidermal growth factor (EGF) were investigated. It was observed that when the hormones are present 24 h prior to the addition of the PAH, they enhance the induction of ECOD activity by benzo[a]pyrene or 3-MC (Ménard, Marion and Denizeau, unpublished observations). This was also seen at the levels of CYP1A1 and CYP1A2 protein and CYP1A1 mRNA. The possible implication of protein phosphorylation on the effect of HC or EGF has been tested in an attempt to understand the mechanisms involved in the modulation of PAH-induced CYP by these factors. Our results demonstrate that phosphorylation is required for the hormone-dependent induction process, and that the protein kinase PKA, but not PKC, could play a role. It has been proposed that the Ah and glucocorticoid receptors are phosphorylated when they exert their function and that the phosphorylation of some factors involved in transactivation by the Ah receptor could be a prerequisite step for both PAH induction and possibly the modulation of PAH induction by glucocorticoids [27].

Another interesting facet of CYP induction in hepatocytes is the possibility of inter-species comparisons. The CYP response can provide a basis for extrapolation of the information obtained in laboratory animals to the human situation. For example, Merrill *et al.* [12] have looked at the effect of polybrominated biphenyls (PBBs) on CYP induction in primary cultures of human and rat hepatocytes. Rat hepatocytes were found to be highly susceptible to CYP induction by PBBs. The lowest concentration of

PBB required for CYP induction in human hepatocytes is 10-1000 times higher than that found in rat hepatocytes, suggesting that humans may be more resistant to these compounds than rats.

26.3 Induction of DNA damage and repair

From a historical perspective, it was the development, in the 1970s, of systems aimed at detecting the carcinogenic potential of chemicals that brought to light the particular usefulness of isolated hepatocytes for investigating the genotoxic activity of xenobiotics. This development eventually generated tremendous interest in cultured hepatocytes as an invaluable cellular model for mechanistic studies in chemical carcinogenesis.

A key paper by Williams [28] presented the basic principles that established primary cultures of hepatocytes as being especially suitable for the detection of genotoxicity. In contrast to most other cells in culture, freshly isolated hepatocytes reproduce well the *in vivo* situation from the point of view of metabolic capacity. As referred to above, they contain the enzymatic machinery necessary to bioactivate genotoxins, a step that is required in most cases for the formation of electrophilic species that bind to the DNA, their ultimate molecular target. This metabolic property obviates the need for an external source of enzyme, such as the liver $9000 \times g$ supernatant, in contrast to what is required in other *in vitro* test systems for genotoxicity. Hepatocytes also contain detoxifying enzymes, allowing for a balance between bioactivation and detoxification pathways, thereby further mimicking what takes place *in vivo*. Another important characteristic of freshly isolated hepatocytes in culture is that they remain quiescent (less that 1% undergoing mitosis). When these cells are incubated with radioactively labelled thymidine, the incorporation background is minimal. Low levels of unscheduled DNA synthesis (UDS) due to DNA excision repair synthesis taking place outside the S phase of the cell cycle, can therefore be measured.

In addition to UDS measurement, other methodological approaches have been developed in hepatocytes to demonstrate the induction of DNA damage by chemicals in exposed cells. Quantification of DNA strand breaks by alkaline elution [29], evaluation of DNA adduct formation by the ^{32}P-post-labeling assay [30] and, more recently, investigation of micronuclei in mitogen-stimulated hepatocytes [31] have all been successfully applied. Moreover, studies focusing on specific DNA adducts and on the links between bioactivation pathways and DNA damage have also provided important information.

Isolated hepatocytes have served as a model to detect the genotoxicity and carcinogenic potential of several environmental agents such as PAHs [32], heavy metals [33], polychlorobiphenyls [34] and nitrosamines. However, nitrosamines, a class of potent carcinogens, have received the most attention. *N*-nitrosodimethylamine (DMN) is a ubiquitous environmental toxin known for its hepatocarcinogenicity. DMN is found in many food products and in polluted air of urban areas. It is also a major toxic component of cigarette smoke. DMN causes the methylation of nucleic acidbases after bioactivation by CYP. The predominant sites of alkylation of both DNA and RNA by activated DMN are the N7 and O6 positions of guanine. O-6-methylguanine (O6-metG) is a highly pro-mutagenic adduct which causes erroneous incorporation of thymidine during DNA replication.

Our laboratory has investigated the molecular lesions due to DMN exposure in rat hepatocyte cultures. We have shown that DMN is positive in the UDS assay specifically designed for sensitive detection of short-patch DNA repair, and that this genotoxic activity is not affected by concomitant exposure to particulate air pollutants such as asbestos fibers [35]. We have also developed antibodies against O6-metG and performed the ultraimmunocytochemical localisation of DMN DNA and RNA adducts in rat hepatocytes. O6-metG-induced immunoreactive sites in hepatocytes exposed to DMN are concentrated in the nucleus and in the rough endoplasmic-rich areas of the cytoplasm. Surprisingly, mitochondria are not significantly labelled, suggesting that alkylation occurs at sites other than the O6 position of guanine, possibly at the nucleophilic N7 position [36]. Moreover, evidence was obtained that hepatocytes contain a cytoplasmic mechanism for repair of the O6-metG lesion, in addition to the nuclear mechanism, presumably attributable to methyltransferase activity. Interestingly, the protective response against guanine alkylation both in the nucleus and cytoplasm appears to be inducible, as it is activated by pre-exposure to the methylating amine [37].

The relationship between *N*-nitrosodialkylamine metabolism and DNA alkylation has also been examined in isolated hepatocytes. It was found that the isoform CYP2E1 is mainly responsible for bioactivation of DMN and DNA methylation [38]. For compounds with longer alkyl chains such as *N*-nitrosodipropylamine, *N*-nitroso-*n*-butyl-*n*-propylamine and *N*-nitrosodibutylamine, it has been demonstrated that CYP2B1 can also participate in the bioactivation of these genotoxins and account for their damaging action on DNA and their hepatocytotoxicity [39].

The enzymology of repair involved in the removal of

nitrosamine-induced DNA lesions is another area that has been investigated in primary cultures of hepatocytes. For this purpose, 4-(methyl-nitrosamino)-1-(3-pyridyl)-1-butanone (NNK), one of the most abundant carcinogens in tobacco smoke, and a well-demonstrated hepatocarcinogen in rats, has been used. By specifically inhibiting enzymes of the excision repair pathways, it was concluded that DNA polymerase α, δ and ε, and to a lesser degree DNA polymerase β (but not DNA topoisomerases I and II) are involved in the repair of pyridyloxobutylated DNA, an adduct formed by exposure to NNK [40]. These results agree with our previous observations which suggested that DNA polymerase α, but not DNA topoisomerase II, participates in the repair of DNA, elicited by the well-known hepatocarcinogen 2-acetylaminofluorene, in cultured hepatocytes [41].

A comprehensive view of the field of genotoxicity evaluation in hepatocytes reveals a salient feature. Most studies that have been carried out with environmental agents tested *in vitro* in cultures of hepatocytes have focused on compounds that act as carcinogens *in vivo*. Using the hepatocyte system to obtain molecular data on the mechanisms of toxicity has undoubtedly clarified the links between genotoxicity and carcinogenicity.

26.4 Induction of metallothionein and cytoprotection

Metallothionein (MT) refers to a family of small proteins that share unique properties. They have a molecular weight ranging from 6 to 10 kDa. MT proteins are formed by a single polypeptide chain generally consisting of 61 amino acids, including 20 cysteines, and devoid of aromatic amino acids and histidine. MT binds copper (Cu) and zinc (Zn) under normal physiological conditions but it has a high potential for binding several nonessential metals such as mercury (Hg) and cadmium (Cd) that are widely distributed, highly toxic, environmental contaminants. Piscator in 1964 was the first to demonstrate that the synthesis of MT is inducible [42]. Although MT synthesis can be stimulated by various agents such as glucocorticoid hormones, growth factors, alkylating agents and physical stresses, metals are the best known and most potent inducers. The detoxification of Cd was the first role attributed to MT. Since then, numerous *in vivo* studies have confirmed the protective function of MT against the toxicity of metals. However, *in vivo* models did not permit the elucidation of the detailed mechanisms of protection by MT. Hepatocytes *in vitro* offered a good tool for this purpose.

The first *in vitro* studies on MT anti-cytotoxic effects were carried out with Cd. Cd is widely used in industry, for example in the manufacture of household objects such as nickel-Cd batteries, in colored pigment in paint, and as a stabilizer in plastics and fungicides. The main source of exposure of the human population to Cd is food of plant origin whose contamination is attributable to the presence of Cd in soils. Hepatic damage is one of the most serious consequences of Cd poisoning. Din and Frazier [43] initiated the use of isolated hepatocytes to study the protective effect of MT on Cd toxicity at the mechanistic level. Subsequently, Klaassen and his collaborators [44] looked at the effect of Zn-induced MT on the toxicity, uptake, and cellular distribution of Cd in rat hepatocyte primary cultures and compared their results to those previously obtained *in vivo*. They were able to show that pretreatment of the hepatocytes with Zn protects the cells against Cd cytotoxicity, in a way similar to that seen in the intact animal [45]. The mechanism for the protective effect of Zn on Cd toxicity is principally attributable to a change in metal subcellular distribution. Indeed, more Cd accumulates in the cytosol and less is found in the nuclear, mitochondrial, and microsomal fractions in Zn-pretreated cells as compared to cells solely exposed to Cd without Zn-pretreatment. Thus, MT induction promotes the sequestration of Cd by MT and a reduction in the amount of Cd associated with critical organelles and proteins.

Like Cd, Hg is a highly toxic environmental contaminant and an MT inducer. Hg has many industrial applications and despite the fact that attempts have been made to control emissions of this metal during processing, pollution of the environment, particularly from mine waste and other waste disposal, is an increasing problem. Isolated rainbow trout hepatocytes have been selected for studying the mechanisms of Hg toxicity, the induction of MT and its role in metal homeostasis in fish [46]. Metal levels were analyzed in cells exposed to a sublethal concentration of Hg. Fish hepatocytes accumulate large quantities of Hg. Furthermore, exposure to Hg leads to enhanced levels of Ca, Cu and Zn within the cells. Data on the molecular distribution of Hg and Cu in the cytosol of Hg-treated trout hepatocytes indicate that, although Hg induces the synthesis of MT, it is found mainly associated with low-molecular-weight (LMW) components. These results suggest that the sequestration of Hg by MT is not the principal mechanism involved in the intracellular distribution and detoxification of this metal in fish hepatocytes. Since Hg has a high affinity for thiol groups, it has been proposed that it binds to glutathione (GSH), a thiol-rich tripeptide, and that this mechanism accounts for protection against Hg cytotoxicity. In addition, in the cytosol of cells exposed to low levels of Hg, the proportion of Cu and Zn associated with the middle-MW components (including MT) is higher, indicating that MT is involved in the regulation of Cu and Zn homeostasis in trout hepatocytes.

The first attempts to demonstrate the protective role of MT focused on metals because of their affinity for the protein, but the concept of cytoprotection by MT has been extended to organic toxins. Cultured rat hepatocytes have been used in our laboratory to demonstrate the protective effect of MT against methylating agents [47]. We have found that pretreatment of hepatocytes with Zn significantly reduces the cytotoxic effect of *N*-methyl-*N*'-nitro-*N*-nitrosoguanine (MNNG) while it does not affect the response to methyl methanesulphonate (MMS) [47]. The synthesis of graded amounts of MT, achieved by incubation with increasing concentrations of Zn or Cu, leads to a high correlation between MT levels and hepatocyte survival. The protection caused by the MT inducers Zn and Cu is not due to the alteration in the activities in the detoxifying enzymes glutathione S-transferase, glutathione peroxidase, catalase, or Cu, Zn-superoxide dismutase [48]. We proposed that the selective protection afforded by MT against MNNG, which does not occur with MMS, can be explained for the most part by the distinct mechanisms of interactions of these two molecules with living cells. Although MNNG and MMS are both considered as direct-acting methylating agents, the metabolism of MNNG is far more complex than is generally assumed and it leads to the formation of various reactive species not observed in the case of MMS [49].

26.5 Alteration of liver-specific functions: protein secretion

The liver is an organ that performs several specific functions including the secretion of plasma proteins such as albumin. In addition, in oviparous species, vitellogenin, a protein that accumulates in the egg yolk as an eventual source of nutrient for the embryo, is also produced by the liver. Albumin and vitellogenin are the two liver-specific proteins that have been the subject of investigation to determine the effects of environmental xenobiotics.

26.5.1 Albumin secretion

A specialized function of the liver is the synthesis and secretion of albumin. Therefore, inhibition of albumin production could serve as a sensitive biomarker for evaluating the hepatocytotoxic potential of environmental chemicals. Efforts towards this objective have concentrated on Cd and Hg.

The liver is one of the major target organs for Cd toxicity *in vivo* where it can alter organ-specific functions. In cultured rat hepatocytes, exposure to

Cd produces a significant inhibition of albumin secretion at concentrations that do not cause any loss of cell viability (up to 0.5 μM Cd for 20 h). It is interesting to note that hepatocytes do not rapidly overcome the inhibitory effects of Cd on albumin production since these effects are maintained even after the cells have been allowed to recover by incubation in Cd-free medium. It appears as though Cd remains firmly bound to critical target sites within the hepatocytes, causing persistence of the metabolic disturbance. Moreover, when the intracellular content of albumin is examined using detection of albumin-reactive sites at the ultrastructural level, immunolabeling is mainly found in the cytoplasm of untreated cells, whereas such sites are practically absent in Cd-exposed hepatocytes. Impairment of albumin production by Cd can take place at the transcriptional and/or translational levels. Inhibition of synthesis is a strong possibility, since it has been demonstrated that Cd inhibits [^{3}H]arginine incorporation into proteins in cultured hepatocytes [50].

The effects of Hg on albumin production have also been investigated in our laboratory. We have shown that Hg inhibits albumin secretion and causes severe morphological anomalies in exposed hepatocytes. The changes are characterized by a decreased number of ribosomes associated with the rough endoplasmic reticulum and circularisation of the latter [51]. As Hg has a very high affinity for sulphhydryl (SH) groups, we have proposed that mercury binds to the SH-rich 60S ribosomal unit, causing the dissociation of ribosomes from the rough endoplasmic reticulum and, as a consequence, a decrease in albumin synthesis.

26.5.2 Vitellogenin secretion

Vitellogenin (VTG) is a yolk protein that is synthesized in the liver of oviparous vertebrates under the influence of estrogens. Estradiol-17β (E2) is the main inducer of VTG. It binds to the cytosolic estrogen receptor (ER) to form an ER/E2 complex that is translocated to the nucleus where it interacts with the estrogen responsive elements of target genes, such as that of VTG, to modulate their transcriptional activity [52]. The alteration of VTG synthesis by environmental chemicals has received particular attention in the last few years. This interest has emerged because a large number of environmental contaminants are now suspected of acting as endocrine disruptors by mimicking or interfering with natural hormones [53]. It has been hypothesized that the capacity of xenobiotics to stimulate VTG synthesis is indicative of estrogenic properties.

Trout hepatocyte cultures, which can perform the biotransformation that is necessary in some cases for the expression of estrogenicity, are a good model for detecting the estrogenic potential of chemicals. This model is based upon the ability of trout hepatocytes to respond to estrogenic stimulation by producing VTG [54]. Flouriot *et al.* [52] have developed a rainbow trout hepatocyte culture system in which the cells maintain stable gene expression of ER and VTG, as *in vivo*. They used this model to show that some environmental pollutants, chosen according to their ability to disturb reproductive functions *in vivo*, modify the expression of the E2-dependent ER and VTG genes. They have shown that chlordecone, nonylphenol, PCB mixtures (Aroclor 1245) and lindane are able to induce ER and VTG mRNA accumulation, whereas pentachlorophenol acts as an antagonist to induction by estradiol. The use of antiestrogens like 4-hydroxytamoxifen and ICI-164,384 prevents the effects of chemicals, suggesting that induction of gene expression is mediated by the ER.

Recently, special attention was given to the effects of aluminium (Al) on VTG induction. Acidification of fresh water leads to an acceleration of the mobilisation of trace metals such as Al, and causes an enrichment of these metals in soft water. This can have serious consequences on aquatic life, especially on fish. By using primary cultures of rainbow trout hepatocytes, Mugiya and Tanahashi [55] have clearly demonstrated that Al impairs hepatic VTG synthesis induced by E2 treatment. The inhibitory effect of Al on VTG can therefore threaten oogenesis due to poor VTG accumulation in the egg yolk.

26.6 Disturbance of ionic homeostasis

In the last decade, there have been intensified efforts to elucidate the mechanisms responsible for the death of cells exposed to toxic agents. It has become clear that perturbations of ionic homeostasis, in particular involving an increase in the intracellular free Ca^{2+} concentration ($[Ca^{2+}]_i$), play an important role in cell death caused by toxic agents. The $[Ca^{2+}]_i$ is regulated within narrow limits, at very low levels in the cytosol (10^{-7}-10^{-8} M) as compared to a relatively high extracellular concentration (10^{-3} M). The regulation of the $[Ca^{2+}]_i$ in most cells is known to take place at three major sites: plasma membrane, endoplasmic reticulum and mitochondria. Exposure to toxic agents can alter these systems and induce a sustained elevation in $[Ca^{2+}]_i$. The latter can in turn activate various Ca^{2+}-dependent degradative enzymes (*e.g.* phospholipases, proteases, endonucleases) and lead to cell death by necrosis. It can also modify the activity of other Ca^{2+}-dependent enzymes such as protein kinases, modulate Ca^{2+}-signalling pathways and cause apoptosis (see Section 26.7).

Isolated trout hepatocytes have been used to study xenobiotic-induced perturbations of ionic homeostasis. We have carried out a detailed investigation with tri-*n*-butyltin (TBT), a biocide widely used in antifouling paints to combat the proliferation of aquatic organisms which encrust ship hulls. TBT ranks among highly toxic contaminants in aquatic ecosystems. The liver is a target for TBT toxicity [56], but the mechanisms involved are poorly understood. Information about the mechanisms of hepatic toxicity of TBT has been obtained *in vitro* in isolated fish hepatocytes by simultaneous assessment of intracellular $[Ca^{2+}]_i$ and pH on a real time basis upon exposure to TBT. Treatment of trout hepatocytes with TBT causes a rapid and sustained elevation of $[Ca^{2+}]_i$ without loss of cell viability, as shown by highly sensitive flow cytometric analysis [57,58]. This change occurs through mobilisation of Ca^{2+} from intracellular stores. In addition to the effects on $[Ca^{2+}]_i$, exposure to TBT induces a rapid and irreversible cytoplasmic acidification explicable by a step resulting from intracellular Ca^{2+} mobilisation and by direct action of TBT on pH-controlling systems. We propose that TBT leads to a more acidotic cell interior by inhibiting the activity of the Na^+/H^+ exchanger, which normally provides an extrusion mechanism for excess protons. The use of isolated trout hepatocytes has demonstrated that imbalances in ionic homeostasis are early events in cell injury by TBT.

Similarly, alterations in intracellular Ca^{2+} and pH have been reported to take place in hepatocytes in the presence of Hg and Cd. Inorganic Hg produces an elevation in cytosolic Ca^{2+} levels in rat and skate hepatocytes [59,60]. In rat hepatocytes, the increase occurs after a delay of several minutes, and is abolished in Ca^{2+}-free medium, suggesting that the Ca^{2+} comes mainly from the extracellular medium. In contrast, in skate hepatocytes, the increase is mediated by the release of Ca^{2+} from internal stores, which are not sensitive to thapsigargin (an ER inhibitor). Persistent cellular acidification due to Cd has also been shown by Koizumi *et al.* [61]. As postulated for TBT, delayed recovery of cytosolic pH after maximal acidification appears to be related to the inhibition of Na^+/H^+ exchange. In addition, induction of MT by Zn or Cd prevents Cd-induced cellular acidification in rat hepatocytes [62]. It has been suggested that Cd binds to a functional sulphhydryl group on the Na^+/H^+ exchanger system.

Ionic disturbances in the cytotoxic response are often associated with the collapse of the mitochondrial membrane potential. It is one of the serious consequences of exposure to toxic agents, as it leads to the depletion of ATP and, ultimately, to cell death. We performed a series of experiments on cultured rat hepatocytes in order to determine the relationship between the dissipation of mitochondrial membrane potential, the decrease in plasma membrane potential, and the loss of cell viability. We found that treatment of rat hepatocytes with Cd causes a rapid drop (within the first 30 min of exposure) in the mitochondrial membrane potential [63]. This event is followed by a decrease in plasma membrane potential and, subsequently, by a decrease in cell viability as observed by the release of lactate dehydrogenase into the extracellular medium. This cascade of events indicates that mitochondrial attack is an early event in the action of Cd on hepatocytes that can trigger a pathway to necrotic cell death. The detailed mechanisms responsible for the alterations in the mitochondrial membrane potential by Cd are not known. It has been proposed that the perturbation of mitochondrial function is linked to the acidification induced by Cd in hepatocytes, as supported by the data of Koizumi *et al.* [61]. However, experiments carried out with isolated mitochondria have shown that an acidic medium has no effect on membrane potential, although such a medium enhances the collapse of membrane potential in the presence of fairly high levels of Cd (>10 μM) [61]. The authors have suggested that the stimulatory effect of the acidic medium on the ability of Cd to dissipate the mitochondrial membrane potential may be related to changes in permeability of the inner membrane.

26.7 Activation of apoptosis signaling pathways

A general description of apoptosis in hepatocytes is given in Chapter 18. Here we focus on the process of apoptosis induced by xenobiotics, which has received very little attention up to now. There are few reports dealing with the induction of apoptosis by environmental contaminants in hepatocytes. Nitrosamines were the first environmental agents tested for their ability to trigger apoptosis in hepatocytes. The DNA damage associated with nitrosamines and resulting in the alkylation of nucleic bases has been extensively studied (for more discussion see Section 26.3). However, as demonstrated by Kamendulis and Corcoran [64,65] in mouse hepatocytes, dimethylnitrosamine (DMN) causes fragmentation of genomic DNA similar to that observed during apoptosis. Furthermore, the use of DNA repair inhibitors enhances DNA fragmentation and toxicity. It has been proposed that failure to repair damaged DNA is a central feature of acute DMN-induced hepatotoxicity. These results obtained *in vitro* agree with those of Ray and collaborators [66], who demonstrated that DMN produces internucleosomal DNA degradation in liver cells of adult male mice *in vivo*, indicating that isolated hepatocytes are a good system for studying apoptosis.

Recently, we performed a series of experiments in freshly isolated trout hepatocytes with the aim of shedding more light on the mechanisms of apoptosis induction by TBT. TBT was recognized as stimulating apoptosis in thymocytes [67], but its apoptogenic action had not been studied in hepatocytes. We found that TBT treatment induces apoptosis in rainbow trout hepatocytes [68]. Moreover, the elevation of cytosolic Ca^{2+} induced by TBT in trout hepatocytes (see Section 26.6) is associated with extensive apoptotic DNA fragmentation, since the cleavage of DNA is prevented by treatment of the cells with the intracellular Ca^{2+} chelator BAPTA-AM. It therefore appears that the increase in intracellular Ca^{2+} is a key event in TBT-mediated apoptosis in trout hepatocytes. Previous studies on physiological apoptosis have pointed to a role for protein kinase C (PKC), an important component of intra-cytoplasmic and nuclear signal transduction pathways, in the regulation of apoptosis. In trout hepatocytes, TBT stimulates PKC γ and δ translocation from the cytosol to the plasma membrane, an event that is followed by the down-regulation of PKC. Inhibition of protein kinase activity in TBT-treated trout hepatocytes leads to enhancement of DNA fragmentation, whereas the phorbol ester, phorbol myristate acetate, a classical activator of PKC, fails to stimulate DNA fragmentation. These observations support the view that down-regulation of PKC is a crucial step in the mechanisms of apoptosis induced by TBT. In addition, proteases such as caspases, a class of enzymes playing a role in the execution phase of apoptosis, are also involved in the death of hepatocytes upon exposure to TBT. Taken together, these *in vitro* data allow a mechanistic model to be proposed for the apoptotic pathways involved in xenobiotic-triggered apoptosis in fish hepatocytes.

Apoptotic cell death is a crucial event in the multistage process of carcinogenesis. It has been proposed that suppression of apoptosis could be an important mechanism of action of tumor promoters. In support of this proposal, an interesting study has shown that the environmental pollutant and rodent liver promoter TCDD inhibits the effects of apoptogenic UV and 2-acetylaminofluorene treatments [69]. These authors have also shown that UV treatment leads to an increase in p53 in the cells and that TCDD suppresses this increase. These data suggest that TCDD blocks the pathway leading to the expression of p53 in response to DNA damage.

The investigations reported so far on the induction of apoptosis by xenobiotics in hepatocytes demonstrate the usefulness of the model at two different levels. First, from a fundamental point of view, there is no doubt that hepatocyte cultures are an invaluable system for unraveling the complexity of apoptotic molecular mechanisms. Second, from an applied point of view, hepatocyte cultures are suitable for detecting the apoptogenic potential of chemicals or for testing for their capacity to inhibit apoptosis in the liver. Because inhibition of apoptosis is associated with the development of cancer, it can be hypothesized that this property should be indicative of carcinogenic potential. Validation of this concept by a well-planned *in vivo-in vitro* experimental strategy could lead to the design of a new test for carcinogens that act though epigenetic (non-genotoxic) mechanisms. This would add a very much needed complement to the classical UDS assay for genotoxicity.

26.8 Conclusion

Isolated hepatocytes have contributed to the field of environmental toxicology in several ways. This review has shown that hepatocytes in culture provide a uniquely versatile tool to further our understanding of the biochemical basis of xenobiotic hepatocytotoxicity. The importance of designing experimental strategies based upon strong links between *in vitro* and *in vivo* approaches has been emphasized. Establishing sound *in vitro-in vivo* correlations is crucial in inter-species comparisons, especially when the purpose is extrapolation to humans. In many of these studies human hepatocytes in culture have been used. Indeed, the tools to evaluate better the impact of accidental exposure of humans to environmental toxins are very limited and cells in culture show great potential for providing solutions to this acute problem.

The aim of protecting the health of humans and that of ecosystems has led to the development of methods to evaluate the risks associated with the presence of toxic substances in the environment. The risk assessment process consists of major steps: (i) hazard identification, which means the demonstration of a toxic potential; (ii) determination of exposure levels; (iii) determination of dose-response relationships; and (iv) risk characterisation [70]. Isolated hepatocytes are very well suited to provide information in an efficient manner at the hazard identification step. Indeed, numerous markers can be used to show hepatocytotoxic properties. Furthermore, information on mechanisms of toxicity derived from hepatocytes can be quite useful in risk characterisation because this step must take into account all the data available in order to build the best scientific judgement possible. The knowledge of mechanisms of toxicity also leads to the definition of biomarkers reflecting exposure levels or toxic effects. The availability of such validated biomarkers is of great value in the enormously challenging task of determining the degree of exposure to specific toxic agents in the environment (step ii). An example of application of this concept is the measurement of DNA adducts as a means of evaluating the "dose" at a critical molecular target. Other endpoints like the induction of MT and that of CYP1A1 are being tested for their usefulness in informing on exposure to metals and aromatic compounds, respectively. Good biomarkers can also help develop reliable dose-response relationships, in particular in humans who cannot be studied directly, as already pointed out above.

As discussed in this review, to date there has been significant development in the application of isolated hepatocytes for use in the field of environmental toxicology. Many efforts in this field can further benefit from this approach, but new areas are also emerging for which the model could be of great value in the future.

References

1. Kedderis GL. Biochemical basis of hepatocellular injury. Toxicol Pathol 1996; 24: 77-83.
2. WHO Environmental Health Criteria 140: Polychlorinated Biphenyls and Terphenyls. 2nd ed, Geneva: World Health Organization, 1993: 682.
3. Vickers AEM, Sipes IG and Brendel K. Metabolism-related spectral characterization and subcellular distribution of polychlorinated biphenyl congeners in isolated rat hepatocytes. Biochem Pharmacol 1986; 35: 297-306.
4. Sawahat T, Olson JR and Neal RA. Identification of metabolites of 2,3,7,8-tetrachlorodibenzo-p-dioxin (TCDD) formed on incubation with isolated rat hepatocytes. Biochem Biophys Res Com 1982; 105: 341-346.
5. KuriHara N, Hori N and Ichinose R. Cytochrome P-450 content and aldrin epoxidation to dieldrin in isolated rat hepatocytes. Pestic Biochem Physiol 1984; 21: 63-73.
6. Reader S, Saint-Louis R, Pelletier E and Denizeau F. Accumulation and biotransformation of tri-n-butyltin by isolated rainbow trout hepatocytes. Environ Toxicol Chem 1996; 15: 2049-2052.
7. Poland A and Knutson J. 2,3,7,8-tetrachlorodibenzo-p-dioxin and related halogenated aromatic hydrocarbons: examination of the mechanism of toxicity. Annu Rev Pharmacol Toxicol 1982; 22: 517-554.
8. Whitlock JP Jr. Genetic and molecular aspects of 2,3,7,8-tetrachlorodibenzo-p-dioxin action. Annu Rev Pharmacol Toxicol 1990; 30: 251-277.
9. Gasiewicz TA, Elferink CJ and Henry EC. Characterization of multiple forms of the Ah receptor: recognition of a dioxin-responsive enhancer involves heteromer formation. Biochemistry 1991; 30: 2909-2916.
10. Schrenk D, Stüven T, Gohl G, Viebahn R and Bock KW. Induction of CYP1A and glutathione S-transferase activities by 2,3,7,8-tetrachlorodibenzo-p-dioxin in human hepatocyte cultures. Carcinogenesis 1995; 16: 943-946.
11. Abdel-Razzak Z, Corcos L, Fautrel A, Campion J-P and Guillouzo A. Transforming growth factor β1 down-regulates basal and polycyclic aromatic hydrocarbon-induced cytochromes P-450 1A1 and 1A2 in adult human hepatocytes in primary culture. Molec Pharmacol 1994; 46: 1100-1110.
12. Merrill J, Beck DJ, Kaminski DA and Li AP. Polybrominated biphenyl induction of cytochrome P450 mixed function oxidase activity in primary rat and human hepatocytes. Toxicology 1995; 99: 147-152.

13. Wölfle D, Becker E and Schmutte C. Growth stimulation of primary rat hepatocytes by 2,3,7,8-tetrachlorodibenzo-p-dioxin. Cell Biol Toxicol 1993; 9: 15-31.
14. Kremers P, Roelandt L, Stouvenakers N, Goffinet G and Thome JP. Expression and induction of drug-metabolizing enzymes in cultured fetal rat hepatocytes. Cell Biol Toxicol 1994; 10: 117-125.
15. Roelandt L, Dubois M, Todaro A, Thome JP and Kremers P. Effect of inducers and PCBs on the cytochrome P450 enzymes in cultured quail hepatocytes. Ecotox Environ Safety 1995; 31: 158-163.
16. Safa B, Lee C and Riddick DS. Role of the aromatic hydrocarbon receptor in the suppression of cytochrome P-450 2C11 by polycyclic aromatic hydrocarbons. Toxicol Lett 1997; 90: 163-175.
17. Dubois M, De-Waziers I, Thomé JP and Kremers P. P450 induction by Aroclor 1254 and 3,3',4,4'-tetrachlorobiphenyl in cultured hepatocytes from rat, quail and man: interspecies comparison. Comp Biochem Physiol C Pharmacol Toxicol Endocrinol 1996; 113: 51-59.
18. Madra S and Smith AG. Induction of cytochrome P450 activities by polychlorinated biphenyls in isolated mouse hepatocytes. Biochem Pharmacol 1992; 44: 455-464.
19. Chun YJ, Koh WS and Yang KH. Suppression of TCDD-induced cytochrome P450 IA1 activity by staurosporine in mouse primary hepatocyte cultures and hepatoma cells. Biochem Molec Biol Interact 1994; 32: 1023-1032.
20. Kennedy SW, Lorenzen A, Jones SP, Hahn ME and Stegeman JJ. Cytochrome P4501A induction in avian hepatocyte cultures: a promising approach for predicting the sensitivity of avian species to toxic effects of halogenated aromatic hydrocarbons. Toxicol Appl Pharmacol 1996; 141: 214-230.
21. Bosveld ATC, Kennedy SW, Seinen W and van den Berg M. Ethoxyresorufin-O-deethylase (EROD) inducing potencies of planar chlorinated aromatic hydrocarbons in primary cultures of hepatocytes from different developmental stages of the chicken. Arch Toxicol 1997; 71: 746-750.
22. Topp RJ and van-Bladeren PJ. Oxidative biotransformation in primary cultures of chick embryo hepatocytes: induction of cytochrome P-450 and the metabolism of benzo[a]pyrene. Arch Toxicol 1986; 59: 150-155.
23. Roelandt L, Todaro A, Thomé JP and Kremers P. Effects of PCBs (Aroclor 1254) on cytochrome-P450 expression and monooxygenase activities in cultured foetal rat hepatocytes. Toxicology 1995; 98: 95-103.
24. Pesonen M and Andersson T. Characterization and induction of xenobiotic metabolizing enzyme activities in a primary culture of rainbow trout hepatocytes. Xenobiotica 1991; 21: 461-471.
25. Pesonen M, Goksoyr A and Andersson T. Expression of P4501A1 in a primary culture of rainbow trout hepatocytes exposed to β-naphthoflavone or 2,3,7,8-tetrachlorodibenzo-p-dioxin. Arch Biochem Biophys 1992; 292: 228-233.
26. Collodi P, Miranda CL, Zhao X, Buhler DR and Barnes DW. Induction of zebrafish (*Brachydanio rerio*) P450 *in vivo* and in cell culture. Xenobiotica 1994; 24: 487-493.
27. Xiao GH, Pinaire JA, Rodrigues AD and Prough RA. Regulation of the Ah gene battery via Ah receptor-dependent and independent processes in cultured adult rat hepatocytes. Drug Metab Dispos 1995; 23: 642-650.
28. Williams GM. Detection of chemical carcinogens by unscheduled DNA synthesis in rat liver primary cell cultures. Cancer Res 1977; 37: 1845-1851.

29. Storer RD, Mckelvey TW, Kraynak AR, Elia MC, Barnum JE et al. Revalidation of the *in vitro* alkaline elution/rat hepatocyte assay for DNA damage: improved criteria for assessment of cytotoxicity and genotoxicity and results for 81 compounds. Mutat Res 1996; 368: 59-101.
30. Holme JA, Bjorge C, Soderlund EJ, Brunborg G, Becher R et al. Genotoxic effects of cyclopenta-fused polycyclic aromatic hydrocarbons in isolated rat hepatocytes and rabbit lung cells. Carcinogenesis 1993; 14: 1125-1131.
31. Muller-Tegethoff K, Kersten B, Kasper P and Muller L. Application of the *in vitro* rat hepatocyte micronucleus assay in genetic toxicology testing. Mutat Res 1997; 392: 125-138.
32. Eberhart J, Coffing SL, Anderson JN, Marcus C, Kalogeris TJ et al. The time-dependent increase in the binding of benzo[a]pyrene to DNA through (+)-anti-benzo[a]pyrene-7,8-diol-9,10-epoxide in primary rat hepatocyte cultures results from induction of cytochrome P450IA1 by benzo[a]pyrene treatment. Carcinogenesis 1992; 13: 297-301.
33. Denizeau F and Marion M. Genotoxic effects of heavy metals in rat hepatocytes. Cell Biol Toxicol 1989; 5: 15-25.
34. Dubois M, Pfohl-Leszkowicz A, Grosse Y and Kremers P. DNA adducts and P-450 induction in human, rat and avian liver cells after exposure to polychlorobiphenyls. Mutat Res 1995; 345: 181-190.
35. Denizeau F, Marion M, Chevalier G and Coté MG. Genotoxicity of dimethylnitrosamine in the presence of chrysotile asbestos UICC B and xonotlite. Carcinogenesis 1985; 6: 1815-1817.
36. Lachapelle M, Fadlallah S, Krzystyniak K, Fournier M, Cooper S and Denizeau F. Colloidal gold ultraimmunocytochemical localization of DNA and RNA adducts in rat hepatocytes. Carcinogenesis 1992; 13: 2335-2339.
37. Lachapelle M, Marion M, Krzystyniak K, Fournier M and Denizeau F. Immunocytochemical evidence for a nuclear and a cytoplasmic O-6-methylguanine repair mechanism in cultured rat hepatocytes. J Toxicol Environ Health 1994; 43: 441-451.
38. Encell L, Foiles PG and Gold B. The relationship between N-nitrosodimethylamine metabolism and DNA methylation in isolated rat hepatocytes. Carcinogenesis 1996; 17: 1127-1134.
39. Shu L and Hollenberg PF. Role of cytochrome P-450 in DNA damage induced by N-nitrosodialkylamines in cultured rat hepatocytes. Carcinogenesis 1996; 17: 569-576.
40. Cloutier JF and Castonguay A. Modulation of DNA repair by various inhibitors of DNA synthesis following 4-(methylnitro-samino)1-1-(3-pyridyl)-1-butanone (NNK) induced DNA damage. Chem Biol Interact 1998; 110: 7-25.
41. Jutras D, Marion M and Denizeau F. The effects of putative DNA repair inhibitors on DNA adduct levels and unscheduled DNA synthesis in rat hepatocytes exposed to 2-acetylaminofluorene. Mutat Res 1989; 216: 35-42.
42. Piscator M. On cadmium in normal human kidneys together with a report on the isolation of metallothionein from livers of cadmium-exposed rabbits. Nord Hyg Tikskr 1964; 45: 76-82.
43. Din WS and Frazier JM. Protective effect of metallothionein on cadmium toxicity in isolated rat hepatocytes. Biochem J 1985; 230: 395-402.
44. Liu J, Kershaw WC and Klassen CD. The protective effect of metallothionein on the toxicity of various metals in rat primary hepatocyte culture. Toxicol Appl Pharmacol 1991; 107: 27-34.

45. Liu J, Kershaw WC and Klaassen CD. Rat primary hepatocyte cultures are a good model for examining metallothionein-induced tolerance to cadmium toxicity. *In Vitro* Cell Dev Biol 1990; 26: 75-79.
46. Gagné F, Marion M and Denizeau F. Metal homeostasis and metallothionein induction in rainbow trout hepatocytes exposed to cadmium. Fund Appl Toxicol 1990; 14: 429-437.
47. Moffatt P, Plaa GL and Denizeau F. Rat hepatocytes with elevated metallothionein expression are resistant to N-Methyl-N'-nitro-N-nitrosoguanidine cytotoxicity. Toxicol Appl Pharmacol 1996; 136: 200-207.
48. Haïdara K, Moffatt P and Denizeau F. Metallothionein induction attenuates the effects of glutathione depletors in rat hepatocytes. Toxicol Sci 1999; 49: 297-305.
49. Moffatt P and Denizeau F. Metallothionein in physiological and physiopathological processes. Drug Metab Rev 1997; 29: 261-307.
50. Beattie JH, Marion M and Denizeau F. The modulation by metallothionein of cadmium-induced cytotoxicity in primary hepatocyte cultures. Toxicology 1987; 44: 329-339.
51. Lachapelle M, Guertin F, Marion M, Fournier M and Denizeau F. Mercuric chloride affects protein secretion in rat primary hepatocyte cultures: a biochemical ultrastructural, and gold immunocytochemical study. J Toxicol Environ Health 1993; 38: 343-354.
52. Flouriot G, Pakdel F, Ducouret B and Valotaire Y. Influence of xenobiotics on rainbow trout liver estrogen receptor and vitellogenin gene expression. J Mol Endocrinol 1995; 15: 143-151.
53. Petit F, Le Goff P, Cravédi JP, Valotaire Y and Pakdel F. Two complementary bioassays for screening the estrogenic potency of xenobiotics: recombinant yeast for trout estrogen receptor and trout hepatocyte cultures. J Mol Endocrinol 1997; 19: 321-335.
54. Pelissero C, Flouriot G, Foucher JL, Bennetau B, Dunogueès J, Le Gac F and Sumpter JP. Vitellogenin synthesis in cultured hepatocytes; and *in vitro* test for the estrogenic potency of chemicals. J Steroid Biochem Molec Biol 1993; 44: 263-272.
55. Mugiya Y and Tanahashi A. Inhibitory effects of aluminium on vitellogenin induction by estradiol-17β in the primary culture of hepatocytes in the rainbow trout *Oncorhynchus mykiss*. Gen Comp Endocrinol 1998; 109: 37-43.
56. Krajnc EI, Wester PW, Loeber JG, Van Leeuwen FXR, Vos JG, Vaessen HAMG and Van der Heijden CA. Toxicity of bis(tri-n-butyltin)oxide in the rat. I. Short-term effects on general parameters and on the endocrine and lymphoïd systems. Toxicol Appl Pharmacol 1984; 75: 363-386.
57. Reader S, Marion M and Denizeau F. Flow cytometric analysis of the effects of tri-n-butyltin chloride on cytosolic free calcium and thiol levels in isolated rainbow trout hepatocytes. Toxicology 1993; 80: 117-129.
58. Reader S, Steen HB and Denizeau F. Intracellular calcium and pH alterations induced by tri-n-butyltin chloride in isolated rainbow trout hepatocytes: A flow cytometric analysis. Arch Biochem Biophys 1994; 312: 407-413.
59. Nieminen AL, Gores GJ, Dawson TI, Herman B and Lemasters JJ. Toxic injury from mercuric chloride in rat hepatocytes. J Biol Chem 1990; 265: 2399-2408.
60. Nathanson MH, Mariwalla K, Ballatori N, and Boyer JL. Effects of Hg^{2+} on cytosolic Ca^{2+} in isolated skate hepatocytes. Cell Calcium 1995; 18: 429-439.

61. Koizumi T, Yokota T, Shirakura H, Tatsumoto H and Suzuki KT. Potential mechanism of cadmium-induced cytotoxicity in rat hepatocytes: inhibitory action of cadmium on mitochondrial respiratory activity. Toxicology 1994; 92: 115-125.
62. Koizumi T, Yokota T, Ohmori S, Kumagai H and Suzuki KT. Protective effect of metallothionein on intracellular pH changes induced by cadmium. Toxicology 1995; 95: 11-17.
63. Martel J, Marion M and Denizeau F. Effect of cadmium on membrane potential in isolated rat hepatocytes. Toxicology 1990; 60: 161-172.
64. Kamendulis LM and Corcoran GB. DNA as a critical target in toxic cell death: enhancement of dimethylnitrosamine cytotoxicity by DNA repair inhibitors. J Pharmacol Exp Ther 1994; 271: 1695-1698.
65. Kamendulis LM and Corcoran GB. Dimethylnitrosamine-induced DNA damage and toxic cell death in cultured mouse hepatocytes. J Toxicol Environ Health 1995; 46: 31-46.
66. Ray SD, Kamendulis LM, Sorge CL and Corcoran GB. Ca^{2+}-activated DNA fragmentation in dimethylnitrosamine-induced hepatic necrosis: Effects of Ca^{2+}-endonuclease and poly(ADP-ribose) polymerase inhibitors in mice. J Pharmacol Exp Ther 1992; 263: 387-394.
67. Raffray M and Cohen GM. Thymocyte apoptosis as a mechanism for tributyltin-induced thymic atrophy *in vivo*. Arch Toxicol 1993; 67: 231-236.
68. Reader S, Moutardier V and Denizeau F. Tributyltin triggers apoptosis in trout hepatocytes: The role of Ca^{2+}, protein kinase C and proteases. Biochim Biophys Acta 1999; 1448: 473-485.
69. Wörner W and Schrenk D. Influence of liver tumor promoters on apoptosis in rat hepatocytes induced by 2-acetylaminofluorene, ultraviolet light, or transforming growth factor β1. Cancer Res 1996; 56: 1272-1278.
70. Suter GW. Defining the field. In: Suter GW, Ed. Ecological Risk Assessment. Chelsea. Michigan: Lewis Publishers, 1993: 3-20.

Chapter 27

The use of isolated hepatocytes for the study of xenobiotic hepatotoxicity

Jian Wu[1], Åke Danielsson[2] and Mark A. Zern[1]
[1]Department of Medicine, Division of Gastroenterology & Hepatology, Jefferson Medical College, Thomas Jefferson University, Philadelphia, Pennsylvania, USA and [2]Department of Medicine, Section for Gastroenterology & Hepatology, Umeå University, Umeå, Sweden

Key words: acetaminophen, apoptosis, bromobenzene, carbon tetrachloride, cytosolic Ca^{2+}, D-galactosamine, dichlorofluorescin diacetate, dimethylnitrosamine, ethanol, glutathione, hepatotoxicity, hepatotoxin, hydroxyl radicals, hydroxynonenal, lipid peroxidation, lucigenin, luminol, malondialdehyde, mitochondrial permeability transition, nitric oxide, peroxynitrite, reactive oxygen species, single strand DNA breakage, superoxide anion, thioacetamide

27.1 Introduction

The liver plays a major role in the metabolism, transformation and elimination of xenobiotics and is frequently injured during these processes, so that isolated hepatocytes are a useful system for studying the metabolism and toxicity of xenobiotics *in vitro* [1]. Commonly hepatotoxins, including carbon tetrachloride (CCl_4), ethanol, acetaminophen (APAP), bromobenzene (BB), dimethylnitrosamine (DMN), thioacetamide (TAA), and D-galactosamine (GalN), display a low degree of direct toxicity to the hepatocytes, whereas more marked toxicity is caused by their intermediate metabolites, such as N-acetyl-*p*-benzoquinone imine (NAPQI) from APAP [2-4] and acetaldehyde from ethanol [5-6]. This chapter describes the use of isolated hepatocytes in the study of the intermediate metabolism of these xenobiotics.

Suspension and primary cultures of hepatocytes are two common tools for the study of toxicity of hepatotoxins *in vitro*. Their advantages and disadvantages in evaluating the effects of various hepatoprotective agents have been summarized previously [7]. In general, the use of hepatocytes in suspension culture is feasible for short term experiments, such as studies of cell viability, induction of single strand DNA breakage or flow cytometry analysis. On the other hand, hepatocytes in primary culture can maintain their biological functions longer than cells in suspension, depending on the substrata on which the cells are seeded, as well as the medium and

M.N. Berry and A.M. Edwards (eds.), The Hepatocyte Review, 455–474.

supplements [8]. Cells in primary culture are widely used for a variety of experiments in the study of toxicology, such as metabolism of xenobiotics, cell viability, apoptosis, intracellular oxidative status, gene expression and transfection, cell proliferation, etc. [3]. Isolated hepatocytes seeded on coverslips and incubated in the same manner as primary cultures are often suitable for microscopic observation, including assays for cytosolic Ca^{2+}, detection of fluorescent signals and immunocytochemical staining [2].

27.2 Hepatotoxins studied with isolated hepatocytes

It is crucial to understand the metabolic characteristics of each individual hepatotoxin, as well as its reactive intermediate metabolites, and the subcellular organelles that are its molecular targets. For the purpose of comparison, the reactive metabolites, forms of hepatic injury, involvement of lipid peroxidation, depletion of glutathione, and generation of reactive oxygen species (ROS) during the metabolism of hepatotoxins are summarized in Table 1. For individual hepatotoxins, the specific requirements for studying their toxicity with isolated hepatocytes are discussed below.

27.2.1 Carbon tetrachloride

CCl_4 is one of the most widely used classical toxins for experimental models of acute and chronic liver injury [9]. Reactive metabolites of CCl_4, *i.e.* trichloromethyl ($\bullet CCl_3$) and trichloromethyl peroxyl ($CCl_3OO\bullet$) radicals, which are generated during the metabolism of CCl_4 by benzoyl peroxide catalysis, or via thermal activation, or by microsomal NADPH-dependent (CYP2E1) biotransformation, are major toxic components and account for almost all the toxicity of the agent [10-11]. Therefore, ethanol may potentiate CCl_4 toxicity by induction of the metabolizing enzymes, such as CYP2E1. Hepatocytes isolated from pyrazole-treated rats were more susceptible to CCl_4 exposure *in vitro* [12]; whereas, hepatocytes isolated from vitamin E-treated rats appeared to be resistant to peroxidative and cytotoxic effects of CCl_4 *in vitro* [13].

For studying the toxicity of CCl_4 to isolated hepatocytes in culture, CCl_4 was not added directly to culture medium due to its insolubility in water and evaporative properties. Instead, it was added to a central well which allowed the toxin to diffuse into the cell suspension. The final concentration of CCl_4 was 0.172 mM, which was in a range shown not to exert a direct solvent effect [13]. A short-term suspension culture of isolated hepatocytes was employed to determine aldehydic products of lipid peroxidation in

Table 1. Toxicity and mechanism of action of common hepatotoxins

Hepatotoxin	Reactive metabolite	Involvement of cytochrome P-450	Generation of reactive oxygen species	Metabolite-protein adduct	Depletion of glutathione	Major forms of liver injury
Carbon tetrachloride (CCl_4) [9,10]	$\cdot CCl_3$, $CCl_3OO\cdot$	NADPH-Cytochrome P-450	Haloalkane radicals	$\cdot CCl_3$-protein adducts GSH-$\cdot CCl_3$ [57]	Yes	Central-lobule necrosis, bridging necrosis of neighbouring lobules; toxicity to isolated hepatocytes.
Ethanol [6,58-59]	Acetaldehyde	CYP 2E1	Hydroxyethyl radical, H_2O_2 [21-22]	Acetaldehyde-adducts, hydroxyethyl-adducts	Yes	Necrosis and steatosis or apoptosis of acinar zone III; toxicity to isolated hepatocytes.
Acetaminophen (APAP) [29]	N-Acetyl-p-benzoquinone imine (NAPQI)	CYP 2E1, 1A2 CYP3A [35]	Undefined	NAPQI-protein (nucleophilic sulfhydryl groups)	Yes	Acute liver injury, central lobule predominance.
Bromobenzene (BB) [42-43]	*p*-Bromophenol	Cytochrome P-450	H_2O_2 [23]	Quinone adducts and epoxide adducts	Yes	Acute or chronic liver injury and carcinogenesis; toxicity to isolated hepatocytes.
Dimethylnitro-samine (DMNS) [45,47]	Alkylating moiety	CYP 2A6, 2E1	Undefined	Alkylation of DNA and proteins	Unknown	Acute central-lobule injury by single dose challenge or fibrosis/cirrhosis and carcinogenesis by long-term exposure.
Thioacetamide (TAA) [48]	Acetylimido-lysine derivative	CYP 2A5	NO	Hydroxynonenal-protein adducts	Yes	Central-lobular apoptosis or necrosis following a single dose of challenge. Fibrosis/ cirrhosis or carcinogenesis by long-term intoxication.
D-Galactosamine (GalN) [3,50]	None	Not involved	None defined	Not defined	Yes, also depletion of uridine triphosphate	Usually combined with LPS to induce acute liver injury.

Reprinted from Table 1, Expert Opinion on Investigational Drugs 1999; 8: 585-607 with permission of Ashley Publications Ltd.

CCl_4-exposed cells [14-15]. This approach may be used to study glutathione depletion, cytosolic Ca^{2+} content and other metabolic alterations in such cells. An alternative method is to dissolve CCl_4 in DMSO, and then add it to the culture medium. This method of dissolving a water-insoluble substance is practical and is often applied in *in vitro* experiments. However, significant evaporation of the toxin may cause problems in controlling its concentration, as well as presenting the risk of becoming a biohazard. In addition, DMSO itself displays significant actions on various aspects of cell functions and viability if the final concentration of DMSO in the medium is more than 2.0%. Therefore, a DMSO control group must be included in experiments when the toxin is dissolved in it [2,16].

Isolated rat hepatocytes in suspension were also employed for evaluating protective effects of honeybee aqueous propolis extract. After cells in suspension culture were pretreated with the extract for 30 min, they were exposed to CCl_4, dissolved in DMSO for a further 30 min [17]. Then, enzyme release in the culture medium, GSH consumption and lipid peroxides in the CCl_4-exposed cells were monitored to determine the cytoprotective and anti-peroxidative effect of the extract [17]. An advantage of *in vitro* experiments with cells in suspension culture is that it is unnecessary to use strict sterile techniques in the preparation of the cells for short term incubation.

27.2.2 Ethanol

Attempts have been made to induce injury of isolated hepatocytes by ethanol treatment; however, the injury is not as apparent as other hepatotoxins used in equivalent concentrations [18]. Treatment of hepatocytes with ethanol at 50 to 150 mM caused a dose- and time-dependent reduction in cellular viability indicated by enhanced leakage of aspartate animotransferase (AST) and decreased phosphatidylcholine synthesis. Addition of vitamin E or ascorbic acid prevented the loss of cell viability in ethanol- or allyl alcohol-treated hepatocytes, suggesting that the ethanol toxicity to hepatocytes is associated with oxidative stress, as shown by other studies [19,20].

Common ROS, such as superoxide radicals ($O_2 \bullet^-$), hydrogen peroxide (H_2O_2) and specific ascorbyl radicals, can be detected during ethanol-induced toxicity in isolated hepatocytes. H_2O_2 production in ethanol-treated hepatocytes was monitored under a confocal microscope by observing conversion of non-fluorescent 2′,7′-Dichlorofluorescin diacetate (DCF-DA) to DCF [21,22]. A microtiter plate method for detection of generation of H_2O_2 in cultured cells by the DCF fluorescent approach was reported recently by our group [23]. With confocal microscopy observation, one can examine the cell morphology and viability, and at the same time, the intensity of weak DCF fluorescent signals at a high magnification can be easily detected and graded from a single cell [21]. Moreover, with laser scanning confocal microscopy, functions of subcellular organelles, such as mitochondria, and lysosomes, can be studied morphologically and even three-dimensional morphology of an individual hepatocyte can be reconstructed [24]. Using two DNA-binding fluorescent dyes, Heochst 33342 and propidium iodide, patterns of chromatin condensation and nuclear fragmentation in a single cell, as well as viability of the cell can be identified at the same time under confocal microscopy [21]. Therefore, this technique is currently used in several toxicological studies of xenobiotics. It was recently confirmed by confocal microscopy that ethanol caused significant apoptosis of isolated hepatocytes, instead of necrosis, via pathways of activation of caspases and cytochrome C [25]. This may be the reason for significant cell death and a lack of enzyme elevation in the culture medium. Agents which inhibit ethanol-derived oxidative stress, effectively attenuate hepatocellular apoptosis or/and necrosis, and inhibitors of intracellular antioxidants exaggerate ethanol-associated hepatocyte apoptosis. Thus, it is conceivable that an oxidant-dependent mechanism is largely involved in the process of ethanol-induced hepatocyte apoptosis [26].

27.2.3 Acetaminophen

Significant acute liver injury with APAP has been frequently reported because of overdose or misuse. In particular, fatal and massive liver necrosis develops in alcoholics after using the drug [27]. The molecular pathogenesis of APAP hepatocellular toxicity has been clarified and is predominantly attributed to the formation of its reactive metabolite, NAPQI, catalyzed by CYP2E1, CYP1A2 and CYP3A4 subtypes of CYP [28]. The reactive NAPQI covalently binds to vital cellular substances, such as proteins, and forms NAPQI-protein adducts [29]. The severity of hepatocellular damage depends on the rate of APAP bio-transformation and formation of NAPQI. The inducers of CYP2E1, such as ethanol or isoniazid, potentiate APAP toxicity. A potent inhibitor of CYP2E1, phenethyl isothiocyanate, protects hepatocytes against APAP toxicity [30]. The covalent binding of NAPQI to proteins is a major contributor to cell toxicity. The early target proteins appear to be those present at highest cellular concentration and with accessible nucleophilic sulphhydryl groups. The more NAPQI formed, the more proteins become targets, and eventually the cellular machinery is overwhelmed. The cell can then no longer function, or there is a disruption of the redox balance within the cell due to the dysfunction of numerous proteins [31]. Depletion of glutathione by reactive NAPQI is a crucial step in the process of toxicity to hepatocytes and is directly related to the severity of the cell damage.

Cultured hepatocytes were employed to compare the toxicity of various APAP metabolites and the difference in its metabolism and toxicity to hepatocytes from various species [32,33]. Cytosolic Ca^{2+} content, evaluated using the Ca^{2+}-binding fluorescent indicators Indo-1 (by scanning laser cytometry) or Fluo-3 (in a fluorescence plate reader) [34] was increased in cultured hepatocytes treated with APAP at 0.5-4.0 mM for 20 h. The increase in cytosolic calcium concentration correlated with formation of plasma membrane blebs and deterioration of cell viability. Pre-exposure to a Ca^{2+} channel blocker, nifedipine, prevented the increase in cytosolic Ca^{2+} levels and the loss of cell viability indicated by LDH leakage and colorimetric MTT [3-(4,5-dimethylthiazol-2-yl)-2,5-diphenyl tetrazolium bromide] assay [34]. The increased cytosolic Ca^{2+} levels are thought to result from impaired Ca^{2+} sequestration activity in both microsomal and mitochondrial compartments [35]. Exposure of mouse hepatocytes to APAP also activates endonuclease, probably mediated by an increased cytosolic Ca^{2+} content, and resulted in formation of a DNA "ladder", an indication of apoptotic changes in the cells [36]. A recent study reported that spontaneous chemiluminescence associated with cell injury was detected in isolated rat hepatocyte suspensions during APAP metabolism, reflecting the occurrence of oxidative stress. APAP at different concentrations affected the hepatocytes in various manners. A low concentration (2 mM), which corresponds to the serum concentration in overdosed animals, caused loss of cell viability due to lipid peroxidation; whereas, APAP at higher concentrations (5-50 mM) protected the hepatocytes due to a chemical antioxidant effect of the unmetabolized APAP that remained in the medium because of the saturation of APAP metabolism [37]. Thus, oxidative stress and associated lipid peroxidation may account for a significant amount of the APAP toxicity [4,38].

27.2.4 Bromobenzene

BB is a classical hepatotoxic agent and its toxicity to the liver is thought to be due to covalent binding of its reactive intermediate metabolites (*p*-bromophenol and *p*-bromophenol-oxide) to vital constituents of the cells. Progressive depletion of cellular glutathione (GSH) is an early event that occurs after exposure to BB, and can be detected 150 min after isolated hepatocytes are exposed to 2.4 mM BB. The GSH content is further depleted 24 h after the exposure. Correspondingly, the cell viability, as indicated by LDH leakage, deteriorates [3]. Using isolated single hepatocytes loaded with Fura-2 and microperifusion techniques, we did not find significant alteration in intracellular Ca^{2+} content during 20 min of treatment with 2.4 mM BB. However, BB at 2.4 mM interfered with Ca^{2+} oscillations induced by phenylephrine (PNLP) in isolated single rat hepatocytes [2]. The oscillations induced by 3.0 μM PNLP were completely blocked during the perifusion with the combination of PNLP and 2.4 mM BB in 58% of the cells studied. The frequency of the remaining PNLP-induced oscillations was markedly reduced by BB [2]. Although the mechanism of the frequency of Ca^{2+} oscillations and its roles in intracellular signalling are not fully understood [39], the BB effect on PNLP-induced Ca^{2+} oscillation may imply that BB or its metabolites interfere with the cellular signalling system that may be involved in the mediation of intracellular or intercellular communications [40].

As indicated by increased malondialdehyde (MDA) levels in BB-treated hepatocytes, lipid peroxidation has been shown to be involved in BB damage to the liver *in vivo* when the GSH content is reduced below a threshold level [41,42]. In a previous study [43], we demonstrated for the first time that BB at 2.4-4.8 mM caused marked single strand DNA breakage, which was determined by a radio-labelling assay [44], in a hepatoblastoma cell line. Water-soluble vitamin E (Trolox C) significantly attenuated the DNA damage. The findings indicated that oxidative insult may be involved in BB-induced single strand DNA breakage since H_2O_2 alone at 400 μM caused marked amounts of single strand DNA breakage [23]. Furthermore, we demonstrated that BB at 2.4 mM caused production of H_2O_2 in the Hep G_2 cells, detected by conversion of DCF-DA to a fluorescein product (DCF) in a microtiter plate [23]. Therefore, it is clear that, in addition to reactive BB metabolites, the generation of ROS, such as H_2O_2 or other oxygen radicals, also contributes to the BB-induced cell injury and DNA damage, and this probably is the reason for the protective effects of anti-oxidative treatment, such as vitamin E or vitamin C [43].

27.2.5 Dimethylnitrosamine

DMN induces hepatic carcinogenesis by chronic intoxication. A single dose of DMN (30 mg/kg, i.p.) led to lethal liver toxicity in shrews and acute central lobular cell injury, but no significant inflammatory reaction and hemorrhage [45]. The mechanism of DMN toxicity is due to recurrent DNA alkylation by an active methylating moiety generated during DMN metabolism. DNA methylation by DMN is thought to be mediated in part by CYP2E1, since the enzyme inhibitor, pyrazole, inhibited the metabolism of DMN and DNA methylation in DMN-treated hepatocytes *in vitro* [46]. It has been shown recently that exposure of isolated mouse hepatocytes to DMN caused marked DNA fragmentation after 6 h, and increased LDH leakage 36 h after the exposure [47]. Thus, both *in vivo* and *in vitro* experiments suggest that DMN-induced apoptosis in parenchymal cells is a major component of the acute liver toxicity.

27.2.6 Thioacetamide

TAA is a well-known hepatotoxin which also induces hepatocyte regeneration and carcinogenesis. An obligate metabolite of TAA binds to proteins with the formation of acetylimidolysine derivatives, which are responsible for TAA-induced hepatotoxicity. The hepatocellular death induced by TAA may occur via both apoptosis and necrosis pathways [48]. Toxicity of TAA has not been studied in isolated hepatocytes. However, hepatocytes isolated from TAA-induced cirrhotic livers displayed less transforming growth factor-β receptor activity than did those from normal rats [49].

27.2.7 D-Galactosamine

GalN is a hepatotoxin that induces liver damage, similar to that seen in human viral hepatitis, by depleting uridine triphosphate (UTP) and other uridine nucleotides, and subsequently diminishes synthesis of mRNA and plasma membrane proteins [50]. However, in contrast to other hepatotoxins, GalN administration leads to increased CYP monooxygenase and UDP-glucuronosyltransferase activity in the kidneys and intestine, respectively, but not in the liver, despite a marked increase in serum ALT levels [51]. Recently, it was reported that the percentage of fragmented DNA in GalN-treated mouse hepatocytes was increased in a dose-dependent manner at 30 min after the exposure. In addition, a ladder-like fragmentation pattern by agarose gel electrophoresis appeared after a 3 h-exposure to 20 mM GalN, after a 6 h-exposure to 10 mM GalN and after a 12 h-exposure to 5 mM GalN, respectively. On the other hand, cytotoxicity indicated by LDH leakage from cultured hepatocytes was detected at 24 h after GalN-treatment. Formation of blebs and apoptotic bodies was observed after 12 h-exposure to 20 mM GalN and after 24 h-exposure to 10 mM GalN. Thus GalN causes hepatocellular death in culture by both necrosis and apoptosis [52]. Moreover, GSH depletion in GalN-treated hepatocytes has been shown to correlate with cellular injury 24 h after rat hepatocytes were exposed to GalN [3]. Therefore, it appears that in GalN-induced toxicity, depletion of GSH, which indicates oxidant stress in the cells, may be a crucial component leading to cell death by necrosis and/or apoptosis.

GalN in combination with tumour necrosis factor-α (TNF-α) induces acute liver injury in a number of species. Kupffer cells, which produce cytokines such as TNF-α and other soluble factors in response to oxidant stress, may be responsible for the second phase of hepatocellular injury caused by hepatotoxins. To understand the role of these cells, a coculture system of Kupffer cells and hepatocytes has been developed [53]. Hepatocytes and Kupffer cells are prepared separately and in most cases, hepatocytes are plated on top of Kupffer cells which are prepared two days earlier than hepatocytes and seeded on collagen-precoated wells [53]. However, depending on the objectives of the studies, it is also feasible to add suspensions of Kupffer cells to hepatocyte cultures [54]. There was no increase in LDH leakage when LPS and complement were added to the medium. A moderate increase was observed with the addition of GalN. A remarkable increase was observed only with the addition of GalN, LPS, and complement to the coculture. These results suggest that Kupffer cells activated by LPS are very important in promoting damage to hepatocytes by complement [55]. Adding hepatocytes seeded on Transwell COL Inserts to cultures of activated peritoneal macrophages is another coculture system used in the study of Kupffer cell and hepatocyte interactions. GalN treatment caused a threefold increase in LDH leakage from the coculture system when compared to the hepatocytes in primary culture subjected to the same treatment [56].

27.3 Assays for Oxidative Stress

27.3.1 Detection of free radicals

Superoxide anion ($O_2^{\bullet -}$) detected by luminol and lucigenin

Luminol chemiluminescence has been used to detect the production of ROS such as $O_2^{\bullet -}$, H_2O_2 and hydroxyl radicals (HO•) by enzymes, as well as from cell and organ systems. It has been useful for examining the kinetics and reaction mechanisms of oxygen radicals. In most cases of luminol chemi-excitation in biological systems, $O_2^{\bullet -}$ is a key intermediate and superoxide dismutase (SOD) diminishes its light intensity. Therefore, luminol chemiluminescence is considered to reflect the generation of $O_2\bullet^-$ in biological systems, such as xanthine oxidase plus xanthine [60]. Actually, luminol is not specific for $O_2\bullet^-$; H_2O_2, HO• and even peroxynitrite, or other oxidants, such as ferricyanide or hypochlorite can provoke its chemiluminescence. Thus, when luminol is applied in studies of xenobotic-induced hepatocyte toxicity, it acts as a general indicator for the generation of various reactive oxygen radicals in the system. Luminol chemiluminescence may be measured in hepatocyte microsomes or homogenates with a luminescence analyser after the cells are preincubated or treated [61].

The chemiluminescence of lucigenin involves the reaction of $O_2\bullet^-$ with a radical form of the luminescent compound. The reactive radical form of lucigenin is generated by univalent reduction, instead of univalent oxidation in the formation of reactive radicals by luminol. $O_2\bullet^-$ is necessary for lucigenin reduction and the luminescent is regarded as a more specific probe for $O_2\bullet^-$ than luminol [62]. As with luminol, lucigenin chemiluminescence is $NADP^+$-dependent and can be detected in microsomes prepared from liver tissue or cultured hepatocytes with various treatments. Moreover, a more specific luminescence probe, MCLA has been utilized for detection of $O_2\bullet^-$ production [63-64], and it should be useful in evaluating oxidative stress in isolated hepatocytes.

Hydrogen peroxide detected by 2′,7′-dichlorofluorescin diacetate

2′,7′-Dichlorofluorescin diacetate (DCF-DA) is often employed to study oxidative burst in a variety of cell types, *e.g.* neutrophils, phagocytes, platelets, pancreatic acinar cells, hepatoma cells, and endothelial cells. The method is based on the fact that DCF-DA, a non-fluorescent compound, is able to diffuse across plasma membranes, and is then cleaved by a cellular esterase to form 2′,7′-dichlorofluorescin (also non-fluorescent) which is trapped within the cells [65]. The latter compound reacts with free radical products, particularly with H_2O_2, and generates the fluorescent 2′,7′-dichlorofluorescein (DCF), which can be detected fluorometrically. Common *in vitro* methods for detection of DCF fluorescence intensity include flow cytometry and fluorophotometry, or observation by laser scanning confocal microscopy [21] of single isolated hepatocytes. By using a multiple well plate, the efficiency in detection of fluorescence intensity may be enhanced [66]. Briefly, after separation, hepatocytes are seeded in 96-well plates with proper precoating. 30 min after DCF-DA is added to the medium at 200 μM, fluorescence intensity from the cells is measured at 501 nm (exciting wavelength) and 521 nm (emission wavelength) in a luminescence spectrometer. This method provides a quick and simple approach for detection of intracellular H_2O_2 production during various treatments and is very sensitive and reliable for the study of oxidant-mediated cell injury.

NO_2• and dinitrosyl iron complex detected by electron paramagnetic resonance

There are two methods for measuring the end products of NO metabolism, nitrate and nitrite. The content of nitrate or nitrite can serve as an indicator of nitric oxide synthase activity. Nitrate in culture medium may be assayed by a commercially available ultraviolet (UV)-method kit from Boehringer-Mannheim, GmbH. Nitrite can be readily determined by the Griess reaction [67] with Griess reagents, sulphanilamide and N-1-naphthyl-ethylenediamine dihydrochloride in H_3PO_4, using sodium nitrite as a reference. The colour can be read at 540 nm, 5 min after all samples and reagents have been added. The assay is very simple and stable, and is widely used in all kinds of samples [68,69]. Generation of NO in isolated hepatocytes can also be monitored simultaneously by detecting signals of the dinitrosyl iron complex (DNIC) using electron paramagnetic resonance (EPR) [70]. NO generated in the hepatocytes binds to an iron-containing molecule, such as DNIC, which gives EPR detectable signals.

Peroxynitrite detected by luminol chemiluminescence

Liver cells generate superoxide (O_2•$^-$) and NO and these two radicals react to form a potent oxidant peroxynitrite anion ($ONOO^-$). Peroxynitrite is an unstable species at physiological pH, protonating to peroxynitrous acid (ONOOH) which spontaneously decomposes to nitrogen dioxide (NO_2•) and HO• at 20-30°C. Therefore, peroxynitrite is speculated to be an important mediator of free radical-dependent toxicity because of its strong oxidizing properties towards different biomolecules, including protein and non-protein thiols, deoxyribose and membrane phospholipids [71]. It has been shown that luminol reacts with peroxynitrite and yields detectable chemiluminescence signals; the signals are greatly enhanced in the presence of bicarbonate [71]. In this case, luminol is oxidized by $ONOO^-$ to form a reactive luminescent radical and the oxidation is also inhibited by SOD. It appears that luminol is more useful in detecting $ONOO^-$ than O_2•$^-$ [71].

27.3.2 Consumption of antioxidative agents

Glutathione is the most important antioxidant in the body, because it reacts with most oxidative radicals directly and can quickly reduce the toxicity of reactive radicals. Consumption or depletion of GSH in isolated hepatocytes reflects the severity of the intoxication of a hepatotoxin and the extent of the oxidative insult. Thus, measurement of GSH in isolated hepatocytes is one of the key indicators for oxidant stress and has been widely applied in studies of toxicology. A number of colorimetric, fluorometric or HPLC methods are available for the measurement of GSH in liver tissue or hepatocyte extracts [19,72]. The most convenient and reliable one we have employed is a patented spectrophotometric kit (GSH-400) from Bioxytech® S. A., Cedex, France, which can be used in both liver homogenate and cultured hepatocytes. Moreover, the assay can be completed in two hours, without requiring any specific equipment [3].

27.3.3 Lipid peroxidation and its products

Lipid peroxidation is a crucial pathologic mechanism by which reactive radicals cause injury. Almost all hepatotoxins discussed earlier in this chapter cause lipid peroxidation and peroxidation of the methylene hydrogen on the carbon between the double bonds of polyunsaturated fatty acids in biomembranes. This in turn leads to formation of lipid hydroperoxides, which decompose into aldehyde products, such as MDA, and 4-hydroxyalkenals (4-HNE). Measurement of MDA and 4-HNE has been widely employed as an indirect evaluation of lipid peroxidation for many years [73]. MDA is a volatile, low molecular-weight, short-chain, 1,3-dicarbonyl compound and a moderately weak acid. It is unstable and easily undergoes oxidative and metabolic degradation. To measure MDA in hepatocytes in suspension or primary culture, MDA is extracted with trichloroacetic acid. In the presence of thiobarbituric acid (TBA) and heat, MDA reacts with TBA and produces an MDA:TBA adduct with red pigment at the highest molar absorptivity, some 5-10 times greater than that of MDA itself [73]. However, other aldehydes may also react with TBA; therefore, the TBA method is called the TBA reactive substance (TBARS) test. Many factors influence the sensitivity of the TBA test of MDA. These include a wide variety of nonlipid-related, TBA-positive materials, such as protein, DNA, bilirubin, etc. Therefore, the TBA test is not a highly sensitive or specific method for detecting and quantifying lipid peroxidation [73]. MDA content in a sample can be assayed more exactly by high pressure liquid chromatography coupled to UV spectrophotometry [73,75-77].

The problems with the TBARS method and the complexity of HPLC for MDA estimation have led to the exploration of more simple methods for the measurement of MDA and 4-HNE in biological tissues. A colorimetric assay (LPO-586) has been developed by Bioxytech S.A., France. The assay takes advantage of a patented chromogenic reagent which reacts with MDA and 4-hydroxyalkenals. The condensation of MDA or 4-HDA with the chromogenic reagent yields a stable chromophore with maximal absorbency at 586 nm wavelength. The assay can be used to detect MDA in liver homogenates or cell extracts. The obvious advantages of the assay are that it is simple to perform and that it excludes nonspecific reactions present in the TBARS method.

27.3.4 Mitochondrial permeability transition

As discussed in an earlier section, the mitochondria play a key role in generation of ATP and regulation of intracellular Ca^{2+} levels. Depletion of ATP and/or GSH and the onset of the mitochondrial permeability transition (MPT) are considered to be important mechanisms leading to cell death [78]. Thus, evaluating inner mitochondrial membrane integrity and monitoring onset of the MPT are two crucial approaches for studying the process of hepatocellular necrosis or apoptosis. For determination of MPT and mitochondrial swelling in hepatocytes, one must prepare the mitochondria in a medium with a low Ca^{2+} concentration [79].

The MPT is measured in mitochondria isolated from liver homogenates or incubated hepatocytes [80]. This method actually determines cyclosporine A-sensitive and [^{3}H]sucrose-accessible mitochondrial space, and is similar to a procedure used for determining permeability of the inner mitochondrial membrane [81]. The swelling of isolated mitochondria can be monitored spectrophotometrically. The change of absorbency at 540 nm over 5 min is recorded as a measure of mitochondrial swelling [81]. To visualize MPT in living isolated hepatocytes, a confocal microscopy technique is required, after loading cells with two fluoresceins, the acetoxymethyl ester of calcein, which gives green fluorescence in the cytoplasm, and tetramethylrhodamine methylester, a cationic dye that generates a red fluorescence signal and electrophoretically accumulates in mitochondria in response to a highly negative mitochondrial membrane potential [82].

27.3.5 Assays for DNA damage: single strand DNA breakage

A number of cytotoxic chemicals induce subcellular damage to many micro-organelles. The nuclear membrane is relatively more resistant to damage caused by toxic agents than either plasma membranes or mitochondria [42]. Damaged DNA is characterized by the formation of single strand breakage (SSDB). Single strand DNA sizes up to $2\text{-}5 \times 10^8$ Dalton can be separated by centrifugation or filtration from intact DNA.

Two methods are available for measurement of SSDB in isolated hepatocytes: one employs radioisotopes, the other alkaline elution. The alkaline elution method requires separation of single strand DNA from the lysed cells by elution and filtration at a high pH value [83,84]. Approximately 10^6 isolated or incubated cells are loaded on a Millipore (mixed esters of cellulose) filter (5 µm pore size), and washed with cold Merchant's solution. The cells are lysed on the filter with a lysing buffer at pH 10.0. Single strand DNA is eluted in the dark with an eluting solution (pH 12.3). At the end of elution, the filter is broken up in the eluting solution. Eluted fractions (single strand DNA) and the filter (intact DNA) are assayed for DNA content by a microfluorometric technique, or by DNA precipitation for quantitation by a spectrophotometer.

The radioactivity labelling method of measuring SSDB requires that hepatocytes are growing, for example as a consequence of mitogen stimulation, and thus take up methyl[^{14}C]thymidine for their DNA synthesis. Cultured cells are pretreated with the isotope (methyl[^{14}C]thymidine) 48 h prior to other treatment. The cells are then exposed to a hepatotoxin, such as bromobenzene, and the damaged DNA is separated by centrifugation [44]. The radioactivity in the supernatant and pellets is determined. The ratio of supernatant/total (supernatant plus pellet) radioactivity is calculated and used as an indicator of single strand DNA breakage. This method is highly sensitive and can be readily used in any cells that have active DNA synthesis [23].

27.4 Summary

Both *in vivo* and *in vitro* approaches are employed to fully understand the pathogenesis of hepatotoxin-induced damage to the liver or to specific cell types. Choosing an *in vivo* or *in vitro* approach depends on the objectives of a study; findings from *in vivo* and *in vitro* studies should be consistent and complementary. For example, in studies of xenobiotic metabolism, acute or chronic injury associated with a hepatotoxin, or protective effects of chemical substances or drugs on the liver, *in vivo* investigations may have advantages in evaluating general effects on the entire liver as well as on other organs. When the objectives of studies are directed at the subcellular or molecular levels, the use of isolated hepatocytes may be more valuable. Finally, cocultures of various liver cell types offer another approach for studying the interactions and the roles of different cell populations in response to xenobiotic intoxication.

Acknowledgements

This work is supported in part by grants from the National Institutes of Health, USA, DK09762 to JW, DK-41875 and AA-06386 to MAZ, and by the Swedish Association of Medical Research (Svenska Sällskapet för Medicinsk Forskning, SSMF) to JW, and the Swedish Medical Research Council (19X-11240) to ÅD.

Jian Wu's present address is: Dept. of Internal Medicine, UC Davis Medical Center. CA.

References

1. Berry MN, Halls HJ and Grivell MB. Techniques for pharmacological and toxicological studies with isolated hepatocyte suspensions. Life Sci 1992; 21: 1-16.
2. Wu J, Danielsson Å, Lindström P, Karlsson K and Sehlin J. The Protective effects of calcium channel blockers on bromobenzene-induced damage to isolated rat hepatocytes. Inhibition of phenylephrine stimulated calcium oscillations. Scand J Gastroenterol 1995; 30: 590-600.
3. Wu J, Söderbergh H, Karlsson K and Danielsson Å. Protective effect of S-adenosyl-L-methionine on bromobenzene- or D-galactosamine toxicity to isolated rat hepatocytes. Hepatology 1996; 23: 359-365.
4. Nelson SD. Molecular mechanisms of the hepatotoxicity caused by acetaminophen. Sem Liver Dis 1990; 10: 267-278.
5. Arnon R, Degli Esposti S and Zern MA. Molecular biological aspects of alcohol-induced liver disease. Alcohol Clin Exp Res 1995; 19: 247-256.
6. Lieber CS. Ethanol metabolism, cirrhosis and alcoholism. Clin Chim Acta 1997; 257: 59-84.
7. Berry MN, Edwards AM and Barritt GJ. Laboratory techniques in biochemistry and molecular biology 21. Isolated hepatocytes: preparation, properties and applications. Amsterdam: Elsevier, 1991: pp. 15-58.
8. Alpini G, Philips JO, Vroman B and Larusso NF. Recent advances in the isolation of liver cells. Hepatology 1994; 20: 494-514.
9. Tamayo RP. Is cirrhosis of the liver experimentally produced by CCl_4 and adequate model of human cirrhosis. Hepatology 1983; 3: 112-120.
10. Faroon O, Derosa CT, Smith L, Mehlman MA, Riddle J. Males Y and Brattin WJ. ATSDR evaluation of health effects of chemicals. 1. Carbon tetrachloride-health effects, toxicokinetics, human exposure and environmental fate. Tox Indus Health 1994; 10: S1-S123.

11. Poli G. Liver damage due to free radicals. Br Med Bull 1993; 49: 604-620.
12. Lauriault VV, Khan S and O'Brien PJ. Hepatocyte cytotoxicity induced by various hepatotoxins mediated by cytochrome P-450IIE1: protection with diethyldithiocarbamate administration. Chem Biol Interact. 1992; 813: 271-289.
13. Danni O, Chiarpotto E, Aragno M, Biasi F, Comoglio A et al. Lipid peroxidation and irreversible cell damage: synergism between carbon tetrachloride and 1,2-dibromoethane in isolated rat hepatocytes. Toxicol App Pharmacol 1991; 110: 216-222.
14. Chiarpotto E, Biasi F, Comoglio A, Leonarduzzi G, Poli G and Dianzani MU. CCl_4-induced of hepatocytes free arachidonate level: pathogenesis and contribution to cell death. Chem Biol Interact 1990; 74: 195-206.
15. Poli G, Dianzani MU, Cheeseman KH, Slater TF, Lang J and Esterbauer H. Separation and characterization of the aldehydic production of lipid peroxidation stimulated by carbon tetrachloride or ADP-iron in isolated rat hepatocytes and rat liver microsomal suspensions. Biochem J 1985; 227: 629-638.
16. Kost DP and Michalopoulos GK. Effect of 2% dimethyl sulphoxide on the mitogenic properties of epidermal growth factor and hepatocyte growth factor in primary hepatocyte culture. J Cell Physiol 1991; 147: 247-280.
17. Mahran LG, el-Khatib AS, Agha AM and Khayyal MT. The protective effect of aqueous propolis extract on isolated rat hepatocytes against carbon tetrachloride toxicity. Drugs Exp Clin Res 1996; 22: 309-316.
18. Lamb RG, Kock JC, Snyder JW, Huband SM and Bush S. An *in vitro* model of ethanol-dependent liver cell injury. Hepatology 1994; 19: 174-182.
19. Sanchez A, Alvarez AM, Benito M and Fabregat I. Cycloheximide prevents apoptosis, reactive oxygen species production, and glutathione depletion induced by transforming growth factor-β in fetal rat hepatocytes in primary culture. Hepatology 1997; 26: 935-943.
20. Glascott PA, Gilfor E, Serroni A and Farber JL. Independent antioxidant action of vitamins E and C in cultured rat hepatocytes intoxicated with allyl alcohol. Biochem Pharmacol 1996; 52: 1245-1212.
21. Kurose I, Higuchi H, Miura S, Saito H, Watanabe N et al. Oxidative stress-mediated apoptosis of hepatocytes exposed to acute ethanol intoxication. Hepatology 1997; 25: 368-378.
22. Reinke LA, Moore DR and McCay PB. Free radical formation in livers of rats treated acutely and chronically with alcohol. Alcoholism Clin Exp Res 1997; 21: 642-646.
23. Wu J, Karlsson K and Danielsson Å. Effects of vitamin E, ascorbic acid and catalase on bromobenzene- and hydrogen peroxide-induced intracellular oxidation and single strand DNA breakage in Hep G_2 cells. J Hepatol 1997; 26: 669-677.
24. Zahrebelski G, Nieminen AL, Al-Ghoul K, Qian T, Herman B and Lemasters JJ. Progression of subcellular changes during chemical hypoxia to cultured rat hepatocytes: a laser scanning confocal microscopic study. Hepatology 1995; 21: 1361-1372.
25. Higuchi H, Adachi M, Kate S, Miura S and Ishii H. Caspase activation system contribute to the process of hepatocyte apoptosis induced by acute ethanol intoxication: cytoprotective role of glutathione. Hepatology 1998; 28: 449A.
26. Higuchi H, Kurose I, Kato S, Miura S and Ishii H. Ethanol-induced apoptosis and oxidative stress in hepatocytes. Alcohol Clin Exp Res 1996; 20: A340-A346.

27. Lieber CS, Robins SJ, Li JJ, DeCarli, Mak KM, Fasula JM and Leo MA. Phosphatidylcholine protects against fibrosis and cirrhosis in the baboon. Gastroenterology 1994; 106: 152-159.
28. Fontana RJ and Watkins PB. Genetic predisposition to drug-induced liver disease. Gastroenterol Clin N Am 1995; 24: 811-838.
29. Gilbson JD, Pumford NR, Samokyszyn VM and Hinson JA. Mechanism of acetaminophen-induced hepatotoxicity: covalent binding versus oxidative stress. Chem Res Toxicol 1996; 9: 580-585.
30. Li Y, Wang EJ, Chen L, Stein AP, Reuhl KR and Yang CS. Effects of phenethyl isothiocyanate on acetaminophen metabolism and hepatotoxicity in mice. Toxicol Appl Pharmacol 1997; 144: 306-314.
31. Pumford NR, Halmes NC and Hinson JA. Covalent binding of xenobiotics to specific proteins in the liver. Drug Metab Rev 1997; 29: 39-57.
32. Holme JA, Hongslo JK, Bjorge C and Nelson SD. Comparative cytotoxic effects of acetaminophen (N-acetyl-p-aminophenol), a non-hepatotoxic regioisomer acetyl-m-aminophenol and their postulated reactive hydroquinone and quinone metabolites in monolayer cultures of mouse hepatocytes. Biochem Pharmacol 1991; 42: 1137-1142.
33. Smolarek TA, Higgins CV and Amacher DE. Metabolism and cytotoxicity of acetaminophen in hepatocyte cultures from rat, rabbit, dog, and monkey. Drug Metab Disp 1990; 18: 659-663.
34. Ellouk-Achard S, Mawet E, Thibault N, Dutertre-Catella H and Thevenin M. Protective effect of nifedipine against cytotoxicity and intracellular calcium alterations induced by acetaminophen in rat hepatocyte cultures. Drug Chemi Toxicol 1995; 18: 105-117.
35. Shen W, Kamendulis LM, Ray SD and Corcoran GB. Acetaminophen-induced cytotoxicity in cultured mouse-hepatocytes: correlation of nuclear Ca^{2+} accumulation and early DNA fragmentation with cell death. Toxicol Appl Pharmacol 1991; 111: 242-254.
36. Shen W, Kamendulis LM, Ray SD and Corcoran GB. Acetaminophen-induced cytotoxicity in cultured mouse hepatocytes: effects of Ca^{2+}-endonuclease, DNA repair, and glutathione depletion inhibitors on DNA fragmentation and cell death. Toxicol App Pharmacol 1992; 112: 32-40.
37. Minamide Y, Horie T, Tomaru A and Awazu S. Spontaneous chemiluminescence production, lipid peroxidation, and covalent binding in rat hepatocytes exposed to acetaminophen. J Pharmaceut Sci 1998; 87: 640-646.
38. Vale JA and Proudfoot AT. Paracetamol (acetaminophen) poisoning. Lancet 1995; 346: 547-552
39. Putney JW. Calcium signaling: up, down, up, down... what's the point? Science 1998; 279: 191-192.
40. Koninck PD and Schulman H. Sensitivity of CaM kinase II to the frequency of Ca^{2+} oscillations. Science 1998; 279: 227-230.
41. Casini AF, Pompella A and Comporti M. Liver glutathione depletion induced by bromobenzene, iodobenzene and diethylaleate poisoning and its relation to lipid peroxidation and necrosis. Am J Pathol 1985; 118: 225-237.
42. Monks TJ and Lau SS. Reactive intermediates and their toxicological significance. Toxicology 1988; 52: 1-53.

43. Wu J, Karlsson K and Danielsson Å. Protective effect of Trolox C, vitamin C and catalase on bromobenzene-induced damage to rat hepatocytes. Scand J Gastroenterol 1996; 31: 797-803.
44. Olive PL, Chan APS and Cu CS. Comparison between the DNA precipitation and alkali unwinding assays for detecting DNA stand breaks and cross-links. Cancer Res 1988; 48: 6444-6449.
45. Shikata N, Oyaizu T, Senzaki H, Uemura Y and Tsubura A. Liver apoptosis after dimethylnitrosamine administration in shrews. Exp Toxicol Pathol 1996; 48: 307-311.
46. Encell L, Foiles PG and Gold B. The relationship between N-nitrosodimethylamine metabolism and DNA methylation in isolated rat hepatocytes. Carcinogenesis 1996; 17: 1127-1134.
47. Kamendulis LM and Corcoran GB. Dimethylnitrosamine-induced DNA damage and toxic cell death in cultured mouse hepatocytes. J Toxicol Environ Health 1995; 46: 31-46.
48. Witzmann FA, Fultz CD, Mangipudy RS and Mehendale HM. Two-dimensional electrophorectic analysis of compartment-specific hepatic protein charge modification induced by thioacetamide exposure in rats. Fund Appl Toxicol 1996; 31: 124-132.
49. Kiso S, Kawata S, Tamura S, Ito N, Takaishi K, Shirai Y, Tsushima H and Matsuzawa Y. Alteration in growth regulation of hepatocytes in primary culture obtained from cirrhotic rat: poor response to transforming growth factor-β 1 and interferons. Hepatology 1994; 20: 1303-1308.
50. Decker K. Mechanisms and mediators in hepatic necrosis. Gastroenterol (Japan) 1993; 28: S20-S25.
51. Omar H el-D, Sanders RA and Watkins JB. Minimal effect of acute experimental hepatitis induced by lipopolysaccharide/D-galactosamine on biotransformation in rats. Biochem Pharmacol 1996; 52: 1921-1924.
52. Tsutsui S, Hirasawa K, Takeda M, Itagaki S, Kawamura S, Maeda K, Mikami T and Doi K. Galactosamine-induced apoptosis in the primary mouse hepatocyte cultures. Exp Toxicol Pathol 1997; 49: 301-306.
53. Hespeling U, Puschel GP, Jungermann K, Gotze O and Zwirner J. Stimulation of glycogen phosphorylase in rat hepatocytes via prostanoid release from Kupffer cells by recombinant rat anaphylatoxin C5a but not by native human C5a in hepatocyte/Kupffer cell co-cultures. FEBS Letters 1995; 372: 108-112.
54. Steinhorn DM and Cerra FB. Comparative effects of lipopolysaccharide on newborn versus adult rat hepatocyte and non-parenchymal cell cocultures. Crit Care Med 1997; 25: 121-127.
55. Matsuo R, Ukida M, Nishikawa Y, Omori N and Tsuji T. The role of Kupffer cells in complement activation in D-Galactosamine/lipopolysaccharide-induced hepatic injury of rats. Acta Medica Okayama 1992; 46: 345-354.
56. McMillan JM and Jollow DJ. Macrophage enhancement of galactosamine hepatotoxicity using a rat hepatocyte culture system. Res Commun Mol Pathol Pharmacol 1995; 88: 327-338.
57. Connor HD, Lacagnin LB, Knecht KT, Thurman RG and Mason RP. Reaction of glutathione with a free radical metabolite of carbon tetrachloride. Mol Pharmacol 1990; 37: 443-451.
58. Clot P, Albano E, Eliasson E, Tabone M, Arico S et al. Cytochrome P4502E1 hydroxyethyl radical adducts as the major antigen in autoantibody formation among alcoholics. Gastroenterology 1996; 111: 206-216.

59. Alboano E, Tomasi A, Goria-Gatti L, Persson JO, Terelius Y et al. Role of ethanol-inducible cytochrome P450 (P450IIE1) in catalysing the free radical activation of aliphatic alcohol. Biochem Pharmacol 1991; 41: 1895-1902.
60. Faulkner K and Fridovich I. Luminol and lucigenin as detectiors for $O_2^{\bullet -}$. Free Rad Biol Med 1993; 15: 447-451.
61. Boelsterli UA, Wolf A and Goldin C. Oxygen free radical production mediated by cocaine and its ethanol-derived metabolite, cocaethylene, in rat hepatocytes. Hepatology 1993; 18: 1154-1161.
62. Klinger W, Lupp A, Karge E, Oelschlager H, Rothley D et al. Ciprofibrate-racemate and enantiomers: effects of a four-week treatment on male inbred Fishcher rats. A biochemical and morphological study. Exp Toxicol Path 1998; 50: 301-314.
63. Tsuchiya M, Seumatsu M and Suzuki H. *In vivo* visualization of oxygen radical-dependent photo emission. Meth Enzymol 1994; 233: 128-140.
64. Gasbarrini A, Pasini P, Nardo B, De Notariis S, Simoncino M, Cavallari A, Roda W, Bernardi M and Roda A. Chemiluminescent real time imaging of post-ischemia oxygen free radicals formation in livers isolated from young and old rats. Free Rad Biol Med 1998; 24: 211-216.
65. Bass DA, Parce JW, Dechatelet LR, Szejda P, Seeds MC and Thomas M. Flow cytometric studies of oxidative product formation by neutrophils: a graded response to membrane stimulation. J Immunol 1983; 130: 1910-1917.
66. Wan CP, Myung E and Lau BSH. Automated micro-fluorometric assay for monitoring oxidative burst activity of phagocytes. J Immunol Methods 1993; 159: 131-138.
67. Green LC, Wagner DA, Glogowski J, Skipper PL, Wishnok JS and Tannenbaum SR. Analysis of nitrate, nitrite, and [^{15}N] nitrate in biological fluids. Anal Biochem 1982; 126: 131-137.
68. Nussler AK, Di Silvio M, Liu ZZ, Geller DA, Freeswick P, Dorko K, Bartoli F and Billiar TR. Further characterization and comparison of inducible nitric oxide synthase in mouse, rat, and human hepatocytes. Hepatology 1995; 21: 1552-1560.
69. Wagner JG, Ganey PE and Roth RA. Nitric oxide is not involved in hepatocyte killing by neutrophils activated by N-formyl-methionyl-leucylphenylalanine or phorbol myristate acetate *in vitro*. Hepatology 1996; 23: 803-810.
70. Sergent O, Griffon B, Morel I, Chevanne M, Dubos MP, Cillard P and Cillard J. Effect of nitric oxide on iron-mediated oxidative stress in primary rat hepatocyte culture. Hepatology 1997; 25: 122-127.
71. Radi R, Cosgrove TP, Bechman JS and Freeman BA. Peroxynitrite-induced luminol chemiluminescence. Biochem J 1993; 290: 51-57.
72. Hissin PJ and Hilf R. A fluorometric method for determination of oxidized and reduced glutathione in tissues. Anal Biochem 1976; 74: 214-226.
73. Janero DR. Malondialdehyde and thiobarbituric acid-reactivity as diagnostic indices of lipid peroxidation and peroxidative tissue injury. Free Rad Biol Med 1990; 9: 515-540.
74. Yao T, Esposti SD, Huang L, Arnon R, Spangenberger A and Zern MA. Inhibition of carbon tetrachloride-induced liver injury by liposomes containing vitamin E. Am J Physiol 1994; 267: G476-G484.
75. Des Rosiers C, Rivest MJ, Boily MJ, Jette M, Carrobe-Cohen A and Kumar A. Gas chromatographic-mass spectrometric assay of tissue malondialdehyde, 4-hydroxynonenal, and other aldehydes after their reduction to stable alcohol. Anal Biochem 1993; 208: 161-170.

76. Lieber CS, Leo MA, Aleynik SJ, Aleynik MK and Decarli LM. Polyenyl phosphatidylcholine decreases alcohol-induced oxidative stress in the baboon. Alcohol Clin Exp Res 1997; 21: 375-379.
77. Lazzarino G, Di Pierro D, Tavazzi B, Cerroni L and Giardina B. Simultaneous separation of malondialdehyde, ascorbic acid, and adenine nucleotide derivatives from biological samples by ion-pairing high performance liquid chromatography. Anal Biochem 1991; 197: 191-196.
78. Susin SA, Zamzami N and Kroemer G. Mitochondria as regulators of apoptosis: doubt no more. Biochim Biophys Acta 1998; 1366: 151-165
79. Walajtys-Rhode E, Zapatero J, Moehren G and Hoek JB. The role of the matrix calcium level in the enhancement of mitochondrial pyruvate carboxylation by glucagon pretreatment. J Biol Chem 1992; 267: 370-379.
80. Lieser MJ, Park JW, Natori S, Jones BA, Bronk SF and Gores GJ. Cholestasis confers resistance to the rat liver mitochondrial permeability transition. Gastroenterology. 1998; 115: 693-701.
81. Pastorino JG, Snyder JW, Hoek JB and Farber JL. Ca^{2+} depletion prevents anoxic death of hepatocytes by inhibiting mitochondrial permeability transition. Am J Physiol 1995; 268: C676-C685.
82. Lemasters JJ, Nieminen AL, Qian T, Trost LC, Elmore SP et al. The mitochondrial permeability transition in cell death: a common mechanism in necrosis, apoptosis and autophagy. Biochim Biophys Acta 1998; 1366: 177-196.
83. Bramilla G, Cavanna M, Parodi S, Sciaba L, Pino A and Lobbiano L. DNA damage in liver, colon, stomach, lung and kidney of balb/c mice treated with 1,2-dimethyhydrazine. Int J Cancer 1978; 22: 174-180.
84. Nizard C and Decloitre F. Lack of DNA single-strand breaks in rat liver cells exposed to 4-acetylaminofluorene, *in vivo* and *in vitro*. Mutation Res 1990; 244: 245-250.

Chapter 28

Human hepatocyte transplantation: biology and therapy

Robert A. Fisher[1] and Stephen C. Strom[2]
[1]*Medical College of Virginia, Virginia Commonwealth University, Division of Transplant Surgery, Richmond, VA, USA and* [2]*University of Pittsburgh, Department of Pathology, Pittsburgh, PA, USA*

Key words: acute liver failure, chronic liver failure, cryopreservation, human hepatocytes, inherited metabolic disorders, hepatocyte transplantation, serum ammonia, transplant site

28.1 Introduction

The biology and technology of hepatocyte isolation and transplantation has been developed over three decades. This chapter which focuses on human hepatocytes, is divided into four sections. First, the methodology and clinically applied preparation, preservation, and banking of human hepatocytes will be described. The subsequent three sections will deal with human hepatocyte transplantation in acute and chronic liver failure and in four characterized primary hepatocellular metabolic disorders. Relevant, but limited animal modelling followed by human data is outlined.

28.2 Human hepatocyte acquisition, isolation and culture

28.2.1 Acquisition

Hepatocyte isolation and transplantation, as with many other significant medical innovations, was the idea of a clinician dealing with a devastating human disease. The concept of infusing cultured liver cells as a future therapy of acute hepatic failure was proposed by the surgeon, Ben Eiseman in 1967 [1]. In 1969, the seminal work of Berry and Friend established the *in situ* collagenase perfusion technique as a reliable procedure for isolating hepatocytes in high cell yield and viability [2]. Seglen subsequently introduced the two-step modification, which improved metabolic performance by decreasing the time during which hepatocytes are exposed to

M.N. Berry and A.M. Edwards (eds.), The Hepatocyte Review, 475–501.

a low Ca^{2+} concentration [3]. The two step isolation technique with some modification of collagenase concentration was adapted by Strom and others in 1982 [4,5] to yield hepatocytes from livers of other species, including human liver, with greater than 60% viability.

Development of human hepatocyte isolation was next influenced in the United States, by the introduction of a system for human liver tissue acquisition. This was stimulated in 1993 by the first report of the use of cryopreserved hepatocytes as a bridging mechanism for maintaining life prior to solid organ transplant [6]. Strom and Fisher in collaboration with a nonprofit tissue bank, the Anatomic Gift Foundation (AGF, Folkston, GA) initiated a nation-wide network in order to educate organ procurement agencies how to identify livers that could be used for cell isolation. In 1994, the United Network of Organ Sharing (UNOS) enacted policy mandating that human livers not used for solid organ transplant be allocated for "bridging" purposes for patients dying of liver failure (Status 1A) [7]. Major reasons that procured human livers are not used for solid allografting include traumatic damage to the organ, steatosis, and donor liver characteristics and procurement perfusion abnormalities considered as placing the transplanted organ at too high a risk of poor or non-function [8,9].

All requirements of UNOS and the FDA, including all donor clinical data collection, strict adherence to viral and serologic testing, "consent for research", and record keeping are followed in the authors' laboratories [10,11]. The procured livers are preserved with University of Wisconsin solution (UW), yielding viable hepatocytes consistently, despite up to 30 hours of ice-cooler storage [12-14]. Because of the critical shortage of donor livers for patients, the number of these organs that might be available for cell isolation has recently become a limitation in the growth of human cellular transplant procedures [15]. In 1997, from 200 specimens of human whole livers not satisfactory for solid organ transplantation or segments resulting from reduction transplant procedures, 75% were suitable for cell isolation. At the University of Pittsburgh and Medical College of Virginia (MCV) laboratories, 230×10^9 viable hepatocytes were isolated primarily from left liver lobe perfusions. Additional sources of human tissue suitable for non-immortalized hepatocyte isolation are liver from resections, and liver specimens from patients with cholestatic liver disease, undergoing hepatectomy prior to transplant. In the future it may be feasible to obtain satisfactory preparations of hepatocytes from the livers of donors whose hearts have stopped beating.

28.2.2 Hepatocyte isolation and culture

Hepatocyte isolation and culture from human liver tissue is conducted as described by Strom *et al.* [16-18]. For cryopreservation, or for use at later times, cells must be cultured. Culture dishes (100 mm in diameter) are prepared by treating with 50 μg of gelatin in sterile water for at least 30 min. The gelatin solution is then removed by aspiration and hepatocytes are added to the culture dishes in Williams' Medium E, Hepatocyte Maintenance Medium, (HMM) (Clonetics) or chemically defined medium (CDM), adapted from Isom and Georgoff [19], supplemented with 10% bovine calf serum. Hepatocytes are allowed to attach for 4 - 16 hours and the cultures are washed twice in serum-free medium to remove dead and unattached cells and maintained thereafter by daily changes with serum-free medium. Cells, plated and cultured as described, are illustrated in Figure 1. Using a similar collagenase isolation methodology, human fetal hepatocytes have been cultured in Leibovitz medium supplemented with 10% fetal calf serum (or 1% human serum) with good proliferation and function [20].

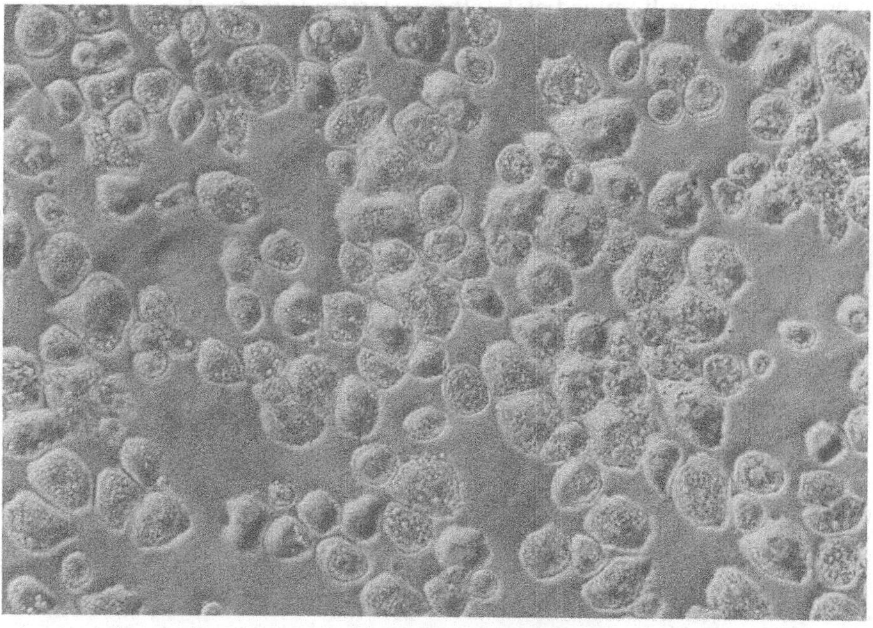

Figure 1. Human hepatocytes on culture plate after isolation

All manipulations of liver tissue are performed in laminar flow culture hoods with surgical sterile technique. Sterility of the cultured cells is assessed by microscopic visual examination of cultures. Gentamicin and

Amphotericin are included in all culture media. Supernatants from cells used for transplantation are cultured for bacterial and fungal analysis.

28.2.3 Cryopreservation

Hepatocytes not used immediately for research or clinical uses can be cryopreserved. Cryopreservation can be conducted immediately following cell isolation (fresh freeze) or cells can be plated as described above, adapted to culture and then cryopreserved. If the cells are cultured first, the cultures are maintained for 24 hours in serum-free medium and changed into CDM supplemented with 2% dimethyl sulphoxide (DMSO) for an additional 24 h. After cryopreservation of more than 300 cases of human hepatocytes we can make some generalisations about the cryopreservation process. It is not always possible to predict which cells will cryopreserve well. In general, the livers with the lowest fat content and the shortest time between cross-clamp and cell isolation are the best. If cold-ischemic times, that is the length of time between when the surgeon begins the flush with UW solution and the time we start the perfusion in our lab, exceed 20 h, the cells will probably not cryopreserve well immediately after isolation. If that is the case we must culture them prior to the cryopreservation process. Figure 2 illustrates the excellent viability, attachment and confluence of human hepatocytes cultured after more than two years of cryopreservation.

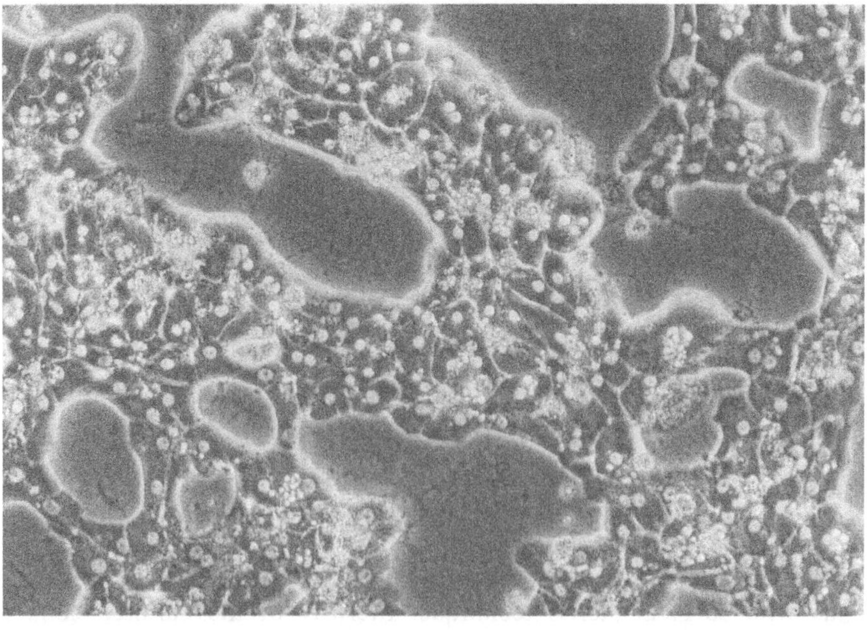

Figure 2. Cultured human hepatocytes after more than 2 years of cryopreservation.

28.2.4 Testing and banking

There are three general tests of hepatocyte viability routinely used in our laboratories. Trypan blue exclusion and plating efficiency of the cells in culture are measured by standard procedures. In addition the expression of basal and inducible CYP genes (measured as immunochemical content and drug metabolism activity) are determined. The above viability data are logged along with donor demographics for banking records. Hepatocytes with plating efficiencies above 10% normally contain basal and inducible drug metabolising enzymes and function well.

Cells isolated and cryopreserved by our techniques express basal levels of CYP genes and their variable expression is induced by exposure to prototypical inducing agents (such as phenobarbital, rifampicin, dexamethosone, and 3-methylcholanthrene (3MC)) [18,21-25]. Ethoxyresorufin-O-deethylase (EROD) and 7-Ethyoxycoumarin-O-deethylase (ECOD) activity is measured in control hepatocyte cultures and compared to cultured hepatocytes exposed for 48 hours to prototypical inducing agents. Separate, duplicate cultures of hepatocytes (60 mM diameter plates) are exposed to the prototypical inducers as described above, and harvested for western blot analysis of the immunoreactive CYP1A and CYP3A as described by Kostrubsky *et al.* [26]. Samples of fresh cells taken immediately after isolation are included in all analyses and the CYP protein content of the cells is quantitated by including different levels of authentic expressed CYP proteins (purchased from Gentest, Inc.) on the western blot along with the samples from that particular experiment. Hepatocytes which demonstrate at least a twofold induction of either the enzymatic activity and/or the CYP protein content in response to the appropriate inducing agent are considered functional. In general, rifampicin, dexamethasone and phenobarbital cause a two to fivefold induction of ECOD activity, and EROD activity is induced four to twenty times by 3MC exposure; however, there is considerable variation between individuals. Western blot results for the cellular content of the specific CYP forms correlate well with the enzymatic levels observed in the ECOD and EROD assays.

Cryopreserved cells have been found to function exactly like the fresh cells with respect to CYP induction [24]. Banked cells from hepatocyte cases that have been cryopreserved are thawed every 6-9 months and are tested for cell attachment and drug metabolism as described above to measure their long-term viability. Cells that are no longer acceptable by the above criteria are discarded.

28.3 Hepatocyte transplantation in acute liver failure

28.3.1 Introduction

Two thousand cases of acute liver failure occur annually in the United States. Acute viral hepatitis accounts for 72% of all cases, with an expected 80% mortality [27]. This human catastrophe is multi-factorial and often complicated by a narrow window of therapeutic options based on timing of presentation to the transplant center. However, the variables for predicting mortality are well established and clinically reproducible, with the etiology of liver failure the most important predictor of survival [28,29]. Cerebral edema, leading to medically recalcitrant cerebral ischemia, is the major pathway to death; [30,31] and infection remains the "Achilles heel", with an incidence of up to 80% for any treatment option [32]. Except for the fortunate 10% who receive orthotopic liver transplant (OLT) [27], the dismal outcome of fulminant hepatic failure is due to hepatocyte loss and not a lack of regenerative capacity of surviving liver cells [33]. The base line replication activity in normal human liver as determined by antibody reaction to nuclear antigens PCNA and Ki-67 is 0.01% to 0.005%. Following hepatotoxic liver injury and hepatitis of viral etiology, hepatocyte replication rates as high as 50% to 60% above baseline are observed, starting at 5 days and continuing for weeks after injury [33-36]. This important feature of human hepatocyte replication, peaking at 5 - 6 days and proceeding over a protracted time frame, is true after a resection stimulus [37,38]. This is surprisingly different from the 20 to 24 h peak regenerative liver cell response following the stimulus of partial hepatectomy in animal models [33].

28.3.2 Value of hepatocyte transplantation

Animal experiments indicate that hepatocyte transplantation can support liver function and hasten host liver regeneration. Chemically induced liver failure in rats, which includes necrosis and functional cell loss, as occurs in the condition of human liver failure, is reversed by hepatocyte fragments and/or isolated viable cells [39-41]. Hepatocyte transplantation rescued all 12 animals when liver cells were implanted less than 28 h following intoxication with D-galactosamine. No animals survived when hepatocytes were transplanted more than 28 h after toxin administration [41]. The survival of rats, subjected to 90% hepatectomy and transplanted with hepatocytes on microcarrier beads into the peritoneal cavity, was 40%

compared with zero survival in controls [42]. A time-dependent efficacy of hepatocyte transplantation was also documented in rats in which liver failure results from partial hepatectomy combined with end-to-side portacaval shunt [43].

28.3.3 Optimal sites for hepatocyte transplantation

Site selection of hepatocyte transplantation in animals has also been thoroughly studied [44]. Analysis indicates that engraftment and subsequent function at the transplant site is superior in the liver, followed by the spleen and peritoneal cavity [44-46]. An experiment measuring the circulating levels of hepatitis B surface antigen (HBsAg) in animals transplanted with hepatocytes procured from transgenic mice, engineered to equally express surface HbsAg, demonstrated at 4 weeks, 4 - 5 times more circulating HBsAg in animals with liver and splenic transplant sites compared to intraperitoneal implants [46]. Hepatocytes transplanted into the spleen display normal morphology, differentiated chord structures, and function for the life of the recipient [47]. Splenic hepatocytes synthesize albumin, store glycogen and control ammonia levels, which were 36% lower in animals with hepatocyte-nidated spleens compared to controls, (even after ammonium chloride challenge) [47]. In a model of ischemic liver injury induced by portal vein and hepatic artery occlusion, gluconeogenesis was significantly increased, and ammonia clearance raised by 60%, with survival times doubled in animals receiving splenic hepatocyte transplants [48]. Animals treated with hepatocyte transplants (after portacaval shunting) were protected from hepatic encephalopathy [49]. Similarly, splenic hepatocyte transplants protected animals infused with ammonia from hyperammonemia-induced encephalopathy and this protection was reversed by splenectomy [50].

28.3.4 Quantity of hepatocytes required for transplantation

The quantity of hepatocytes required for transplant success has also been examined. In rats subjected to 90% hepatectomy, almost all of which die without transplantation, the transplantation of 1 - 2% of the liver mass prevented encephalopathy following ammonia infusion, as well as resulting in long term survival of 60% of the rats [42,50]. The same volume of hepatocytes after implantation has yielded a tenfold better survival in chemically induced liver failure models, even though only 10-20% of cells survived [39-41,43]. In transgenic mice bearing a urokinase gene with albumin promoter, 1×10^4 hepatocytes repopulated the test animal's entire

Table 1A. Acute liver failure hepatocyte recipients - clinical details

Patient #	Etiology Liver Disease	Age Y/M	Organ Systems Failed *	Grade Encephalo -pathy
4. (sp)	Dilantin toxicity	27y	L, P, B, R	4
5.	Hepatitis B (-delta)	28y	L, P, B	4
6. (sp)	Herpetic encephalitis Valproate toxicity	37y	L, B	4
7. (sp)	Idiopathic	23y	L, B, P	3
8. (sp)	Hepatitis B (-delta)	43y	L, B	4
10. (sp)	Acetaminophen toxicity	26y	L, B	4
+11. (sp)	Trisegmentectomy for liver adenocarcinoma	69y	L, B, P	3→4
+12. (L)	Idiopathic	5m	L, B	1
13. (L)	Hepatitis B (-delta)	37y	L, B	1

All data points pre-cell transplant (trx) reflect day prior to hepatocyte trx. Patients. #1, #2, #3, #9 data not included. Patients had chronic liver disease with acute decompensation.
*Abbreviations used in Tables: L, liver; P, pulmonary; C, cardiac; B, brain; (sp), splenic cell trx; (L), Liver cell trx; OLTx, orthotopic solid liver trx.
+Patients treated at University Hospital of Cincinnati or Cincinnati Children's Hospital.

Table 1B. Acute liver failure recipients – function, treatment and outcome

Patient	PT (sec)	Bilirubin (mg/dl)	% Total hepatocyte mass	Cell no. / % Viability	Outcome
4	19	8.3	0.01	2.8×10^7 / 86%	OLT (d10) Full recovery
5	25	20.3	0.07	1.7×10^8 / 56%	OLT (d3) Full recovery
6	100	2.8	0.06	1.2×10^8 / 70%	(d5) *
7 #1	100	21.1	0.14	2.9×10^8 / 72%	OLT (d5)**
#2				1×10^8 / 82%	
8	16.9	16.4	0.2	7.3×10^8 / 83%	OLT (d1) Full recovery
10	60	4.0	0.6	1.2×10^9 / 69%	OLT (d2) Full recovery
11 #1	16	16.8	0.25	3.6×10^8 / 82%	(d2)***
#2				5.3×10^8 / 78%	
12	29.3	14.7	0.5	1.8×10^8 / 95%	OLT (d1) Full recovery
13	100	11.9	0.4	8.8×10^8 / 96%	No OLT$^+$ Full recovery

*gram negative and herpetic pulmonary sepsis, cardiopulmonary death; gram positive sepsis **multi-organ failure, death; ***cardiopulmonary death. $^+$ discharged home (d10).

liver [51]. This very important, repeatedly demonstrated, efficacy of a surprisingly small number of hepatocytes can be understood when we take into account the possible 10^4 times increase in hepatocyte replication incited by the liver failure stimulus [33-35]. Transplantation of 10^7-10^9 hepatocytes can bolster the immediate cellular volume loss, while providing additional regenerative cellular volume needed to overcome the basic defect of human liver failure. Liver failure in the recipient and the placement of hepatocytes in sites other than the liver have been reported to stimulate the regenerative response [52-55]. In the future, to reduce the need for large hepatocyte volumes, there is the prospect of using pharmacological inducers to stimulate a higher level of hepatocellular protein or enzyme production to meet the requirements of individual patients [56].

The solid organ corollary of hepatocyte transplantation, auxiliary partial orthotopic liver transplant (OLT), yielded a 68% host liver regeneration rate in 30 patients with fulminant hepatic failure [57]. A similar liver volume of 0.6% to 0.8% donor graft to recipient body weight ratios as employed in living human donor liver transplantation has also been successful in providing life-assuring liver function [58]. This solid mass of liver tissue of 360-500 g, when extrapolated into cell volume for a 60 kg human, would be equivalent to 7×10^{10} to 1×10^{11} hepatocytes. This cell volume is achievable by cellular transplantation.

28.4 Clinical experience of hepatocyte transplantation for acute liver failure

A summary of our initial experience for hepatocyte transplantation into the spleen has been reported [59,60]. Table 1 summarizes the MCV experience of 11 hepatocyte infusions in 9 patients with acute liver failure (encephalopathy within 8 weeks of onset of illness [27]). Patients 4 and 5 correspond to those similarly numbered in the our previous report [59].

28.4.1

Patients 10 and 11 (Table 1) illustrate that the numbers of hepatocytes that can be safely infused is no greater than 3 to 4×10^8 when the spleen is normal in size (100-150 g) and not hypertrophied as in chronic liver disease. Infusion of 1×10^9 hepatocytes created splenic vein thromboses followed by splenic infarction requiring distal pancreatectomy and splenectomy, fortunately with complete recovery post-OLT in patient 10. Even with extension of the hepatocyte infusion over 4-6 h, as in patient 11, there was

Table 2. Control patients with acute liver failure

Patient #	Etiology of Liver Disease	Age Y	Organ Systems Failed	Grade Encephal-opathy	PT (sec)	bilirubin (mg/dl)	Length ICU support (days)	Cause of death
2.	Traumatic vascular injury	55	L, P, R	3	20	24	3	Sepsis cardiopulmonary failure
3.	Voltaren toxicity	67	L, P, R, B	4	40	15.3	2	Brain stem herniation
4.	Valproate toxicity	46	L, P, C, B	4	27	2	1	Brain stem herniation sepsis
5.	Idiopathic	42	L, B	3	35	25.8	7	Brain stem herniation sepsis
7.	Aldomet toxicity	36	L, B, P	4	50	39.9	1	Brain stem herniation

All data points reflect admission date to MCV or VCU Hospitals.
Patient Controls #1, #6 not included here. Patients had chronic and acute liver disease.

angiographic evidence of diminished venous outflow leading to decreased arterial inflow.

In the acute liver failure patient group with coma, followed by cell transplant, the mean decrease of intracranial pressure of 43%, with concomitant normalisation of cerebral perfusion pressures, corresponded to a decrease in serum ammonia levels that averaged 67% after cell transplant [59]. This ammonia clearance was also demonstrated by Habibullah and associates who used human fetal hepatocytes in treating fulminant hepatic failure [61]. We use ammonia clearance as a marker of hepatocyte function and base additional cell infusions on this. Cardiopulmonary stability is also noted with declining ammonia levels; this allows reduction or withdrawal of cardiac pressor drug support.

In our experience of acute liver failure, 8 of 11 infusions in 7 of the 9 patients have utilized hepatocytes cryopreserved for 2 months to 2 years. These hepatocytes nidate in the spleen and also travel to the liver, and are able to organize and function as fresh hepatocytes, as we previously reported [6,59]. The finding that cryopreserved human hepatocytes form organized structures, and function as well as fresh cultured hepatocytes, parallels findings in animal models [47,62-64].

28.4.2 Clinical outcome for patients who did not receive cell transplantation

Table 2 characterizes the 5 patients (of the total 25 acute liver failure patients evaluated) who were considered cell transplant candidates but did not receive cells due to lack of consent in patients #2 and #3, death within 24 hours of transfer to MCV and/or uncontrolled sepsis (patients #4, #5, #7). Of 7 patients transplanted with hepatocytes as a bridge to OLT, six patients were successfully bridged with 1 post-OLT death due to sepsis (Table 1).

In the 4.5 year time frame during which 25 patients have been evaluated at MCV, 13 of 16 (81%) of patients with fulminant liver failure have enjoyed a 1 year survival after OLT. Finally, patients #8, #12, and #13 reflect the primary author's emerging belief that the liver allows for a safer placement of a larger cell volume then can be placed in the spleen of this group of patients, who exhibit falling transaminases indicative of resolving cell death and acute injury. In addition, earlier cell transplantation, using the grade of encephalopathy as an indicator, is likely to improve the chances of obtaining the maximum number of days of replication stimulus and thus allow an adequate volume of hepatocyte regeneration in order to overcome the principal pathophysiology of acute liver failure, cell loss.

28.5 Hepatocyte transplantation in chronic liver failure

Chronic liver disease patients receive the majority of liver transplant allografts (OLTx) but also have the greater risk of death while on the waiting list due to sheer volume of patients and lack of donors [65]. Like the acute liver failure patient, the chronic liver failure patient suffers from hepatocellular loss; but in contast to the situation in acute disease, the rest of the organism is not normal or healthy prior to decompensation. The therapies such as Transjugular Intrahepatic Portal Systemic Shunt or surgical shunts, used to treat the complications of the shrinking fibrotic liver, such as portal hypertension and ascites, can contribute to worsening encephalopathy and portal vein thrombosis in 4 to 10 % despite halting bleeding [66-68]. These bridges to transplant provide no cellular mass with potential for replication. It is not surprising that animal models mimicking the diverse features of human chronic liver disease are problematic [69-71]. The portacaval shunt concept used in the human to treat complications of cirrhosis, when applied in an end-to-side fashion in the rat creates a chronic liver disease with the encephalopathy, wasting, and neurobehavioral abnormalities similar to the human disease [72-74].

28.5.1 Intrasplenic hepatocyte transplantation in chronic liver disease

As previously well documented in the rat, syngeneic hepatocytes transplanted into the spleen, whether cryopreserved or fresh, can hepaticize the spleen and survive functionally for up to 18 months [45,47,75,76]. Intrasplenic hepatocyte transplantation in the portacaval shunt rat model of chronic liver failure corrects hepatic encephalopathy and reverses the neurobehavioral disorder as well as correcting ultrastructural alterations of the corpus striatum [49,73,77]. The concept that the spleen in the chronic liver failure model is a safe and effective vessel for hepatocyte survival has been further strengthened by the 8 month survival of structurally normal autotransplanted hepatocytes in the spleens of monkeys with portacaval shunts [78]. Based on this extensive preclinical work, Mito and Kusano performed 10 hepatocyte autotransplants, using the recipients' own cirrhotic left lateral segments to extract hepatocytes; and then directly inoculated 1.7×10^7 to 6×10^8 hepatocytes into their spleens. These human cellular transplants in patients with chronic liver disease were combined with an open common hepatic artery ligation. Clinical improvement occurred in 2 patients who demonstrated intrasplenic hepatocyte uptake of

Technetium-99 m, but artery ligation, rather than hepatocyte transplantation was judged causal for the relief of ascites and encepholopathy [79,80].

The Fisher-MCV protocol, which has been used in 15 hepatocyte transplants in 13 patients (patients #1, #2, #3 and #9 with chronic liver disease previously described [59]), differs from that of Mito and Kusano, in its use of allogeneic hepatocytes isolated from non cirrhotic liver tissue. These cells are then infused by a transfemoral intra-arterial asplenic infusion and a direct or transvenous portal vein infusion technique without laparoscopic or open surgical procedures. Human hepatocytes extracted from non-cirrhotic liver tissue, when cultured even after cryopreservation, attach and expand as we have illustrated in Figures 1 and 2 in the first section of this chapter. In our experience, use of the isolation techniques for cirrhotic human liver as described by Mito and Kusano [79,80] has yielded cellular cultures of hepatocytes forming encapsulated nodular structures, shown in Figure 3, and this may be an important difference. In the four chronic liver disease patients treated with cellular transplants, spleen size has measured 800 g to 1200 g (unlike the normal spleen size of 100-120 g in the patients

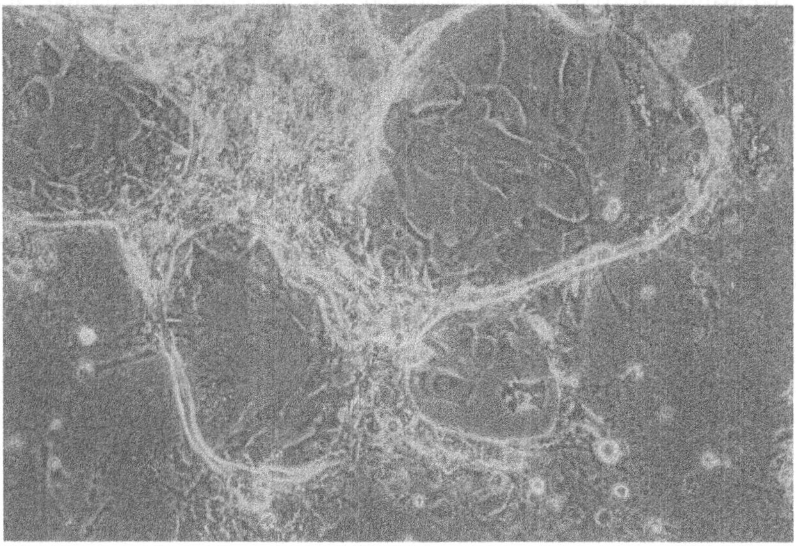

Figure 3. Culture plate of human hepatocytes, isolated from regenerating nodules of cirrhotic liver tissue, forming encapsulated structures.

Figure 4. Topographical view of colonies of hepatocytes in the spleen of chronic patient #2.

with acute liver failure), providing the necessary red pulp volume to safely allow up to 6×10^8 hepatocytes to be infused at one time. No evidence of pulmonary embolism, hemodynamic instability, or infectious contamination has been recorded with the intra-arterial splenic infusion pioneered at MCV. In the chronic liver disease patient where the liver is shrunken and cirrhotic, isolated human hepatocytes can nidate in the spleen and produce liver specific proteins (*i.e.* α_1-antitrypsin), reduce serum ammonia, and assume normal hepatocyte structural characteristics. Figure 4 is a topographical view of the hepatocyte colonies previously illustrated at higher power magnification (× 250) in chronic patient #2 [59].

28.5.2 Use of immunosuppression

In our clinical series, the majority of cellular allogeneic transplants have been performed under cyclosporin (Neoral/Sandimmune, Novartis, East Hanover, New Jersey) immunosuppression. Two patients received tacrolimus (Prograf, Fujisawa, Deerfield, Il). This follows the experience of successful allogeneic hepatocyte transplantation in the rat spleen with primary cyclosporin immunosuppression [81]. Bumgardner and colleagues utilized a functional model of splenic hepatocyte transplantation of donor hepatocytes from mice transgenic for human α_1-antitrypsin, and demonstrated the weak humoral, and strong T-cell mediated, rejection of allogeneic hepatocytes [82,83]. Our initial immunologic studies of human patients with liver failure, bridged with human allogeneic hepatocytes under cyclosporin immunosuppression, did not show an increase in recipient immunoreactivity to cell and solid organ donor or third party lymphocytes. These results at 3 and 6 months after hepatocellular bridging to OLT were no different to concurrent matched OLT patients who did not receive cell transplantation [84].

28.5.3 Case history

A chronic liver failure patient with alcoholic cirrhosis and poorly controlled encephalopathy, on maximal tolerable lactulose, and with intractable ascites and hydrothorax requiring frequent paracentesis, was transplanted with 6×10^8, 80% viable hepatocytes. This 62 year old male was found to have clotted splenic, mesenteric, and portal veins not surgically correctable for OLT. He consented to treatment involving hepatocyte transplant for encephalopathy management, combined with percutaneous peritoneovenous shunt placement [85] for ascites control, and hemodialysis for renal failure 3 times a week. This patient had serum ammonia levels of 89 μmol/l despite lactulose treatment every 4 hours before cell transplant. After cell transplant the serum ammonia levels fell to 20-36 μmol/l and remained at this level on a general diet without lactulose. This ammonia control and freedom of encephalopathy ended when the patient, who was discharged home 11 days after cellular transplant, voluntarily discontinued dialysis and died from pulmonary fluid overload and hyperkalemia 33 days after splenic hepatocyte transplant.

In keeping with the auto-hepatocyte transplant experience of Mito and colleagues [79,80], we found that increasing allogeneic hepatocellular activity in the spleen of this patient was detectable with Technetium-99m diisopropyl-iminodiacetic acid. Activity rose, from zero at day 2 after cell transplant, to easily visible and increasing hepatocellular activity at days 9 and 23 post hepatocyte transplant (Figure 5). This recorded increase in hepatocellular activity could have improved encephalopathy and ammonia clearance but had no positive effect on synthetic factor production, bilirubin control, or renal failure.

In human chronic liver failure, earlier timing of hepatocyte transplant and the role of repeated hepatocyte volume infusion to enhance hepatocellular functional mass, appear to be the next relevant issues to be addressed in order to make hepatocellular splenic transplant clinically relevant to the long waiting time until OLT.

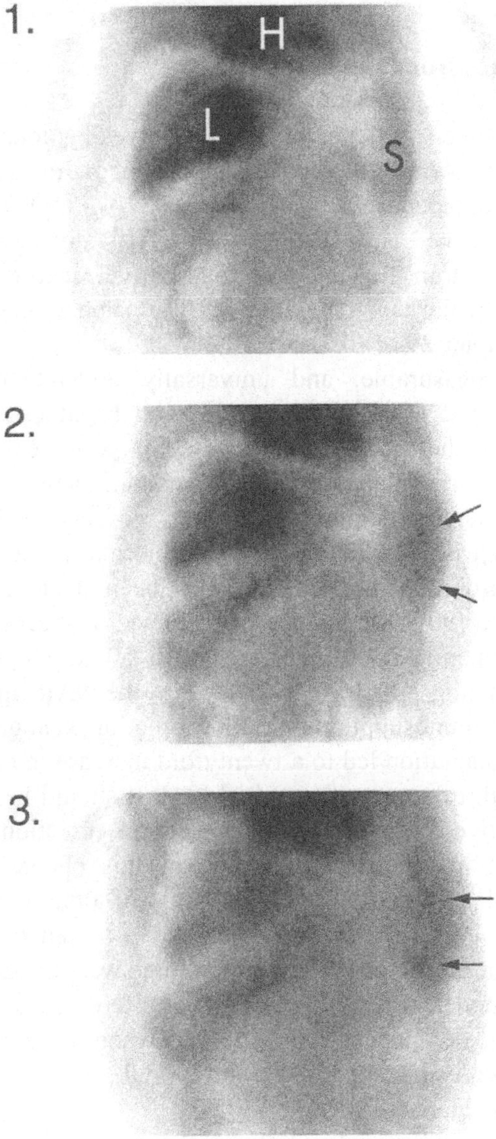

Figure 5. Serial Technetium - 99m DISIDA scan images at the same perfusion phase of study in chronic liver failure patient #9. Images are taken at 1: day 2; 2: day 9; 3: day 23 post splenic hepatocyte transplant. Arrows in frames 2 and 3 point to hepatocellular activity in the spleen. (H = heart, L = liver, S = spleen)

28.6 Hepatocyte transplantation for metabolic disorders

28.6.1 Evidence from animal models

There have been extensive animal investigations of hepatocyte transplantation as potential therapy for inborn errors of hepatic metabolism where genetic defects result in absence, or impaired function of hepatocellular proteins. This section will specifically address research on isolated hepatocyte transplantation of cells natively expressing the gene of interest in 100% of the donor population. The science of gene therapy will be detailed in Chapter 29.

The readily measurable, and universally understood correlation of albumin (mRNA and protein) production with hepatocellular volume and function, has been the measure of success for peritoneal, intraportal, and splenic hepatocyte transplantation in the analbuminemic rat [86-89]. This model has repeatedly demonstrated the profound effect of 10^6-10^7 hepatocytes, which is only about 1% of the normal rat total liver mass. A fifteenfold increase in serum albumin has been achieved by intraportal cellular transplant [90]. Intrasplenic hepatocyte isotransplantation in the Nagase Analbuminemic rat (NAR) created a 50 to 80 mg/dl increase in serum albumin, lasting 5 to 16 months [89]. In the NAR model treated with tacrolimus for immunosuppression, allogeneic or xenogeneic hepatocyte intrasplenic transplantation led to a twentyfold increase in albumin levels for 4 weeks as a result of xenograft survival [91]. It should be emphasized that when a regenerative stimulus, (as is present in resection or liver failure models discussed in the earlier sections of this chapter) was added to hepatocyte transplantation, long term elevation of albumin levels to 60% of normal occurred [92]. The effectiveness of a regenerative stimulus in overcoming a limit of safe cell transplant volume was further corroborated in the Shionogi rat model, which involves a fatal osteogenic disorder associated with ascorbic acid deficiency. Rats survived when a regenerative stimulus was combined with intrasplenic or intraportal, but not intraperitioneal, hepatocellular transplants [93].

In the larger-animal model, involving the use of allogeneic hepatocyte transplantation to lower the cholesterol level in the Watanabe hyperlipidemic rabbit, a regenerative stimulus, prior to administration of hepatocytes representing only 2% to 5% of the total liver cell mass of the recipient's native liver, resulted in a prolonged reduction of 40% in serum cholesterol [94,95]. In another study with the Watanabe rabbit, the use of normal xenogeneic hepatocytes for transplantation, resulted again in a 40% decrease in serum cholesterol despite data implying that only 0.1% to 0.5% of the actual liver mass of hepatocytes survived [96].

28.6.2 Transplantation of human hepatocytes

The autotransplantion of hepatocytes (2×10^8) from a left lateral liver segment after *ex vivo* LDL receptor gene transduction, resulted in a 20% reduction in serum cholesterol for 18 months in a human patient with type 1 familial hypercholesterolemia [97]. Unfortunately, these results were not duplicated in a later trial with 5 patients by Grossman, *et al.* [98]. For treatment of familial hypercholesterolemia, no transplants of human allogeneic hepatocytes naturally expressing the LDL receptor have yet been reported.

Results of allogeneic splenic and portal venous transplantation of normal human hepatocytes expressing α_1-antitrypsin and ornithine transcarbamoylase (OTC) in enzyme deficient patients have been reported [59,60]. In both circumstances a small number of hepatocytes convincingly normalized ammonia levels. An increase of α_1-antitrypsin of 34% above baseline, expected from plasma support, with an increase in the M phenotype bands in a ZZ phenotype patient, occurred under the regenerative stimulus of liver failure together with cyclosporine immunosuppression [6,59]. In a 5 year old child without vascular access for solid liver transplantation and with no detectable OTC on liver biopsy, 1×10^9 hepatocytes infused intraportally, under tacrolimus immunosuppression, resulted in a liver tissue level of 0.3 to 0.5% of normal liver OTC on biopsy 28 days post hepatocyte transplant [60]. In contrast to the patient with α_1-antitrypsin deficiency, the regenerative stimulus in the patient with OTC deficiency occurred in the absence of liver failure or resection of liver parenchyma.

28.6.3 The Gunn rat model

One of the most extensively studied models of hepatocyte transplantation for overcoming an inborn error of hepatic metabolism has involved efforts to correct bilirubin conjugation and excretion in the Gunn rat. The defect of bilirubin uridine diphosphate glucuronosyltransferase in the Gunn rat serves as the model for the Crigler-Najjar syndrome type 1 in humans [86,90,99,100]. In this Gunn rat model, segmental liver transplantation of 12% of the whole liver volume can normalize hyperbilirubinemia [101]. In an elegant set of experiments, normal, untreated, congenic rats (Wistar Shi) were used as the source for whole organs (by OLT), or as donors of fetal or adult hepatocytes, for enzyme replacement therapy in Gunn rats, and the long term (12 months) effects on bilirubin correction were studied [102]. In the OLT group, total bilirubin normalized by 2 months, with the bilirubin glucuronide fractions normalising by 4 months post transplant. The intrasplenic transplanted fetal tissue and adult congenic hepatocytes (10^7 hepatocytes representing 1-2% of the whole liver hepatocellular volume) reduced total bilirubin by 30% by 2 months with continued and further reduction to <4 mg/dl total bilirubin at 12 months compared to <0.5 mg/dl total bilirubin by 2 months in the OLT group. Furthermore, by 4 months, Gunn rats transplanted with cellular or fetal tissue exhibited bilirubin glucuronides of 28-36% of total bilirubin [102]. Intraportal hepatocyte transplantation in the Gunn rat of comparable liver cell volumes has been shown to decrease mean serum bilirubin levels by 51% at 28 days following transplantation [90]. In the EHBR Eizai rat model, comparable to the human Dubin-Johnson syndrome, hepatocyte transplantation of 2% of the total liver cellular volume resulted in a 50% decrease in conjugated bilirubin for 120 days [103]. These studies are important in relation to yet another facet of cellular biology: the correction of a defect in canalicular transport can only occur if transplanted hepatocytes migrate, incorporate, and make functional connections to the biliary transport system.

28.6.4 Human hepatocyte transplantation in Crigler-Najjar syndrome

In the human, the nature of the genetic mutation giving rise to Crigler-Najjar syndrome type 1 was revealed following the cloning of the human bilirubin UDP-glucuronosyltransferase complementary DNA [104] and gene [105]. No cases of human disease have been reported without mutation in the gene form responsible for bilirubin glucuronidation, UGTIAI [105,106]. The simple nature of the defect in Crigler-Najjar syndrome type 1, involving a single gene, together with the results of animal studies, has led to the hope that this disease could be amenable to a gene or cell therapy. In a human 10 year old child, awaiting OLT, with deficient bilirubin glucuronidation activity unaffected by phenobarbital therapy, an intraportal infusion of 7.5×10^9 hepatocytes (representing 5% of the native total hepatocyte volume) was performed with tacrolimus immunosuppression. Using the methods detailed in the first section of this chapter, hepatocytes were isolated by Strom and colleagues from a cadaveric donor having normal bilirubin glucuronosyltransferase activity. By day 7 after the infusion, serum bilirubin concentration had declined from a starting level of 26 mg/dl to 21.5 mg/dl. The patient was then started on a phenobarbital regimen with the intent of increasing the level of UGTIAI expressed in the transplanted hepatocytes. The effectiveness of this treatment was demonstrated by the further decline in serum concentration of unconjugated bilirubin to 11-14 mg/dl. The bilirubin level has remained at this stable level for more than 6 months following the cell transplant [107]. Although the risk of kernicterus is reduced under these conditions, further diminution of the serum bilirubin levels to normal range (<1 mg/dl) would be desirable. This last human cell transplant case highlights three valuable facts: 1) Intraportal cell transplant with careful portal pressure monitoring allows safe hepatic seeding with a larger volume of hepatocytes than can safely be placed in a human spleen of normal size. 2) Hepatocytes isolated by the techniques of Strom *et al.* [59,60] have long term viability and metabolic function [107]. 3) Pharmacologic induction of cells *in vivo*, detailed as part of the standard transplant protocol by our group, [59] has clinical utility.

28.7 Conclusion

The scientific community has devoted years of work on hepatocellular isolation, preservation, transport, and ideal cell transplant number and location. We are learning that one key to overcoming the barrier of time and number of hepatocytes needed to treat disease, is greater understanding of the mechanisms influencing the expression and/or activity of hepatocellular enzymes (*i.e.* phenobarbital induction of CYP and UGTIAI gene expression).

Acknowledgements

The authors express our deepest gratitude to Melissa Thompson and Kenneth Dorko for their consistently superb hepatocyte isolation; to Joe Ritter, Ph.D., for his excellent comments and additions to the metabolic section; to Jaime Tisnado, M.D. and Mel Fratkin M.D. for their never faltering radiologic support; and to Mrs. Joanne Fisher for her eternal patience.

References

1. Eiseman B. Treatment of liver failure. In: Read AE, Ed. Colston Papers No. 19 The Liver. London: Butterworths, 1967: 279-285.
2. Berry MN and Friend DS. High yield preparation of isolated rat liver parenchymal cells. A biochemical and fine structural study. J Cell Biol 1969; 43: 506-520.

3. Strom SC, Jirtle RL, Jones RS, Rosenberg M, Novicki NA, Irons G, McLain JR and Michalopoulos G. Isolation, culture and transplantation of human hepatocytes. J National Cancer Institute 1982; 68: 771-778.
4. Guguen-Guillouzo C, Campion JP, Brissot P, Glaise D, Launois B et al. High yield preparation of isolated human adult hepatocytes by enzymatic perfusion of the liver. Cell Biol Int Rep 1982; 6: 625-628.
5. Habibullah CM, Hassan SI, Ayesha Q, Zaman TU, Baig MMA and Devi BG. Fetal liver perfusion - a new approach. Surg Res Commun 1990; 8: 87-90.
6. Fisher RA, Strom SC, Thompson MT et al. Human hepatocyte transplantation - A life bridge in terminal liver failure. Hepatology. 1993; 18: Suppl: 66A.
7. U.N.O.S. Allocation of livers for other methods of hepatic support. 1994; Policy 3.6.10.
8. Ploeg RJ, D'Alessandro AM, Knechtle SJ, Stegall MD, Pirsch JD et al. Risk factors for primary dysfunction after liver transplantation: a multivariate analysis. Transplantation 1993; 55: 807-813.
9. Marino IR, Doyle HR, Aldrighetti L, Doria C, McMichael J et al. Effect of donor age and sex on the outcome of liver transplantation. Hepatology 1995; 22: 1754-1762.
10. U.N.O.S. Minimum information for liver offers. June 1997; Policy 3.6.9.
11. FDA. Human tissue intended for transplantation and guidance for screening and testing of donors of human tissue intended for transplantation; availability. Federal Register 1997; 62: 40429-40537.
12. D'Alessandro AM, Kalayoglu M, Sollinger HW, Hoffmann RM, Pirsch JD et al. Experience with Belzer UW cold storage solution in human liver transplantation. Transplant Proc 1990; 22: 474-476.
13. Sorrentino D, VanNess K, Ribeiro I and Miller CM. Functional and morphologic features of isolated hepatocytes preserved in University of Wisconsin solution. Hepatology 1991; 14: 331-339.
14. Belzer FO. Clinical organ preservation with the UW solution. Transplantation 1989; 47: 1097-1098.
15. Lake JR. Hepatocyte transplantation. N Engl. J Med 1998; 338: 1463-1465.
16. Strom SC, Dorko K, Thompson MT, Pisarov LA and Nussler AD. Large scale isolation and culture of human hepatocytes. In Utilization Therapeutiques des ilots de Langerhans et des Hepatocytes. Editions de Inserm. 1997.
17. O'Connell JF, Cox S, Buontempo P, Skelton A, Pisarov LA, Dorko K and Strom SC. Primary human hepatocyte cultures for the study of Hepatitis C Virus (HCV). In: Methods in Molecular Medicine. Chapter 55, Humanna Press, Totowa, NJ, 1997.
18. Strom SC, Pisarov LA, Dorko KD, Thompson MT, Schuetz EG and Schuetz JD. The use of human hepatocytes to investigate the regulation of cytochrome P450 genes, Meth Enzymol 1996; 272: 388-401.
19. Isom HC and Georgoff I. Quantitative assay for albumin-producing liver cells after simian virus 40 transformation of rat hepatocytes maintained in chemically defined medium. Proc. Natl Acad Sci (USA). 1984; 81: 6378-6382.
20. Bisell DM and Tilles JG. Radioimmunoassay of serum albumin produced by human embryonic liver in monolayer culture. J Cell Biol 1972; 53: 819-823.
21. Monteith D, Ding JH, Chen YT, Michalopoulos G and Strom SC. Induction of cytochrome P(I)-450 specific RNA and benzo(a)pyrene metabolism in isolated human hepatocytes. Toxicology and Applied Pharm. 1990; 105: 460-471.
22. Schuetz EG, Schuetz JD, Strom SC, Thompson MT, Fisher RA, Molowa DT, Li D and Guzelian PS. Regulation of human liver cytochromes P450 in the family 3A in primary and continuous culture of human hepatocytes. Hepatology. 1993; 18: 1254-1262.

23. Kocarek TA, Schuetz EG, Strom SC, Fisher RA and Guzelian PS. Comparative analysis of cytochrome P450 3A induction in primary cultures of rat, rabbit and human hepatocytes. Drug Metab Disp 1995; 23: 415-421.
24. Schuetz EG, Schuetz JD, Thompson MT, Fisher RA, Madaraga JR and Strom SC. Phenotypic variability in the induction of P-glycoprotein mRNA by aromatic hydrocarbons in primary human hepatocytes. Mol Carcinogenesis 1995; 12: 61-65.
25. Schuetz JD, Schuetz EG, Trotassery JV, Guzelian PS, Strom SC and Sun D. Identification of a novel dexamethasone responsive region in the human CYP3A5 gene and its activation in human and rat liver cells. Molecular Pharm 1996; 49: 63-72.
26. Kostrubsky VE, Strom SC, Wood SG, Wrighton SA, Sinclair PR and Sinclair JF. Ethanol and isopentanol increase CYP3A and CYP 2E in primary cultures of human hepatocytes. Arch Biochem Biophys 1995; 322: 516-520.
27. Lee WM. Acute liver failure. New Eng J of Med 1993; 329: 1862-1872.
28. O'Grady JG, Alexander GJM, Hayllar KM and Williams R. Early indicators of prognosis in fulminant hepatic failure. Gastroenterology 1989; 97: 4439-4445.
29. Hughes RD, Wendon J and Gimson AES. Acute liver failure. Gut 1991; suppl: S86-S91.
30. Williams R and Gimson AES. Intensive liver case and management of acute hepatic failure. Dig Dis Sci 1991; 36: 820-826.
31. Blei AT. Cerebral edema and intracranial hypertension in acute liver failure: distinct aspects of the same problem. Hepatology 1991; 13: 376-379.
32. Rolando N, Harvey F, Brahm J, Philpott-Howard J, Alexander G et al. Prospective study of bacterial infection in acute liver failure: an analysis of fifty patients. Hepatology 1990; 11: 49-53.
33. Wolf HK and Michalopoulos GK. Hepatocyte regeneration in acute fulminant and nonfulminant hepatitis: A study of proliferating cell nuclear antigen expression. Hepatology 1992; 15: 707-713.
34. Koukoulis G, Rayner A, Tan KC, William SR and Portmann B. Immunolocalization of regenerating cells after submassive liver necrosis using PCNA staining. J Pathol 1992; 166: 359-368.
35. Kayano K, Yasunaga M, Kubota M, Takenaka K, Mori K et al. Detection of proliferating hepatocytes by immunohistochemical staining for proliferating cell nuclear antigen (PCNA) in patients with acute hepatic failure. Liver 1992; 12: 132-136.
36. Gerdes J, Li L, Schlueter C, Duchrow M, Wohlenberg C et al. Immunobiochemical and molecular biologic characterization of the cell proliferation-associated nuclear antigen Ki-67. Am J Path 1991; 138: 867-873.
37. Fausto N and Mead JE. Regulation of liver growth: protooncogenes and transforming growth factors. Lab Invest 1989; 60: 4-13.
38. Franenvilla A, Panella C and Polimeno L. Hormonal and enzymatic parameters of hepatic regeneration in patients undergoing major liver resections. Hepatology 1990; 12: 1134-1138.
39. Sutherland DER, Numata M, Matas AJ, Simmons RL and Najarian JS. Hepatocellular transplantation in acute liver failure. Surgery 1977; 82: 124-132.
40. Sommer BG, Sutherland DER, Matas AJ, Simmons RL and Najarian JS. Hepatocellular transplantation for the treatment of d-galactosamine-induced acute liver failure in rats. Transplant Proc 1979; 11: 578-584.

41. Baumgartner D, La Plante-O'Neill PM, Sutherland DER and Najarian JS. Effects of intrasplenic injection of hepatocytes, hepatocyte fragments and hepatocytes culture supernatants on D-galoctosamine-induced liver failure in rats. Eur Sur Res 1983; 15: 129-135.
42. Demetriou AA, Reisner A, Sanchez J, Levenson SM, Moscioni AD et al. Transplantation of microcarrier-attached hepatocytes into 90% partially hepatectomized rats. Hepatology 1988; 8: 1006-1009.
43. Cuervas-Mons V, Cienfuegos JA, Maganto P, Golitsin A, Eroles G et al. Time related efficacy of liver cell isografts in fulminant hepatic failure. Transplantation 1984; 38: 23-25.
44. Gupta S and Chowdhury JR. Hepatocyte transplantation: back to the future. Hepatology 1992; 15: 156-162.
45. Mito M, Kusano M, Onishi T, Saito T and Ebata H. Hepatocellular transplantation. Gastroenterol Jpn 1978; 13: 480-490.
46. Gupta S, Vemuru RP, Lee CD, Yerneni PR, Aragona E et al. Hepatocytes exhibit superior transgene expression after transplantation into liver and spleen compared with peritoneal cavity or dorsal foot pad: Implications for hepatic gene therapy. Human Gene Therapy 1994; 5: 959-967.
47. Mito M, Ebata H, Kusano M, Onishi T, Saito T et al. Morphology and function of isolated hepatocytes transplanted into rat spleen. Transplantation 1979; 28: 499-505.
48. Takeshita K, Ishibashi H, Suzuki M and Kodama M. Hepatocellular transplantation for metabolic support in experimental acute ischemic liver failure in rats. Cell Transplant 1993; 2: 319-324.
49. Ribeiro J, Nordlinger B, Ballet F, Cynober L, Coudray-Lucas C et al. Intrasplenic hepatocellular transplantation corrects hepatic encephalopathy in portacaval-shunted rats. Hepatology 1992; 15: 12-18.
50. Schumacher IK, Okamota T, Kim BH, Chowdhury NR, Chowdhury JR et al. Transplantation of conditionally immortalized hepatocytes to treat hepatic encephalopathy. Hepatology 1996; 24: 337-343.
51. Rhim JA, Sandgren EP, Degen JL, Palmiter RD and Brinster RL. Replacement of diseased mouse liver by hepatic cell transplantation. Science 1994; 263: 1149-1152.
52. Lamers WH, Been W, Charles R and Moorman AFM. Hepatocytes explanted in the spleen preferentially express carbomoylphosphate synthetase rather than glutamine synthetase. Hepatology 1990; 12: 701.
53. Strain AJ, Ismail T, Tsubouchi H, Arakaki N, Hishida T et al. Native and recombinant HGF are highly potent promoters of DNA synthesis in both human and rat hepatocytes. J Clin Invest 1991; 87: 1853.
54. Tomiya T, Nagoshi S and Fujiwara K. Significance of serum human growth factor levels in patients with hepatic failure. Hepatology 1992; 15: 1.
55. Wolf HK, Zarnegar R and Michalopoulos GK. Localization of hepatocyte growth factor in human and rat tissues: an immunohistochemical study. Hepatology 1991; 14: 488.
56. Strom SC, Pisarov LA, Dorko K, Thompson MT, Schuetz JD et al. Use of human hepatocytes to study P450 gene induction. Meth Enzymol 1996; 272: 388-401.
57. Chenard-Neu MP. Auxiliary liver transplantation: Regeneration of the native liver and outcome in 30 patients with fulminant hepatic failure. A multicenter European study. Hepatology 1996; 23: 1019-1027.
58. Lo CM, Sheung TF, Chan JKF et al. Minimum graft volume for successful Adult-to-Adult living donor liver transplantation for fulminant hepatic failure. Transplantation (Brief Communications) 1996; 62: 696-698.

59. Strom SC, Fisher RA, Thompson MT, Sanyal AJ, Cole PE et al. Hepatocyte Transplantation as a bridge to orthotopic liver transplantation in terminal liver failure. Transplantation 1997; 63: 559-569.
60. Strom SC, Fisher RA, Rubinstein WS, Barranger JA, Towbin RB et al. Transplantation of human hepatocytes. Transplant Proc 1997; 29: 2103-2106.
61. Habibullah CM, Syed IH, Qamar A and Taher-Uz Z. Human fetal hepatocyte transplantation in patients with fulminant hepatic failure. Transplantation 1994; 58: 951-952.
62. Maganto P, Cienfuegos JA, Santamaria L, Rodriguez V, Eroles G et al. Auxiliary liver by transplanted frozen-thawed hepatocytes. J Surg Res 1990; 48: 24-31.
63. Maganto P, Cienfuegos JA, Santamaria L, Eroles G, Andres S et al. Cryopreservation and transplantation of hepatocytes: an approach for culture and clinical application. Cryobiology 1988; 25: 311-322.
64. Fuller BJ, Lewin J and Sage L. Ultrastructural assessment of cryopreserved hepatocytes after prolonged ectopic transplantation. Transplantation 1983; 35: 15-18.
65. Weisner RH. Current indications, contraindications, and timing for liver transplantation. In: Busuttil RW and Klintmalm GB, Eds. Transplantation of the liver. Philadelphia: WB Saunders Company, 1996: 71-84.
66. Freedman AM, Sanyal AJ, Tisnado J, Shiffman ML, Luketic VA et al. Results with percutaneous transjugular intrahepatic portasystemic stent-shunts for control of variceal hemorrhage in patients awaiting liver transplantation. Transplant Proc 1993; 25: 1087-1089.
67. Henderson JM, Gilmore GT, Hooks MA, Galloway JR, Dodson TF et al. Selective shunt in the management of variceal bleeding in the era of liver transplantation. Ann Surg 1992; 216: 248-255.
68. Sanyal AJ, Freedman AM and Purdum PP. TIPS-associated hemolysis and encephalopathy. Annals of Int Med 1992; 117: 443-444.
69. Mullen KD and McCullough AJ. Problems with animal models of chronic liver disease: suggestions for improvement in standardization. Hepatology 1989; 9: 500-503.
70. Bengtsson F, Bugge M, Vagianos C, Jeppsson B and Nobin A. Brain serotonin metabolism and behaviour in rat with CCl_4-induced cirrhosis. Res Exp Med 1987; 187:429-438.
71. Martin J, Cuervas-Mons V, Colas A, Diz M and Tormo J. Intrasplenic hepatocellular transplantation as a hepatic support measure in cirrhotic dogs. Transplant Proc 1987; 19: 3930-3931.
72. Lauterbough BH, Sautter V, Preisig R and Bircher J. Hepatic functional deterioration after portacaval shunt in the rat. Gastroenterology 1976; 71: 221-227.
73. Bengtsson F, Gage FH, Jeppsson B, Nobin A and Rosengren E. Brain monomine metabolism and behavior in portacaval-shunted rats. Exp Neurol 1985; 90: 21-35.
74. Gitlin N, Lewis DC and Hinkley L, The diagnosis and prevalence of subclinical hepatic encephalopathy in apparently healthy, ambulant, non-shunted patient with cirrhosis. J Hepatol 1986; 3: 75-82.
75. Ebata H, Kusano M, Onishi T, Saito T and Mito M. Liver regeneration utilizing isolated hepatocytes transplanted into the rat spleen. Surg Forum 1978; 29: 338-340.
76. Kusano M, Ebata H, Onishi T, Saito T and Mito M. Transplantation of cryopreserved isolated hepatocytes into the rat spleen. Transplant Proc 1981; 13: 848-854.
77. Mariani P, Coudray-Lucas C, Baudrimont M, Ribeiro J, Legendre C et al. Glutamine metabolism and neuropathologic disorder in experimental hepatic encephalopathy: effect of transplanted hepatocytes. Surgery 1996; 120: 93-99.
78. Kasai S, Sawa M, Kondoh K, Ebata H and Mito M. Intrasplenic hepatocyte transplantation in mammals. Transplant Proc 1987; 19: 992-994.

79. Mito M, Kusano M and Kawaura Y. Hepatocyte transplantation in man. Transplant Proc 1992; 24: 3050-3051.
80. Mito M and Kusano M. Hepatocyte transplantation in man. Cell Transplantation 1993; 2: 65-74.
81. Makowka L, Lee G, Cobourn CS, Farber E, Falk JA and Falk RE. Allogeneic hepatocyte transplantation in the rat spleen under Cyclosporin immunosuppression. Transplantation 1986; 42: 537.
82. Bumgardner GL, Li J, Heininger M, Ferguson RM and Orosz CG. *In vivo* immunogenicity of purified allogeneic hepatocytes in a murine hepatocyte transplant model. Transplantation 1998; 65: 47-52.
83. Bumgardner GL, Heininger M, Li J, Xia D, Parker-Thornburg J et al. A functional model of hepatocyte transplantation for *in vivo* immunologic studies. Transplantation 1998; 65: 53-61.
84. Fisher RA, Kimball PM, Thompson M et al. An immunologic study of liver failure patients bridged with hepatocyte infusions. Transplantation 1998; 65: S45.
85. Valentine RJ and Turner WW. "Peel away" technique for Denver shunt insertion. Surgery 1984; 96: 937-939.
86. Demetriou AA, Whiting JF, Feldman D, Levenson SM, Chowdhury NR et al. Replacement of liver function in rats by transplantation of microcarrier-attached hepatocytes. Science 1986; 233: 1190-1192.
87. Ohta T, Ogawa K and Nagase S. Increase in albumin mRNA by repeated intrahepatic transplantation of Fischer 344 rat hepatocytes into the liver of congenic analbuminemic rats. Biochem Biophys Res Commun 1993; 194: 601-609.
88. Rozga J, Holzman M, Moscioni AD, Fujioka H, Morsiani E and Demetriou AA. Repeated intraportal hepatocyte transplantation in analbuminemic rats. Cell Transplant 1995; 4: 237-243.
89. Nozawa M, Ikebukuro H, Otsu I et al. Hepatic support by hepatocyte transplantation in congenitally metabolic diseased rats. Transplant Int 1992; 5(suppl 1): S211-S213.
90. Holzman MD, Rozga J, Neuzil DF, Griffin D, Moscioni AD and Demetriou AA. Selective intraportal hepatocyte transplantation in analbuminemic and Gunn rats. Transplantation 1993; 55: 1213-1219.
91. Kokudo N, Horimoto H, Ishida K, Takahashi S and Nozawa M. Allogeneic hepatocyte and fetal liver transplantation and xenogeneic hepatocyte transplantation for Nagase's analbuminemic rats. Cell Transpl 1996; 5: S21-S22.
92. Moscioni AD, Rozga J, Chen S, Naim A, Scott HS et al. Long-term correction of albumin levels in the Nagase analbuminemic rat: repopulation of the liver by transplanted normal hepatocytes under a regeneration response. Cell Transplantation 1996; 5: 499-503.
93. Onodera K, Ebata H, Sawa M, Katoh K, Kasai S, Mito M and Nozawa M. Comparative effects of hepatocellular transplantation into the spleen, portal vein, or pertioneal cavity in congenitally ascorbic acid biosynthetic enzyme-deficient rats. Transplantation Proc 1992; 24: 3006-3008.
94. Wang J, Pollak R and Bartholomew A. Sustained reduction of serum cholesterol levels following allotransplantation of parenchymal hepatocytes in Watanabe rabbits. Transplantation Proc 1991; 23: 894-895.
95. Eguchi S, Rozga J, Lebow T, Chen SC, Wang CC et al. Treatment of hypercholesterolemia in the Watanabe rabbit using allogeneic hepatocellular transplantation under a regeneration stimulus. Transplantation 1996; 62: 588-593.

96. Gunsalus JR, Brady DA, Coulter SM, Gray BM and Edge AS. Reduction of serum cholesterol in Watanabe rabbits by xenogeneic hepatocellular transplantation. Nature Med 1997; 3: 48-53.
97. Grossman M, Raper S, Kozarsky K, Stein EA, Engelhardt JF et al. Successful *ex vivo* gene therapy directed to the liver in a patient with familial hypercholesterolemia. Nature Genetics 1994; 6: 335-341.
98. Grossman M, Rader DJ, Muller WM, Kolansky DM, Kozarsky K et al. A pilot study of *ex vivo* gene therapy for homozygous familial hypercholesterolemia. Nature Med 1995; 1: 1148-1153.
99. Groth CG, Arborgh B, Bjorken C, Sundberg B and Lundgren G. Correction of hyperbilirubinemia in the glucuronosyltransferase-deficient rat by intraportal hepatocyte transplantation. Transplantation Proc 1977; 9: 313-319.
100. Vroemen JPAM, Buurman WA, Heirwegh KPM, van der Linden CJ and Kootstra G. Hepatocyte transplantation for enzyme deficiency disease in congenic rats. Transplantation 1986; 42: 130-135.
101. Asonuma K, Gilbert JC, Stein JE, Takeda T and Vacanti JP. Quantitation of transplanted hepatic mass necessary to cure the Gunn rat model of hyperbilirubinemia. J Pediatric Surg 1992; 27: 298-301.
102. Kokudo N, Otsu I, Okazaki T, Takahashi S, Sanjo K et al. Long term effects of intrasplenically transplanted adult hepatocytes and fetal liver in hyperbilirubinemic Gunn rats. Transpl Int 1995; 8: 262-267.
103. Hamaguchi H, Yamaguchi Y, Goto M, Misumi M, Hisama N et al. Hepatic biliary transport after hepatocyte transplantation in the Eizai hyperbilirubinemic rats. Hepatology 1994; 20: 220-224.
104. Ritter JK, Crawford JM and Owens IS. Cloning of two human liver bilirubin UDP - glucuronosyltransferase cDNAs with expression in COS-1 Cells. J Biol Chem 1991; 266: 1043-1047.
105. Ritter JK, Chen F, Sheen YY, Tran HM, Kimura S, Yeatman MT and Owens IS. A novel complex locus UGTI encodes human bilirubin, phenol, and other UDP-glucuronosyltransferase isozymes with identical carboxyl termini. J Biol Chem 1992; 267: 3257-3261.
106. Gantla S, Bakker CT, Deocharan B, Thummala NR, Zweiner J et al. Splice-site mutations: a novel genetic mechanism of Crigler-Najjar Syndrome Type 1. Am J Hum Genet 1998; 62: 585-592.
107. Fox IJ, Chowdhury JR, Kaufman SS, Goertzen Z, Chowdhury NR et al. Treatment of the Crigler-Najjar Syndrome Type 1 with hepatocyte transplantation. N Engl J Med 1998; 338: 1422-1426.

Chapter 29

Ex vivo hepatocyte-directed gene therapy

Steven E. Raper and A. David Moscioni
Institute for Human Gene Therapy, University of Pennsylvania Medical Center, 607BRBII/III 421 Curie Blvd, Philadelphia, PA 19104-6160, USA

Key words: adeno-associated virus, allogeneic hepatocytes, autologous hepatocyte transplantation, baculovirus, gene delivery, herpes simplex I virus, hepatocyte proliferation, inherited metabolic disease, lentiviral vectors, murine leukemia virus

29.1 Introduction

The liver has emerged as a prime target organ for the evolving field of somatic cell gene therapy because of the many inherited and acquired diseases that require innovative treatment [1-2]. *Ex vivo* gene therapy was the initial strategy used to achieve correction of inherited defects by the introduction of adenosine deaminase into the bone marrow of patients with severe combined immunodeficiency [3]. Using this approach, autologous target cells are harvested from a donor, genetically modified *in vitro*, and then transplanted back into the host. This is the strategy which will be discussed in detail in this chapter. For example, in the bone marrow, pluripotent stem cells have been genetically modified and then induced to proliferate after being returned to the host [4]. Early excitement for liver-directed applications of the *ex vivo* approach subsided with the realisation that hepatocyte transduction was inefficient, with expression of the genetically modified vector DNA occurring only rarely. A number of new and innovative methods to enhance transgene integration and engraftment of transduced hepatocytes are currently being studied. In this chapter, we will review the work that has led to present hepatic gene therapy protocols, primarily in the context of an *ex vivo* approach. We will review both the delivery systems that are currently most promising and the vectors that are likely to arise from this technology.

M.N. Berry and A.M. Edwards (eds.), The Hepatocyte Review, 503–520.

29.2 Ex-vivo gene therapy: the human clinical trial

Liver-directed *ex vivo* gene therapy for humans was first used to treat the rare disorder familial hypercholesterolemia [5]. In this trial, LDL receptor cDNA was retrovirally transduced into autologous hepatocytes. The left lateral segment of the liver containing the targeted hepatocytes was removed from the patient. These cells were plated in culture and transduced with retrovirus containing the therapeutic LDL receptor gene. The genetically modified cells were harvested and transplanted into the patient from whom they had been removed. A persistent, clinically detectable effect was demonstrated in three of five recipients, although total correction of serum cholesterol was not achieved in any [6]. More importantly, the procedures were conducted safely and without any noted morbidity or mortality [7].

Although recombinant retroviruses are the preferred substrate for transducing genes into dividing cells *in vitro*, the strategy of transplanting retrovirally transduced hepatocytes has several limitations. As demonstrated by the now completed clinical trial, far less than 1% of the liver mass is taken over by these transplanted cells [6-9]. To date, overexpression of the transduced gene has not completely compensated for the vast majority of uncorrected cells. In the human trial in which patients with familial hypercholesterolemia were treated, compensation was observed to occur as increased LDL receptor expression and decreases in serum cholesterol were noted, despite engraftment levels of less than 1% [5,6]. It was decided that a higher degree of engraftment must be achieved before proceeding with this treatment modality in humans.

29.3 Allogeneic versus autologous cells

In the human clinical trial, autologous cells were used [5-7]. This brings up an important decision when developing *ex vivo* liver-directed gene therapy protocols. Autologous, allogeneic, and possibly xenogeneic hepatocytes may all be suitable donor cells. The main advantage of using autologous cells is to avoid the need for immunosuppression following transplantation, however this approach is invasive, requiring major surgery. Allogeneic or xenogeneic hepatocytes are more available, and might even be cryopreserved to allow hepatocyte banks to be created. The use of non-self donor cells has the requirement of immunosuppression. This requirement was demonstrated in the Watanabe heritable hyperlipidemic (WHHL) rabbit. The WHHL rabbit has a mutation in its LDL receptor gene, rendering the receptor completely dysfunctional [10]. When allogeneic hepatocytes were used in non-immunosuppressed WHHL rabbits, correction of the serum cholesterol could be demonstrated for only 7-10 days, after which rejection resulted in loss of the transplanted cells [11].

When immunosuppression is instituted during allogeneic hepatocyte transplantation, better gene function is seen. Isolated cryopreserved human liver cells have been attached to collagen-coated microcarriers and injected intraperitoneally into Gunn rats deficient in hepatic bilirubin uridine diphosphoglucuronosyltransferase or Nagase analbuminemic rats. Subgroups of both rat strains were made immunodeficient by crossing them with athymic strains with an inherited T-cell deficiency. No histologic evidence of rejection was seen in the athymic rats, while immunocompetent rats did manifest evidence of rejection. In both groups, there was evidence of partial correction of both genetic defects [12]. Another promising method of avoiding an immune response is to modify the transplanted cells in order to reduce their immunogenicity. This was done with ultraviolet-B irradiation and short term culture. Hepatocytes isolated from Wistar RHA rats were subjected to UV-B irradiation then injected intraportally into Nagase analbuminemic rats. Plasma albumin concentration increased significantly [13]. Presently, the immune system remains a formidable obstacle for allogeneic cell transplantation, and the issue of transgene immunity has not been completely resolved. Immunosuppressive agents such as cyclophosphamide, cyclosporine A and tacrolimus can prevent rejection and lead to stable effects. However, these agents are not without risks. Decreasing the amount of immunosuppression leads to less recipient morbidity, but may also inhibit the function of the transplanted cells.

29.4 Hepatocytes as vector targets

The hepatocyte is a differentiated cell capable of limited proliferation *in vitro*. Although it is possible to isolate relatively pure fractions of hepatocytes in culture, they cannot be passaged and they are difficult to maintain for extended periods of time. The hepatocyte begins to lose differentiated function within several days of isolation. This necessitates quick and efficient viral transduction and rapid transplantation. The liver is thought to provide the most physiologic site for hepatocyte transplantation. The extracellular matrix of the liver is most conducive to supporting the hepatocyte, and only in the liver would hepatocytes be able to develop polarity and secrete bile. Questions concerning the amount of cells that can be transplanted need to be addressed. Using a rabbit model, 2×10^8 cells were safely transplanted into a 2 kg animal. In 200-g rats, intraportal infusion of 2×10^7 cells was well tolerated [14]. The 5 patients in the human trial (weight range: 20-50 kg) received between 1×10^9 and 3.2×10^9 cells, without any associated morbidity [7]. The transplanted cell mass did, in many instances, cause acute self-limited increases in portal venous pressure which were felt to be related to occlusion of sinusoids. Therefore, approximately 1×10^8 cells/kg appears to be an appropriate number. As the number of cells transplanted increases, complications such as portal hypertension, venous thrombosis and even pulmonary embolism might occur [15,16].

29.4.1 Methods of delivery

The portal vein can be accessed using different strategies, including inferior mesenteric vein (IMV) injections, umbilical vein injections, direct injections into the spleen, or even cannulation of jejunal venous tributaries. IMV infusions were first performed in mongrel dogs using 9.6 Fr Hickman catheters secured with 3 vicryl ties [17]. Ten days later, the catheter was pulled and the abdomen was explored for intraperitoneal pathology. No hemorrhage, catheter migration, or portal thrombosis was observed. This technique was then repeated in baboons in preparation for a human trial for the treatment of familial hypercholesterolemia [18]. Transduced autologous hepatocytes were reinfused via an indwelling catheter placed into the IMV at initial laparotomy. No acute portal hypertension or long-term evidence of portal vein thrombosis was detected angiographically. Biochemical and hematologic parameters remained within normal limits.

In the initial human trial for familial hypercholesterolemia, 4 of the 5 patients had their IMV cannulated for cell infusion (1 patient had a left gastric vein cannulation) without any associated perioperative hemorrhage, or portal vein thrombosis [7]. Only transient elevations in portal vein pressure were observed during the cell infusions. Despite these promising results in pre-clinical trials and in the initial clinical trials, several potential complications must be anticipated when cannulating the inferior mesenteric vein. Catheter infection might be a problem. Portal vein occlusion or thrombosis is another potential life threatening complication, which could lead to variceal hemorrhage or ascites, or which may complicate subsequent attempts at orthotopic liver transplantation. This has been observed during infusion of crude pancreas preparations into the portal vein [19]. Infusion of highly purified preparations of islet cells appears to result in a much lower incidence of portal vein thrombosis [20]. The incidence of portal vein thrombosis following infusion of hepatocytes should be lower because the infusate contains a purified suspension of cells. Autologous cells should result in a lower incidence of thrombosis than allogeneic cells by virtue of avoiding the immune response, which may contribute to this complication.

Although not yet used for hepatocyte transplantation, the umbilical vein remnant in the falciform ligament can be cannulated to access the portal system. Islet cells have been transplanted into the liver using this technique [18]. Patients for whom this procedure is inappropriate are usually those with anatomic anomalies, and include patients with a potentially scarred-down vein, or who have had prior surgical division of the falciform ligament, or who may have hepatofugal flow through this vessel, such as patients with cirrhosis and portal hypertension.

Intrasplenic transplantation of HBsAg-producing G7 HBV transgenic rat hepatocytes into livers of recipient rats has been reported. Transgenic cells could be seen in the intrahepatic portal spaces and over half of the transplanted cells permanently engrafted into the liver [21]. Intrasplenically engrafted hepatocytes were shown to function in a liver-specific manner by the maintenance of differentiated liver gene expression [22]. Intrasplenic human hepatocyte transplantation has also been reported in a clinical setting [23] (see also Chapter 28). Cannulation of jejunal venous branches for hepatocyte infusion may present a feasible approach with minimal morbidity. Finally, as interventional radiologists continue to develop better techniques to access the arterial system, the hepatic artery might be used more readily for hepatocyte transplantation.

29.5 Vectors suitable for *ex vivo* liver-directed gene therapy

An ideal vector would cause no toxicity, transduce only the target cell population *in vivo*, and integrate in a site-specific manner at the site of the abnormal gene, thus restoring normal function. A large variety of vectors has been studied as suitable candidates for liver-directed gene therapy. No single vector currently is ideal. Many of the properties of currently existing vectors are presented in Table 1. Those vectors most commonly used in *ex vivo* studies are described in the following Sections.

Table 1. Characteristics of commonly used vectors

Vector	Source	Nucleic acid	Integration occurs	Proliferation required for integration	Period of expression	Size of insert †kb
Retrovirus	Murine leukemia virus	RNA	Yes	Yes	Permanent	4
Adeno-associated virus	Human parvo virus	SS DNA*	Yes, site specific on chromosome 19	No	Permanent	4
Herpes simplex virus type 1	Human herpes virus	DS DNA	Yes	No	With latency, permanent	10
Lentivirus	Human immuno-deficiency virus	RNA	Yes	No	Permanent	4
Baculovirus	Insect virus	DS DNA	No	No	At least 6 months	4

* - SS = single stranded, DS double stranded. † - kb = kilobase

29.5.1 Murine leukemia retroviral vectors

The retrovirus is a commonly used vector for piloting liver-directed gene delivery applications *ex vivo* and *in vivo*. Retroviruses are popular because of their relatively efficient entry into cells and integration into the host genome. Retroviral particles are composed of an RNA genome encapsulated into a complex virus particle containing both viral and cellular components. The virus enters cells primarily on the basis of interactions between the viral envelope protein and receptors for the virus on the recipient cell [24]. Once internalized, the viral RNA is converted to a double stranded DNA sequence and this proviral DNA is integrated into the host chromosome by means of an integrase protein [25]. Integrated proviral DNA is then propagated as the host cell replicates [26]. Viruses used in somatic gene transfer experiments have been rendered replication defective. These recombinant viruses are produced in two stages [27,28]. Initially, a packaging cell line is used to produce viruses that contain only the desired therapeutic RNA. A packaging cell is generated by transfecting a mutated wild-type murine retroviral genome into a cell permissive for virus replication. This genome has a partial genomic deletion, called the psi sequences, in the 5' untranslated region. The psi sequences enable the transcript to be incorporated and packaged into virions. The psi-deleted genome still expresses all proteins necessary to assemble a virion, namely gag, pol and env. A second retrovirus is introduced into the packaging cell line, which contains RNA molecules with the desired therapeutic or marker gene. This second virus retains the psi sequence so that transcribed RNA molecules can be packaged into viral protein coats.

Gene delivery using retroviruses results in long-term expression because they integrate into cell chromosomes. The use of retroviruses for *in vivo* gene transfer has been hampered by two major limitations. The first is the need for cell proliferation since retroviruses only integrate into dividing cells [29]. The second limitation is that it is possible to generate only relatively low titres of Moloney-based amphotropic retroviral vectors, in the range of 10^5-10^7 colony forming units (cfu) per ml. Hepatocytes are quiescent cells, so that most *in vivo* delivery attempted to date has required a 76% partial hepatectomy or administration of a liver-damaging agent prior to injection of the retrovirus into the portal vein [30-32].

One problem encountered during retrovirus production is that the retroviral particles are relatively labile when compared to other vectors used for gene therapy. A significant loss of infectivity can occur during purification. Titres of infectious retroviral particles are usually in the range of 10^6 to 10^7 particles/ml from media harvested from packaging cell lines. Centrifugation of large volumes of cell culture supernatant has recently shown promise as a technique used to concentrate retroviruses and overcome the limitation of low retroviral titres. These concentrated virus suspensions, containing 7×10^7 cfu per rat, have been used to deliver the E. coli β-galactosidase gene into up to 25% of hepatocytes [33] More recent evidence for a direct correlation between viral titre and effective liver transduction efficiency was provided in a report showing that two injections of high titre unconcentrated retrovirus (5×10^8 cfu/rat) into rats resulted in the transduction of 45% of the hepatocytes [34].

Attempts to stimulate proliferation in the absence of hepatic resection have met with some promise, and bear directly on the use or retroviral vectors to transduce hepatocytes *in vitro*. Hepatocyte growth factor (HGF) is known to play a primary role in liver regeneration [35-37]. Follistatin blocks the inhibition of DNA synthesis by activin A, a member of the TGF-β superfamily in rat liver parenchymal cells [38]. Intraportal administration of a single dose of follistatin has accelerated liver regeneration in rats after 70% hepatectomy and was more efficient at promoting liver regeneration than HGF [39-41]. The ability to stimulate hepatocyte proliferation *in vivo* and *in vitro* by single or combination growth factors can be expected to lead to better transduction of hepatocytes by retroviruses.

Other alternative strategies have also had some success in improving *in vivo* gene transfer of retrovirally transduced hepatocytes. Following partial hepatectomy, there is an increase in retroviral transduction efficiency, as cellular regeneration occurs [42,43]. Intraportal infusion of retrovirus during either asanguineous perfusion or hepatic artery in-flow occlusion has also increased viral transduction [44,45]. Finally, *in vivo* regional delivery of retroviral vectors either prior to or after a partial hepatectomy, will increase the amount of hepatocytes infected by the retrovirus [46-49]. Further investigation will define the role of hepatectomy and blood flow manipulation on *ex vivo* gene therapy.

The promoter used in the retroviral vectors has also been studied as a potential factor affecting hepatocyte transduction. The Moloney murine leukemia virus long terminal repeat (LTR) promoter has been linked to the murine RNA polymerase II (Pol II) promoter, with human alpha-1-antitrypsin (hAAT) as the reporter gene. After intraportal infusion, serum hAAT levels were significantly higher using the Pol II promoter [50]. *In vivo* hAAT expression has also been studied using a retroviral vector containing the liver-specific albumin promoter-enhancer. Serum levels of hAAT were fivefold to tenfold higher using the albumin promoter than levels obtained using either the Pol II or endogenous LTR promoter [51].

29.5.2 Adeno-associated viral vectors

Adeno-associated virus (AAV) is a human parvovirus that is being considered for use in gene therapy protocols. This small non-enveloped virus contains a 4.7-kb single stranded DNA genome that encodes regulatory and structural genes [52]. Recombinant forms of AAV (rAAV) have been developed as vectors by substituting viral open reading frames with therapeutic minigenes [53]. The advantages of AAV include stable and efficient integration of viral DNA into the cellular genome, a broad host range of cultured cell types, ability to transduce growth arrested cells, high titers of vector produced in tissue culture, and a physically stable virion that can be concentrated [53,54].

Little has been done to date with respect to *ex vivo* gene therapy, however, the *in vivo* data amassed to date is instructive. When rAAV vector was injected into the liver via the portal vein, <0.01% of hepatocytes were transduced. However, upon co-administration with an adenoviral vector containing both the E1 and E4 regions, transduction efficiencies increased to 10-25%. *In vitro*, transduction was also increased significantly when an adenoviral vector expressing early regions 1 and 4 was co-transfected with rAAV [55].

Recombinant adeno-associated virus vectors (AAV) were prepared in high titre (10^{12} to 10^{13} particles per ml) for the expression of human factor

IX (F.IX) after *in vivo* transduction of murine hepatocytes. Injection of AAV-CMV-F.IX (expression from the human cytomegalovirus IE enhancer/promoter) into the portal vein of adult mice resulted in no detectable human F.IX in plasma, but in mice injected intravenously as newborns with the same vector, expression was initially 55 to 110 ng/ml. The expression in the liver was mostly transient, and plasma levels decreased to undetectable levels within 5 weeks. However, long-term expression of human F.IX was detected by immunofluorescence staining in 0.25% of hepatocytes 8 to 10 months postinjection. A second vector in which F.IX expression was controlled by the human EF1α promoter was constructed and injected into the portal vein of adult C57BL/6 mice at a dose of 6.3×10^{10} particles. This resulted in therapeutic plasma levels (200-320 ng per ml) for a period of at least 6 months, whereas no human F.IX was detected in plasma of mice injected with AAV-CMV-F.IX. Doses of AAV-EF1α-F.IX of 2.7×10^{11} particles resulted in plasma levels of 700-3,200 ng/ml. Liver-derived expression of human F.IX from the AAV-EF1α-F.IX vector was confirmed by immunofluorescence staining. The loss of expression with the former vector was probably caused by suppression of the CMV promoter, because polymerase chain reaction data showed no substantial loss of vector DNA in mouse liver [56].

A separate study assessed the feasibility of transferring the insulin gene into liver cells of diabetic mice using a defective AAV vector. The rat insulin gene was packaged into a defective AAV vehicle (AAV-INS). Successful AAV-mediated transfection and expression of the insulin gene into hepatocytes was demonstrated by immunocytochemistry and reverse transcriptase-polymerase chain reaction (RT-PCR) [57]. After AAV-mediated transfection of the insulin gene into hepatocytes, glucose in the medium was significantly reduced for up to 5 days. After direct injection of AAV-INS into liver parenchyma of diabetic mice, successful transfection was demonstrated by RT-PCR, and blood glucose was significantly decreased for at least 6 days. Taken together, these studies suggest that recombinant AAV vectors may be used to transfer therapeutic genes into hepatocytes *in vitro* and *in vivo*.

29.5.3 Herpes simplex 1 viral vectors

Herpes simplex virus-1 (HSV-1) is a 150-kb double-stranded DNA virus that is known to infect a variety of mammalian cell types. Defective HSV amplicons can be produced in culture that are deleted of most replication and packaging proteins [58]. Using these principles, defective HSV-1 vectors have been developed and used to transfer genes into differentiated neurons

[59-60]. HSV-1 also exhibits tropism for hepatocytes, suggesting a role for these vectors in liver-directed gene therapy [61]. HSV-1 vectors containing the Escherichia coli lacZ reporter gene are able to transduce both mouse and human hepatocytes in culture with high efficiency [62,63]. In addition, this vector was shown to transduce a high percentage of human hepatocytes (40%) in suspension culture, thus eliminating the problems encountered during hepatocyte culture under standard conditions [63]. When transduced murine hepatocytes were delivered to congenic recipients intrasplenically, lacZ expression persisted for 2 weeks [62]. Delivery of this vector to rats *in vivo*, by isolated hepatic perfusion, resulted in lacZ expression at 72 h (1-3% of hepatocytes) without associated histological signs of tissue damage or inflammation [63]. Advantages of this vector for liver-directed gene therapy may include high-efficiency transduction of non-replicating cells, effective transduction of cells in suspension, and the ability to accommodate larger fragments of genetic material.

Using features of AAV to augment deficiencies of herpes simplex virus type 1 (HSV-1), a helper virus-free HSV/AAV hybrid amplicon vector was used to extend transgene expression in hepatocytes *in vivo*. HSV-1 amplicon or HSV/AAV hybrid amplicon vectors were created to express reporter genes from different transcriptional regulatory sequences and packaged using a helper virus-free system. To determine relative transduction efficiencies, vector stocks were tested on four different cell lines, including hamster kidney (BHK21) and human lung (Hs913T) fibroblasts, and mouse (G6Pase-/-) and human (NPLC) hepatocytes. After *in vivo* injection of vector stocks into mouse liver, tissue sections were examined for reporter gene expression and cellular inflammatory response. Blood samples were collected to measure serum transaminase levels as a biochemical index of liver toxicity. Expression of a reporter gene from liver-specific promoter sequences was consistently more effective in hepatic cells compared with fibroblasts, whereas the opposite was true when using an HSV-1 immediate-early promoter. Expression in hepatocytes *in vivo* was markedly longer from HSV/AAV hybrid vector than HSV-1 amplicon vectors. The number of transduced cells (approximately 2% of all hepatocytes) remained stable over 7 days after injection of HSV/AAV hybrid vector, whereas no transduced cells were detected 7 days after gene transfer with standard HSV-1 amplicon vector. The rapid decline in reporter gene expression from standard amplicons was not solely caused by a cell-mediated immune response, as it also occurred in the immunodeficient mouse strain RAG-2. Hepatocyte toxicity and cellular inflammatory effects associated with HSV/AAV hybrid vector-mediated gene transfer were minimal, and readministration of vector stock proved equally effective in naive mice and in animals that received a first vector dose 4 weeks earlier. HSV/AAV hybrid amplicon vectors support gene expression *in vivo* for considerably longer than do traditional HSV-1 amplicon vectors, and need to be tested in an *ex vivo* gene therapy setting [64].

29.5.4 Lentiviral Vectors

Lentiviral vectors are RNA-based, retroviral vectors such as HIV that are able to integrate into the genome of non-proliferating cells. This characteristic gives this vector an advantage over other retroviral vectors, in that transduction efficiency of non-proliferating cells such as hepatocytes, myofibrils and hematopoetic cells will be much higher. Recently, a lentiviral vector was shown to be more effective than a murine leukemia-based retroviral vector in providing improved transduction of rat fibroblasts blocked in the cell cycle, or in human primary macrophages, *in vitro* [65]. These authors demonstrated a similar effect *in vivo* with terminally differentiated neurons. Because lentiviral vectors can be pseudotyped with other coat proteins, such as vesicular stomatitis virus G glycoprotein or influenza hemagglutinin, they can selectively transduce a broad range of tissues and cell types. Sustained expression of green fluorescent protein (GFP), used as a surrogate for therapeutic protein, was observed for more than 22 weeks after *in vivo* transduction into the liver. Similar long-term expression (more than eight weeks) was observed in transduced muscle. In contrast, little or no GFP could be detected in liver or muscle transduced with the Moloney murine leukemia virus, a prototypic retroviral based vector under similar conditions. At a minimum, 3-4% of the total liver tissue was transduced by a single injection of $1\text{-}3 \times 10^7$ infectious units (I.U.) of recombinant HIV vector. Furthermore, no inflammation due to recruitment of lymphocytes could be detected at the site of injection. Animals previously transduced with a lentiviral vector can be efficiently re-infected with lentiviral vectors. A requirement for lentiviral accessory proteins to establish efficient transduction *in vivo* is tissue dependent and deserving of further investigation [66].

29.5.5 Baculoviral Vectors

The baculovirus, *Autographa californica*, nuclear polyhedrosis virus (AcNPV) is a new potential vector for liver directed gene therapy. This virus is widely used as a vector for expression of foreign genes in insect cells [67]. It has also been studied in mammalian cells, where neither significant gene expression nor viral DNA replication could be observed [68]. However, a recent discovery suggests that this vector may in fact be useful for gene therapy in humans. Recombinant AcNPV containing a heterologous promoter (CMV) and the luciferase reporter gene is efficiently taken up by both human and rabbit hepatocytes *in vitro* by an endosomal pathway, resulting in a high level of luciferase activity [69]. Mouse hepatocytes and many other human cell types tested had very low levels of luciferase activity. Using the viral polyhedrin promoter, no activity was seen in the human hepatocytes. Stable expression with the CMV promoter-luciferase construct was observed in resting hepatocytes for over 2 weeks. The data suggest that baculoviral vectors may also be advantageous for liver-directed gene therapy. *In vitro* efficiency of gene transfer is comparable to retrovirus. *In vitro* expression level is comparable to adenovirus. These vectors appear to be specific for human hepatocytes but do not integrate into the host genome. Endogenous baculoviral genes are most likely not expressed in mammalian cells due to restricted functions of their promoters. Future work testing hepatocyte selectivity and immune responses *in vivo* may prove fruitful.

29.6 A new concept: selective repopulation by corrected hepatocytes

The liver is unique in its capacity to regenerate, and regardless of the type of vector used to transduce the hepatocyte *in vitro*, once returned to the recipient, further proliferation of the genetically corrected cells is a desirable outcome. A pioneering discovery was made by Sandgren, in the laboratory of Dr. Ralph Brinster, with the observation that a proportion of transgenic mice containing the albumin-urokinase-type plasminogen activator gene (Alb-uPA) survived expression of the usually lethal construct [70]. The rationale for construction of this transgene was to create high levels of urokinase-type plasminogen activator (uPA) that would be expressed only in the liver by linking it to the liver-specific albumin promoter. Mice expressing the uPA gene would be susceptible to a severe bleeding diathesis. The uPA-expressing hepatocytes were found to be at a selective proliferative disadvantage. In the surviving mice, a gradual decrease in the level of plasma uPA activity was accompanied by a restoration of clotting function that was complete between 1 and 2 months. It was discovered that DNA rearrangements within a subset of hepatocytes abolished transgene expression. The uPA-transgene-deficient cells selectively proliferated relative to uPA-expressing hepatocytes and repopulated the entire liver. This initial discovery demonstrated the remarkable developmental potential of liver cells. The data suggest that a few corrected or normal hepatocytes could potentially salvage and restore the entire liver if the internal environment were suitable.

The Brinster group further studied this idea when they introduced foreign cells into a diseased liver to see if transplanted cells could also repopulate the liver. Once again using the Alb-uPA transgenic mouse, normal hepatocytes were isolated from adult mice and transplanted. The transplanted liver cell populations replaced up to 80% of the diseased recipient liver, demonstrating the feasibility of liver cell transplantation as a method to replace lost hepatic parenchyma [71]. Mouse hepatocytes have also been transduced *in vivo* with an adenovirus expressing uPA [72]. This resulted in significant liver damage and death. Concurrently, a retrovirus expressing a marker gene was delivered to hepatocytes *in vivo*. The hepatocyte destruction and regeneration stimulated by the recombinant adenovirus led to a fivefold to tenfold greater transduction efficiency of the retrovirus. After 4 weeks, the livers appeared normal. Thus, this two-virus system offers an alternative to partial hepatectomy to achieve permanent levels of hepatocyte transgene expression.

Further refinement of the concept of *in vivo* selection of corrected hepatocytes to improve the efficiency of liver gene transfer using recombinant retrovirally transduced hepatocytes was shown by recent work from the laboratory of Grompe. Hereditary tyrosinemia type-1 is a recessive liver disease caused by deficiency of fumarylacetoacetate hydrolase (FAH). It was found that in a transgenic FAH knockout mouse, as few as 1,000 transplanted wild-type hepatocytes were able to repopulate the mutant liver, demonstrating the strong competitive growth advantage of transduced hepatocytes. The data suggest that adult mouse hepatocytes can divide at least 18 times *in vivo* [73]. After reporting the use of hepatocyte transplantation and *in vivo* gene therapy for the correction of a mouse model of Hereditary Tyrosinemia Type I, Grompe and his collaborators subsequently demonstrated repopulation of fumarylacetoacetate hydrolase (FAH)-deficient livers with *ex vivo* gene therapy using retrovirally transduced hepatocytes. Correction of the disease phenotype was achieved by retrovirally transducing cultured FAH-deficient hepatocytes *ex vivo*, followed by transplantation and selective repopulation. Treated mice were phenotypically normal and had corrected plasma amino acid levels and liver function tests. The results demonstrate that efficient hepatic repopulation using *ex vivo* genetically manipulated hepatocytes is feasible [74]. To test further the upper limit of mouse hepatocyte regenerative capacity, serial transplantation of hepatocytes in the fumarylacetoacetate hydrolase deficiency murine model of liver repopulation was performed. Hepatocytes from adult donors were serially transplanted in limiting numbers 6 times and resulted in complete repopulation during each cycle. This corresponds to a minimal number of 69 cell doublings or a 7.3×10^{20}-fold expansion. No evidence for abnormal liver function or altered hepatic architecture was found in repopulated animals [75].

29.7 Summary

Currently, transduction of retroviral vectors into autologous hepatocytes may be considered to be the best approach for *ex vivo* liver directed gene therapy. The retroviral vectors have been proven to be safe, and the surgical procedures minimally morbid. Recent advances in production techniques have yielded higher titre vector, which will increase the efficiency of *ex vivo* hepatocyte transduction. Current limitations remain the relatively few corrected hepatocytes that can be transplanted back into the host and the as yet incomplete understanding of the requirement for proliferation.

The ultimate goal in human applications of *ex vivo* liver-directed gene therapy would be to transduce efficiently - and without toxicity - hepatocytes with the selected gene *in vitro* with a safe method of cell transplantation. These cells would be stimulated to thrive in an environment that would lead to repopulation of the liver by the transduced cells, as has been shown to occur by the work of Brinster and Grompe. The degree of cellular proliferation needs to be sufficient to achieve clinical correction of the particular disease, but appears within the capabilities of most mature hepatocytes, provided the correct selective pressure can be brought to bear. Much of the work presented in this chapter demonstrates the great strides that have been made in trying to reach this goal. However, much investigation is still needed before *ex vivo* liver-directed gene therapy is an accepted treatment modality for inherited liver diseases that require permanent correction.

Acknowledgements

This work was funded in part by NIH RO1 DK 47811 (SER, ADM).

References

1. Horwitz AL. Inherited hepatic enzyme defects as candidates for liver-directed gene therapy. Curr Top Microbiol Immunol 1991; 168: 185-200.
2. Scriver CR, Beaudet AL, Sly WS et al, Eds. Metabolic Basis of Inherited Diseases. New York: McGraw Hill Information Services, 1989.

3. Mullen CA, Snitzer K, Culver KW et al. Molecular analysis of T lymphocyte-directed gene therapy for adenosine deaminase deficiency: long-term expression *in vivo* of genes introduced with a retroviral vector. Human Gene Therapy 1996; 7: 1123-1129.
4. Williams DA, Lemischka IR, Nathan DG et al. Introduction of new genetic material into pluripotent hematopoietic stem cells of the mouse. Nature 1984; 310: 476-480.
5. Grossman M, Raper SE, Kozarsky K et al. Successful *ex vivo* gene therapy directed to the liver in a patient with familial hypercholesterolemia. Nature Genet 1994; 6: 335-341.
6. Grossman M, Rader DJ, Muller D et al. A pilot study of *ex vivo* gene therapy for homozygous familial hypercholesterolemia. Nature Med 1995; 1: 1148-1154.
7. Raper SE, Grossman BS, Rader DJ et al. Safety and feasibility of liver-directed *ex vivo* gene therapy for homozygous familial hypercholesterolemia. Ann Surg 1995; 223: 116-126.
8. Cepko CL, Roberts BE and Mulligan RC. Construction and applications of a highly transmissible murine retrovirus shuttle vector. Cell 1984; 37: 1053-1062.
9. Choudhury JR, Grossman M, Gupta S et al. Long-term improvement of hypercholesterolemia after *ex vivo* gene therapy in LDLR-deficient rabbits. Science 1991; 254: 1802-1805.
10. Goldstein JL, Kita T and Brown MS. Defective lipoprotein receptors and atherosclerosis. Lessons from an animal counterpart of familial hypercholesterolemia. N Engl J Med 1983; 309: 288-296.
11. Wilson JM, Johnston DE, Jefferson DM et al. Correction of the genetic defect in hepatocytes from the Watanabe heritable hyperlipidemic rabbit. Proc Natl Acad Sci USA 1988; 85: 4421-4425.
12. Moscioni AD, Roy-Choudhury J, Barbour R et al. Human liver cell transplantation: Prolonged function in athymic-Gunn and athymic-analbuminemic hybrid rats. Gastroenterology 1989; 96: 1546-1551.
13. Patel A, Hardy MA, Chowdhury NR, Wajsman R, Sandoval M et al. Long-term correction of genetic defect of liver function in rat by transplantation of liver cells after ultraviolet irradiation. Mol Biol Med 1989; 6: 187-196.
14. Holzman MD, Rozga J, Neuzil DF, Griffin D, Moscioni AD and Demetriou AA. Selective intraportal hepatocyte transplantation in analbuminemic and Gunn rats. Transplantation 1993; 55: 1213-1219.
15. Ponder K, Gupta S, Leland F, Darlington G, Finegold M et al. Mouse hepatocytes migrate to liver parenchyma and function indefinitely after intrasplenic transplantation. Proc Natl Acad Sci USA 1991; 88: 1217-1221.
16. Peeters MJ, Lieber A, Perkins J et al. Method of multiple portal vein infusions in mice: Quantitation of adenovirus-mediated hepatic gene transfer. Biotechniques 1996; 20: 278-285.
17. Grossman M, Wilson JM and Raper SE. A novel approach for introducing hepatocytes into the portal circulation. J Lab Clin Med 1993; 121: 472-478.
18. Grossman M, Raper SE and Wilson JM. Transplantation of genetically-modified autologous hepatocytes into non-human primates: Feasibility and short-term toxicity. Human Gene Therapy 1992; 3: 501-510.
19. Memsic L, Busuttil RW and Traverso LW. Bleeding esophageal varices and portal vein thrombosis after pancreatic mixed-cell autotransplantation. Surgery 1984; 95: 238-242.
20. Scharp DW, Lacy PE, Santiago JV, McCullough CS, Weide LG et al. Results of our first nine intraportal islet autografts in type 1 insulin-dependent diabetic patients. Transplantation 1991; 51: 76-85.
21. Gupta S, Aragona E, Vemuru RP et al. Permanent engraftment and function of hepatocytes delivered to the liver: implications for gene therapy and liver repopulation. Hepatology 1991; 14: 144-149.

22. Maganto P, Traber PG, Rusnell C, Dobbins WO, Keren D and Gumucio JJ. Long term maintenance of the adult pattern of liver specific expression for P-450b, P-450e, albumin and α-fetoprotein genes in intrasplenically transplanted hepatocytes. Hepatology 1990; 11: 585-592.
23. Mito M, Kusano M and Kawaura Y. Hepatocyte transplantation in man. Transplant Proc 1992; 24: 3050-3051.
24. Varmus H and Swanstrum R. Replication of retrovirus. In: Weiss R, Teich N, Varmus H and Coffin J, Eds. RNA tumor viruses. New York: Cold Spring Harbor Press, 1984: 369-512.
25. Mulligan RC. Gene transfer and gene therapy: Principles, prospects, and perspective. In: Lindsten J and Petersson U, Eds. Etiology of human disease at the DNA level. New York: Raven Press, 1991: 143-189.
26. Miller DG, Adam MA and Miller AD. Gene transfer by retrovirus vectors occurs only in cells that are actively replicating at the time of infection. Mol Cell Biol 1990; 10: 4239-4242.
27. Mulligan RC. The basic science of gene therapy. Science 1993; 260: 926-932.
28. Raper SE and Wilson JM. Cell transplantation in liver-directed gene therapy. Cell Transplantation 1993; 2: 381-400.
29. Wilson JM, Jefferson DM, Choudhury JR, Novikoff PM, Johnston DE and Mulligan RC. Retrovirus-mediated transduction of adult hepatocytes. Proc Natl Acad Sci USA 1988; 85: 3014-2018.
30. Grossman M, Raper SE and Wilson JM. Towards liver-directed gene therapy: Retrovirus-mediated gene transfer into human hepatocytes. Som Cell Mol Genet 1991; 17: 601-607.
31. Roe T, Reynolds TC, Yu G and Brown PO. Integration of murine leukemia virus DNA depends on mitosis. EMBO J 1993; 12: 2099-2108.
32. Ferry N, Duplessis O, Houssin D, Danos O and Heard JM. Retroviral-mediated gene transfer into hepatocytes *in vivo.* Proc Natl Acad Sci USA 1991; 88: 8377-8381.
33. Cardoso J, Branchereau S, Jeyaraj PR, Houssin D, Danos O and Heard JM. In situ retrovirus mediated gene transfer into dog liver. Human Gene Therapy 1993; 4: 411-418.
34. Lieber A, Peeters MJ, Meuse L et al. Adenovirus-mediated urokinase gene transfer induces liver regeneration and allows for efficient retrovirus transduction of hepatocytes *in vivo*. Proc Natl Acad Sci USA 1995; 92: 6210-6214.
35. Bowles NE, Eisensmith RC, Mohuiddin R, Pyron M and Woo SL. A simple and efficient method for the concentration and purification of recombinant adenovirus for increased hepatocyte transduction *in vivo*. Human Gene Therapy 1996; 7: 1735-1742.
36. Kitten O, Cosset FL and Ferry N. Highly efficient retrovirus- mediated gene transfer into rat hepatocytes *in vivo*. Human Gene Therapy 1997; 8: 1491-1494.
37. Matsumoto K and Nakamura T. Role of HGF as a pleiotropic factor in organ regeneration. In: Goldberg ID and Rosen EM, Eds. Hepatocyte growth factor-scatter factor (HGF-SF) and the c-met receptor. Basel: Birkhauser Verlag, 1993: 225-249.
38. Roos F, Ryan AN, Chamow SM, Bennett GL and Schwall RH. Induction of liver growth in normal mice by infusion of hepatocyte growth factor/scatter factor. Am J Physiol 1995; 268: G380-G386.
39. Fujiwara K, Nagoshi S, Ohno A, Hirata K, Ohta Y et al. Stimulation of liver growth by exogenous human hepatocyte growth factor in normal and partially hepatectomized rat liver. Hepatology 1993; 18: 1443-1449.
40. Yasuda H, Mine T, Shibata H, Eto Y, Hasegawa Y, Takeuchi T, Asano S and Kojima I. Activin A: an autocrine inhibitor of initiation of DNA synthesis in rat hepatocytes. J Clin Invest 1993; 92: 1491-1496.

41. Kogure K, Omata W, Kanzaki M, Zhang YQ, Yasuda H, Mine T and Kojima I. A single intraportal administration of follistatin accelerates liver regeneration in partially hepatectomized rats. Gastroenterology 1995; 108: 1136-1142.
42. Kogure K, Zhang YQ, Kanzaki M, Omata W, Mine T and Kojima I. Intravenous administration of follistatin: delivery to the liver and effect on liver regeneration after partial hepatectomy. Hepatology 1996; 24: 361-366.
43. Ishiki Y, Ohnishi H, Muto Y, Matsumoto K and Nakamura T. Direct evidence that hepatocyte growth factor is a hepatotropic factor for liver regeneration and has a potent antihepatitis effect *in vivo*. Hepatology 1992; 16: 1227-1235.
44. Miller DG, Adam MA and Miller AD. Gene transfer by retrovirus vectors occurs only in cells that are actively replicating at the time of infection. Mol Cell Biol 1990; 10: 4239-4242.
45. Hatzoglou M, Lamers W, Bosch F, Wynshaw-Boris A, Clapp DW and Hanson RW. Hepatic gene transfer in animals using retroviruses containing the promoter from the gene for phosphoenolpyruvate carboxykinase. J Biol Chem 1990; 265: 17285-17293.
46. Ferry N, Duplessis O, Houssin D, Danos O and Heard JM. Retroviral-mediated gene transfer into hepatocytes *in vivo*. Proc Natl Acad Sci USA 1991; 88: 8377-8381.
47. Rettinger SD, Ponder KP, Saylors RL, Kennedy SC, Hafenrichter DG and Flye MW. *In vivo* hepatocyte transduction with retrovirus during in-flow occlusion. J Surg Res 1993; 54: 418-425.
48. Moscioni AD, Rozga J, Neuzil DF, Overell RW, Holt JT and Demetriou AA. *In vivo* regional delivery of retrovirally mediated foreign genes to rat liver cells: Need for partial hepatectomy for successful foreign gene expression. Surgery 1993; 113: 304-311.
49. Ferry N, Duplessis O, Houssin D, Danos O and Heard JM. Retroviral-mediated gene transfer into hepatocytes *in vivo*. Proc Natl Acad Sci USA 1991; 88: 8377-8381.
50. Rettinger SD, Kennedy SC, Wu X, Saylors RL, Hafenrichter DG et al. Liver-directed gene therapy: Quantitative evaluation of promoter elements by using *in vivo* retroviral transduction. Proc Natl Acad Sci USA 1994; 91: 1460-1464.
51. Kay MA, Li Q, Liu TJ, Leland F, Toman C, Finegold M and Woo SL. Hepatic gene therapy: persistent expression of human α 1-antitrypsin in mice after direct gene delivery *in vivo*. Human Gene Therapy 1992; 3: 641-647.
52. Hermonat PL, Labow MA, Wright R, Berns KI and Muzyczka N. Genetics of adeno-associated virus: isolation and preliminary characterization of adeno-associated virus type 2 mutants. J Virol 1984; 51: 329-339.
53. Hermonat PL and Muzyczka N. Use of adeno-associated virus as a mammalian DNA cloning vector: transduction of neomycin resistance into mammalian tissue culture cells. Proc Natl Acad Sci USA 1984; 81: 6466-6470.
54. Kotin RM. Prospects for the use of adeno-associated virus as a vector for human gene therapy. Human Gene Therapy 1994; 5: 793-801.
55. Cheung AK, Hoggan MD, Hauswirth WW and Berns KI. Integration of the adeno-associated virus genome into cellular DNA in latently infected human Detroit 6 cells. J Virol 1980; 33: 739-748.
56. Nakai H, Herzog RW, Hagstrom JN, Walter J, Kung SH et al. Adeno-associated viral vector-mediated gene transfer of human blood coagulation factor IX into mouse liver. Blood 1998; 91: 4600-4607.
57. Sugiyama A, Hattori S, Tanaka S, Isoda F, Kleopoulos S et al. Defective adeno-associated viral-mediated transfection of insulin gene by direct injection into liver parenchyma decreases blood glucose of diabetic mice. Horm Metab Res 1997; 29: 599-603.
58. Frenkel N. Defective interfering herpesvirus. In: Nahmias AJ and Dowdel WR, Eds. The human herpes viruses: an interdisciplinary perspective. New York: Elsevier, 1981: 91-120.

59. Geller AI and Breakefield XO. A defective HSV-1 vector expresses Escherichia coli β-galactosidase in cultured peripheral neurons. Science 1988; 241: 1667-1669.
60. Johnson PA, Yoshida K, Gage FH and Friedmann T. Effects of gene transfer into cultured CNS neurons with a replication-defective herpes simplex virus type 1 vector. Mol Brain Res 1992; 12: 95-102.
61. Dienes HP, Ramadori G, Falke D and Thoenes W. Electron microscopic observations on primary hepatocyte cultures infected with herpes simplex virus types I and II. Virchows Arch 1984; 46: 321-332.
62. Lu B, Gupta S and Federoff H. *Ex vivo* hepatic gene transfer in mouse using a defective herpes simplex virus-1 vector. Hepatology 1995; 21: 752-759.
63. Fong Y, Federoff HJ, Brownlee M, Blumberg D, Blumgart LH and Brennan MF. Rapid and efficient gene transfer in human hepatocytes by herpes viral vectors. Hepatology 1995; 22: 723-729.
64. Fraefel C, Jacoby DR, Lage C, Hilderbrand H, Chou JY et al. Gene transfer into hepatocytes mediated by helper virus-free HSV/AAV hybrid vectors. Molecular Medicine 1997; 3: 813-825.
65. Naldini L, Blomer U, Gallay P, Ory D, Mulligan R, Gage FH, Verma IM and Trono D. *In vivo* gene delivery and stable transduction of nondividing cells by a lentiviral vector. Science 1996; 272: 263-267.
66. Kafri T, Blomer U, Peterson DA, Gage FH and Verma IM. Sustained expression of genes delivered directly into liver and muscle by lentiviral vectors. Nature Genetics 1997; 17: 314-317.
67. Luckow VA and Summers MD. Baculovirus expression in insect cells. Biotechnology 1988; 6: 47-55.
68. Carbonell LF, Klowden MJ and Miller LK. Baculovirus-mediated expression of bacterial genes in dipteran and mammalian cells. J Virology 1985; 56: 153-160.
69. Hofmann C, Sandig V and Jennings G. Efficient gene transfer into human hepatocytes by baculovirus vectors. Proc Natl Acad Sci USA 1995; 92: 10099-10103.
70. Sandgren EP, Palmiter RD, Heckel JL, Daugherty CC, Brinster RL and Degen JL. Complete hepatic regeneration after somatic deletion of an albumin-plasminogen activator transgene. Cell 1991; 66: 245-256.
71. Rhim JA, Sandgren EP, Degen JL, Palmiter RD and Brinster RL. Replacement of diseased mouse liver by hepatic cell transplantation. Science 1994; 263: 1149-1152.
72. Lieber A, Vrancken Peeters MJ, Meuse L, Fausto N, Perkins J and Kay MA. Adenovirus-mediated urokinase gene transfer induces liver regeneration and allows for efficient retrovirus transduction of hepatocytes *in vivo*. Proc Natl Acad Sci USA 1995; 92: 6210-6214.
73. Overturf K, Al-Dhalimy M, Tanguay R, Brantly M, Ou CN, Finegold M and Grompe M. Hepatocytes corrected by gene therapy are selected *in vivo* in a murine model of hereditary tyrosinaemia type I. Nature Genetics 1996; 12: 266-273, [published erratum appears in Nature Genetics 1996; 12: 458].
74. Overturf K, Al-Dhalimy M, Manning K, Ou CN, Finegold M and Grompe M. *Ex vivo* hepatic gene therapy of a mouse model of Hereditary Tyrosinemia Type I. Human Gene Therapy 1998; 9: 295-304.
75. Overturf K, Al-Dhalimy M, Ou CN, Finegold M and Grompe M. Serial transplantation reveals the stem-cell like potential of adult mouse hepatocytes. Am J Path 1997; 151: 1273-1280.

Chapter 30

A review of bioartificial liver development from an artificial organ engineering perspective

Mark D. Smith, M. Helen Grant, Rod B. Cousins and John D.S. Gaylor
Bioengineering Unit and Dept. Chemical & Process Engineering, University of Strathclyde, Glasgow, UK

Key words: bioartificial liver, hepatocyte bioreactor design, hepatic failure, human hepatocytes, liver support, monolayers, oxygen delivery, plasma perfusion, porcine hepatocytes, self-assembling aggregates

30.1 Introduction

Fulminant hepatic failure (FHF) and acute episodes in chronic liver failure are associated with high mortality [1-3]. Organ transplantation is the only treatment acknowledged to improve patient survival when compared with intensive care [4,5]. There is a shortage of donors worldwide [6-8] resulting in approximately 50% mortality amongst patients awaiting transplants [9]. Furthermore, OLT renders graft recipients dependent upon immunosuppression [10,11]. US hospital charges average $250,000 for liver transplants [12]. It was estimated that in the USA in 1996, there were approximately 1,700 emergency liver transplants with 1,000 patients dying while waiting for a transplant. A further 1,000 patients with FHF or sudden onset of acute liver failure (ALF) were either not listed for transplant or died before being listed [13].

The immediate goal for liver support therapy is to maintain patients until a donor organ becomes available. However, a more optimistic expectation would be to reduce the number of transplants by sustaining patients with acute liver failure until their damaged livers regenerated sufficiently [14].

Many therapies have been evaluated, viz. hemodialysis with hemofiltration [15,16], plasma exchange [17,18], hemoperfusion [19,20] and cross circulation [21]. However, none has found universal approval. Treatment of acute liver failure patients with adsorbents via hemo- or plasma perfusion may prove to be worthwhile in combination with other modalities [14,22].

M.N. Berry and A.M. Edwards (eds.), The Hepatocyte Review, 521–541.

Failure of support measures has been attributed to the fact that they cannot replace all the metabolic and synthetic operations of the natural liver [9,12,14]. Whilst this is reasonable, in this review we suggest that the goal for novel therapies should be replacement of the regulatory role of the natural organ. We clarify this distinction by contrasting the list of hepatocyte functions which are normally given in reply to the question "what is the function of the liver?" with the answer supplied by Gumucio and Chianale. They suggested that "the function of the liver is to regulate the concentration of solutes in the terminal hepatic venule and in bile" [23].

Many consider the bioartificial liver (BAL) to be the most promising potential therapy for acute liver failure. In this approach, living liver-derived cells are maintained in a bioreactor which is incorporated into an extracorporeal blood circuit.

30.2 History

The BAL concept first appeared over 40 years ago [24]. Later, in 1963, Nosé and co-workers reported the first clinical treatments with a bioartificial liver based upon a dialyser containing canine liver tissue [25]. Experiments with cultured liver-derived cells were reported in 1975 when Wolf and Munkelt detailed bilirubin conjugation in a simple hepatocyte bioreactor incorporating rat hepatoma cells [26].

Matsumura and colleagues were the first group to report clinical treatment with a hepatocyte bioreactor [27]. Using a filtration module filled with cryopreserved rabbit hepatocytes, they treated one patient with acute liver failure arising from an inoperable cancer. There was anecdotal evidence of improvement in the patient during two sessions of therapy.

A large trial of a liver support system was reported in 1989 [28]. Fifty-nine patients received complex therapy including sessions of hemoperfusion through activated carbon and a small mass of porcine hepatocytes. Improved survival in this treated group when compared with a similar number of patients receiving normal intensive care therapy was noted. However, there were no data to suggest that this improvement could be attributed to the BAL component.

30.2.1 Recent development and use of "generic" BALs

The first BAL system to be evaluated at more than one clinic was the Hepatix™ ELAD developed by Sussman and Kelly [29,30]. This device was based upon a commercial hollow fiber filtration cartridge. Human hepatoma cells were seeded into the extracapillary space. Anhepatic dogs, and dogs with acetaminophen-induced liver failure, were treated. Some animals survived longer than controls. Clinical use of this device was subsequently reported by the developers [31-33] and by workers at King's College Hospital in London [34,35]. In the first report of a controlled trial of a bioartificial liver device, McCarthy and co-workers described the treatment of 24 patients with acute liver failure. There was no significant improvement in survival in patients receiving treatment with the bioartificial liver system in addition to standard therapy, when compared with that in control patients. This result was tentatively attributed to there being a higher than expected rate of survival in the controls.

The most widely reported BAL system is that developed by Demetriou and Rozga [36,37]. Again, this system is based upon a commercial hollow fiber filtration cartridge but with the extracapillary space seeded with primary porcine hepatocytes. The developers have published clinical results [38-42]. More recently, this system has been commercialized and is now owned by Circe Biomedical Inc. and known as the HepatAssist 2000™. The results of phase 1 clinical trials in the USA and France have now been reported [43-45] and phase 2 trials are underway. This body of work represents the biggest effort so far to develop a commercial, clinically proven bioartificial liver support system. There have been striking results, notably from recent applications in patients with advanced acute liver failure. Some of those treated achieved a full recovery after BAL treatment without a subsequent transplant. Also, the reduction in intracranial pressure which has been recorded in patients following treatment with this system is generally accepted as being significant. However, due to the complex nature of HepatAssist therapy, it is difficult to quantify the contribution of the hepatocyte bioreactor.

Another device based upon a commercial hollow fiber filtration cartridge has been reported by Nyberg and co-workers [46-51]. In this bioreactor, primary hepatocytes are entrapped within a collagen gel in the intracapillary compartment. *In vitro* studies and testing in rabbit and dog models of liver failure have been reported. Hepatocytes in the bioreactor reportedly remain viable and functional in terms of albumin and urea synthesis for 9 days if aggregates are used. In the rabbit model of liver failure, biochemical function of the reactor was noted whilst in dogs with galactosamine-induced liver failure, animals treated with the bioreactor survived longer than controls.

30.2.2 "Second generation" hollow fiber BALs

The foregoing devices have been based upon commercial hollow fiber filtration cartridges. This is an obvious choice in terms of cost and development time. However, purpose-built designs that may prove to be more effective have emerged during the last 5 years. These novel configurations are referred to in this review as Second Generation (SG) systems. This nomenclature is adopted in order to distinguish those purpose-built hepatocyte bioreactors from the generic devices previously used and to suggest that after 40 years, we are now in a new phase of BAL design.

Currently, the SG BAL which is at the most advanced stage is that invented by Gerlach in Berlin. This system is based on a novel multi-compartment reactor configuration and represents a departure from the use of parallel bundle hollow fiber cartridges or hemoperfusion columns. In Gerlach's bioreactor, primary porcine hepatocytes are cultured as small aggregates which are reported to form "tissue like structures" between multiple sets of capillary membranes. The development and testing of this system has been widely reported [52-57]. These studies indicated that hepatocytes could be maintained in the bioreactor for up to 6 weeks. Hepatectomized pigs survived significantly longer when treated with the bioreactor than control pigs treated with a placebo system. More recently, the device has been used to treat a small series of patients in Berlin [58].

Flendrig, Chamuleau and colleagues in Amsterdam have incorporated a hollow fiber membrane oxygenator in their SG BAL. Porcine hepatocytes are cultured as small aggregates in a non-woven polyester mesh, wound with a mat of microporous, gas permeable hollow fibers [59-61]. Testing of this device demonstrated *in vitro* performance for 4 days and rats with ischemic livers, treated with a small scale device, survived twice as long as control animals. This system has also been tested in a pig model of liver failure, however, at the time of writing, the data were not published.

30.2.3 "Second generation" flat-sheet BALs

The SG bioartificial livers developed by Bader and co-workers in Hannover [62-64] and by our own group in Glasgow [65-68] are discussed together in this section since they share some features. Bader's design and the Strathclyde system are based upon flat sheet reactor technology. In Bader's reactor, a modified bi-laminar sheet membrane oxygenator is employed. The Strathclyde bioreactor is based upon stacked plate technology similar to that used in flat sheet hemodialysers. but hollow fiber oxygenation is integrated into each reactor plate. Bader's device has been tested *in vitro* and has demonstrated an array of biochemical functions. Our own system has been subjected to *in vitro* testing for hepatic functional markers and modelling of cell attachment and fluid dynamics. It has recently been scaled-up for evaluation in a large-animal model of liver failure. The fact that neither of these two systems has yet reached the clinic might be attributed to increased manufacturing difficulties associated with this type of device. However, developers of this type of device should take encouragement from early membrane blood oxygenators in which surface areas of 25 m^2 were achieved with flat plate configurations (the estimated required surface area for monolayer hepatocyte culture BALs is approximately 10 m^2).

30.3 Theoretical considerations concerning replacement of liver function

This section considers the provision of "liver function" by BAL devices. Argument is confined to those areas which may be influenced by device design. For acute liver failure therapy, the most immediate target for a BAL is reversing, or preventing development of complications, especially hepatic encephalopathy. We introduce this section with two statements:

1. It is not known which blood solutes must be regulated in order to reverse the complications of liver failure.
2. It is not known at what rate essential processes must proceed.

However, we can make some educated assumptions about which processes may be critical and hence infer which biochemical pathways should be maintained in hepatocytes intended to provide liver support. Until such time as the essential reactions are defined, we can only make broad assumptions about the required mass transfer kinetics in hepatocyte bioreactors.

When the more optimistic target of complete recovery without subsequent transplantation is considered, the picture becomes more complex since the necessary conditions for liver regeneration must also be provided. This would likely require a balance between growth stimulators and inhibitory factors [14]. However, this topic lies beyond the scope of this review.

Which pathways are critical? Those compounds for which an increase in circulating concentration is associated with the pathogenesis of liver failure (particularly hepatic encephalopathy) include: ammonia, aromatic amino acids, benzodiazepines, bile acids, bilirubin, cytokines, endotoxins, fatty acids and mercaptans. At the same time, there are compounds which are depleted in the plasma of liver failure patients. These include: blood clotting proteins, fibronectin and growth factors other than HGF.

The urea cycle is the main mechanism by which the circulating concentration of ammonia is regulated. Conversion of ammonia to glutamine (reversible) is also quantitatively significant. In the acinus, ammonia regulation can be considered as a two stage process. A large capacity upstream process associated with zone 1 (periportal) hepatocytes can dramatically reduce the concentration of both ammonia and glutamine in acinar blood via conversion to urea. A second more sensitive downstream process regulates the concentrations of ammonia and glutamine by converting some remaining ammonia to glutamine.

This suggests a requirement for reproduction of aspects of acinar architecture and development of appropriate solute gradients in BAL devices if regulated ammonia metabolism is to be achieved. It may also illuminate the question of why, if ammonia is a significant "liver failure toxin" does high flux dialysis fail to show an improvement in ALF patients. Aside from the likelihood of synergistic effects of several "toxic" compounds [69], it could be that "stripping" ammonia is not desirable. Rather, circulating concentrations of ammonia and glutamine should be normalized via appropriate regulation. This may also contribute to restoration of the normal balance between branched chain- and aromatic amino acids.

Other implicated compounds such as benzodiazepines, bile acids and bilirubin are modified in the liver via phase I and/or phase II metabolism. Many groups acknowledge that retention of CYP activity and conjugation enzyme activity is a key requirement of hepatocytes in BAL devices. The literature describes many hepatocyte culture models which can enhance these functions. However the possible benefits of particular culture configurations in maintaining particular liver-specific pathways could be negated if poor fluid dynamics and mass transfer characteristics arise in BAL devices.

Concentrations of benzodiazepines, bile acids and bilirubin may also be regulated by a BAL which operates under acinar conditions. The acinar distribution of the enzymes associated with metabolism of these compounds is heterogeneous, with zone 3 cells having the greatest concentration and activity of both CYP and conjugation enzymes. The findings of some *in vitro* studies indicate that both CYP and conjugation activity are better expressed when the oxygen partial pressure of the culture is akin to that in acinar zone 3 [70,71]. We may assume that reproduction of the hepatocyte cell surface PO_2 gradients found in the acinus will maximize device output in terms of these functions.

It is not possible to extrapolate this argument further, or to discuss the consequences of this design approach on other markers of hepatic function that may be significant. The extent of plasma protein binding of solutes of interest will affect the extent to which reactor design influences clearance. Fatty acids and mercaptans can be effectively removed by adsorbents such as activated carbon. BAL devices relying upon hepatocyte-rich cultures are unlikely to be effective in removal of endotoxin. Specific adsorbent regimes can be used to eliminate endotoxin though perhaps in the future, cultured Kupffer cells in BAL devices may have a role in this respect. It is unlikely that BAL devices will prove capable of appropriate regulation of cytokines, clotting factors or growth factors. These are subject to interactions, balances and feedback involving various tissues and cell types. The release of such factors by the damaged natural organ will make this more difficult. Direct infusion of purified clotting factors can be used to treat coagulopathy with good results and, in clinical practice a contribution from the BAL is probably not required.

30.4 BAL design

The principal design considerations for BAL devices are:

1. cell source
2. culture model
3. bioreactor configuration
4. mode of operation

The primary goal in BAL design is clinical efficacy. Secondary requirements arise from economic, manufacturing and ethical considerations. These include: biocompatibility, reliability and cost-effectiveness in both manufacture and operation, satisfaction of emerging regulatory requirements and scalability/modularity. The various approaches adopted by workers in the field are discussed in the subsequent chapters. Here, we will discuss these topics in general terms.

30.4.1 Sources of cells

In selecting a cell source workers face two choices, between primary cells or cell lines, and between human cells or xenogenic cells. Researchers may use primary non-human cells and accept the attendant costs and difficulties of repeated cell isolations providing they can satisfy quality requirements. This approach is beset by ethical doubts and uncertainties regarding the possibility for transmission of infectious agents. Such issues may soon be covered by legislation [72]. There is no present or foreseeable source of sufficient human tissue for this purpose.

At present, no clinically safe cell line (human or xenogenic) has demonstrated stable, adequate expression of the hepatocyte phenotype to be considered appropriate for use in a BAL.

Most workers have elected to use primary porcine hepatocytes. (So far, the only BAL device incorporating a cultured cell line which has been used clinically is the Hepatix ELAD [31-35]). The rationale for the selection of pigs as the source animal for primary cells is less obvious. Clinical familiarity with pigs exists and there are physiological similarities with humans. Yet, we are not certain whether porcine hepatocytes can provide essential human liver function. Differences in expression of P450 isoenzymes and/or differences in conjugation activity between humans and pigs could be significant. One report has identified a potential immunological problem associated with the use of porcine hepatocytes [73].

Recent developments have fuelled debate on the suitability of porcine hepatocytes as the cellular component of BAL devices. Evidence for the existence of transmissible agents (including retroviruses) harbored by pigs, and which might infect humans via BAL therapy, has led to some ethical committees advising against such practice. The FDA and the European Commission are considering related legislation. Developments in transgenic husbandry have, on the other hand, offered the possibility of sustained supply of pathogen-free pig livers which might prove suitable for transplantation into humans. Some argue that this foretells the demise of the BAL concept. However, BAL developers should not be deterred since, if BAL treatment can negate the requirement for transplants in some cases, then clinical and economic incentives for selecting BAL therapy in appropriate patients will be strong.

A criticism cited against use of primary cells is that the time elapsing between identifying a patient requiring BAL therapy and provision of a device ready for clinical use will be too long. Two groups have taken different routes to solving this logistical problem and both claim to have found a working solution. Circe Biomedical Inc. state that they have developed cell cryopreservation, thawing and seeding protocols that enable them to prepare a device for clinical use within an acceptable time frame. Gerlach and co-workers in Berlin consider that they can maintain their BAL devices for several weeks without appreciable decline in measured function. Consequently, they have adopted a protocol based on stand-by devices being continuously available in the clinic. Further details of these two approaches are provided in the following chapters.

30.4.2 Culture model

In BAL design, culture models may be defined as the functional sub-unit of devices. In other words, BAL devices are realized by the scale-up of culture models in a bioreactor [74]. The functional subunit of the natural organ is the hepatic acinus. Evidently, a culture model which replicates some of the facets of the acinus is necessary. Various culture models are used in BAL developments. Some aim to replicate the gross structure of the acinus while others attempt to reproduce the surroundings of the hepatocytes. It is possible to address both aspects in a culture model. Whilst culture models can be established in simple static environments (culture flasks or Petri dishes) their measured performance in such environments may not be extrapolated to bioreactor performance. Therefore, enhanced expression of specific functions achieved in static configurations (collagen sandwiches, encapsulated or entrapped hepatocytes or spheroid cultures [75-78]), cannot automatically be anticipated in BAL devices. We recall the distinction between "liver function" and the array of "hepatocyte functions" and suggest that, for replacement of liver function, it will be necessary to elicit particular hepatocyte functions from cells according to their location within the BAL with respect to the direction of blood or plasma flow. Hence, the mass transfer and fluid dynamic characteristics of the bioreactor should be considered together with the culture model in order for the device to work appropriately. That is to say, the culture model ought to be scaled up as a series of parallel elements, each of which provides essential liver function.

There are three general culture models employed in current BAL devices:

1. Hepatocyte monolayers
2. Self-assembling hepatocyte aggregates
3. Pre-formed hepatocyte aggregates or spheroids

Hepatocyte monolayers

BAL designs based upon hepatocyte monolayers enable reproduction of acinar geometry since the acinus is essentially a single cell thick plate structure. For example, axial gradients can be effectively regulated, diffusion distances minimized and multiple parallel units can be stacked. There are two disadvantages associated with this model. First, the reactor must provide an attachment surface for each of the cells and second, specialized manufacturing techniques are required since suitable commercial bioreactors are not available.

Self-assembling aggregates

BAL devices based upon self-assembling hepatocyte aggregates have been realized recently. A suspension of hepatocytes is seeded into the bioreactor and the cells subsequently form aggregates within the device. This technique has the advantage of reducing the surface area required for attachment. It might dispense altogether with the requirement for an attachment surface though current devices make some provision for cell attachment or immobilisation. This approach has a significant disadvantage: it is difficult to control the size of the aggregates. Aggregates are often described as "tissue-like structures". In the natural organ, the hepatocytes are arranged in mainly single cell layers and layers more than two hepatocytes thick almost never occur. Multiple cell layers are unavoidable in aggregates. Radii of the order of 50 to 250 μm are common. These dimensions can be correlated to a "thickness" of 3 to 15 cell layers from the outer edge of a spherical aggregate to its core. Aggregates of this size give rise to two significant problems in BAL devices. First, essential nutrients, especially O_2, can be depleted at the core rendering the cells there quiescent or dead. Second, it is difficult to develop appropriate solute gradients across the bioreactor when large localized gradients exist within the aggregates themselves.

Pre-formed aggregates

Most argument about self-assembling aggregates holds for pre-formed aggregates. However, whilst for self-assembling aggregates, the internal structure of the bioreactor is the only means of controlling the size of the aggregates, pre-formed aggregates can be processed prior to seeding. Hence, the latter offers two potential advantages. First, means of regulating the size of the aggregates seeded into the bioreactor are available. This may be attempted during the aggregation process or resultant aggregates could be sieved to an appropriate size range prior to seeding. Second, there is the opportunity to encapsulate the aggregates. If the size of aggregates was well controlled and the encapsulation process was efficient in terms of guaranteeing that cells were completely covered by a sufficiently thin gel layer, then this approach would, potentially, enable relatively simple bioreactors to be used. However, use of pre-formed aggregates also has a disadvantage when compared with use of post-seeding aggregation: these cell processing steps require time and, when using primary cells, it is advisable to seed them as soon after isolation as possible.

30.4.3 Bioreactor configuration

The configurations used in, or incorporated into, BAL designs are:
1. Packed bed
2. Parallel bundle hollow fiber filtration cartridge
3. Hollow fiber oxygenator
4. Flat sheet filtration / gas exchange module

Some BAL devices are based upon these generic configurations. However, SG devices are modifications of generic devices or hybrids of different generic configurations.

Packed bed devices

The packed bed is the simplest configuration used in BAL devices. An example is a hemoperfusion column packed with encapsulated hepatocyte aggregates. It is difficult to incorporate the monolayer culture model into this type of reactor and, if post-seeding aggregation is employed, the aggregation process cannot be allowed to progress unhindered in this configuration as the voids between the aggregates form the perfusate channel. Immuno-protection can only be conferred if the cells are encapsulated. Therefore, pre-formed encapsulated aggregates are most commonly used. The packed bed has no provision for specific oxygen delivery. However, this type of reactor is readily scalable and commercial modules are available.

Parallel bundle hollow fiber filtration cartridge

This is the most widely employed configuration in BAL devices. This widespread use can be attributed to the fact that cartridges are available in a range of sizes and with a range of filtration properties. They are extensively used as bioreactors for other cell types including hybridomas and offer integral immuno-protection if appropriate membranes are used. However, they make no provision for oxygen delivery, monolayer culture models cannot be incorporated, mass transfer limitations are imposed by the filtration membrane and the development of appropriate axial solute gradients is very difficult to achieve. Gerlach's SG bioreactor attempts to address this latter problem by incorporating multiple sets of hollow fibers. It is operated with 100% transmembrane convective flow so the solute gradients develop across the interfiber distance rather than along the fiber length.

Hollow fiber oxygenators

These are structurally similar to hollow fiber filtration cartridges so the possibilities and limitations, as far as culture models and scaling are concerned, are similar. However, commercial oxygenators typically feature a dense winding of small diameter fibers making the configuration unsuitable for extracapillary cell culture. The Amsterdam SG bioreactor provides an elegant solution to this problem by winding larger diameter fibers together with a sheet of cell-supporting mesh, thereby increasing the extracapillary volume and reducing shear rates. Obviously, hollow fiber oxygenators make provision for specific oxygen delivery which is advantageous. An attendant drawback is that integral immuno-protection is sacrificed.

Flat sheet filtration modules

Flat sheet filtration modules offer an advantage over other generic configurations in that monolayer culture models can be accommodated. They are also scalable. However, scale-up requires specialized manufacturing and modules suitable for use as BAL devices, or which can be easily adapted for this purpose, are not commercially available. Furthermore, the large surface areas required will make BAL devices based on this configuration bulky. Immuno-protection can be incorporated into the flat sheet configuration. However, there is no provision for specific oxygen delivery. Of course, if hydrophobic membranes are used then dedicated oxygen delivery can be provided at the expense of immuno-protection. The SG BAL devices developed by Bader and by the Strathclyde group represent modified flat sheet configurations. Both feature purpose-designed integral oxygenators.

30.4.4 Mode of operation

Three considerations merit discussion here:

1. Whole blood versus plasma perfusion
2. Single pass versus recirculation
3. Extent of intermission in therapy

Whole blood versus plasma perfusion

Clinicians in general would prefer BAL systems to operate with whole blood perfusion. A plasma separation stage is likely to exacerbate hemodynamic instability in the patients and also results in a more complex system (at least two extra pumps, a plasma filter and additional monitoring are required). Whole blood perfusion offers the theoretical advantage of improved oxygen supply to the hepatocytes. However, there is then complete reliance upon immuno-protection being provided by the bioreactor. Therefore, some configurations do not allow use of whole blood. This may be considered for bioreactors incorporating encapsulated cells or for those based on filtration modules. This is an important consideration since membranes or cell capsules which form a reliable barrier to cells, cell fragments and immunoglobulins will also offer significant transfer resistance to low molecular weight proteins (for example, albumin) which can be essential solute-binding proteins. Such membranes or capsules also have limited diffusive permeability to dissolved oxygen. Therefore, increased availability of oxygen arising from blood hemoglobin, can be negated by mass transfer limitations in bioreactors with integral immunoisolation. Some BAL designs are reliant upon an upstream plasma separation membrane as the only immunological barrier between the patient and the cells in the bioreactor. It is not yet clear whether or not this will prove to be an adequate level of immuno-protection for BAL systems using xenogenic hepatocytes. Another problem with whole blood perfusion is that the perfusate flow rate is limited by the need to keep shear rates below those resulting in hemolysis. Consequently, the ultrafiltration rate, and hence mass transfer performance, are limited in whole blood perfused filtration cartridges. This means that the available flow for single-pass plasma perfused bioreactors is similarly limited by the rate of plasma filtration in the upstream separator.

Single pass or recirculation

There appears to be no advantage gained by recirculation in blood perfused devices. Our analyses predict that oxygen transfer is limiting for a BAL device based on a typical hollow fiber ultrafiltration cartridge perfused with venous blood [79,80]. We identified device inlet PO_2 and membrane oxygen permeability as the two main limiting parameters. For BAL devices supplied with arterial blood, our theory predicts that anticipated cellular oxygen requirements could be met. Arterial blood access requires more difficult surgery than venous access and is generally not desirable. There may, therefore, be a role for an upstream oxygenator in whole blood BAL treatment. Oxygen debt can also be predicted to occur for plasma-perfused BAL devices. Assuming that the available plasma flow resulting from the separation phase is 20-30% of the extracorporeal blood flow rate, then we arrive at an available plasma flow of 10-60 ml per min. If the plasma flow is equilibrated with pure oxygen at atmospheric pressure, it will only carry sufficient oxygen to meet the requirement of approximately 5×10^9 hepatocytes at basal metabolic activity under ideal transfer conditions [79-82]. This represents a drop in PO_2 across the device of 760 mmHg which is not desirable and is, in any case, practically unattainable. One might expect that high flow rate recirculation of plasma via an oxygenator incorporated into the extracorporeal circuit would alleviate such oxygen debt. However, we predict that if the PO_2 gradient across the BAL device is to be kept within the physiological range (e.g. 100 mm Hg to 0.5 mm Hg, noting that within the hepatic acinus, the maximum PO_2 is actually 65 mmHg), then recirculating flow rates of 1.6 litres per minute would be required to sustain 5×10^9 hepatocytes at basal metabolic rate. This indicates the requirement for specific oxygen delivery within plasma-perfused BAL bioreactors. It follows that the decision of whether or not to operate such systems in single pass or recirculation mode will be based on other considerations. One parameter which could be modified via recirculation is the axial solute gradient.

Continuous or intermittent treatment with the BAL

It seems likely that patients would benefit more from continuous than from intermittent therapy. However, for practical reasons, therapy will be intermittent. The duration of intermission is discussed here. We consider, for now, that BAL therapy will be employed as a bridge to transplantation. Hence, the duration of required therapy will be in the range of a few hours to a few days. The main practical reason for interrupting therapy will be to allow replacement of system components. Therefore, if a component other than the bioreactor fails first, the shortest possible intermission in therapy will be preferred in order to maximize the therapeutic effect of the system. The therapeutic life span of blood-perfused hepatocyte bioreactors is not likely to be significantly greater than that of other system components since it is susceptible to clotting and fouling problems. So we can state that, for blood-perfused systems, intermittent therapy is a consequence of the short life expectancy of the hepatocyte bioreactor. Cost, ease of replacement and rate of remission in the patient will influence the duration of the intervals between therapy. However, in plasma-perfused systems, the plasma filtration cartridge has an operational life of only a few hours. So here, the degree of intermission becomes a function of the ratio of the life expectancy of the bioreactor to that of the plasma separator. As this ratio increases, it will become necessary to rapidly replace spent plasma filters.

It should be noted that use of discrete microporous membrane oxygenators in BAL systems might also influence this aspect of therapy since these devices tend to fail because of pore wetting after a few hours of operation.

30.5 Conclusions and outlook

We have drawn conclusions, from an artificial organ engineering perspective, about the extent to which designs are capable of functioning in the manner of the natural organ. We make a distinction between "liver function" and "hepatocyte functions". A design which theoretically offers liver function must be populated with cells capable of providing the required hepatic functions. We did not discuss reported cell functionality for two reasons. First, different groups often investigate different markers of hepatic function and, where common markers are studied, different analytical methods and/or differing units are often reported. This makes meaningful comparison of reported functional data impossible. Second, factors associated with the cell source and the culture model are significant. Cell harvesting methods and the formulation of culture media used will have an effect on subsequent device performance. Also, whilst particular static culture models may influence hepatocyte behaviour, their corresponding effect in BAL devices is poorly understood. However, we may draw some conclusions. The most obvious differences arise inevitably between the two classes of device:

1. Generic devices offer rapid development and reduced initial costs. SG devices involve more extensive development programs and may invoke higher costs. Attendant manufacturing difficulties may also be expected.
2. SG devices are purpose-designed with optimized performance being the main target. Generic bioreactors appear to be adopted as a means of culturing a large mass of hepatocytes in an available device readily accommodated by existing hardware.

It is probably not a coincidence that all of the SG BAL devices are plasma-perfused devices with specific integral oxygenation. No generic device makes this latter provision. Simple modelling demonstrates that plasma-perfused devices require integral oxygenation for desirable operating conditions. The modelling required to calculate the availability of oxygen to hepatocytes in blood-perfused devices is more complex. Our analyses predict that the diffusive permeability of an immunoisolation barrier (for example, a 100,000 nominal molecular weight cut-off hydrophilic membrane or an alginate capsule) to oxygen can limit cellular oxygen uptake in devices perfused with venous blood. Have SG developers recognized these limitations and therefore elected to perfuse their devices with plasma? This offers improved mass transfer if direct contact between plasma and hepatocytes occurs. Are SG developers reckoning upon there being no immunological problems associated with this mode of therapy? They may also be monitoring future prospects for supply of suitable human cell lines.

Devices should be capable of regulating the circulating concentrations of target solutes. But, how do we demonstrate this? Let us consider BAL bioreactors as black boxes at steady state. Fully described system operating conditions and simultaneous quantification of input concentrations of target solutes and outlet concentrations of these solutes (or reaction products arising from them) would be a basis for meaningful reporting of performance data. This is not a panacea but it would enable inter-device comparisons if it was adopted as a standard method. Standard "BAL challenge" media containing appropriate concentrations of relevant "liver failure solutes" could be defined for use in such tests. From resultant data, we might determine the extent to which each BAL model requires to be scaled in order to provide adequate regulatory capacity.

To date, there are no published data on any BAL device which demonstrate that the design is scalable. For BAL devices to be appropriately scaled, characterisation of the fluid dynamics and reaction kinetics of the functional element of the reactor is required [83]. The BAL could then be scaled to suit predicted loading as a set of parallel elements. Although, at present, we are unable to predict the load placed upon a BAL device by a patient.

Perhaps evaluation of the efficacy of BAL designs in regulating plasma biochemistry should be considered as a priority in bioartificial liver research. We may learn a great deal from standardized experiments. For example, we could identify promising designs worthy of continued development and, by comparing results from such experiments with clinical findings, we may be able to elucidate what is required of BAL devices in order to reverse the complications of liver failure and reduce the requirement for liver transplantation.

The demise of the BAL concept has been predicted in the past and, at the time of writing, is once again being discussed. Furthermore, it is now being suggested that tissue-engineered livers may be a realistic future prospect [84]. However, it has taken more than 40 years for us to achieve the present state of the art. Purpose designed hepatocyte bioreactors are now emerging and understanding of the pathogenesis of liver failure continues to improve. We should not discount the advantages of temporary extracorporeal therapy which, when allied to the prospect of the emergence of suitable cell lines, will make BAL treatment attractive. There is life in the old BAL yet.

References

1. O'Grady JG, Alexander GJM, Hayllar KM and Williams R. Early indicators of prognosis in fulminant hepatic failure. Gastroenterology 1989; 97: 439-445.
2. Tygstrup N and Ranek L. Assessment of prognosis in fulminant hepatic failure. Sem Liv Dis 1986; 6: 129-137.
3. Hughes RD, Wendon J, and Gimson AES. Acute liver failure. Gut 1991; Supplement: s86-s91.
4. Edmond JC, Aran PP, Whitington PF, Broelsch CE and Baker AL. Liver transplantation in the management of fulminant hepatic failure. Gastroenterology 1989; 96: 1583-1588.
5. Gerlach JC, Ziemer R and Neuhaus P. Fulminant liver failure: Relevance of extracorporeal hybrid liver support systems. Int J Artif Organs 1996; 19: 7-13.
6. Bismuth H, Samuel D, Gugenheim J, Castaing D, Bernau J et al. Emergency liver transplantation for fulminant hepatitis. Ann Intern Med 1987; 143: 337-341.
7. Iwatsuki S, Stieber AC, Marsh JW, Tzakis AG, Todo S et al. Liver transplantation for fulminant hepatic failure. Transplant Proc 1989; 21: 2431-2443.
8. Vickers C, Neuberger J, Buckels J, McMaster P and Elias E. Transplantation of the liver in adults and children with fulminant hepatic failure. J Hepatol 1988; 7: 143-150.
9. Kamlot A, Rozga J, Watanabe FD and Demetriou AA. Review: Artificial liver support systems. Biotechnology and Bioengineering 1996; 50: 382-391.
10. Kalayoglu M, Stratta RJ, Hoffman RM and Belzer FO. Quadruple immunosuppresive therapy for liver transplantation. Transpl Proc 1988; 20: 524-528.
11. McDiarmid SV, Klintmalm GB and Busuttil RW. FK506 conversion for intractable rejection of the liver allograft. Transplant Int 1993; 6: 305-312.
12. Sussman NL and Lake JR. Treatment of hepatic failure - 1996: Current concepts and progress towards liver dialysis. Am J Kidney Dis 1996; 27: 605-621.
13. Circe-Biomedical-Inc., Registration Statement under the Securities Act of 1933 (USA). Form S1. Sec. File Number 333-29751. Document filed with Securities and Exchange Commission, Washington DC, 20549, USA. 1997, Circe Biomedical Inc., 1 Ledgemont Centre, 128 Spring Street, Lexington, MA, 02173, USA.
14. Hughes RD and Williams R. Use of bioartificial and artificial liver support devices. Sem Liv Dis 1996; 16: 435-444.
15. Matsubara S, Okabe K, Ouchi K, Miyakazi Y, Yajima Y et al. Continuous removal of middle molecules by hemofiltration in patients with acute liver failure. Crit Care Med 1990; 18: 1331-1338.
16. Sarnatskaya VV, Nikolaev VG, Osipova LA, Ivanov AI, Maslenny VN et al. Peritoneal dialysis for the removal of protein-bound markers of hepatic insufficiency. Artificial Organs 1993; 17: 828-836.
17. Kondrup J, Almdal T, Vilstrup H and Tygstrup N. High volume plasma exchange in fulminant hepatic failure. Int J of Artific Org 1992; 15: 669-676.
18. Yoshiba M, Sekiyama K, Iwamura Y and Sugata F. Development of reliable artificial liver support (ALS) - plasma exchange in combination with hemodiafiltration using high performance membranes. Dig Dis Sci 1993; 38: 469-476.
19. Chang TMS. Haemoperfusion Over microencapsulated adsorbent in a patient with hepatic coma. The Lancet 1973; ii: 1371-1372.

20. O'Grady, J, Gimson AES, O'Brien CJ, Pucknell A, Hughes RD et al. Controlled trials of charcoal hemoperfusion and prognostic factors in fulminant hepatic failure. Gastroenterology 1988; 94: 1186-1192.
21. Eiseman B, Liem DS and Raffucci F. Heterologous liver perfusion in treatment of hepatic failure. Ann Surg 1973; 162: 329-345.
22. Schafer DF. Desperate Appliances: A short review of therapies for fulminant hepatic failure. Blood Purification 1993; 11: 158-162.
23. Gumucio JJ and Chianale J. Liver Cell Heterogeneity and Liver Function. In: Arias IM et el., Eds. The Liver: Biology and Pathobiology, Second Edition. Raven Press Ltd. New York. 1998: 931-947.
24. Sorrentino F. Prime richerche per la realizzatione di un fegato artificiale. Chir Patol Sper 1956; 4: 1401-1414.
25. Nose Y, Mikami J and Kasai Y. An experimental artificial liver utilizing extracorporeal metabolism with slices or granulated canine liver. Trans Am Soc Artif Intern Organs 1963; 9: 358-362.
26. Wolf CFW and Munkelt BE. Bilirubin conjugation by an artificial liver composed of cultured cells and synthetic capillaries. Trans Amer Soc Artif Int Organs 1975; 21: 16-27.
27. Matsumura KN, Guevera GR, Huston H, Hamilton WL, Rikimaru M et al. Hybrid bioartificial liver in hepatic failure: Preliminary clinical report. Surgery 1987; 101: 99-103.
28. Margulis MS, Erukhimov EA, Andreiman LA and Viksna LM. Temporary organ substitution by hemoperfusion through suspension of active donor hepatocytes in a total complex of intensive therapy in patients with acute hepatic insufficiency. Resuscitation 1989; 18: 85-94.
29. Sussman NL, Chong MG, Koussayer T, He D-E, Shang TA et al. Reversal of fulminant hepatic failure using an extracorporeal liver assist device. Hepatology 1992; 16: 60-65.
30. Kelly JH, Koussayer T, He D-E, Chong MG, Shang TA et al. Assessment of an extracorporeal liver assist device in anhepatic dogs. Artificial Organs 1992; 16: 418-422.
31. Sussman NL and Kelly JH. Improved liver function following treatment with an extracorporeal liver assist device. Artificial Organs 1993; 17: 27-30.
32. Sussman NL, Finegold MJ, Barish JP and Kelly JH. A case of syncytial giant-cell hepatitis treated with an extracorporeal live assist device. Am J Gastroent 1994; 89: 1077-1082.
33. Sussman NL, Gislason GT, Conlin CA and Kelly JH. The Hepatix extracorporeal liver assist device: Initial clinical experience. Artificial Organs 1994; 18: 390-396.
34. Ellis AJ, Hughes RD, Wendon JA, Dunne J, Langley PG et al. Pilot-controlled trial of the extracorporeal liver assist device in acute liver failure. Hepatology 1996; 24: 1446-1451.
35. McCarthy M, Ellis AJ, Wendon JA, Heaton N, Rela M et al. Use of extracorporeal liver assist device and auxilliary liver transplantation in fulminant hepatic failure. Eur J Gastroent Hepatol 1997; 9: 407-412.
36. Rozga J, Williams F, Roo M-S, Neuzil DF, Giorgio TD et al. Development of a bioartificial liver: Properties and function of a hollow-fiber module inoculated with liver cells. Hepatology 1993; 17: 2558-265.

37. Rozga J, Holzman MD, Roo M-S, Griffin DW, Neuzil DF et al. Development of a hybrid bioartificial liver. Ann Surg 1993; 217: 502-511.
38. Rozga J, Podesta L, Le Page E, Hoffman A, Morsiani E et al. Control of cerebral oedema by total hepatectomy and extracorporeal liver support in fulminant hepatic failure. The Lancet 1993; 342: 898-899.
39. Neuzil DF, Rozga J, Moscioni AD, Roo M-S, Hakim R et al. Use of a novel bioartificial liver in a patient with acute liver insufficiency. Surgery 1993; 113: 340-343.
40. Rozga J, Podesta L, Le Page E, Morsiani E, Hoffman A et al. A bioartificial liver to treat severe acute liver failure. Ann Surg 1994; 219: 538-546.
41. Demetriou AA, Rozga J, Podesta L, Le Page E, Morsiani E et al. Early clinical experience with a hybrid bioartificial liver. Scand J Gastroenterol 1995; 30 (Suppl. 208): 111-117.
42. Watanabe FD and Demetriou AA. Support of acute liver failure patients with a bioartificial liver. J Clin Apheresis 1996; 11: 138-142.
43. Chen SC, Mullon C, Kahaku E, Watanabe F, Hewitt W et al. Treatment of severe liver failure with a bioartificial liver. Ann NY Acad Sci 1997; 831(Bioartificial organs): 350-360.
44. Watanabe FD, Mullon CJ-P, Hewitt WR, Arkadopoulos N, Kahaku E et al. Clinical experience with a bioartificial liver in the treatment of severe liver failure. A phase 1 clinical trial. Ann Surg 1997; 225: 484-494.
45. Bismuth H. New developments in clinical use of a bioartificial liver. In: Crepaldi G, Demetriou AA and Muraca M. eds. Bioartificial Liver Support: The Critical Issues. Italy: C.I.C. Edizioni Internationali, 1997:
46. Nyberg SL, Shatford RA, Peshwa MV, White JG, Cerra FB et al. Evaluation of a hepatocyte entrapment hollow fiber bioreactor: A potential bioartificial liver. Biotechnol Bioeng 1993; 41: 194-203.
47. Nyberg SL, Platt JL, Shirabe K, Payne WD, Hu W-S et al. Immunoprotection of xenocytes in a hollow fiber bioartificial liver. ASAIO Journal 1992; 38: M463-M467.
48. Nyberg SL, Payne WD, Amiot B, Shirabe K, Remmel RP et al. Demonstration of biochemical function by extracorporeal xenohepatocytes in an anhepatic animal model. Transplantation Proc 1993; 25: 1944-1945.
49. Sielaff TD, Hu MY, Amiot B, Rollins MD, Rao S et al. Gel-entrapment bioartificial liver therapy in galactosamine hepatitis. J Surg Res 1995; 59: 179-184.
50. Wu FJ, Friend JR, Lazar A, Mann HJ, Remmel RP et al. Hollow fiber bioartificial liver utilizing collagen entrapped porcine hepatocyte spheroids. Biotechnol Bioeng 1996; 52: 34-44.
51. Hu W-S, Friend JR, Wu FJ, Sielaff T, Peshwa MV et al. Development of a bioartificial liver employing xenogenic hepatocytes. Cytotechnology 1997; 23: 29-38.
52. Gerlach JC, Kloppel K, Muller C, Smith MD and Neuhaus P. Hepatocyte aggregate culture technique for bioreactors in hybrid liver support systems. Int J Artific Org 1993; 16: 843-846.
53. Gerlach J, Schnoy N, Smith MD and Neuhaus P. Hepatocyte culture between woven capillary networks: A microscopy study. Artificial Organs 1994; 18: 226-230.
54. Gerlach, J, Encke J, Hole O, Muller C, Ryan CJ et al. Bioreactor for a larger scale hepatocyte *in vitro* perfusion. Transplantation 1994; 58: 984-988.
55. Fuchs M, Gerlach J, Encke J, Smith MD, Neuhaus P et al. Amino acid metabolism by hepatocytes in a hybrid liver support bioreactor. Int J Artific Org 1994; 17: 663-669.

56. Fuchs M, Gerlach J, Unger J, Encke J, Smith MD et al. α-keto acid metabolism by hepatocytes cultured in a hybrid liver support bioreactor. Int J Artific Org 1994; 17: 554-558.
57. Gerlach J, Trost T, Ryan CJ, Meissler M, Hole O et al. Hybrid liver support system in a short term application on hepatectomized pigs. Int J Artific Org 1994; 17: 549-553.
58. Gerlach JC. Development of a hybrid artificial liver: A review. Inte J Artific Org 1996; 19: 645-654.
59. Flendrig LM, la Soe JW, Jorning GGA, Steenbeek A, Karlsen OT et al. *In vitro* evaluation of a novel bioreactor based on an integral oxygenator and a spirally wound nonwoven polyester matrix for hepatocyte culture as small aggregates. J Hepatol 1997; 26: 1379-1392.
60. Flendrig LM, te Velde AA and Chamuleau RAFM. Semipermeable hollow fibre membranes in hepatocyte bioreactors: A prerequisite for a successful bioartificial liver? Artificial Organs 1997; 21: 1177-1181.
61. Flendrig LM, Di Florio E, Mancini A, Vacca C, Evangelista A et al. *In vivo* investigation of a novel bioartificial liver (BAL) based on porcine hepatocytes. Hepatology 1997; 26(4 pt.2 SS): 168A.
62. Bader A, Knopp E, Boeker K, Schuettler W, Oldhafer K et al. A novel bioreactor design for *in vitro* reconstruction of *in vivo* liver characteristics. Artificial Organs 1995; 19: 368-374.
63. Bader A, Knopp E, Fruehauf N, Crome O, Boeker K et al. Reconstruction of liver tissue *in vitro*: geometry characteristic of flat bed, hollow fiber and spouted bed bioreactors with reference to the in vivo liver. Artificial Organs 1995; 19: 941-950.
64. Bader A, Fruehauf N, Steinhoff G and Haverich A. Development of a bioartificial liver by reconstructing 3-D cell plates and sinusoidal analogues. Artificial Organs 1997; 21: 526.
65. Smith MD, Airdrie I, Cairns D, Cousins RB, Ekevall E et al. A novel hybrid artificial liver with integral membrane oxygenation: Theory and initial *in vitro* evaluation. In: International Congress on Hepatocytes: Applications in cell biology, toxicology and medicine. Kluwer Academic Publishers, Germany, 1996.
66. Smith MD, Airdrie I, Cairns D, Courtney JM, Cousins RB et al. Characterisation of a hybrid artificial liver bioreactor with integral membrane oxygenation. Artificial Organs, 1997; 21: 531.
67. Ekevall E, Smith MD, Airdrie I, Grant MH, Cousins RB et al. A modular hepatocyte bioreactor with integral membrane oxygenation: In vitro testing. Int J Artific Org 1997; 20: 529.
68. Smith MD, Airdrie I, Courtney JM, Cousins RB, Ekevall E et al. Development and characterization of a hybrid artificial liver bioreactor with integral membrane oxygenation. In: Crepaldi G, Demetriou AA and Muraca M, Eds. Bioartificial Liver Support: The Critical Issues. Italy: C.I.C. Edizioni Internationali, 1997: 27-35.
69. Zieve L, Doizaki WM and Zieve FJ. Synergism between mercaptans and ammonia or fatty acids in the production of coma: A possible role for mercaptans in the pathogenesis of hepatic coma. J Lab Clin Med 1974; 83: 16-28.
70. Nishikawa M, Uchino J, Matsushita M, Takahashi, M, Taguchi K et al. Optimal oxygen tension conditions for functioning cultured hepatocytes *in vitro*. Artificial Organs 1996; 20: 169-177.
71. Bhatia SN, Toner M, Foy BD, Rotem A, O'Neil KM et al. Zonal liver cell heterogeneity: effects of oxygen on metabolic functions of hepatocytes. Cellular Engineering 1996; 1: 125-135.

72. UK government clears animal to human transplant trials under strict guidelines. In: Clinica. 1998: 5.
73. Takahashi M, Ishikara H, Takahashi C, Nakajima Y, Matsushita M et al. Immunologic considerations in the use of cultured porcine hepatocytes as a hybrid artificial liver. ASAIO Journal 1993; 39: M242-M246.
74. Gerlach JC. Long-term liver cell cultures in bioreactors and possible application for liver support. Cell Biol Toxicol 1997; 13: 349-355.
75. Joly A, Desjardins J-F, Fremond B, Desille M, Campion J-P et al. Survival, proliferation and functions of porcine hepatocytes encapsulated in coated alginate beads: a step towards a reliable bioartificial liver. Transplantation 1997; 63: 795-803.
76. Stockman HBAC, Tompkins RG and Berthiaume F. Expression of long-term liver-specific function by adult rat hepatocytes cultured on microcarriers. Tissue Engineering 1997; 3: 267-279.
77. Kern A, Bader A, Pichlmayer R and Sewing K-F. Drug metabolism in hepatocyte sandwich cultures of rats and humans. Biochem Pharmacol 1997; 54: 761-772.
78. Hansen LK, Hsiao C-C, Friend JR, Wu FJ, Bridge GA et al. Enhanced morphology and function in hepatocyte spheroids: a model of tissue self assembly. Tissue Engineering 1998; 4: 65-74.
79. Smith MD, Cairns DE, Veitch R, Cousins RB et al. Analysis of oxygen transfer in hollow fibre hepatocyte bioreactors. Artificial Organs 1997; 21: 531.
80. Smith MD. A hybrid artificial liver with integral membrane oxygenation: Theory, developmental studies and prototype testing. In: PhD Thesis, Bioengineering Unit, University of Strathclyde, Glasgow, UK, 1998: pp205.
81. Myers JD. The hepatic blood flow and splanchnic oxygen consumption of man - Their estimations from urea production or bromosulpthalein during catheterisation of the hepatic veins. J Clin Invest 1947; 26: 1130-1137.
82. Rotem A, Toner M, Tompkins RG and Yarmush ML. Oxygen uptake in cultured rat hepatocytes. Biotechnol Bioeng 1992; 40: 1286-1291.
83. Catapano G and De Bartolo LD. Importance of the kinetic characterization of liver cell metabolic reactions to the design of hybrid liver support devices. Int J Art Org 1996; 19: 670-676.
84. Griffith LG, Wu B, Cima MJ, Powers MJ, Chaignaud B et al. *In vitro* organogenesis of liver tissue. Ann NY Acad Sci 1998; 831(Bioartificial Organs): 382-397.

Chapter 31

Development and testing of a bioartificial liver

Jacek Rozga and Achilles A. Demetriou
Liver Support Research Laboratory, Department of Surgery, Cedars-Sinai Medical Center, Los Angeles, California, USA

Key words: artificial liver, artificial organs, hepatocyte aggregates, clinical trial, cryopreservation, fulminant hepatic failure, hepatic growth factor, hollow-fiber bioreactor, plasma perfusion, spheroids, TGF-β, transcription factors

31.1 Introduction

The development of an artificial liver is a major challenge because the liver controls, or has an effect on, almost every aspect of metabolism, blood coagulation and fibrinolysis, the immune system, endocrine responses and detoxification. It is believed that devices utilising intact viable hepatocytes have the greatest potential to provide liver support at this level of complexity. In recent years, advances in hepatocyte isolation and culture techniques, better understanding of hepatocyte-matrix interactions, availability of new biomaterials, and improved hollow-fiber technology have resulted in the development of such hepatocyte-based systems. Some of these bioartificial livers (BAL), including the one developed by our group, are currently being tested in the clinical setting [1-5]. In patients with severe acute hepatic failure, their role is to provide physiologic support and to "buy time" until a new liver is procured and to render patients stable and more likely to survive transplantation. There will also be patients who suffer from severe, but potentially reversible liver injury, in whom BAL therapy can potentially serve as the definitive treatment.

During the past two decades, two important advances took place which made possible the construction of devices containing isolated functional hepatocytes. One was the realisation of the importance of cell-cell and cell-matrix interactions in maintaining hepatocyte differentiation, and another was the development of improved cell culture techniques utilising serum-free defined media enriched with hormones and growth factors [6]. Various types of matrices were developed, including rat-tail collagen, Matrigel®, and solid matrix either alone or coated with extracellular matrix

M.N. Berry and A.M. Edwards (eds.), The Hepatocyte Review, 543–559.

constituent(s) [7]. Our own experience with transplantation of hepatocytes attached to collagen-coated dextran microcarriers [8] was a natural step preceding use of this same matrix-cell preparation in a hollow-fiber bioreactor [9].

The critical issues that must be addressed by any program aiming at the development of a hepatocyte-based liver support system are as follows:

1. Selection of hepatocyte donors.
2. Harvest, processing and storage of mammalian hepatocytes.
3. BAL (bioreactor) design.
4. BAL perfusion technique.
5. Hepatocyte and BAL testing *in vitro* and *in vivo* in animals with experimentally-induced liver failure.
6. Development of an adequate clinical study and care protocols.

In this chapter, we will briefly describe our own approach to these problems and comment on ideas and practical solutions which are favoured by other investigators. The results of a Phase I clinical study of our hybrid bioartificial liver will be summarized as well.

31.2 Bioartificial liver: a single group experience

31.2.1 Source of hepatocytes

Ideally, normal human hepatocytes should be used in the BAL: as mentioned, however, there is a shortage of cadaveric organs. The development of a primary human hepatocyte cell line is another alternative [10], but suitable candidates are as yet unavailable for use in human patients. An extracorporeal device which utilized cells from a clonally-derived hepatoma cell line of human origin has been developed and clinically tested. Early trials in dogs with acetaminophen-induced liver failure as well as treatments in a limited number of liver failure patients appeared promising [11,12]. However, results of a prospective controlled study performed in two groups of liver failure patients - those who had a potential for recovery and in those listed for emergency transplantation - were negative [4]. One of the major problems with this approach is the potential risk of leakage of tumor cells and their products into the patient's circulation [13]. Additionally, the establishment of a transformed cell line, using media conditioned by the HepG2 tumor cell line, was reported [10]. The consequences of both events in a patient can be magnified after the patient receives a transplant and becomes exposed to immunosuppression. In this context, the idea of constructing abioartificial liver utilising human hepatocyte progenitors deserves attention (Reid L., personal communication). It remains to be seen, however, whether cloning of large numbers of hepatocytes with subsequent full differentiation from a limited pool of committed hepatocyte progenitors is feasible and if the concept itself is applicable on an industrial scale. As a result of biosafety issues and because primary mammalian hepatocytes outperform all current liver cell lines [13], most investigators contemplate the use of primary porcine hepatocytes. First, pigs are widely used in medical research due to their similar physiology to man. As such, pathogen-free conditions have already been established for their breeding and maintenance. Second, unlike smaller mammals such as rabbits or rats, one adult pig liver can provide enough hepatocytes for several patient treatments. Third, isolated porcine hepatocytes have a tendency towards formation of cell aggregates. The latter may be of particular importance because cells arranged into spheroids regain polarity, establish cell-to-cell contact and recreate other elements of the liver environment. Several groups of investigators, including ourselves, succeeded in creating liver cell aggregates expressing liver-specific functions for prolonged periods of time and, reportedly, at higher levels than cells grown in a two-dimensional environment [14,15]. In most current BAL devices, the cells should be used as soon as possible, probably not later than 24-48 hours after isolation because of a progressive loss of differentiation in culture.

31.2.2 Hepatocyte isolation

Portal vein collagenase perfusion, as developed by Berry and Friend [16], remains the generally accepted method of hepatocyte isolation. In our laboratory, the need for large-scale production of isolated porcine hepatocytes has led to the development of a highly efficient and reproducible method utilising a filtration chamber and an automated cell washer [17]. Under ketamine anesthesia (30 mg/kg, i.v.), the abdomen of a young (10-12 kg) pig is entered through a midline incision, the hepato-duodenal ligament is dissected and all its structures ligated and divided except the portal vein which is cannulated with silicone tubing. The liver (300-400 g) is perfused *in situ* with 3 L of a medium comprising 2 mM EDTA, 140 mM NaCl, 5 mM KCl, 0.8 mM $MgSO_4$, 1.6 mM Na_2HPO_4, 0.4 mM KH_2PO_4, 25 mM $NaHCO_3$ solution at 300 ml/min. Then, it is resected and placed in a sterile basin which contains 0.05-0.1% collagenase solution prepared in Ca^{2+}-enriched buffer. The solution is recirculated through the liver (300 ml/min), following passage through silicone tubing submerged in an oxygen-saturated water bath at 39-40°C. The collagenase-digested liver tissue is hand crushed, suspended in a large volume of ice cold Dulbecco's Modified Eagle's Medium/10% Bovine Calf Serum and passed through a stack of steel meshes of decreased porosity (400 μ - 200 μ - 100 μ) in a custom-made chamber attached to a rotary shaker. Graded filtration allows rapid collection of approximately 1 l of liver cell suspension. Subsequent hepatocyte enrichment and purification is carried out using a cell washer (blood cell processor, COBE 2991). With this method, over 2×10^{10} liver cells can be obtained in less than 60 min by processing 250-300 g of the digested liver tissue. Initial viability of cells is greater than 90% on both the Trypan blue exclusion test and reduction of the tetrazolium salt 3(4,5-dimethyl-thiazoyal-2-yl)2,5-diphenyl tetrazolium bromide (MTT). On transmission electron microscopy, hepatocyte purity is $98.7 \pm 1.0\%$ and therefore, we do not use any additional purification techniques. Hepatocytes appear morphologically intact, showing viability of plasma membrane and intracellular organelles. Also when cultured either on collagen-coated dextran microcarriers or collagen-coated plates, hepatocytes retained differentiated functions (maintenance of microsomal 7-ethoxycoumarin-*O*-deethylase activity, a CYP-dependent enzyme, maintenance of mitochondrial succinate dehydrogenase activity) for several days [9,15]. It is also important to note that this method minimizes the risk of contamination during cell processing.

31.2.3 Solid support and matrix

As mentioned above, various types of matrices can be used to anchor hepatocytes and induce differentiation. We chose collagen-coated Cytodex 3 microcarriers (Pharmacia) because the total surface area available for cell attachment is greatly increased and because after inoculation into the extra-fiber compartment of the hollow-fiber bioreactor, the microcarrier-cell aggregates do not attach to the fiber wall thereby allowing free convection of plasma. Recently, by using only a small number of collagen-coated microcarriers (1.6 g per 1×10^9 liver cells) as a nidus for induction of cell aggregation, we have created liver cell clusters (MILCA: matrix-induced liver cell aggregates). After 3 hours in culture, hepatocytes in MILCA regain polarity, re-establish gap junctions and form distinct bile canaliculi [15] (Figure 1).

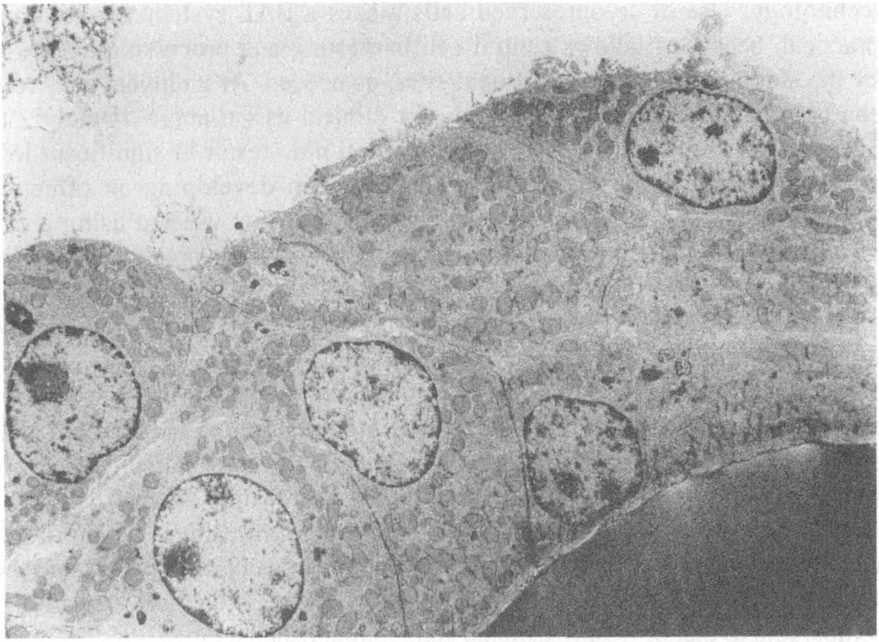

Figure 1. Electron micrograph depicting a cluster of freshly isolated pig hepatocytes attached to the surface of the collagen-coated microcarrier. Hepatocytes appear to have normal junctional complexes, organelles and bile canaliculi with microvilli (3,800 ×).

31.2.4 Hepatocyte function

The functional effect of the BAL depends on the metabolic performance of cells seeded in the bioreactor. Many tests are available for quality control, including measurement of protein synthesis, ureagenesis, oxygen uptake, CYP content and activity (*e.g.* lidocaine metabolism), drug metabolic assays, clearance techniques, and others. In our laboratory, we use a repertoire of tests which includes Trypan blue exclusion, MTT reduction, cell attachment to matrix, oxygen consumption, urea synthesis and measurement of diazepam, lidocaine, 7-ethoxycoumarin and resorufin metabolism [9,15].

31.2.5 Hepatocyte storage by cryopreservation

One of the crucial aspects of BAL technology is availability of cells. It becomes particularly important in the clinical setting where treatment of patients with fulminant hepatic failure (FHF) is carried out as an emergency at any hour. This and many other problems related to hepatocyte storage and distribution are in our view best solved by use of cryopreservation technology. Use of cryopreserved cells makes a BAL system versatile and practical, because it allows central cell harvesting and processing, followed by the shipping of cells to treatment sites, as needed. At a clinical site, cells can be processed easily and be ready for clinical use in approximately 2 to 3 hours. Although in general cryopreservation may result in significant loss of viability and function, we have succeeded in developing an effective method for hepatocyte cryopreservation. Cells are first washed using a cell conditioning and cryopreservation medium (CCCM) that contains adenosine, a calcium channel blocker, free oxygen radical scavengers (allopurinol and glutathione) and sufficient metabolic substrates to maintain cell viability and function throughout the cryopreservation and recovery processes. Before being placed in a programmed freezer, the cells are passed through a series of proprietary buffered CCCM/10% dimethylsulphoxide washes with progressively increasing osmolality, whereas upon rapid thawing, they are washed using solutions with progressively decreasing osmolality. As a result, morphological abnormalities (blebbing, swelling) are minimal; viability of cells upon thawing consistently remains at or above 80% and the ability of cryopreserved cells to metabolize diazepam and ethoxyresorufin is comparable to that of freshly isolated cells (unpublished data).

31.3 Features of the bioartificial liver

31.3.1 How many hepatocytes are needed in a BAL?

It has been suggested that as little as 10% of total liver mass is compatible with life, which would correspond to $\sim 1.5 \times 10^{10}$ hepatocytes. The actual number of cells used by various investigators in hepatocyte-based systems varies from 0.5×10^8 to 2.0×10^{10}. We should note, however, that even in the severely injured liver, there is usually a certain amount of residual liver function and as little as 3-5% of additional hepatocyte mass in a BAL could potentially make a difference.

31.3.2 Hepatocyte bioreactor

Numerous bioreactors have been tested as prototypes of a bioartificial liver, including commercially available filters, membrane oxygenators and plasma separators. None of those, however, were able to provide an adequate environment for liver cells and to ensure an effective bi-directional mass transport. Bioreactors containing bundles of small-diameter polysulphone or cellulose-acetate tubes, sealed at each end, were developed for protein ultrafiltration, membrane oxygenation and mass tissue cultures [18]. Wolf and Munkelt [19] were the first to use them as a prototype of an artificial liver. We followed their lead and chose a macroporous (0.2 μ) hollow-fiber bioreactor, because it is simple and reproduces, to some extent, the liver microarchitecture. For example, when the bioreactor is perfused axially, the fiber wall acts as fenestrated endothelial layer that separates the lumen of the liver sinusoid from the space of Disse. In both instances, the pressure gradient along the fiber/sinusoid long axis forces convection of plasma (Starling flow) through the extrafiber space. Thus, the basic principles of bi-directional mass transport in the liver parenchyma and the hollow-fiber bioreactor are similar. Except for Nyberg [20], all investigators utilising hollow-fiber bioreactors inoculate cells in the extra-fiber space, while culture medium, blood, or plasma is pumped through the fiber lumen.

Blood vs. Plasma Perfusion. Proponents of whole blood perfusion argue that plasma separation complicates liver support treatment, increases the extracorporeal priming space and raises the cost of the procedure. On the other hand, whole blood perfusion requires heparinisation which, in a coagulopathic patient, increases the risk of bleeding. In addition, it may result in hemolysis, platelet depletion, fiber clotting and system failure with cell loss. We, and other advocates of plasma perfusion, have stressed that plasma perfusion can be carried out safely for long periods of time and, more importantly, that it eliminates direct cell-cell interactions between host blood cells and xenogeneic liver cells. Additionally, by excluding the BAL from the primary blood circuit, its versatility and safety are enhanced.

31.3.3 BAL design

The design of the current BAL system (Figure 2) has evolved from a series of *in vitro* and *in vivo* experimental animal and pilot clinical studies carried out over a period of several years [5,7,8,21]. Since 1994, the BAL has been tested in an FDA-approved clinical trial under the brand name HepatAssist™ (Circe Biomedical, Inc., Lexington, MA). It is a hybrid system, where the function of the hepatocyte bioreactor is enhanced and/or supplemented by the column filled with particles of activated cellulose-coated charcoal (Adsorba 300C; Gambro, Hechingen, Germany). Blood is removed from a patient at 90-100 ml/min through a double-lumen catheter placed in the superficial femoral vein. Blood is then separated into blood cells and plasma using a plasma separator (Spectra, Cobe, Lakewood, CO). Plasma enters the BAL system at 50 ml/min; the high-flow plasma recirculation loop consists of a reservoir, an Adsorba 300C column, an oxygenator (Capiox 308; Terumo Inc., Tokyo, Japan), a heater (Temp Marg,

Marquest, Lakewood, CO) and a hollow-fiber module (Circe Biomedical, Inc, Lexington, MA). The total extracorporeal volume is about 600 ml. The extra-fiber space of the cell module (200 ml) is filled with five billion viable porcine hepatocytes aggregated into MILCA by collagen-coated dextran microcarriers (Cytodex 3; Pharmacia, Piscataway, NJ). Plasma circulates through the fibers at 400 ml/min. As a safety measure, pressure in the system is monitored to alert the operator if an unexpected leak or blockage were to occur. Sodium citrate is used as an anticoagulant to prevent thrombosis within the circuit. Plasma ionised calcium levels are measured hourly during BAL treatments to prevent ionised hypocalcemia due to the calcium-chelating effect of citrate. Calcium chloride infusion is initiated at the start of BAL treatment. Recently, we have introduced a more efficient hollow-fiber module consisting of polysulphone fibers (Circe Biomedical,). After return from the BAL, plasma and blood cells are reconstituted and returned to the patients by way of a double-lumen venous catheter.

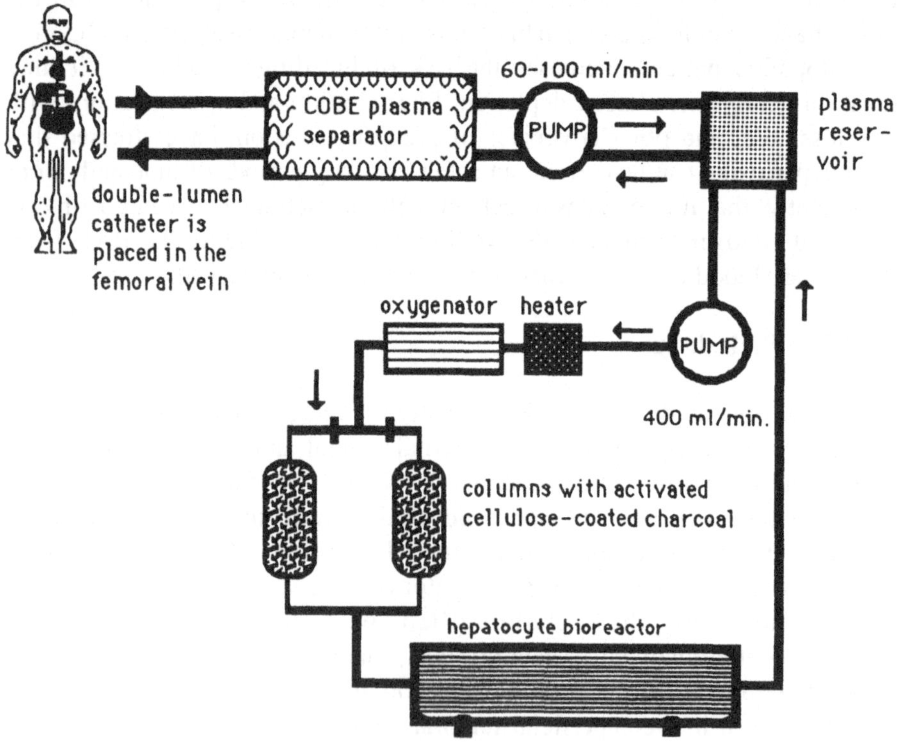

Figure 2. The bioartificial liver (BAL) system incorporating a hollow-fiber bioreactor inoculated with five billion viable porcine hepatocytes aggregated into MILCA by collagen-coated dextran microcarriers. The function of the hepatocyte bioreactor is enhanced and/or supplemented by the column filled with activated cellulose-coated charcoal. The high-flow plasma recirculation loop also includes an oxygenator and a heater .

31.3.4 Bioartificial liver testing in experimental settings

Assessing BAL function is a major challenge, because there is no single "global" marker for quantifying liver function. During *in vitro* testing, most authors measure liver specific functions such as synthesis of glucose, urea, albumin, and drug metabolism (CYP activity). We performed a large number of such experiments using both stationary and perfusion culture systems and demonstrated that microcarrier-attached hepatocytes, either alone or aggregated into spheroids (MILCA) can synthesize albumin, conjugate bilirubin, respond to proliferative stimuli, utilise oxygen and carry out CYP-specific functions (*e.g.* metabolism of cyclosporin A, 19-nor-testosterone, diazepam, lidocaine, ethoxyresorufin, 7-ethoxycoumarin) at rates similar to those seen *in vivo* [5,7-9,21].

31.4 BAL testing in the clinical setting

When the BAL is used in patients, the relevance of data generated from the use of tests such as galactose elimination and measurement of lidocaine metabolite levels is not clear, because it is not possible to distinguish the effect of the BAL from that of the native liver. In our experience, a relatively reliable indicator of the BAL metabolic activity is extraction of oxygen by the hepatocyte hollow-fiber cartridge during perfusion. This is carried out routinely at hourly intervals by measuring PO_2 samples that are being taken simultaneously at the inlet and outlet of the cell bioreactor.

31.4.1 Patients and methods

We treated three groups of patients. Group I (24 patients) fulfilled all diagnostic criteria of FHF and were listed for emergency liver transplantation. Group 2 (3 patients) had undergone orthotopic liver transplantation (OLT) and developed primary graft nonfunction in the immediate postoperative period; they were also candidates for urgent re-transplantation. Group 3 (10 patients) presented with acute exacerbation of an underlying chronic liver disease process and were not candidates for transplantation because of active substance abuse, sepsis, and concurrent medical problems. In Group 1, all patients but one were in stage 4 encephalopathy (deep coma) and exhibited signs of brainstem dysfunction and decerebration because of intracranial hypertension. Patients in Group 2 were in stage 3 encephalopathy and in Group 3, all patients were in stage 4 encephalopathy with the exception of two patients who were in stage 3.

31.4.2 Liver Support Unit

The clinical Liver Support Unit team consists of physicians (hepatology, nephrology, critical care, pulmonology, neurology, neurosurgery, transplantation, infectious diseases), nurses, pharmacists, nutritionists, physical and respiratory therapists. The research team comprises experimental surgeons, neurophysiologists, molecular biologists, hepatic physiologists, and engineers. In addition to using a computerized on-line clinical data collection system (Care Vue 9000 Clinical Information System; Hewlett-Packard, Andover, MA), standardized diagnostic and therapeutic clinical protocols were developed and implemented to allow meaningful and accurate data collection and analysis. Immediately on admission, patients undergo rapid evaluation. Invasive hemodynamic monitoring (peripheral arterial catheter and pulmonary artery catheter) is instituted after rapid blood product infusion to partially correct the coagulopathy. Patients are intubated endotracheally when in stage 4 encephalopathy and are placed on ventilator support. In patients with clinical signs of intracranial hypertension, a subdural intracranial pressure (ICP) monitor is placed at the bedside for ICP and cerebral perfusion pressure (CPP) monitoring, the head of the bed is elevated, the room is kept dark and quiet and standard medical therapy is instituted (hyperventilation, mannitol, lactulose). Standard medical therapy also includes correction of electrolyte (*i.e.* hypocalcemia), metabolic (*i.e.* hypoglycemia, lactic acidemia), respiratory, coagulation, and hemodynamic abnormalities. Renal dialysis and filtration is used, as needed. Vital signs, urinary output, and central hemodynamic parameters are monitored continuously. Neurologic monitoring and assessment is carried out in all patients with clinical evidence of brainstem dysfunction. This consists of hourly clinical examination, serially assessing brainstem dysfunction by using the comprehensive level of consciousness score (CLOCS) system, determination of the Glasgow coma score, and continuous monitoring of ICP and CPP. Liver function tests, coagulation tests, complete blood cell count, serum electrolytes, ammonia, lactate, blood urea nitrogen, and creatinine levels are determined at the start, midpoint, and end of each BAL treatment. Arterial gases are determined serially.

31.4.3 BAL treatment

Each BAL treatment lasts 6 hours and is preceded by an hour-long control period of plasma separation only. The treatment endpoint in the first two groups of patients was survival. Secondary endpoints included neurologic improvement accompanied by a decrease in ICP and successful "bridge" to transplantation. Depending on the donor organ availability, our patients received 1 to 5 BAL treatments.

31.4.4 Results

With one exception, all treatments were completed as scheduled. During one treatment, the patient experienced significant hypotension and neurologic deterioration after 2 hours and treatment was discontinued. In the remaining patients, no technical problems were encountered during plasma separation and BAL perfusion (*e.g.* embolisation, thrombosis, plasma clotting, bleeding) and no adverse reactions to porcine hepatocytes were observed. Patients remained hemodynamically stable throughout the treatment period. Despite contact with "toxic" plasma, viability of freshly isolated hepatocytes at the end of a treatment, was high (>75%). In the case of cryopreserved hepatocytes, a steep decrease in cell viability (<50%) was noted at the end of a treatment.

Survival

In Group 1 patients, seventeen were bridged successfully to OLT, experienced full neurologic recovery and were discharged from the hospital. Five patients with acetaminophen-induced FHF, although listed for transplantation according to King's College criteria, improved after BAL treatments and recovered fully without undergoing OLT. One patient with FHF due to ischemic liver injury after a heat stroke underwent five BAL treatments and showed neurologic recovery; however, severe ischemic necrotising pancreatitis subsequently developed and he was removed from the transplant list and died 21 days later due to multiorgan failure and sepsis.

All three Group 2 patients were bridged successfully to retransplantation, experienced full recovery, and were discharged from the hospital.

All Group 3 patients enjoyed transient clinical improvement after BAL treatment. However, only two patients were able to recover enough native liver function to survive the acute-on-chronic failure; these two patients later became candidates for OLT. The remaining eight patients died 1 to 21 days (mean 7.1 days) after their last BAL treatment from variceal bleeding, sepsis and multiorgan failure.

Neurologic effects of the BAL

Patients in Group 1 experienced remarkable neurologic improvement with reversal of the decerebrate state after BAL treatments. Brainstem function improved, as judged by the higher CLOCS scores (Table 1). There was a significant reduction in ICP with a concomitant increase in CPP (Table 1). This can be better appreciated if one examines more closely patients whose ICP was >25 mm Hg; in these, the effect of BAL treatment was dramatic (ICP: 31 ± 7 to 13 ± 10 mm Hg, p<0.002).

Table 1. Neurologic effects of the BAL in Group 1 FHF patients. Abbreviations: ICP - intracranial pressure, CPP - cerebral perfusion pressure, GCS - Glasgow Coma Score, CLOCS- Comprehensive Level of Consciousness Score

	ICP (mmHg)	CPP (mmHg)	GCS	CLOCS
pre-BAL	17.0 ± 1.5	70.1 ± 2.4	6.8 ± 0.4	27.0 ± 1.2
post-BAL	11.0 ± 1.1	75.1 ± 1.8	7.4 ± 0.4	32.0 ± 1.0
P*	0.0002	0.04	0.01	0.0001

* Paired Student "t" Test; Values: Mean ± S.D.

It was difficult to quantitate the neurologic impact of BAL treatment in patients in Group 2 because they remained under the influence of medications used during recent liver transplantation. Patients in Group 3 experienced transient neurologic improvement after BAL treatments, manifested primarily by an increased responsiveness and improvement in the CLOCS score (pre-BAL 29.7 ± 2.3 vs. post-BAL 34.0 ± 1.7; p<0.001).

Laboratory data

The effects of BAL treatment on liver function tests, metabolic and renal function, and hematologic and coagulation parameters are listed in Table 2 and 3. A decrease in the activity of transaminases, bilirubin levels and blood ammonia were the most common findings. The effect of BAL treatment on prothrombin time was difficult to assess, because most patients had already been treated with fresh frozen plasma and clotting factor concentrates.

Immune response

At this time, because we use only short-term treatments and because fulminant hepatic failure patients are immunosuppressed, use of xeno-hepatocytes does not appear to cause problems. In addition, when the BAL is used as a bridge to liver transplantation, development of immunological reaction to porcine antigens is prevented by post-transplant immunosuppression. However in the future, with multiple BAL treatments,

Table 2. Effect of Bioartificial Liver (BAL) treatment on laboratory parameters in Group 1 FHF patients

	pre-BAL	post-BAL	P*
Glucose (mg/dl)	126 ± 5	175 ± 7	0.001
Lactate (mmol/l)	4.4. ± 0.7	4.2 ± 0.6	0.2
Ammonia (μmol/l)	159 ± 8	133 ± 6	0.001
Total Bilirubin (mg/dl)	17.9 ± 1.5	14.6 ± 1.2	0.001
Dir. Bilirubin (mg/dl)	8.5 ± 1.0	6.4 ± 0.6	0.001
Indir.Bilirubin (mg/dl)	9.3 ± 0.8	7.9 ± 0.6	0.001
AST (U/l)	1255 ± 261	879 ± 148	0.002
ALT (U/l)	1075 ± 184	674 ± 120	0.001
Alk. Phosph. (U/l)	116 ± 16	90 ± 5	0.001
Uric acid (mg/dl)	4.3 ± 0.4	2.6 ± 0.2	0.001
Creatinine (mg/dl)	1.5 ± 0.2	1.1 ± 0.1	0.001
BUN (mg/dl)	18.9 ± 2.5	17.0 ± 2.0	0.001
Prothrombin Time (s)	21.1 ± 0.7	22.0 ± 0.5	0.11

* Paired Student "t" Test; Values: Mean ± S.D.

Table 3. Effect of Bioartificial Liver (BAL) treatment on metabolic parameters in Group 3 patients

	pre-BAL	post-BAL	P*
Glucose (mg/dl)	141 ± 19	171 ± 11	0.001
Lactate (mmol/l)	5.7 ± 1.1	5.6 ± 0.9	0.3
Ammonia (μmol/l)	173 ± 31	131 ± 15	0.06
Total Bilirubin (mg/dl)	26.0 ± 2.7	21.6 ± 2.2	0.001
Dir. Bilirubin (mg/dl)	12.8 ± 1.5	10.3 ± 1.2	0.001
Indir.Bilirubin (mg/dl)	11.9 ± 1.8	10.2 ± 1.5	0.002
AST (U/l)	692 ± 374	739 ± 409	0.5
ALT (U/l)	349 ± 126	281 ± 114	0.06
Uric acid (mg/dl)	4.1 ± 0.5	2.6 ± 0.1	0.001
Creatinine (mg/dl)	2.8 ± 0.3	2.2 ± 0.2	0.001
BUN (mg/dl)	38.5 ± 3.8	35.5 ± 3.2	0.004
Prothrombin Time (s)	22.6 ± 1.2	21.4 ± 0.7	0.2

* Paired Student "t" Test; Values: Mean ± S.E.M.

for longer periods of time, in patients with more chronic forms of liver failure, the use of primary xenogeneic hepatocytes may not be possible. Thus there is a need for developing well-differentiated hepatocyte lines of decreased immunogenicity and/or the development of immunoisolation methods which do not impair transport of nutrients and other factors between cells and perfusate.

31.5 Conclusions

Based on the preliminary data presented here, BAL treatment appears to be an effective bridge to liver transplantation [5]. Further appropriately controlled trials are currently underway to confirm this observation. Certainly, the ultimate goal of a liver support treatment is to treat patients early enough and repeatedly to allow them to regenerate their livers and recover normal function. In this respect, five patients in Group 1 fell in this category and had full recovery without a transplant.

31.6 Bioartificial liver: future perspectives

There have been recent technological advances in cell isolation, culture and cryopreservation, improved knowledge on hepatocyte-matrix interactions, availability of new biomaterials, hollow-fiber bioreactors and better understanding of flow and mass transport across membranes. All these have resulted in the development of a new generation of liver-assist devices. However, none of the current designs appears to be optimal. One of the areas where further research efforts are particularly needed includes studies to identify the humoral, cellular and subcellular mechanisms responsible for functional and morphologic recovery of the diseased liver and whether the cell-based BAL therapy interferes in any way with these mechanisms. This raises a seemingly naive question on how does such a BAL work? A common view is that it is a source of liver-specific activity which complements the residual function of the diseased liver. This notion has never been challenged because, as mentioned earlier, no morphological or functional tests exist that enable discrimination between the BAL effects and function of residual hepatocytes. In theory, the observed beneficial effects can be attributed to either BAL cell function, to improved function of the native liver, or both. We have initiated studies on these mechanisms and our preliminary data suggest that after massive destruction of hepatic tissue, blood levels of both positive and negative regulators of hepatocyte proliferation and several proinflammatory factors increase dramatically.

These changes in growth factor and cytokine profiles were found to interfere not only with proliferation of hepatocytes in the native liver, but also with mechanisms regulating expression of liver-specific genes. In particular, we found that:

1. In low-density cultures of primary adult rat hepatocytes stimulated by mitogen (hepatocyte growth factor), upregulation of STAT-3 expression is associated with down-regulation of C/EBP, HNF1 and HNF4 expression. In similar culture conditions, addition of transforming growth factor β1 (TGF-β1 helped maintain normal STAT3, C/EBP, HNF1 and HNF4 expression and down-regulated hepatic growth factor (HGF) receptor c-met expression [22-24].
2. *In vivo* in FHF rats, and in rats subjected to 90% hepatectomy, blood levels of HGF, TGF-β1 and IL-6 were greatly elevated and liver remnants showed suppression of both proliferation-associated (STAT3, AP-1) and liver-enriched (C/EBPα,β, HNF1, HNF4) transcription factors and organ-specific genes (*e.g.* PEPCK) as well as early up-regulation of SOCS1 and PIAS3 transcription factors (unpublished). At the same time, residual livers showed delayed HGF and c-met expression and no signs of hepatocyte proliferation (negative bromo-deoxy-uridine (BrdU) and mitotic indices) [25-27].
3. In FHF rats transplanted ectopically (spleen) with limited (2% liver mass) numbers of syngeneic hepatocytes, blood levels of both HGF and TGF-β1 were significantly lower than in sham-transplanted FHF rats and liver remnants showed accelerated HGF and c-met expression and signs of vigorous regeneration (positive BrdU uptake, mitotic index of 3% at 72 hours) [26].
4. In FHF rats, a pulse 4-hour treatment with the bioartificial liver (BAL) resulted in lowering of blood TGF-β1 levels and re-appearance of STAT3, AP-1, C/EBP, HNF1 and HNF4 expression in the native liver. Additionally, liver remnants showed accelerated HGF and c-met expression [28].
5. In FHF patients, livers showed no c-met expression [29] and blood HGF and TGF-β1 levels were highly elevated (unpublished data).

These data suggest that the HGF/TGF-β1 ratio in the circulating blood plays an important role in maintaining an adequate balance between hepatocyte proliferation and differentiation. Furthermore, it appears that increased blood TGF-β1 levels have an inhibitory effect on HGF-induced hepatocyte proliferation not only through the down regulation of the HGF gene, as shown by other investigators, but also through the suppression of HGF receptor c-met proto-oncogene expression [23]. At the same time, increased blood TGF-β1 levels help maintain liver specific functions in

HGF-stimulated hepatic tissue. Based on these data we are proposing a notion that after massive destruction of hepatic tissue, abnormal (elevated) blood levels of both positive and negative growth regulators inhibit proliferation and liver specific functions of residual hepatocytes. Hepatocyte-based support, in turn, seems to improve liver function and to facilitate liver regeneration by correcting humoral, cellular and molecular aberrations resulting from massive loss of functional hepatic tissue. These data represent the first step towards identification of specific humoral, cellular and/or molecular aberrations responsible for loss of liver regeneration and suppression of residual liver function in the diseased liver. We believe that knowledge of the requirements for recovery of hepatic functional reserve and liver growth (repair) will help design novel, rational and specific liver support strategies.

References

1. Margulis MS, Erukhimov EA and Andreiman LA. Temporary organ substitution by hemoperfusion through suspension of active donor hepatocytes in a total complex of intensive therapy in patients with acute hepatic insufficiency. Resuscitation 1989; 18: 85-94.
2. Gerlach J, Trost T and Ryan CJ. Hybrid liver support system in a short term application on hepatectomized pigs. Int J Artif Organs 1994; 17: 549-53.
3. Li AP, Barker G and Beck D. Culturing of primary hepatocytes as entrapped aggregates in a packed bed bioreactor: a potential bioartificial liver. In Vitro Cell Dev Biol 1993; 29A: 249-54.
4. Ellis AJ, Hughes RD, Wendon JA and Williams R. Pilot-controlled trial of the extracorporeal liver assist device in acute liver failure. Hepatology 1996, 24: 1446-1451.
5. Watanabe FD, Mullon CJ-P, Hewitt WR, Arkadopoulos N, Kahaku E et al. Clinical experience with a bioartificial liver in the treatment of severe liver failure. A phase I clinical trial. Ann Surg 1997, 225: 484-494.
6. Reid LM and Rojkind M. New techniques for culturing differentiated cells: reconstituted basement membrane rafts. Meth Enzymol 1989; 58: 263-278.
7. Rozga J and Demetriou AA. Artificial Liver. Evolution and future perspectives. ASAIO Journal 1995; 41: 831-837.
8. Demetriou AA, Whiting J and Levenson AM. New method of hepatocyte transplantation and extracorporeal liver support. Ann Surg 1986; 204: 259-271.
9. Rozga J, Williams RF, Moscioni AD, Giorgio T, Ro MS, Daniel Neuzil, Hakim R and Demetriou AA. Development of a Bioartificial Liver: Properties and function of a hollow-fiber module inoculated with liver cells. Hepatology 1993; 17: 258-265.
10. Roberts EA, Letarte M, Squire J and Yang S. Characterization of human hepatocyte cell lines derived from normal liver tissue. Hepatology 1994; 19: 1390-1399.
11. Sussman NL and Kelly JH. Improved liver function following treatment with an extracorporeal liver assist device. Artif Organs 1993; 17: 27-30.
12. Sussman NL, Chong MG, Koussayir T, He DE, Shong TA et al. Reversal of fulminant hepatic failure using an extracorporeal liver assist device. Hepatology 1992; 16: 60-65.

13. Nyberg SL, Remmel RP and Mann HJ. Primary hepatocytes outperform HepG2 cells as the source of biotransformation functions in a bioartificial liver. Ann Surg 1994; 220: 59-67.
14. Gerlach J, Schnoy N and Smith MD. Hepatocyte culture between woven capillary networks. A microscopy study. Artif Organs 1994; 18: 226-230.
15. Kong LB, Chen S, Demetriou AA and Rozga J. Matrix-induced liver cell aggregates (MILCA) for bioartificial liver use. Int J Artif Organs 1995; 18: 43-49.
16. Berry MN, Edwards AM and Barritt GJ. Isolated hepatocytes: preparation, properties and applications. In: Burdon RH and van Knippenberg PH, Eds. Laboratory Techniques in Biochemistry and Molecular Biology. Amsterdam: Elsevier, 1991; 21: 1-58.
17. Morsiani E, Rozga J, Scott H, Lebow L, Moscioni AD et al. Automated liver cell processing facilitates large scale isolation and purification of porcine hepatocytes. ASAIO Journal 1995; 41: 155-161.
18. Knazek RA, Gullino PM, Kohler PO and Dedrick R. Cell culture on artificial capillaries: An approach to tissue growth *in vitro*. Science 1972; 178: 65-66.
19. Wolf CFW and Munkelt BF. Bilirubin conjugation by an artificial liver composed of cultured cells and synthetic capillaries. Trans Am Soc Artif Intern Organs 1975; 21: 16-27.
20. Nyberg SL, Shatford RA, Peshwa MV, White JM, Cerra FB and Hu W-S. Evaluation of a hepatocyte-entrapment hollow-fiber bioreactor: A potential bioartificial liver. Biotechnol Bioeng 1993; 41: 194-203.
21. Arnaout WS, Moscioni AD, Barbour RL and Demetriou AA. Development of bioartificial liver: Bilirubin conjugation in Gunn rats. J Surg Res 1990; 48: 379-382.
22. Lilja H, Blanc P, Eguchi S, Arkadopoulos N, Margulies J et al. Fetal rat hepatocytes: Isolation, characterization and transplantation in the Nagase Analbuminemic rats. Transplantation 1997; 64: 1240-1248.
23. Lilja H, Suh KS, Neuman T, Demetriou AA and Rozga J. Transforming growth factor β1 down-regulates HGF induced expression of c-Met in primary rat hepatocyte cultures (submitted).
24. Lilja H, Kamohara Y, Suh KS, Neuman T, Demetriou AA and Rozga J. Transforming growth factor β1 maintains expression of liver enriched transcription factors in mitogen-stimulated primary rat hepatocyte cultures. Surg Forum 1998, 49: 176-178.
25. Eguchi S, Kamlot A, Ljubimova J, Hewitt W, Demetriou AA and Rozga J. Fulminant hepatic failure in rats: survival and effects on blood chemistry and liver regeneration. Hepatology 1996; 24: 1452-1459.
26. Eguchi S, Lilja MH, Kamlot A, Middleton Y, Demetriou AA and Rozga J. Loss and recovery of liver regeneration in rats with fulminant hepatic failure. J Surg Res 1997; 72: 112-122.
27. Kamohara Y, Lilja H, Neuman T, Demetriou AA and Rozga J. Activity of proliferation-associated transcription factors in rats with surgically induced liver failure. Surg Forum 1998, 49: 157-158.
28. Suh KS, Lilja H, Eguchi S, Arkadopoulos N, Detry O et al. Bioartificial liver treatment in rats with fulminant hepatic failure: effect on function and regeneration of the native liver. J Surg Res 1999; 85: 243-250.
29. Ljubimova JY, Petrovic LM, Arkadopoulos N, Blanc P, Geller SA and Demetriou AA. Lack of hepatocyte growth factor receptor (c-met) expression in fulminant hepatic failure livers before transplantation. Dig Dis Sci 1997; 42: 1675-1680.

Chapter 32

The Berlin extracorporeal liver support system

Benedikt Busse and Jörg Gerlach
Chirurgische Klinik, Virchow-Klinikum, Humboldt-Universität, Berlin, Germany

Key words: acute liver failure, hollow-fiber bioreactor, hybrid liver support, liver cell coculture, orthotopic liver transplantation, porcine hepatocytes

32.1 Introduction

Developments in medical biotechnology and engineering of liver cells and tissues have made attractive the concept of temporary extracorporeal hybrid liver support for clinical use. In contrast to implantable hepatocyte systems [1] for therapy of chronic diseases, this chapter focuses on liver support systems for temporary extracorporeal therapy [2] in acute liver failure (ALF).

An introduction to this topic is given in Chapter 30. Here we describe the bioreactor for the Berlin Extracorporeal Liver Support System and relate this development to the state of the art. The latter is not easy, since publications on biotechnology-based therapies for ALF use different terms for the description of systems. We believe that the use of terms in the field should distinguish between systems containing additional non-biological components as a therapeutic tool (*e.g.* charcoal columns, resin adsorbers) and systems which depend on biological functions.

Demetriou *et al.* introduced the term "bio-artificial liver" to describe the combination of liver cells in a system with an additional artificial detoxification component [3]. The term "cartridge" was used for the apparatus, containing hepatocytes, cultivated on microcarriers. The term "hybrid system" has been applied by Chang *et al.* in 1988 for therapeutic devices utilising biological tissue [4]. Pappas in the same year used the term "hybrid liver" to describe liver cells cultivated in a bioreactor, and utilized in extracorporeal perfusion systems [5]. Since tissue engineering aims to enable systems for the support of liver functions including detoxification, we are using the term "bio-artificial" for combined systems and the term "hybrid" for biological systems. The term "cartridge" is employed for devices utilising liver cells in conventional culture techniques and the term "bioreactor" is used for systems utilising liver cells engineered to liver tissue structures.

M.N. Berry and A.M. Edwards (eds.), The Hepatocyte Review, 561–573.

For the development of a liver support system (LSS), two complementary basic approaches have to be considered. On the one hand, a high level of biochemical and physiological activity is thought to be necessary to counter the deteriorating course of patients in ALF with progressive hepatic coma. Use of primary liver cells makes necessary substantial effort to stabilize and maintain cellular functions. The challenge for bio-engineering and tissue engineering is to mimic physiological conditions and optimize the cellular microenvironment.

On the other hand, the feasibility, scale-up and clinical logistics of the device have to be considered. Once a patient is evaluated for treatment, a long time of preparation is not possible due to the dramatic progression of the disease and complications.

Both considerations have in common the fact that our knowledge of the mechanisms of the disease and associated complications (namely progressive encephalopathy, hyperdynamic cardiovascular insufficiency, renal failure and increased risk of infection) is limited. The success of orthotopic liver transplantation proves that biologic treatments are effective whereas no benefits for survival have been shown for otherwise highly effective artificial detoxification methods. It is the hope and challenge of hybrid liver support to fill the gap between these two therapeutic concepts.

One of the most important open questions relates to the magnitude of the appropriate cell mass. Whilst clinical and experimental studies of partial hepatectomy have shown the possibility of survival after reduction of the liver mass to one fifth, the necessary cell mass for a LSS is not known. Extrapolation of the hepatectomy data is too optimistic since the patient with ALF presents already with dramatic disturbances in metabolic state compared to the compensated state of a patient after partial hepatectomy, performed to remove liver metastasis. Furthermore, the reduction in mass transfer of a LSS, compared to the native liver, should be mitigated by an efficient and biochemically active cell mass. Here we will describe the compromise between the demands of bio-engineering and cell physiology as well as clinical and logistic requirements worked out during our own developments.

32.2 Considerations for the development of the culture model

Conventional culture techniques lead to a remarkable alteration in the environmental conditions of liver cells in comparison to their *in vivo* situation. Electron microscopy investigations have demonstrated that major differences in the ultrastructure of hepatocytes can be observed when cultured in monolayers. Important findings are the endocytosis of the junctional complexes with secondary effects on the orientation of the cytoskeleton and cell organelles [6]. If a culture model could support high cell densities [7] as well as a free three-dimensional cell re-arrangement, a more physiological microenvironment of hepatocytes could be established. Cell-to-cell aggregation [8] would enable the reconstruction of junctional complexes, as well as re-orientation of the cytoskeleton and cell organelles *in vitro* [6]. Furthermore a liver tissue-like formation of parenchymal and non-parenchymal cells is thought to support cellular orientation and differentiated function. A culture technique in which the different cell populations can reorganize themselves into three-dimensional structures may result in a hepatocyte microenvironment more akin to the physiological situation along the liver sinusoids.

Most conventional culture models present important differences in mass exchange features, compared to the liver. They rely on diffusion for substrate exchange, while hepatocytes *in vivo* operate under continuous perfusion conditions. Moreover, the natural zonation within a liver lobule results in gradients of oxygen, carbon dioxide and metabolites and different metabolic functions that can be identified along the liver sinusoids. In conventional culture models, gradients and vectors for metabolite transfer are often greatly distorted when compared to the *in vivo* situation. A culture technique which allows the cells to operate under more physiological conditions, *e.g.* convection with low metabolite gradients, may result in a cell micro-environment more closely related to the situation in the liver sinusoids.

32.3 Background to our development

To enable the use of primary hepatocytes for a LSS with a high cell mass (approximately half or one-third of the liver mass of an adult human), techniques for large scale cell isolation had to be developed [9-12]. A principal factor in the maintenance of differentiated hepatocyte function *in vitro* is liver cell immobilisation [13,14]. In addition, the immobilisation technique appears as a problematical task if the cultivation of several hundred grams of liver cells is intended. Furthermore a bioreactor module must allow the introduction of a large cell mass into an extracorporeal circuit for therapeutic use. Monolayer cultivation of hepatocytes allows a cell density of 3×10^5 cells per cm^2. Scale-up of monolayer cultures for 500 g of cells would call for an adhesion surface of about 100 m^2. Surface area can be enlarged by use of microcarriers. By using microcarriers *i.e.* a cell density of 10^6 cells per ml bioreactor volume is possible (dependent on the microcarrier volume). In contrast by high density immobilisation between capillary membranes, more than 10^7 cells can be used per ml bioreactor volume. For a cell mass of 10^{10} hepatocytes, considered as a minimum for clinical application, this results in a construction of a 10 l reactor for microcarrier culture and a 1 l reactor for high cell density immobilisation. Cell-to-cell aggregate formation [15] provides a technique of cell immobilisation that allows reduction of artificial substrates and total volume of the culture model. From the clinical point of view it is important to decrease the volume of the culture model to reduce the patient's extracorporeal plasma compartment. But the spontaneous aggregation of the cells induces new problems. In particular, appropriate nutrition and oxygenation of larger cellular clusters has to be guaranteed. The deconstruction of the vascular system of the liver during enzymatic dissociation calls for alternative solutions to enable sufficient mass exchange and oxygenation.

Artificial flat sheet or hollow-fiber membranes may act as a substrate for immobilisation of cell aggregates [16]. In addition to cell immobilisation [17], the arrangement of membranes included in a bioreactor construction is able to define the size of the cellular aggregates. Relevant membrane properties are shape, chemical structure and permeability. Because of a number of theoretical considerations [18], we chose not to use conventional dialyser cartridges for extracorporeal perfusion. Instead we decided to construct a bioreactor adapted to the demands of liver cells. The bioreactor uses interwoven hollow fiber membranes to create a three dimensional framework over which self-assembling hepatocyte aggregates are distributed. The more complex design needs some explanation. The bioreactor may be considered as three separate parallel hollow-fiber bundles. Two sets of hydrophilic, high molecular weight cut-off (~300,000 Da) hollow fiber membranes are incorporated into the bioreactor. By closing one end of each set, plasma entering the reactor via one set must enter the extracapillary space before leaving the reactor via the second set of membranes. In other words, 100% of the plasma flow is transferred by convection across two sets of membranes. The third set of hollow fibers is composed of hydrophobic membranes which are used for supply of oxygen and removal of surplus CO_2. A proprietary technique for combining the three sets of hollow fiber membranes results in a complex extracapillary framework. The periphery of the bioreactor is sealed by polyurethane potting. This produces a central region in which the hepatocytes are distributed upon the membrane framework. The extracapillary space is,

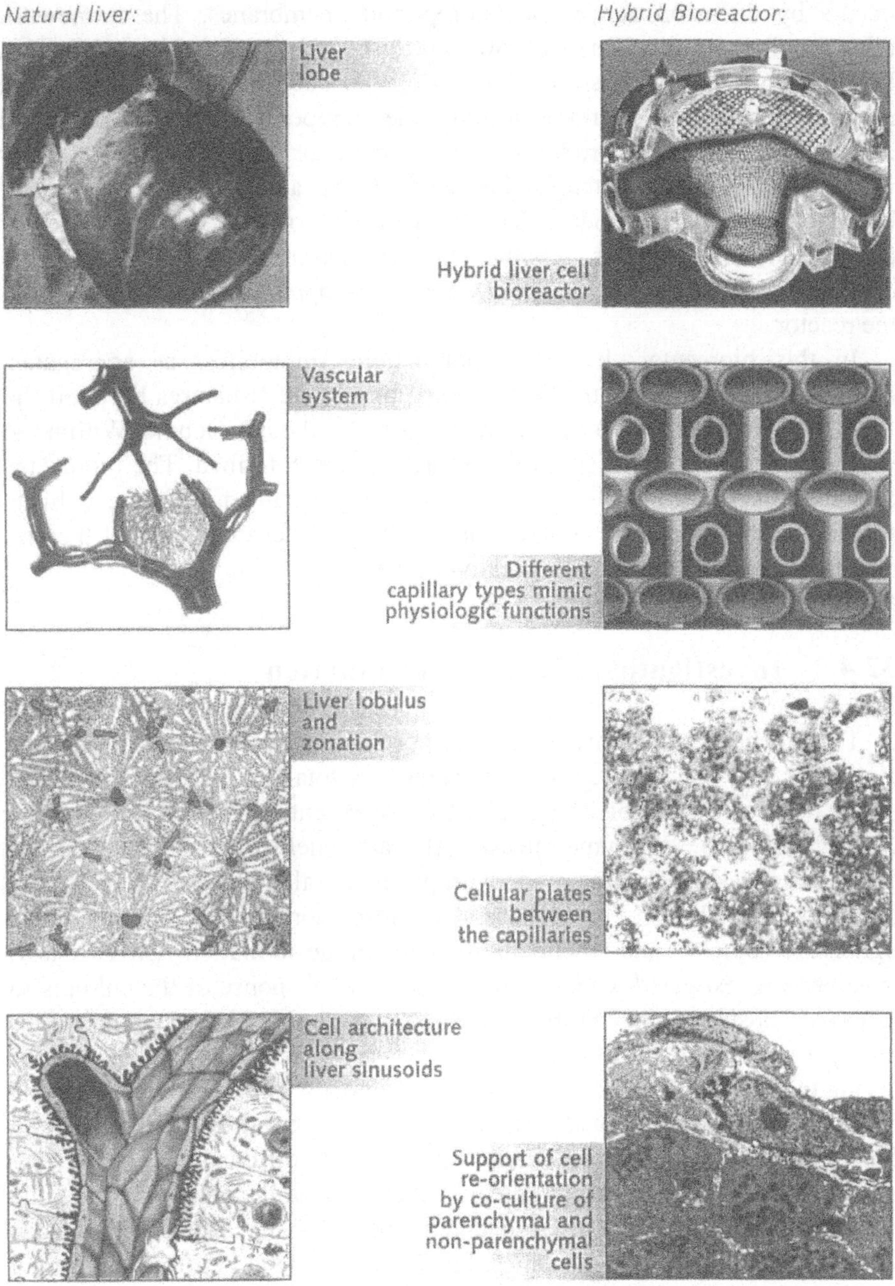

Figure 1. From the liver *in vivo* to a hybrid reactor *in vitro*: A bioreactor for the maintenance of differentiated hepatic functions in hybrid systems

therefore, perfused with plasma (or culture medium) whilst gas exchange occurs by diffusion across the hydrophobic membranes. The framework allows a close localisation of inflow-, outflow- and oxygenation-membranes. Due to the hollow fiber size, distances of 400-800 μm are produced in the extracapillary space. Therefore a multitude of repetitive units for perfusion and oxygenation are produced. The bioreactor design allows further independent capillary systems to be added. Hence, additional functions could be integrated into the module. The incorporation of dialysis membranes or the use of adsorbents is feasible. The utilisation of water-impermeable capillaries as a heat exchanger, might permit cryopreservation of cells within the reactor.

In this bioreactor, liver cell populations immobilize as aggregates, produce their own biomatrix and re-form tissue-like structures between the capillaries. The smallest repetitive unit resembles the liver lobule. Within the cellular aggregates, liver sinusoidal structures are reformed. The bioreactor design and cellular morphologic behaviour are explained in Figure 1. In its actual size the bioreactor contains up to 600 g of hepatocytes, and thereby maintains the cell mass of a human liver lobe.

32.4 Investigation of bioreactor function

The biochemical activity of primary porcine liver cell cultures within the bioreactor was followed up to eight weeks. A total of 25 cultures with an initial liver cell mass of 210 g to 680 g (representing $1.2\text{-}4.0 \times 10^{10}$ cells) were evaluated. Enzyme-release (lactate dehydrogenase, aspartate aminotransferase, glutamate dehydrogenase), albumin secretion, urea formation, metabolism of glucose, galactose, sorbitol and amino acids, lignocaine uptake and monoethylglycinxylidide-formation (MEGX-test) have been investigated. Our studies included the response of the cultures to different substrate concentrations and rates of supply. The data showed a highly reproducible time course, with an initial peak of enzyme release during the first 48-72 h, accompanied by a temporary disturbance in nitrogen metabolism. This initial phase was followed by a period of stable metabolic function for up to three weeks. After three weeks, differentiated hepatocyte functions started to decrease gradually. At the end of the culture periods, the bioreactor housings were opened and biopsies of the cellular aggregates were taken for morphologic studies. Freshly isolated pig hepatocytes formed a tissue-like structure in the culture model. Scanning electron microscopy shows that hepatocytes spontaneously formed aggregates in a three-dimensional structure between and on the surface of artificial capillaries. Transmission electron microscopy revealed that even after seven weeks of *in vitro* perfusion the cell ultrastructure remained similar to that of the hepatocyte *in vivo*. Golgi complexes, active membrane processes, reorganisation of cell junctions and bile canaliculi-like intercellular spaces were demonstrated. Cell organelles which depend on an intact cytoskeleton, such as the endoplasmic reticulum and the Golgi apparatus, were well maintained.

The results of the studies demonstrate that primary hepatocyte cultures can be maintained over a period of several weeks *in vitro* and that they reorganize themselves to give tissue-like structures within the artificial system. The cells spontaneously form aggregates on, and between, the capillaries and reorganize into a three-dimensional array. Rebuilt cell junctions and the reconstitution of bile canaliculi-like structures were demonstrated in all reactors under investigation.

32.5 Development of perfusion system for extracorporeal hybrid liver support

To allow a therapeutic use of the liver cell cultures for temporary extracorporeal hybrid liver support, a perfusion system with respect to clinical standards had to be developed. Consideration of technical problems due to direct hemoperfusion of the bioreactor resulted in the introduction of a continuous plasma separation unit between the patient and the bioreactor circuit. As a consequence of plasma separation before bioreactor perfusion, the following beneficial features can be achieved:

1. A uniform perfusion of the cell cultures is obtained.
2. Platelet aggregation within the bioreactor can be avoided leading to a prolongation of the life cycle of the device.
3. Direct cellular immunological interactions between the patients specific and non-specific immunocompetent cells and the heterologous cell cultures can be avoided.
4. The running of an extracorporeal blood circuit and a parallel bioreactor circuit allows an easy connection and disconnection of the two circuits without producing stasis in either circuit (reduction of technical complications).

Animal experiments were performed in order to develop a system for plasma perfusion [19,20] and to enable [21-23] the first clinical applications of the bioreactor. An overview of the device is given in Figure 2. Up to August 1998, the Berlin Extracorporeal Liver System [24] had been used in eight patients listed for urgent liver transplantation. All applications (7-46 h) were successful in bridging to transplantation and all patients survived.

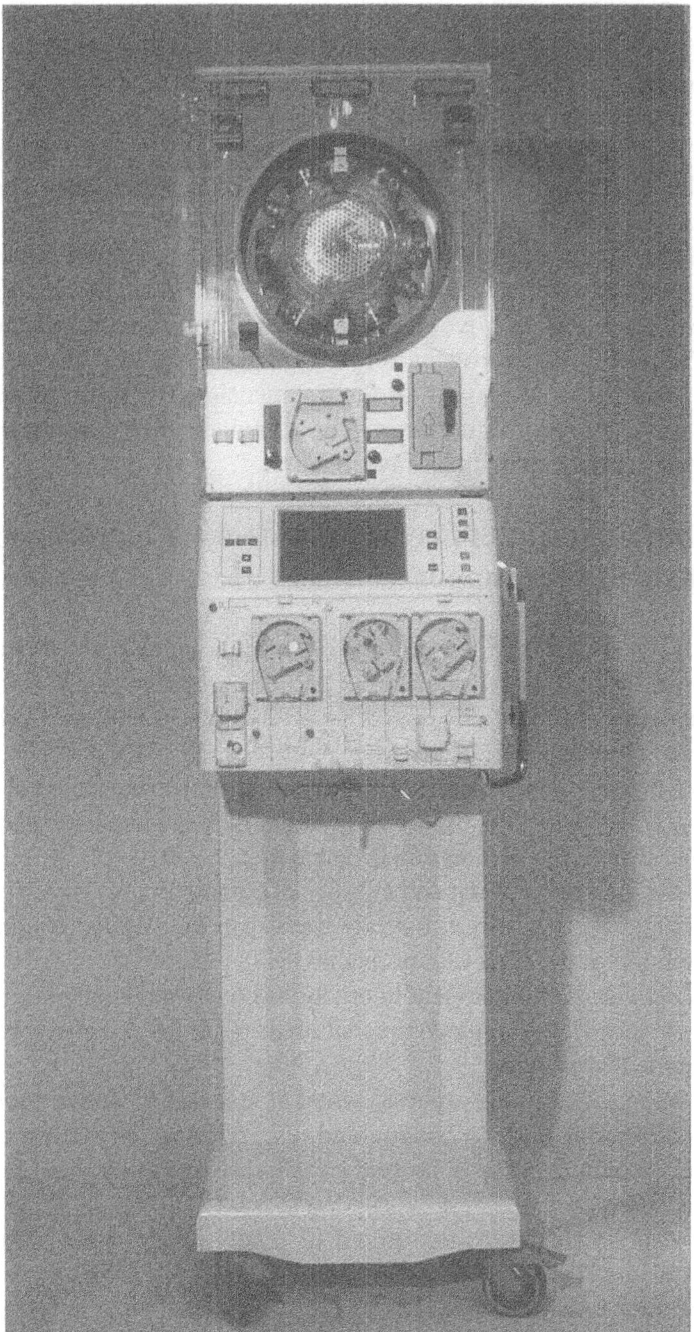

Figure 2. (above) The Berlin extracorporeal liver support system. (top right) Bioreactor perfusion unit. (bottom right) Plasma separation unit.

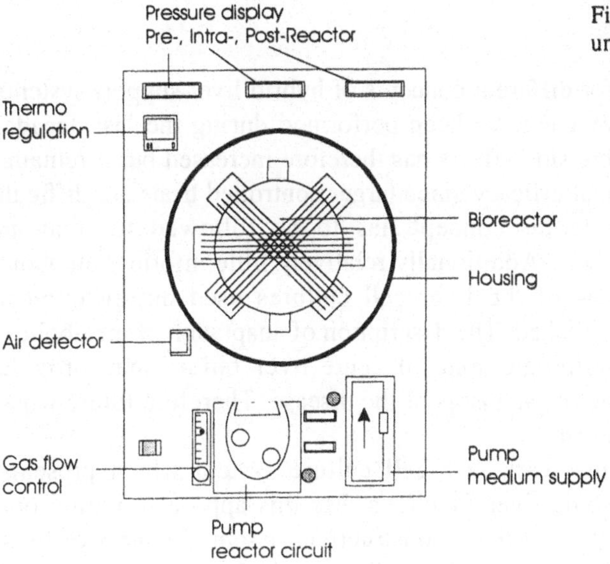

Figure 2b: Bioreactor perfusion unit (add-on module).

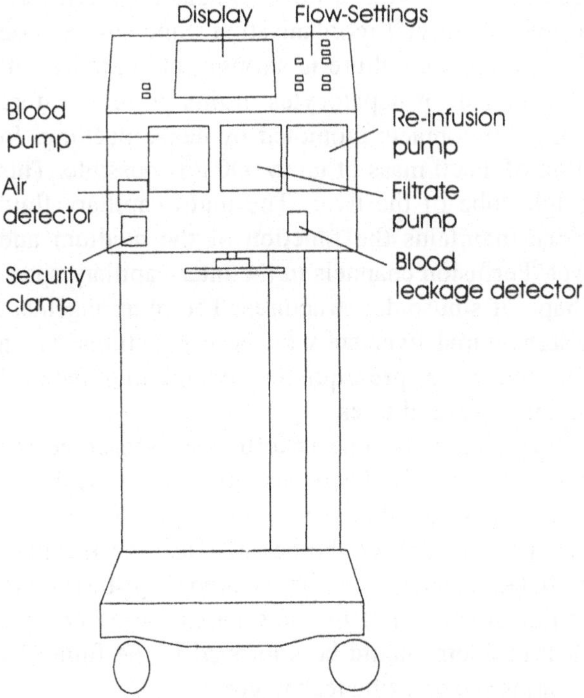

Figure 2c: Plasma separation unit.

32.6 Discussion

Initial clinical trials for different concepts of hybrid liver support systems and bioartificial liver devices have been performed during the last decade. The knowledge of possible side effects has therefore increased but it remains difficult to evaluate clinical efficacy since larger controlled trials are difficult to design and a multi-center concept has to be followed to treat an appropriate range of patients. Additionally relatively little information about the biochemical functions of the liver cell cultures used during clinical applications is actually published. The description of adaptation of metabolic activity in the pathophysiologic state of acute liver failure may provide information on pathogenic mechanisms of the disease. Therefore future work in this field is of high interest.

A prerequisite for this work is a cell culture system which presents features close to physiologic liver functions. For this approach, within our own model, it was thought that the reconstruction of some features of liver architecture within the reactors should contribute to the long term stability of hepatocytes. The woven capillaries provide a three-dimensional artificial framework as a macroenvironment for cell reorganisation. Since expression of differentiated cell functions in tissue culture is strongly affected by cell density [7,25], the microenvironment of hepatocytes in this present culture model may partly result from cell contacts promoted by dense packing. In the actual model, a cultivation of a cell mass of up to 600 g is possible. This could be compared to a single lobe of the liver. The multi-capillary fluid distribution within the device maintains the function of the medium and small size vessels of the liver. Perfusion channels in the inter-capillary space reach the dimension and shape of sinusoidal structures. The re-arrangement of parenchymal and non-parenchymal liver cells in these structures was a finding which was interpreted as a prerequisite for maintenance of biochemical activity for a period of several weeks.

The morphology of the surviving cells appears to be well preserved but non-surviving cells could also be detected. This phenomenon could be due to a less than optimal bioreactor design: the diameter of the capillaries used is relatively large in comparison to the size of the hepatocytes and the liver sinusoids. Therefore, the distances between the packed capillaries are (depending on the localisation of the cells) in some cases increased to a factor of 2-5. These technical problems could be improved in the future by using specially adapted capillaries and weaving techniques.

Initial clinical experience in experimental transplantation surgery [26] provides information that the cell mass of a liver lobe should be sufficient to support patients in acute liver failure. The developed bioreactor in its actual size incorporates 300-600 g of cells, which can be compared to that of the liver lobe. However, it is unlikely that this quantity of *in vitro* culture enables qualitative and quantitative organ functions equivalent to that of the same cell mass in a liver lobe. Furthermore the mass exchange between the patient and the device is not comparable to the physiologic blood supply of the liver. Clinical investigations have to show whether a sufficient clinical function can be achieved simply by adjusting the cell mass, or whether a radical redevelopment of the liver support devices is necessary.

As seen in other Chapters in this book, the biological approach to liver support with primary cells has its general technical limitations. The foremost is the gradual loss of cell metabolism with time in culture. Much work appears to be required to enable a half-life of liver cells *in vitro* comparable to that of liver cells *in vivo*. Half-life of biochemical activity of the cells influences the logistics of hybrid liver support. To avoid further critical damage of the cells, a cryopreservation step to facilitate distribution and handling of the devices was excluded from our protocol concept. The cross-over from the organ to culture and from in-vitro culture to therapeutic use are critical points for metabolic adaptation. Another problem which may influence the logistics of clinical application of primary hepatocytes is their need for time for recovery after the trauma of cell isolation. This may also be relevant after the trauma of thawing and reoxygenation in case of utilising frozen cells. An initial metabolic disorder following cell isolation was observed [27] which normalizes after the reorganisation of the cells to tissue-like structures in the bioreactors. During this process of 24-48 h, cell function is impaired. Considering the time required after cell isolation and the half-life of cell function *in vitro*, a time window for possible clinical applications of bioreactors may be defined.

A further finding of our investigations was that detection of viability of the initial cell suspension, as measured by Trypan blue exclusion, does not allow an accurate prediction of biochemical activity. Therefore, a critical evaluation of every single cell culture before its clinical use is indicated. These considerations suggest that stand-by cultivation of liver cells is required, to enable on-demand availability of biochemically active devices and to allow screening of the pre-treatment biochemical activity. From the clinical and practical point of view, the concept of centralized cell culture laboratories and distribution and transport of the devices (and logistical support) to the patient in different clinics has to be evaluated. The vulnerability of function of primary liver cell cultures demands sophisticated physiologic support during all periods of device handling. Whether the higher demands of logistics will be accepted depends on the evaluation of possible clinical benefits. The problems due to the limited life span of primary liver cells and the control of possible transmission of infectious diseases or immunological risks bring permanent liver cells of human origin into the center of interest. However, the enormous research efforts required in the past to prolong the maintenance of differentiated cellular functions of primary liver cells, suggest that a lot more work remains to be done. Unsolved problems of long-term in-vitro hepatocyte cultivation will also effect the work on natural or artificially immortalized cell lines, as well as the developments in adult or embryonic hepatic stem cell cultivation.

References

1. Chang TM. Applications of artificial cells in medicine and biotechnology. Biomater Artif Cells Artif Organs 1987; 15: 1-20.
2. Gerlach J, Ziemer R and Neuhaus P. Fulminant liver failure: relevance of extracorporeal hybrid liver support systems. Int J Artif Organs 1996; 19: 7-13.
3. Demetriou AA, Whiting J, Levenson SM et al. New method of hepatocyte transplantation and extracorporeal liver support. Ann Surg 1986; 204: 259-271.
4. Chang TM. The role of biotechnology in bioartificial or hybrid artificial cells and organs. Biomater Artif Cells Artif Organs 1988; 16: v-vi
5. Pappas SC. Fulminant hepatic failure and the need for artificial liver support. Mayo Clin Proc 1988; 63: 198-200.
6. Berry MN, Edwards AM and Barritt GJ. Isolated hepatocytes preparation, properties and applications. In: Burdon RH and van Knippenberg PH, Eds. Laboratory techniques in biochemistry and molecular biology. Amsterdam: Elsevier, 1991: 1-460.
7. Clayton DF, Harrelson AL and Darnell-JE. Dependence of liver-specific transcription on tissue organization. Mol Cell Biol 1985; 5: 2623-2632.
8. Koide N, Sakaguchi K, Koide Y et al. Formation of multicellular spheroids composed of adult rat hepatocytes in dishes with positively charged surfaces and under other nonadherent environments. Exp Cell Res 1990; 186: 227-235.
9. Gerlach JC, Brombacher J, Courtney JM and Neuhaus P. Nonenzymatic versus enzymatic hepatocyte isolation from pig livers for larger scale investigations of liver cell perfusion systems. Int J Artif Organs 1993; 16: 677-681.
10. Gerlach J, Kloppel K, Schon MR et al. Comparison of pig hepatocyte isolation using intraoperative perfusion without warm ischemia and isolation of cells from abattoir organs after warm ischemia. Artif Organs 1993; 17: 950-953.
11. Gerlach JC, Brombacher J, Kloppel K, Schnoy N and Neuhaus P. Comparison of four methods for mass hepatocyte isolation from pig and human livers. Transplantation 1994; 57: 1318-1322.
12. Gerlach J, Brombacher J, Smith M and Neuhaus P. High yield hepatocyte isolation from pig livers for investigation of hybrid liver support systems: influence of collagenase concentration and body weight. J Surg Res 1996; 62: 85-89.
13. Gerlach J, Kloppel K, Schauwecker HH, Tauber R, Muller C and Bucherl ES. Use of hepatocytes in adhesion and suspension cultures for liver support bioreactors. Int J Artif Organs 1989; 12: 788-792.
14. Gerlach J, Vienken J, Walker P and Affeld K. Computer aided time-lapse video analysis of hepatocyte morphology during adhesion to cellulose membranes. Int J Artif Organs 1990; 13: 365-369.

15. Gerlach JC, Kloppel K, Muller C, Schnoy N, Smith MD and Neuhaus P. Hepatocyte aggregate culture technique for bioreactors in hybrid liver support systems. Int J Artif Organs 1993; 16: 843-846.
16. Gerlach J, Stoll P, Schnoy N and Bucherl ES. Membranes as substrates for hepatocyte adhesion in liver support bioreactors. Int J Artif Organs 1990; 13: 436-441.
17. Gerlach JC, Schnoy N, Vienken J, Smith M and Neuhaus P. Comparison of hollow fibre membranes for hepatocyte immobilisation in bioreactors. Int J Artif Organs 1996; 19: 610-616.
18. Gerlach JC, Encke J, Hole O, Muller C, Ryan CJ and Neuhaus P. Bioreactor for a larger scale hepatocyte *in vitro* perfusion. Transplantation 1994; 58: 984-988.
19. Gerlach J, Trost T, Ryan CJ et al. Hybrid liver support system in a short term application on hepatectomized pigs. Int J Artif Organs 1994; 17: 549-553.
20. Gerlach J, Jorres A, Trost O et al. Side effects of hybrid liver support therapy: TNF-α liberation in pigs, associated with extracorporeal bioreactors. Int J Artif Organs 1993; 16: 604-608.
21. Janke J, Gerlach J, Kardassis D, Bohmer C and Rossaint R. Effect of a hybrid liver support system on cardiopulmonary function in healthy pigs. Int J Artif Organs 1997; 20: 570-575.
22. Gleissner M, Bornemann R, Stemerowicz R, Meissler M, Neuhaus P and Gerlach JC. Immunoisolation of hybrid liver support systems by semipermeable membranes. Int J Artif Organs 1997; 20: 644-649.
23. Bornemann R, Smith MD and Gerlach JC. Consideration of potential immunological problems in the application of xenogenic hybrid liver support. Int J Artif Organs 1996; 19: 655-663.
24. Gerlach JC. Development of a hybrid liver support system: a review. Int J Artif Organs 1996; 19: 645-654.
25. Yuasa C, Tomita Y, Shono M, Ishimura K and Ichihara A. Importance of cell aggregation for expression of liver functions and regeneration demonstrated with primary cultured hepatocytes. J Cell Physiol 1993; 156: 522-530.
26. Gubernatis G, Pichlmayr R, Kemnitz J and Gratz K. Auxiliary partial orthotopic liver transplantation (APOLT) for fulminant hepatic failure: first successful case report. World J Surg 1991; 15: 660-665.
27. Gerlach JC, Fuchs M, Smith MD et al. Is a clinical application of hybrid liver support systems limited by an initial disorder in cellular amino acid and α-keto acid metabolism, rather than by later gradual loss of primary hepatocyte function? Transplantation 1996; 62: 224-228.

Chapter 33

The bioartificial liver of the Academic Medical Center at Amsterdam

Robert A.F.M. Chamuleau and Leonard M. Flendrig
Laboratory of Experimental Internal Medicine, Academic Medical Center, University of Amsterdam, The Netherlands

Key words: acute liver failure, bioartificial liver, bioreactor, porcine hepatocytes

33.1 Introduction

There is an urgent need for an effective liver support system to bridge patients with hepatic failure to liver transplantation or to a stage where their own liver is able to regenerate. In the past most attempts to reach this goal were focused on blood detoxification, based on the assumption that liver failure could be reversed if the associated toxins were removed from the circulation of the patient. Although improvement of the neurological status of patients has been reported, none achieved long-term survival [1]. It was therefore concluded that an effective liver support system should be able to perform the liver's multiple synthetic and metabolic functions, including detoxification and excretion. The most logical approach to this problem is the introduction of active functioning hepatocytes. At present, the bioartificial liver is thought to hold the greatest promise for successful temporary liver support. The heart of such an extracorporeal system consists of a bioreactor: a cell culture device which comprises well nourished and oxygenated, viable, hepatocytes immobilized on a mechanical support and separated from the patient's blood circulation by semi-permeable membranes. This paper reports on the principle and the *in vitro* and *in vivo* results of a highly original new design of bioartificial liver support developed at the Academic Medical Centre of the University of Amsterdam, the so called Academic Medical Centre Bioartificial Liver (AMC-BAL).

M.N. Berry and A.M. Edwards (eds.), The Hepatocyte Review, 575–583.

33.2 Features of the Academic Medical Center bioartificial liver

33.2.1 Design

Our goal was to devise a cell culture system which allows individual perfusion of high density cultured, well oxygenated hepatocytes resulting in low diffusional gradients (as in the intact liver), which are essential for optimal mass transfer and cell function. Furthermore, we considered that the device needed to be based on cell aggregates, which are known to maintain many of the cyto-architectural characteristics found *in vivo* and exhibit higher and more prolonged functional activity than hepatocytes cultured in monolayers. For these reasons our bioreactor consists of a spirally wound non-woven polyester matrix, *i.e.* a sheet-shaped three-dimensional framework for hepatocyte immobilisation and spontaneous self-aggregation, and of integrated hydrophobic hollow-fiber membranes for decentralized oxygen supply and CO_2 removal. A schematic drawing of a transverse and longitudinal cross-section of the AMC-BAL is shown in Figure 1.

Microscopic examinations revealed that hepatocytes cultured in this polyester fabric at a density of 20-40 $\times$ 10^6 cells/ml, formed tiny irregularly shaped aggregates (3 to 20 cells) with space in between the aggregates (Figure 1, lower left). In this way, medium (or plasma *in vivo*) can be perfused along the aggregates and make contact with almost every hepatocyte, as in the intact liver.

33.2.2 *In vitro* results

The *in vitro* results have been described in detail in 1997 by Flendrig *et al* [2]. In summary: urea synthesising capacity of hepatocytes in the bioreactor was twice that of hepatocytes in monolayer cultures, indicating the efficacy of this approach. Other functions like protein synthesis, galactose elimination, metabolism of amino acids, and lactate/pyruvate ratios were well maintained over the investigated period of 3 days. In addition, monitoring of lidocaine elimination up to 14 days was maintained at levels at least 70% of those at the first day. Transmission electron micrographs showed viable, well differentiated cells (bile canaliculi, tight junctions, mitochondria, and desmosomes) after 4 days of culturing. Flow sensitive Magnetic Resonance Imaging revealed that the bioreactor construction ensured medium flow through all parts of the device irrespective of its size.

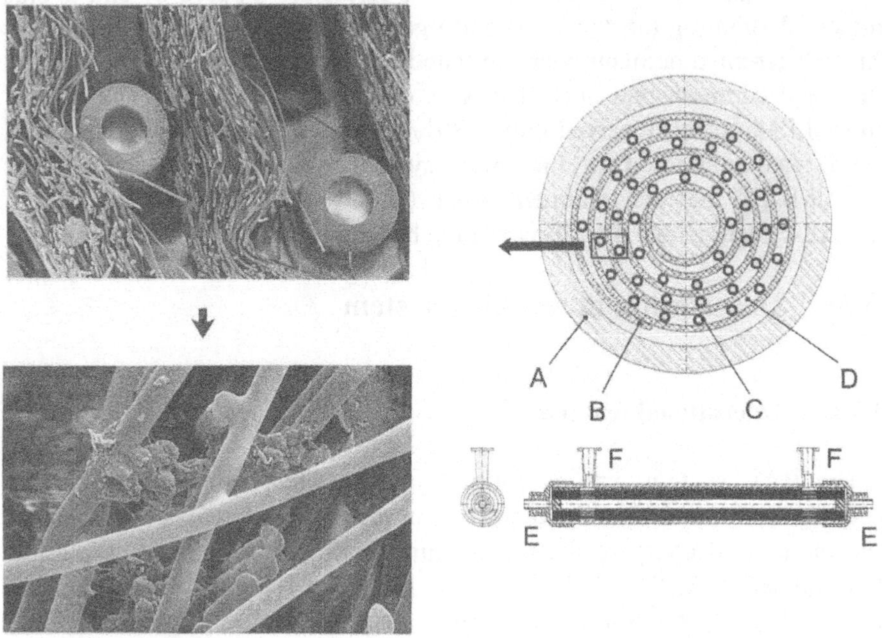

Figure 1. The University of Amsterdam bioreactor. Shown are schematic drawings of a transverse and longitudinal cross-section of the bioreactor (right), a scanning electron micrograph of a small section of the cell culture compartment (upper left), and a scanning electron micrograph of hepatocyte aggregates attached to a polyester scaffold (lower left, diameter polyester fibers: 13 μm). The system is composed of a polysulfon dialysis housing (A) comprising a three-dimensional nonwoven polyester fabric (B) for high density hepatocyte culture as small aggregates, and hydrophobic polypropylene hollow-fiber membranes (C, external diameter: 630 μm) for oxygen supply and CO_2 removal. The combination of the polyester matrix and the oxygenation tubing creates a third compartment (D). These channels are used to perfuse the plasma of the recipient through the bioreactor which can make direct contact with the hepatocytes in the polyester fabric. Plasma is perfused through the bioreactor via the side ports (F). The integrated oxygenator of the bioreactor is connected to the gas supply (5% CO_2 in air) via the endcaps (E). The homogenous distribution of the oxygenation hollow-fibers throughout the bioreactor compartment ensures every hepatocyte has an oxygen source within its direct surroundings.

It was concluded that the AMC-BAL showed encouraging efficiency. The device is easy to manufacture and easily scaled-up to the liver mass required for possible short-term support of patients in hepatic failure.

Protocol for testing device *in vivo*

The next step was to evaluate the efficacy and safety of the device, incorporating porcine hepatocytes, in rats with irreversible acute liver failure. Complete liver ischemia (LIS) was surgically induced by end-to-side portacaval shunting (at day T -3) and ligation of the hepatic artery (day 0). Three different treatments were initiated 30 min after ligating the hepatic artery and common bile duct, and were continued until death. The control group (LIS-Control) received only a 20% glucose infusion, a second control group was connected to the BAL system without hepatocytes (LIS No-Cell-BAL) and the treatment group (LIS Cell-BAL) was connected to the BAL-system with 4.4×10^8 viable pig hepatocytes.

33.3 Description of working system

33.3.1 Operational features

The BAL system was as follows: it consisted of a plasma circuit including the bioreactor and a blood circuit. Blood was pumped (1.5 ml/min) from the carotid artery of the rat to a mini hollow-fiber membrane plasma filter (Fresenius SSP1003-P, Germany, number of fibers: 80, membrane surface 90 cm^2, membrane material: polypropylene [AKZO-NOBEL, Wuppertal, Germany], membrane pore size: ~0.5 μm, Fresenius, St. Wendel) for continuous separation of plasma from the blood. As plasma could not be withdrawn from the plasma filter at a higher rate than 0.3 ml/min without inducing hemolysis, a high flow plasma loop (5 ml/min) was introduced to reduce the formation of substrate and metabolite gradients inside the bioreactor. After recirculation through the bioreactor, the treated plasma was pumped through an outlet filter (Mediakap 5, membrane material: mixed cellulose acetate/cellulose nitrate, membrane pore size: ~0.2 μm, Microgon, Laguna Hills, CA) to remove possible cell debris. This plasma was then reunited with the blood cells from the plasma filter and given back to the rat via the jugular vein. The BAL-system had a total extracorporeal volume of 25 ml (blood circuit 5 ml, plasma circuit 20 ml). A computerized pressure monitoring system was included to monitor the mean arterial blood pressure of the rat, to check the condition of the two filters regarding membrane fouling, and to detect obstruction of the arterial and venous blood lines and catheters. Additional safety features involved controls for gas flow, temperature, and balanced plasma exchange.

33.3.2 *In vivo* results

BAL treatment significantly improved survival time in recipients with liver ischemia (Figure 2). In addition, the onset of hepatic encephalopathy was significantly delayed and the mean arterial blood pressure significantly improved (not shown). Significantly lower levels of ammonia (at T=8h: Controls 1800 ± 200 μM Cell-BAL 800 ± 80 μM, and No-Cell BAL 1300 ± 120 μM $p<0.05$) and lactate (Cell-BAL 1.5 mM, No-Cell BAL 6 mM) indicated metabolic activity of the porcine hepatocytes in the bioreactor. Furthermore, this study reports on the first attempt to monitor the viability of the hepatocytes in the bioreactor during xenogeneic BAL treatment. This was accomplished using a new species-specific enzyme immunoassay for the determination of porcine α-glutathione S-transferase (α-GST), a marker for hepatocellular damage. Low pig α-GST levels suggested that our bioreactor was capable of maintaining hepatocyte viability during treatment (14 h). For more details see Flendrig *et al* [3].

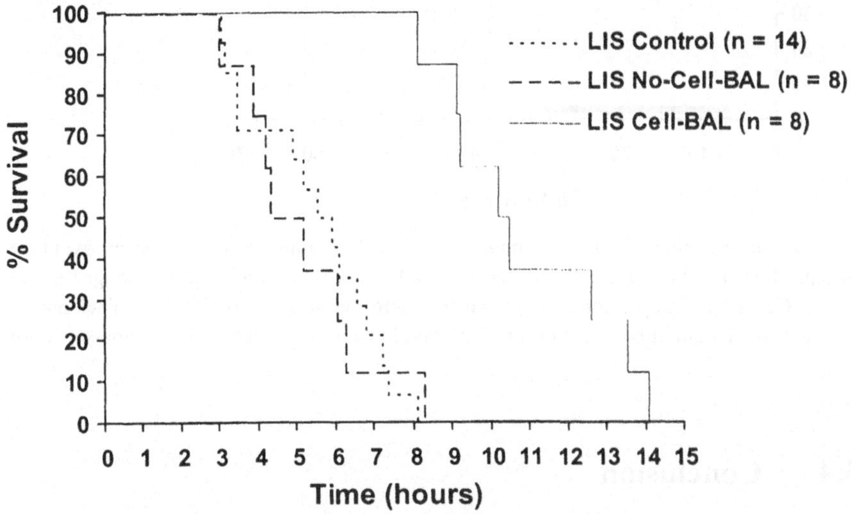

Figure 2. Percentage of rats surviving as a function of time after inducing liver ischemia. (t = 0 h). Kaplan-Meier Log-Rank test; $p < 0.001$, LIS Cell-BAL versus LIS Control and LIS No-Cell-BAL; $p = 0.8931$, LIS Control versus LIS No-Cell-BAL.

Finally we demonstrated that the scaled-up version of our bioartificial liver based on 14×10^9 viable hepatocytes was able to support 35 kg pigs with complete liver ischemia [4]. To our knowledge, this was the first time that BAL treatment significantly prolonged life in large animals (35 kg) with complete liver ischemia: (treated animals: 51.1 ± 3.4 h; controls: 33.1 ± 3.0 h; longest survivor 63 h, Figure 3). See also [6,7]. In addition, arterial blood ammonia (Cell BAL 600 μM, No-Cell BAL 1700 μM, $p<0.05$) and total bilirubin levels (Cell BAL 3 mg/dl and No-Cell BAL 5.2 mg/dl, $p<0.05$) decreased significantly, indicating metabolic activity of the porcine hepatocytes in the bioreactor. Microscopic examination of the hepatocytes in the bioreactor revealed that the viability was well maintained during the first 12 hours of treatment, but decreased to around 50% by the end of the therapy (not shown). Probably more hepatocytes are required to reduce the metabolic stress per hepatocyte.

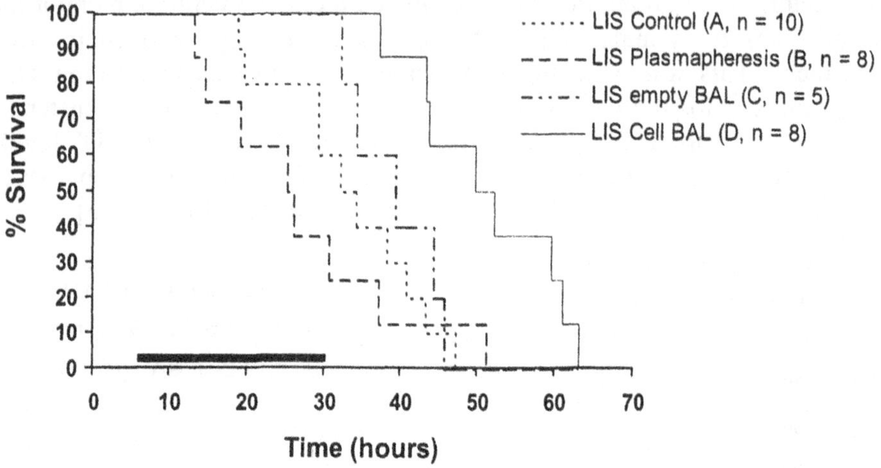

Figure 3. Survival time of pigs after induction of liver ischemia. Shown are the survival times of pigs after induction of liver ischemia (at t = 0 h). Life was significantly prolonged in the LIS Cell-BAL group compared to the three control groups ($p < 0.027$). The black bar represents the length of time that groups B,C, and D were connected to the plasma-separator.

33.4 Conclusion

The present data comprise a series of studies in which we evaluated the efficacy of our novel design bioreactor to support hepatocytes and prolong life in animals with acute liver failure. The efforts of recent years have resulted in a bioreactor design that can be produced by any dialysis Company in any size, and is capable of supporting hepatocytes at high density with better functional activity than hepatocytes in monolayer cultures. For the first time in bioartificial liver research, statistically significantly prolonged survival times have been obtained in animals (rat and pig) with complete liver ischemia; one of the most acute models of hepatic failure. We believe that our novel BAL has excellent qualities to support patients with acute liver failure. However, before a phase I and a multi-center randomized controlled clinical trial can be started some of the following questions have to be solved.

33.5 Future questions

Characteristic for all BAL-systems is the lack of a bile excretory system. We have shown [2] that porcine hepatocytes cultured in our bioreactor form bile canaliculi-like structures. This implies that conjugated bile-acids and bilirubin can be excreted into the circulation during in vivo application. This may result in high levels of bile-salts, which might limit the long term function of the hepatocytes in the bioreactor [7]. A possible solution could be the introduction of anion exchange resins and/or silica-based strong anion exchange cartridges for selective adsorption of bile acids and bilirubin [8].

The use of a xenogeneic hepatocyte source for the clinical application of a BAL may present cellular and humoral immunological problems. This could result in either destruction of the hepatocytes in the BAL or in hypersensitivity reactions in the patient receiving BAL treatment [9,10]. Proper immuno-isolation is therefore essential, which might be achieved by using semi-permeable membranes. The macroporous membranes in our *in vivo* studies [3,4] were chosen to allow the hepatocytes to function in nearly whole plasma contact, thereby closely resembling physiological liver cell perfusion. Cellular immunological complications were circumvented, but no humoral immunological protection was provided. The latter has the following implications for the clinical application of the BAL. Firstly, patients should be screened for human IgM against porcine hepatocytes to avoid an early malfunction of the bioreactor during BAL treatment [11]. Secondly, our group has demonstrated that rats infused with supernatant of cultured pig hepatocytes formed antibodies against pig hepatocyte products [9]. As a result, immunological complications may arise in the recipient if the BAL treatment is repeated after one week or more. A possible humoral reaction could be serum sickness a type III hypersensitivity reaction. For these reasons we conclude that bioartificial liver support based on xenogeneic hepatocytes in humans should not be used longer than 6-7 days.

In March 1997, Patience *et al.* [12] reported that cell free (0.45 μm filtered) C-type virus particles produced by pig kidney cell lines (PK-15 and MPK), were able to infect human kidney 293 cell lines *in vitro*. Co-cultivation of the PK-15 cells with human cells resulted in a broader range of human cell infection, including lung and muscle cells. At present it is not clear what the impact of this findings will be for xenogeneic BAL treatment. However, the first studies on possible transfection of porcine endogenous retrovirus into human cells are reassuring [13,14].

Final remark: a much safer approach could be the use of immortalized but non-tumorigenic human hepatocytes having the same phenotype and function as primary cells. Recently, attempts have been made to obtain such immortalized human liver cells [15]. These hepatocytes, cultured in our bioreactors, are currently being investigated to support pigs with acute liver failure.

References

1. Sussman NL, Gislason GT and Kelly JH. Extracorporeal liver support. Application to fulminant hepatic failure. J Clin Gastroenterol 1994; 18: 320-324.
2. Flendrig LM, La Soe JW, Jörning GG, Steenbeek A, Karlsen OT, Bovée WMMJ, Ladiges NCJJ, te Velde AA and Chamuleau RAFM. *In vitro* evaluation of a novel bioreactor based on an integral oxygenator and a spirally wound non-woven polyester matrix for hepatocyte culture as small aggregates. J Hepatol 1997; 26: 1379-1392.
3. Flendrig LM, Chamuleau RAFM, Maas MAW, Daalhuisen J, Hasset B, Kilty CG, Doyle S, Ladiges NCJJ, Jörning GGA, la Soe JW, Sommeijer D and te Velde AA. Evaluation of a novel design bioartificial liver in rats with complete liver ischemia: treatment efficacy and species-specific α-GST detection as a first attempt to monitor hepatocyte viability. J Hepatol 1999; 30: 311-320.
4. Flendrig LM, Di Florio E, Mancini A, Mezza E, Ceriello A, Mazzone M et al. *In vivo* evaluation of a novel bioartificial liver in a surgical model of acute liver failure in the pig. Preliminary results. In: Crepaldi G, Demetriou AA and Muraca M. Bioartificial liver support systems. The critical issues. Rome: CIC Edizioni Internationali, 1997: 42-47.
5. Rozga J, Holzman MD, Ro M-S, Griffin DW, Neuzil DF, Giorgio T, Moscioni AD and Demetriou AA. Development of a hybrid bioartificial liver. Ann Surg 1993; 217: 502-511.
6. Sheil AGR, Sun J, Mears DC, Waring M, Woodman K, Johnston B, Horvat M. Watson J, Kuotalistras N and Wang L. Positive biochemical effects of a bioartificial liver supprt system (BALSS) in a porcine fulminant hepatic failure (FHF) model. Int J Artif Organs 1998; 21: 43-48.
7. Pazzi P, Morsiani E, Muraca M, Vilei MT, Rozga J and Demetriou AA. Metabolic characterization of bioartificial liver: focus on bile production and excretion. In: Crepaldi G, Demetriou AA and Muraca M. Bioartificial liver support systems. The critical issues. Rome: CIC Edizioni Internationali, 1997: 42-47.
8. Morimoto T, Matsushima M, Sowa N, Ide K and Sawanishi K. Plasma adsorption using bilirubin adsorbent materials as a treatment for patients with hepatic failure. Artif Organs 1989; 13: 447-452.
9. te Velde AA, Flendrig LM, Ladiges NCJJ and Chamuleau RAFM. Immunological consequences of the use of xenogeneic hepatocytes in a bioartificial liver for acute hepatic failure. Int J Artif Organs 1997; 20: 229-233.
10. te Velde AA, Flendrig LM, Ladiges NCJJ and Chamuleau RAFM. Possible immunological problems of bioartificial liver support. Int J Artif Organs 1997; 20: 418-421.
11. Takahashi M, Ishikura H, Takahashi C, Nakajima Y, Matsushita M et al. Immunologic considerations in the use of porcine hepatocytes as a hybrid artificial liver. ASAIO J 1993; 39: M242-M246.

12. Patience C, Takeuchi Y and Weiss RB. Infection of human cells by an endogenous retrovirus of pigs. Nature Med 1997; 3: 282-286.
13. Heneine W, Tibell A, Switzer WM, Sandstrom P, Rosales GV et al. No evidence of infection with porcine endogenous retrovirus in recipients of porcine islet-cell xenografts. Lancet 1998; 352: 695-699.
14. Patience C, Patton GS, Takeuchi Y, Weiss RA, McClure MO et al. No evidence of pig DNA or retroviral infection in patients with short-term extracorporeal connection to pig kidneys. Lancet 1998; 352: 699-701.
15. Fourneau I, Depla E, van Pelt J, Crabbé T, Creses E et al. Development and characterization of immortalized human hepatocyte lines and the application in a bioartificial liver device. In: Crepaldi G, Demetriou AA and Muraca M. Bioartificial liver support systems. The critical issues. Rome: CIC Edizioni Internationali,1997: 62-69.

Chapter 34

Performance of a flat membrane bioreactor utilizing porcine hepatocytes cultured in an extracellular matrix

Loredana De Bartolo[1] and Augustinus Bader[2,3]
[1]Research Institute on Membranes and Modelling of Chemical Reactors, IRMERC-CNR, c/o Department of Chemical and Materials Engineering, University of Calabria, Rende, Italy, [2]GBF, Gesellschaft für Biotechnologische Forschung Braunschweig, [3]Leibniz Laboratorien für Biotechnologie und künstliche Organe, Medizinische Hochschule, Hannover, Germany

Key words: ammonia elimination, cell plates, culture in bioreactor, detoxification functions, diazepam metabolism, DNA in bioreactor, drug biotransformation, extracellular matrix, flat-membrane bioreactor, gas-permeable membranes, hepatocyte isolation, hybrid liver support device, oxygen supply, porcine hepatocytes, rate of albumin synthesis, rate of urea synthesis, scaled-up device, synthetic functions, tissue architecture

34.1 Introduction

To be clinically successful, an extracorporeal bioartificial liver (BAL) should allow maintenance of liver specific functions *in vitro* and achievement of high density cell culture for minimising the size of the BAL. The module must be designed to ensure optimum conditions for the mass transfer of nutrients and oxygen that are required for cell survival and metabolism. On this basis, it is generally agreed that there is room for improvement in bioreactor design. The majority of the current bioreactor designs do not fully meet these requirements. This could result in a situation in which BALs perform well for limited periods of time after which the hepatocytes deteriorate rapidly. In this respect, we consider that the impact of bioreactor design on hepatocyte viability and functions has been grossly undervalued. It is our view that when designing a BAL, special attention should be paid to providing the architectural basis for reconstructing a proper cellular microenvironment that ensures the highest and prolonged functional activity of hepatocytes cultured in the bioreactor. Tissue architecture and parenchymal cell morphology are crucial for the proper functioning of hepatocytes within the organ *in situ*. In the liver, hepatocytes are organized three-dimensionally into cellular plates that are identified as cords in a

M.N. Berry and A.M. Edwards (eds.), The Hepatocyte Review, 585–595.

cross-sectional view. In these plates hepatocytes exhibit a distinctive epithelial polarity, as well as a strongly developed cell-cell communication structures including bile canaliculi and tight junctions. Thus, hepatocytes must interact with other cells as well as with chemically complex substrata to sustain viability and functions. This organisation allows them to obtain an adequate supply of oxygen and nutrients.

34.2 Development of a flat-membrane bioreactor

It has been well demonstrated that extracellular matrix induces both cell polarity and tissue organisation. In fact, it was reported, recently, that the culture of hepatocytes within extracellular matrix in a three-dimensional system amplifies their specific metabolic activities and cell stability: hepatocytes in the extracellular matrix culture maintained a stable level of albumin and stable ratios of transcription and rates of protein secretion [1,2] (see also Chapter 8).

To imitate the organisation of liver cells *in vivo*, a flat membrane bioreactor (FMB) for hepatocyte culture has been developed [3-9], in which cells are cultured within two layers of collagen supported on flat sheet oxygen-permeable polymeric membranes. In such a FMB, hepatocytes are organized in a simple and repetitive way: cells are cultured as individual plates. The membranes permit unlimited oxygen uptake by the hepatocytes and the right geometry for the attachment of hepatocytes as monolayers. In BALs, where oxygen is solely transported by diffusion, meeting the cell oxygen metabolic needs requires that the bioreactor provides an oxygen path length shorter than about 100-150 μm [10]. In the FMB, the mean diffusion distance from the oxygen-permeable membranes to liver cells is about 20 μm. Embedding of hepatocytes within two layers of collagen improves the maintenance of liver specific functions of cells. Another important goal was to develop a bioreactor that could be scaled-up to incorporate sufficient cell mass for possible therapeutic liver support. Porcine hepatocytes are probably one of the best xenograft candidates with regard to differentiated metabolic functions and high-yield availability. In this study, we investigated the *in vitro* performance of the FMB using pig liver cells in terms of maintenance of viability and liver specific functions of cells cultured in the bioreactor.

34.3 Porcine hepatocyte isolation

Liver cells were isolated from female pigs weighing from 20 to 25 kg. The liver was pre-perfused *in situ* with 1 l of Ringer solution, consisting of

0.147 mM NaCl, 4 mM KCl and 2.24 mM $CaCl_2$ at 4°C. After the pre-perfusion step, the liver was removed from the animal and was placed in an ice-cold sterile Ringer solution. Liver pieces of 150 g were resected from the whole liver and perfused in single-pass at 80 ml/min with 1 l of buffer I consisting of 142 mM NaCl, 6.7 mM KCl, 10 mM HEPES and 0.25 mM EGTA (Sigma, St. Louis, MO, USA) at 37°C. This was followed by perfusion with 1 l of buffer II containing 142 mM NaCl, 6.7 mM KCl, 10 mM HEPES at 37°C. Thereafter, 200 ml of buffer III containing 142 mM NaCl, 6.7 mM KCl, 10 mM HEPES and 0.7 g/l $CaCl_2$, supplemented with 130 U/ml collagenase type IV (Biochrom, Berlin, Germany) was recirculated for 18 min at 70 ml/min. Upon disruption of the liver capsule, cells were liberated in ice cold buffer and resuspended twice after sedimentation, as described previously [8]. Viability of the hepatocytes ranged between 90% and 95% as assessed by Trypan blue exclusion.

34.4 Hepatocyte culture

34.4.1 Culture medium

The isolated hepatocytes were incubated in Williams' E medium (Biochrom, Berlin, Germany) supplemented with 5% (v/v) fetal bovine serum, prednisolone 2.1 μM, glucagon 0.38 μM (Novo, Mainz, Germany), insulin 0.169 U/ml (Hoechst, Frankfurt, Germany), penicillin 200 U/ml and streptomycin 137 μM (Biochrom, Berlin, Germany). This medium was chosen as a result of preliminary experiments.

34.4.2 Seeding of hepatocytes

Freshly isolated hepatocytes were seeded in the FMB on gas-permeable surfaces pre-coated with rat-tail collagen type I (prepared according to the method of Elsdale and Bard [11]) Seeding of the cells was performed at an ambient oxygen concentration of 20%. Bioreactors without hepatocytes were used as controls. The cells and controls were incubated at 37°C in a 5% CO_2:20% O_2 atmosphere with 95% relative humidity for the first 24 h and thereafter in a 5% CO_2:10% O_2 atmosphere for the remainder of the culture time. The perfusion of the FMB was initiated at 9 ml/h in single-pass, 24 h after cell loading. Media samples were collected daily on the inlet and outlet stream to determine product formation and substrate clearance. After 48 h, cultures were overlaid with a second layer of collagen gel (see Chapter 8). Thereafter, the culture was continued under serum-free conditions.

34.4.3 Assessment of hepatocyte morphology and function

The morphology and the concentration of cells cultured inside the FMB were assessed by inverted light microscopy for each day of culture. The survival of hepatocytes was assessed by DNA measurements, which together with morphological observations provide a measure of viability. To measure liver-specific functions of the isolated hepatocytes, we investigated albumin synthesis, ammonia elimination and urea synthesis, as well as xenobiotic detoxification and drug transformation abilities. The ability of the hepatocytes to eliminate ammonia and to synthesize urea were established by culturing the cells in the presence of 1 mM ammonia. Albumin secretion was measured in the samples by immunometric method (ELISA) as previously described [6]. For DNA analysis, cells were detached from the bioreactor membrane, which had been precoated with collagen, by addition of 0.05% collagenase followed by incubation at 37°C for 30 min. The cell pellets were suspended in a solution containing 0.1 M Tris, 1 mM EGTA, 0.1% SDS. Fluorescence of samples was measured after the addition of 0.1 μg/ml bisbenzimide H 33258 Fluorochrome Fa (Calbiochem, Germany) by fluorometer.

Diazepam metabolism by hepatocytes was determined by estimating diazepam elimination and the formation of metabolites in presence of 10 μg/ml diazepam in the culture medium. diazepam is metabolized through N-desmethylation and C3-hydroxylation to N-desmethyldiazepam (Nordiazepam) and 3-hydroxydiazepam (temazepam) respectively, with either or both being converted to oxazepam. diazepam and metabolite concentrations were assessed by HPLC analysis (Merck/Hitachi, Darmastadt, Germany).

Experimental metabolic rates are reported as mean ± standard error of the mean of 6 replicated cultures for each time point from 3 different isolations. The statistical significance of the experimental results was established according to the Student's t-test.

34.5 Bioreactor construction

The bioreactor consists of a multitude of stackable flat membrane modules, each having an oxygenating surface area of 1150 cm^2 (Figure 1) [8]. Up to 50 modules can presently be run in parallel mode. Each module is seeded with 2×10^8 cells, which results in a total seeded number of 1×10^{10} cells. We presently use up to 20 modules (4×10^9 cells).

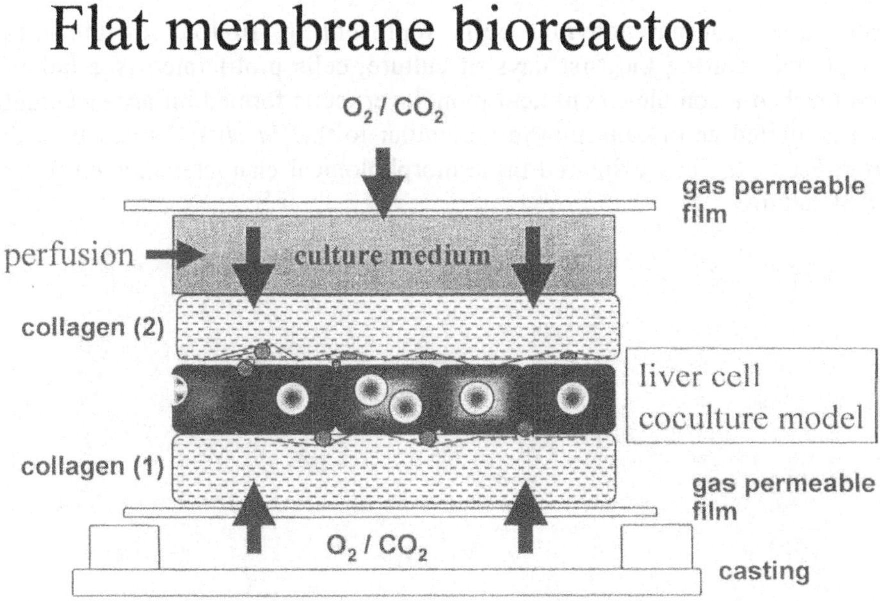

Figure 1. Scheme of Flat Membrane Bioreactor.

The hepatocytes were separated from membrane support and media by 10-20 μm of extracellular matrix. Medium and cells were oxygenated by diffusion of humidified air across the membrane support. A microporous membrane that separates the cell compartment from the medium compartment is generally used when the FMB is used at high flow rate in order to protect cells from shear forces and to control mass transfer of different solutes. In these experiments the FMB was perfused with serum-free medium at a flow rate of 9 ml/h. The bioreactor was prepared for cell culture by coating the polytetrafluorethylene membrane (DuPont) with rat tail collagen. After seeding and attachment of the hepatocytes, a second layer of a matrix was placed on top. Under these operative conditions, we characterized the fluid dynamics of the FMB by tracer experiments, in order to calculate the rates of metabolic reactions.

34.5.1 Morphology and viability of isolated hepatocytes in the bioreactor

Freshly isolated pig liver cells were prepared as small aggregates. When loaded in the FMB, isolated cells rapidly attached and formed small clusters. After 1 day, hepatocytes re-established cell to cell contacts with neighbouring cells and exhibited a polygonal shape. In the cell culture there were also non-parenchymal cells representing 10-15% of the total population. During the first days of culture, cells proliferated (see below) and reached a complete confluent monolayer: cells formed biliary canaliculi and exhibited an organisation more similar to that *in vivo*. As can be seen from Figure 2, cells exhibited these morphological characteristics until day 18 of culture.

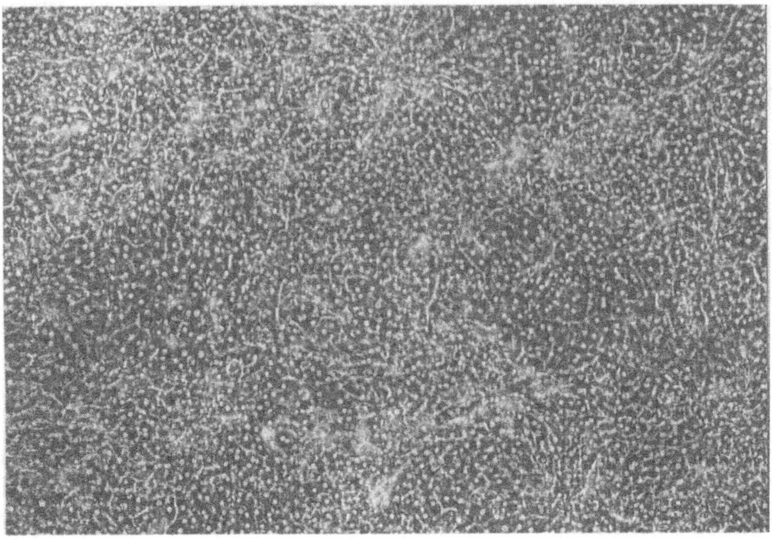

Figure 2. Inverse light microscopy of isolated hepatocytes cultured inside the FMB after 18 days of culture (100×)

The morphological observation with regard to the increase of cell number on the surface with time is confirmed by measurement of DNA content in the FMB over time. As is shown in Figure 3, DNA concentration peaked 48 h after cell seeding reaching a 1.4-fold value compared with cells 2 h after isolation. Thereafter, the DNA concentration remained constant for 10 days, decreased at about 10% per day from day 12, and then decreased more rapidly during the last days of culture, reaching a low value of 0.5 $\mu g/cm^2$.

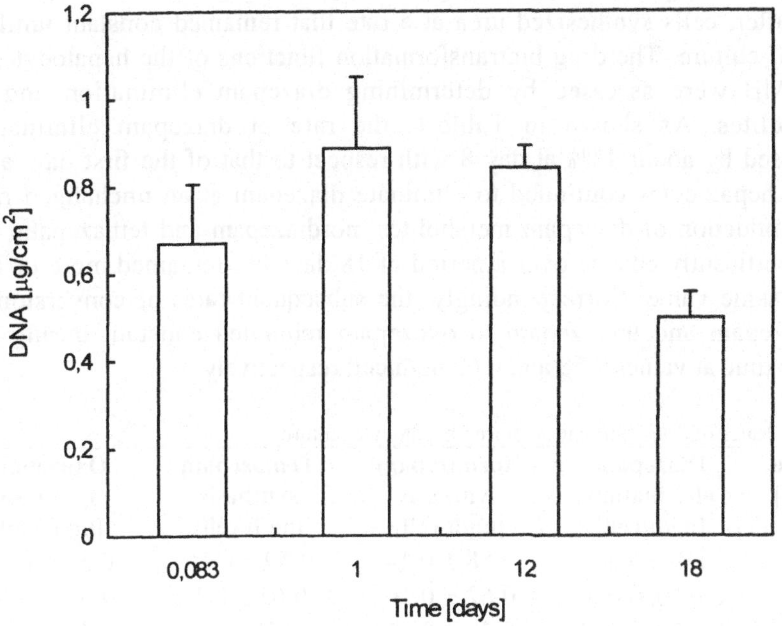

Figure 3. Fluorometric analysis of DNA content of hepatocytes cultured in the FMB

34.5.2 Liver specific functions of isolated hepatocytes in the bioreactor

Albumin production has long been used as an indicator of liver synthetic functions. To evaluate the performance of the FMB, each bioreactor module was monitored for a period of 18 days for albumin synthesis. The ability of hepatocytes to synthesize albumin increased during the first days of culture. Higher rates of albumin synthesis were obtained at day 9 and remained at a value of 1.41 pg/h/cell until day 18 of culture. However, at this time, the value of albumin production remained significantly higher than during the first days of culture.

Since levels of ammonia are considered to play a role in hepatic encephalopathy, we tested the bioreactor's ability to eliminate ammonia and to synthesize urea. Isolated hepatocytes eliminated ammonia present in the culture medium, at a rate of about 56 pg/h/cell during culture time and the rate of ammonia elimination of cells was constant over the experimental period. The rate of urea synthesis of the hepatocytes in the FMB consistently increased during the initial culture period from 22.6 ng/h/cell to 28 ng/h/cell. Thereafter, cells synthesized urea at a rate that remained constant until 18 days of culture. The drug biotransformation functions of the hepatocytes in the FMB were assessed by determining diazepam elimination and its metabolites. As shown in Table 1, the rate of diazepam elimination decreased by about 43% at day 8 with respect to that of the first day, after which hepatocytes continued to eliminate diazepam at an unchanged rate. The production of diazepam metabolites, nordiazepam and temazepam, did not significantly change over a period of 18 days but remained more or less at the same value. Correspondingly, the subsequent rates of conversion of nordiazepam and temazepam to oxazepam remained constant during the culture time at values 0.58 and 0.62 ng/h/cell respectively.

Table 1. Rates of drug elimination during hepatocyte culture

Time [day]	Diazepam elimination [ng/h/cell]	Nordazepam synthesis [ng/h/cell]	Temazepam synthesis [ng/h/cell]	Oxazepam synthesis [ng/h/cell]
1	1.1 ± 0.03	0.58 ± 0.14	0.69 ± 0.16	0.51 ± 0.07
8	0.63 ± 0.036	0.62 ± 0.16	0.63 ± 0.1	0.52 ± 0.09
18	0.64 ± 0.043	0.58 ± 0.11	0.65 ± 0.15	0.49 ± 0.06

34.6 Oxygen and nutrient supply

Considering that achieving the equivalent of 10% of the mass of the adult human liver requires about 1.5-2 × 10^{10} hepatocytes to be used in a BAL [12,13], it is evident that the treatment of hepatic failure requires high density hepatocyte culture. In many bioreactors the scale-up to such high densities gives rise to the formation of clusters of cells. In these large aggregates the reduced rate of supply of oxygen and nutrients may impair the metabolism and viability of the hepatocytes. In the liver the problem of oxygen and nutrient supply to the hepatocytes is averted by their arrangement in cell plates, with sinusoidal structures located on both sides. Our device has been designed to take advantage of this principle to the greatest degree possible. In the FMB hepatocytes are cultured within two layers of collagen between flat sheet oxygen permeable membranes as individual plates. In this way, each individual cell has its own membrane support and thereby an optimal situation for oxygen supply.

34.7 Scale-up of the bioreactor

For potential treatment of acute human fulminant hepatitis, a scale-up in bioreactor size and activity is required. The significantly increased demand for functionally active cells can be met by changing to larger animal sources. From recent investigations it seems that porcine hepatocytes are temporarily capable of replacing human liver functions in an extracorporeal BAL. As can be deduced from Figure 1, scaling-up the FMB implies increasing the number of stackable flat-membrane modules. This does not affect the individual bioreactor configuration, since the dimensions and the thickness of respective flat membrane modules remain the same. Furthermore, the bioreactor is constructed of materials that withstand the high thermal stress of autoclaving: it can be steam sterilized. This procedure is biologically much safer than the commonly used ethylene-oxide sterilisation that may cause sensitisation and allergic reaction in patients.

34.8 Conclusions

In vitro studies demonstrated that the performance of a scaled-up FMB using porcine hepatocytes, is stable for the duration of culture and compares well with that of other systems described in the literature. As shown in Figure 2, isolated hepatocytes, cultured in the FMB, reconstitute many of the features of the liver *in vivo*: cells exhibited a polyhedral shape and became arranged in a three-dimensional structure with formation of biliary canaliculi. The results concerning DNA content in each FMB module suggest that porcine hepatocytes cultured in the bioreactor proliferated and maintained their viability for 18 days. The finding of Taguchi *et al.*, [14], who measured the DNA content of hepatocytes cultured between two collagen layers in a multi-plated type of bioartificial liver suggest that cells maintain their surface density in the bioreactor for about 5 days of culture. In our long-term culture experiments we observed only a modest decrease (26%) from the initial level of DNA and this was apparent only after many days (day 18) in culture. Likewise, biochemical and drug metabolism of hepatocytes, cultured in a full-scale FMB, were sustained for almost three weeks.

Various functions of hepatocytes *in vivo* are regulated by cell-cell contact and the stable functions are not maintained unless the cells from tissue receive complex regulation from their microenvironment [15,16]. Specific metabolic functions of hepatocytes in terms of albumin synthesis, ammonia elimination and urea synthesis are maintained for the investigated culture time. Albumin production increased from 1 pg/h/cell to 1.45 pg/h/cell in the culture time and remained at this value until 18 days of culture. FMBs sustained detoxification properties in terms of ammonia elimination (56 μg/h/cell) over 18 days. Also the urea synthesising capacity of hepatocytes in the FMB increased in the culture time by 24% and remained high for the 18 days of the experiment. Likewise the drug-biotransforming ability of hepatocytes within the FMB was consistently maintained over the experimental period with high rates of metabolic reactions. The observed rates of the metabolic reactions for hepatocyte biochemical and drug metabolism described here are comparable with those reported by other authors [14,17,18]. The important feature of our study is that hepatocytes cultured in a full-scale FMB maintained these rates of metabolic reactions at high values until 18 days in culture.

These results demonstrate that a FMB sustained viability and specific metabolic functions of hepatocytes over a prolonged culture period. The scale-up of the FMB was assessed and it could be demonstrated that the device design, aimed at the reconstruction of the specific tissue architecture of the liver supports the expression of tissue specific functions. These characteristics suggest a potential application in human therapy. *In vivo* experiments in animal models and clinical trials will be performed in order to evaluate the full potential of the FMB as a bioartificial liver.

Acknowledgements

We thank Dr. Tanja Bestman of Leibniz Institute for Biotechnology and Artificial Organs, Medizinische Hochschule Hannover, Germany for assisting in DNA measurements. The Alexander von Humboldt Stiftung (Bonn, Germany) is gratefully acknowledged for providing funding to Dr. De Bartolo for this collaboration.

References

1. Dunn JCY, Tompkins RG and Yarmush ML. Hepatocyes collagen gel sandwich: evidence for transcriptional and translational regulation. J Cell Biol 1992; 116: 1043-1053.
2. Bader A, Rinkes IHB, Closs IE, Ryan CM, Toner M et al. A stable long-term hepatocyte culture system for studies of physiologic processes: Cytokine stimulation of acute phase response in rat and human hepatocytes. Biotechnol Prog 1992; 8: 219-225.
3. Knop E, Bader A, Böker K, Pichlmayr R and Sewing KF. Ultrastructural and functional differentiation of hepatocytes under long-term culture conditions. Anatomical Record 1995; 242: 337-349.

4. Bader A, Knop E, Böker K, Frühauf N, Schüttler W et al. A novel bioreactor design for *in vitro* reconstruction of *in vivo* liver characteristics. Artif Organs 1995; 19: 368-374.
5. Bader A, Knop E, Fruhauf N, Crome O, Boker K et al. Reconstruction of liver tissue *in vitro*: geometry of characteristic flat bed, hollow fiber, and spouted bed bioreactors with reference to the *in vivo* liver. Artif Organs 1995; 19: 941-950.
6. Bader A, Knop E, Böker KHW, Crome O, Frühauf N et al. Tacrolimus (FK 506) metabolism in primary rat hepatocytes depends on extracellular matrix geometry. NS Arch Pharm 1996; 353: 461-473.
7. Bader A, Knop E, Kern A, Boker K, Frühauf N, Crome O et al. 3-D coculture of hepatic sinusoidal cells with primary hepatocytes-Design of an organotypical model. Exp Cell Res 1996; 226: 223-233.
8. Bader A, De Bartolo L and Haverich A. Initial evaluation of the performance of a scaled-up flat membrane bioreactor (FMB) with pig liver cells. In: Crepaldi G, Demetriou AA and Muraca M Eds. Bioartificial Liver: the Critical Issues. Rome: CIC International Editions, 1997: 36-41.
9. Bader A, Frühauf N, Zech K, Haverich A and Borlak J. Development of a small scale bioreactor for drug metabolism studies maintaining hepatospecific functions. Xenobiotica 1998; 28: 815-825.
10. Catapano G, De Bartolo L, Lombardi CP and Drioli E. The effect of oxygen transportresistances on the viability and functions of isolated rat hepatocytes. Int J Artif Organs 1996; 19: 61-71.
11. Elsdale T and Bard J. Collagen substrata for studies on cell behavior. J Cell Biol 1972; 54: 626-637.
12. Kelly JH and Sussman NL. The hepatix extracorporeal liver assist device in the treatment of fulminant hepatic failure. ASAIO J 1994; 40: 83-85.
13. Kamlot A, Rozga J, Watanabe FD and Demetriou AA. Artificial liver support systems. Biotech Bioeng 1996; 50: 382-391.
14. Taguchi K, Matsushita M, Takahashi M and Uchino J. Development of a bioartificial liver with sandwiched-cultured hepatocytes between two collagen gel layers. Artif Organs 1996; 20: 178-185.
15. Koike M, Matsushita M, Taguchi K and Uchino J. Function of culturing monolayer hepatocytes by collagen gel coating and coculture with nonparenchymal cells. Artif Organs 1996; 20: 186.
16. Nakamura T, Yoshimoto K, Nakayama Y, Tomita Y and Ichihara A. Reciprocal modulation of growth and differentiated functions of mature rat hepatocytes in primary culture by cell-cell contact and cell membranes. Proc Natl Acad Sci 1983; 80: 7229-7233.
17. Wu FJ, Friend JR, Lazar A, Mann HJ, Remmel RP, Cerra FB and Hu W-S. Hollow fiber bioartificial utilizing collagen-entrapped porcine hepatocyte spheroids. Biotech Bioeng 1996; 52: 34-44.
18. Ono S, Hatanaka T, Miyazawa S, Tsutsui M, Aoyama T, Gonzalez FJ and Satoi T. Human liver microsomal diazepam metabolism using cDNA-expressed cytochrome P450s: role of CYP2B6, 219 and the 3A subfamily. Xenobiotica 1996; 26: 1155-1166.

Index

Note: page numbers in bold indicate Chapter key words

The manufacturer's authorised representative in the EU is Springer Nature Customer Service Centre GmbH, Europaplatz 3, 69115 Heidelberg, Germany. If you have any concerns regarding our products, please contact ProductSafety@springernature.com

Printed and bound by CPI Group (UK) Ltd, Croydon, CR0 4YY
30/09/2026
02225436-0014